WHEELMAKING

WOODEN WHEEL
DESIGN & CONSTRUCTION

WHEELMAKING

WOODEN WHEEL
DESIGN and CONSTRUCTION

OVAL TENON-FORMING MACHINE.

LARGE WHEEL-BOXING MACHINE.

CARRIAGE-WHEEL BOXING MACHINE.

WHEEL-SCREWING MACHINE.

Compiled by the
**Carriage Museum
of America**

Edited by
Don Peloubet,
Wheelwright

SPOKE-TENONING MACHINE.

SPOKE-FACING MACHINE.

HOSLER'S SPOKE DRIVER.

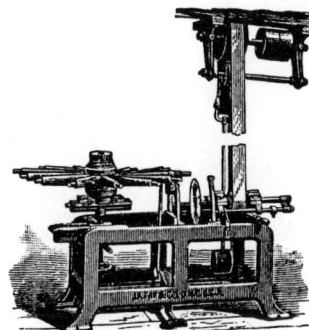

ROUND WHEEL-TENONING MACHINE.

ASTRAGAL PRESS
MENDHAM, NEW JERSEY

Published by
The Astragal Press
P.O. Box 239
Mendham, NJ 07945-0239

Copyright 1996 by the Carriage Museum of America

The material in this book was derived from the reference collection of the Carriage Museum of America, P.O. Box 417, Bird-in-Hand, PA 17505. Tel: 717-656-7019.

Manufactured in the United States of America

Library of Congress Catalogue Card Number: 96-86030

ISBN 1-879335-73-5

TABLE OF CONTENTS

INTRODUCTION

For many people, wheelmaking is perceived as a lost art; its practitioners, artisans of a rural society. While wheelwrights in the early years of the nation might have fit this description, those practicing their trade through the 19th century lived in an environment of rapid technological development and experienced a transformation of their craft. The articles that follow offer the reader a look at an industry in transition during the last decades of the 19th century.

Until 1850, the wheelwright was still working with hand tools, using methods that had been employed for centuries. The trade was primarily in support of the ever-growing, but still exclusive, carriage industry. (The rural wheelwright shop was involved in making the entire vehicle, while the urban wheelwright worked in carriage factories.) The labor-intensive processes used by the wheelwright greatly limited the number of vehicles that could be made; it would take a wheelwright six days (with an average work day of 10 hours) to make a set of wheels. Therefore, carriages were unavailable to the average person. As already noted, however, both the carriage and wheelmaking industries were on the verge of a transformation that would alter the economics, and accessibility, of carriages for the next 40 to 50 years.

Actually, significant developments in wheelmaking had been taking place starting in the mid-1830's and continuing through the century. The wheel industry patented approximately 8000 designs for wheels, wheel parts, and machinery from 1790 to 1910, including the invention of the spoke turning lathe in 1841, the development of the hub mortising machine in 1858, solid rubber tires in 1868, the cold tire-setter in 1870, and fluted tire bolts in 1890. (The above are the approximate dates these developments were patented, although it often took years before they were actually adopted.) The invention of the patent hub and the development of technology to steam-bend felloes (rims) at mid-century were even more significant. All these advancements in wheelmaking played a large role in the growing success of the carriage industry in the latter part of the century. Nothing played so significant a role in the transformation of the industry, however, as the advent of power machinery and the mechanization of the wheelmaking process.

By the end of the Civil War, the use of hand methods had virtually ceased, as wheelmaking evolved into large wheelmaking factories, and machines transformed the wheelwright into a type of assembler. These factories were fitted with power machinery, manufactured by such major machine makers of the day as Bentel & Margedant, Defiance, J.A. Fay, and The Egan Co. The machines used were usually belted to line-shafts driven by steam or gas engines. This development made finished wheels and wheel parts in large quantities readily available for the first time, spurring tremendous growth in the carriage industry. By 1890, with the advent of mass-production methods, the carriage industry, and wheel-making, reached their zenith, bringing costs within the reach of the average consumer.

On the following pages you will find a compilation of articles on wheelmaking from late 19th century trade journals published in America for the carriage industry. Most of the articles are from the two leading trade journals of the period, *The Hub* and *The Carriage Monthly*. *The Coach-makers' Magazine* was first published in New York in 1858; in 1871 the name was changed to *The Hub*. *The Coach-makers' International Journal* was founded in Philadelphia in 1865; in 1873 the name became *The Carriage Monthly*. These journals were aimed at those involved in the manufacture of pleasure carriages, business vehicles, and sleighs. They encouraged an active interchange of ideas, in the belief that this would lead to as perfect a vehicle as could be made.

We hope that readers will enjoy this book, not only for its historical value, but for the wealth of technical information that it provides.

Don Peloubet, Wheelwright
Lake Hiawatha, New Jersey

I
Wheelmaking

WHEELMAKING

As we read the articles on wheelmaking, we begin to comprehend the complexity of the processes involved. The discussions and disagreements contained in this and following chapters cover all aspects of wheelbuilding, from the types and properties of wood required for various parts of the wheel and the seasoning of hub blanks, through the tiring process, and everything in between. This interchange of ideas in journals subscribed to by carriage and wheelmakers of the period is one of the principal reasons the carriage industry attained such a high level of quality in the late 19th century.

The majority of articles in this chapter, which presents an overview of wheelmaking, were written by Howard M. Dubois, an outstanding lecturer, educator, editor, and patent holder, as well as a prominent Philadelphian wheelmaker. Dubois' essay on "Applied Mechanics in Wheelmaking" was published in *The Hub,* one chapter each month from May 1878 through August 1879. During that period, his Jones-Lewis award-winning essay, starting on page 14, was also published in the same journal. Another essay included here, "Whys and Wherefores in Wheelmaking," was delivered by Dubois in 1882 to a class in carriage drafting at the Carriage Builder's Technical School in New York City.

It was the opinion of Dubois that if the wooden wheel was properly made, it was sufficiently flexible; the newer, patented wheels, he felt, were too rigid. This reluctance to accept the newly developing technology of the period is repeated throughout this book by many authors writing on the full range of wheelmaking subjects.

Jared Maris, a wheelmaker, editor, and inventor, also wrote prolifically on wheel building. As evident in this and following chapters, Maris held strongly different attitudes from those of Dubois on many, if not most, aspects of wheelmaking.

One theme recurring throughout the articles in this chapter is the ever-present problem faced by wheelwrights: the expansion and contraction of wood in response to moisture in the air. This problem affected the tightness of the iron tire and hence the amount of dish. Oiling the hub and rim sections, sometimes soaking the completed wheel in linseed oil for several days, and painting, all helped combat the effects of weather.

Another problem discussed here was that of smoothness. Though production grew rapidly, triggered by the incredible variety and increased mechanization of wheelmaking machinery, the resulting finish on the product was not as smooth as that left by the spokeshave, and additional sanding and polishing machines were needed to smooth out the ridges left by the shaping operations.

One of the most striking observations to be drawn from the writings in this and following chapters is their authors' passionate feelings about the craft and the commitment each felt to his individual method of wheelmaking.

WHEEL-BUILDING.

MESSRS. EDITORS : Wheel-building being a specialty with me, I write this article on wheel-building ; hoping some of my rules may be of benefit to some who have had but little experience in this line, I shall here give them in the plainest manner I know of.

Of what Material shall our Hub be?

I prefer hickory-elm, some know it by the name of rock-elm. It is hard, (so are our spokes,) does not easily split, and holds a spoke firmer than any other timber I know of. I do not think that because a dealer gives you what he calls elm hubs, you have the desired thing, for I have seen some that were not any better than poplar would make. Good hickory-elm hubs are heavy. I will say as regards gum hubs, I have used but few, and therefore I am no judge as regards their merits.

How large shall our Hub be in Diameter to the Spoke?

The mortise in the hub must be as deep as the spoke is wide at the shoulder. If we now take a $3\frac{1}{4}$-inch hub, (I have here drawn a diagram by way of illustrating my meaning,) and space off sixteen mortises, scant $\frac{3}{8}$ of an inch in width, we find our mortises will come together at barely one inch in depth, and this is as small a hub as I would dare use for a medium inch spoke. This rule I apply to all hubs, whether $\frac{3}{4}$ or 3-inch spoke. It is an important point that the spoke shall have sufficient length of tenon to keep it firmly in place. We must not for the sake of appearances have our hub so small in diameter as to endanger the stability of our wheel. (This is the point where the patent wheel has acquired its reputation.) For an oak spoke in heavy wheels, I want a hard, solid, dry oak hub. The hub must be of as near the same hardness as the spoke to hold well. Locust I have no use for. Having selected my hubs, I now try the front end to see if it is rounding. If so, dress it so that my staff will fit up to it perfectly. Having chosen a set with the mortises small enough to allow me to true up to the sized wheel I want, and taper I give, I now place my hub in the driving-stand with the staff on and the gauge set exactly in height where the shoulder of the tenon for the felloe will be.

Shall our Front and Hind Wheel have the same Dish drove in them?

I practice giving front wheels (when the difference in size of front and hind are six inches or more) a trifle more dish than the hind ones, because hind wheels will dish more in setting of tire than the front will, and I want my axles of the same length, set on a plumb spoke. Now, taking a narrow straight-edge and placing it in the front end of the mortises, if my straight-edge is in line with my gauge, set to $\frac{1}{8}$ of an inch dish for one-inch spoke, $\frac{3}{16}$ for $1\frac{1}{2}$ inch spoke and $\frac{1}{4}$ for 2-inch staggered mortises, plain, $\frac{3}{8}$ for 2-inch and $\frac{1}{2}$ for $2\frac{1}{2}$ to 3-inch. I shall now have to taper my mortises to suit my rule, that the mortises must taper smaller in length a full $\frac{1}{16}$ of an inch to every inch in depth, no matter what its size is ; the mortise being $\frac{15}{16}$ at the outside of the hub for a 1-inch spoke, $1\frac{7}{8}$ inches for 2-inch and $2\frac{15}{16}$ inches for a 3-inch spoke. I give these sizes, it being easy to get the ones that are between, having these as a guide. I shall now have to

cut the corners of the mortises and take a portion off the side corner of the mortises, so that the shoulder of the spoke will be snugly down, and prevent the shoulder from splitting off. As for steaming hubs, I don't like it. My light hubs must be dry. Your Philadelphia correspondent's views on green hubs I agree with. On heavy wheels, $1\frac{3}{4}$-inch spoke and upward, with an oak hub, I steam them, because spokes of this size having side drive from full $\frac{1}{32}$ on $1\frac{1}{4}$ to $\frac{3}{32}$ on 3-inch, dressed off on the side, slightly dovetailing—or, to make my meaning plainer, a spoke for $\frac{3}{4}$-inch mortise would be $\frac{13}{16}$ in thickness at point of the tenon and plumb $\frac{3}{4}$ next to the shoulder. The hubs are softer, drive easier, and keep the glue hot, and when dry shrink down on the dovetailed tenon ; but I will here remark, I have some grave doubts of its being the best way, even on heavy work, and would like to hear from some experienced brother chip on this subject. Hubs do not shrink away and leave the shoulder of the spoke, for I have experimented thus far on steamed hubs ; but does the glue not lose its tenacity and hardness by being kept moist so long ? and is not our wheel liable to be rim-bound by the hub shrinking ? and how shall we make allowance for it ? The spokes must be dry, and to be certain of it all my spokes are kept in dry kiln from two to three weeks before using. Having selected the size I want—second-growth hickory for light wheels, A No. 1 white-oak for heavy, (I prefer oak for heavy wheels on account of its being better able to stand poor care and very little paint, which is the way truck-work is treated generally,)—I now dress the tenon equally on the sides until it fits fully, for all light hubs without permanent spoke-bands on. For truck-work I follow the rules I have given. I examine my spokes and reject all that are sprung, or not of quality I want, or wormy. Do not, for the sake of quickness, only dress the tenon on one side : if you do, the spoke will not stand in line with the center when drove. As for a sprung spoke, I would just as well have none in. I shall next true the face of my spokes, then having a pattern ready made to taper the back of my tenon with.

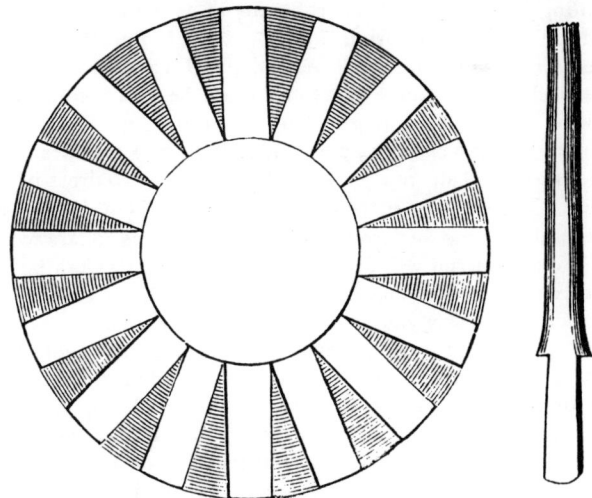

FIG. I.   FIG. II

What Drive shall my Spoke have?

The tenon must taper $\frac{1}{16}$ smaller in width to every inch in length. By comparing this with the rule I give for the mortise, it will be seen that the tenon will bind more at the point than it does at the shoulder. I have followed these rules for years, and my spokes never work in the hub. I taper my tenons, chamfer the ends, trim and smooth the spokes ready to drive ; next heat up the kettle of sand to put the tenon end of our spoke in, and by so doing prevent the glue in the mortise from becoming chill-

continued next page

Wheelmaking

ed before I have the spoke home. In heavy work, if you steam the hub, no need of warming the spokes. In driving, the spokes must dish to the rules I have given for the cutting of the front end of the mortises—⅛ for one-inch spoke, etc., etc. In sawing off my spokes I leave a full eighth of an inch too long, so that I can saw off easily what projects of the tenon, when the felloe is on. I set my adjustable hollow auger so that it will cut a tenon a little larger than the bit will bore in the felloe. I now cut my tenons, being very careful to have them in line with my spoke. Untrue tenons cause sprung spokes and breaking of tenon at the shoulder. Now taking a safe-edged wood file, (smooth-edged,) I cut the tenons on face and back in an oval form, just enough to prevent splitting my rims, and by this means dispensing with wedges and still have the rim held firmly up to the shoulder of spoke. I follow the same plan on heavy work, bent rims, with the difference of not making the tenon oval, because oak rims will drive on tight enough to hold up to the shoulder without checking, as hickory will.

Of what Material shall the Rims be?

Having a hard hub, a hard spoke, I now want firm material in my rims, and for light work in hickory we have it—firmness and elasticity to bound back to its place from the blows it receives while in use. I shall select a dry, hard, straight-grained set, two inches longer in diameter than the wheel is to be. I have had trouble to get the joints high enough with rims bent to the right size being cut off so near the required length. In rims two inches higher they are long enough to avoid splits from nailing on the binding strips, and will save yourself the trouble of pulling out the ends to make a high joint.

What Size shall my Rim be in Depth?

It should never be any smaller in depth than the spoke is wide at the shoulder. For hard service I prefer them from ⅟₁₆ to ⅛ inch heavier. As regards width for tire, it is rather difficult to give. I make them generally the same as the axle-arm is in size. These rules apply to all bent rims, light or heavy. I make my holes for tenon in diameter half the size of tire in width. When making the tire same as the axle-arm, I would say to all wheelers, tenoning machines are truer than any of us can be, to cut tenons and bore holes in our rims, and it is vitally important that our holes as well as tenons shall be true. If I have to use the bit-stock and bit, the gauge has to be used, being careful that the mark on the rim is no longer than the bit will cut out in boring; if it is longer, the rim has a check already started. I bore a dowel hole for as small an iron pin as will hold the rim to dress it in the center. The weakest place in our rim must be cut away as little as possible. Now be sure that your rim inside is square with the holes, and your rim will fit to the shoulder snug as can be when you drive it on. In regard to rounding the rims, it does not matter much which way you do it; whether you take one of your rejected spokes, cut it off in length the same as the spoke in the wheel is, cut a short tenon on, a little small, and mark around your hole for your spotting; in this way you can round up before putting on, or drive your rim on; fit your joints a little open on the tire side. Be sure your rim is up all round; and I can't describe it better than one of your correspondents does—just enough open to draw a sheet of paper through; now saw off your tenons close, mark around your spokes, dress off face side of rim to straight-edge, square the tire side with it, (I see your correspondent from Yonkers, N. Y., recommends dressing the rim on tire side square to the plumb spoke; but suppose we put on a set of felloes 4 inches wide on this plan, ¹-inch dish to the spoke, I don't believe that our friend J. L. H. Mosier could put on the tire to fit,) gauge for thickness, dress off the back, number the rims, drive off and round up in vise; in driving on again by this plan the rim may not set up as snug as it did at first, but you have it fitted, and when the tire is put on it will come all right. The tenon of spoke must be flush with the outside of rim. In heavy work, 2-inch spoke and upward, with oak rim, I leave from ⅟₁₆ of an inch to ⅛ opening in joint, in wheels heavier than 2¼ spoke. Sawed

felloes should always be used, taken out of oak plank, with the chins toward the sap. I prefer timber that is close-grained and rather soft, but thoroughly dry. Timber of this kind is not so liable to chin out, and it holds the tire better than timber that is hard. As regards making patterns to get your felloes out by, your English correspondent has saved me the trouble to write. I will add that in boring dowel-holes put them as close to the outside of felloe as you can with safety; in broad tread, four inches and over, it is better to *double dowel.* If our friend across the big pond were here, he would form a different opinion about the thumb rule or sawed felloes being rare'; but I will say this much, if a wheeler can make a good, true, sawed-felloe wheel, one with bent rims won't bother him very much.

I will add, in concluding this article, if you use hubs already mortised to the size you want, (especially staggered mortises,) do not expect to make a good set of wheels four inches lower than the hubs are intended for; if you do make them, when the tire expands, as it will do in time from the continual pounding it receives, the wheel will go backward. My excuse for mixing in heavy wheels with light ones is, that I have to build wheels from 1-inch spoke ⅞ tire, to 3-inch spoke and 4-inch tire, and I judge that other of your readers have to do the same thing, and a few rules are needed as well by the builder of truck-wheels as by the one who builds for a trotting wagon. As for these rules to govern in every case, I should be foolish to think so; but for generality of work they are my guide, and I know from experience they make good wheels. If I am in error in some part of my article, I shall consider it a favor if some of our experienced wheelers will point it out. I expect to citicise others when I think they are in error, and in this way by giving our manner of building wheels to one another and criticising each other in a friendly spirit, we shall be materially benefited. I want to build better wheels next year than I do this; my motto is "Excelsior," and I think it is so with a good many of our craft.

MEMPHIS, TENN.   R. KETTLEWELL.

The Hub, May 1873

Wheelmaking

APPLIED MECHANICS IN WHEEL-MAKING.

Chapter I.—Ancient Art and Modern Timber.

IN the year 1850 Prof. Agassiz opened in Boston the burial case of the oldest artificer who ever visited this country ; his name was "Got-thothi-aunkh," a master mechanic who died 2,800 years ago, and from the many pictures of wheels found on the stones of the temples of Thebes, in Egypt, where he wrought, we think he may be classed as one of the oldest wheel-makers on record ; at any rate it is a wonderful fact, that the wheel of his age compares favorably with the wheels used in some parts of Europe to-day, and the technical terms used to denote the several parts are identical with those in use at the present time.

Where can we find another art with such an unchanging record, and this stability comes from the fact that the ancients recognized the laws of mechanics, and everything made in accordance with those laws can never change in principle, however varied the applications to suit the ever changing wants of mankind.

The greatest changes from this ancient standard have been made by the mechanics of this country, who have carried wheel-making to the extreme of lightness and symmetry. In making such sweeping changes in proportions as we have made in the States, each mechanic has formed, as it were, proportions of his own for building wheels ; these are often so widely at variance with each other and the laws of mechanical construction, that the want of some standard, offering a practical reason for each dimension, should meet with some favor, and more criticism. To this end we propose to devote a series of papers, beginning with this one, in which we will endeavor to collect the experience of many of the best wheel-makers of this country, aided by our own experience of more than twenty years in the preparation and designing of material for wheels.

We will begin by a description of the qualities and varieties of the timber so peculiarly adapted to light wheel-making, known to the trade as hickory, and to botanists as the "Carya" (*Juglandea*). It is classed among the walnut family on account of the leaves and flowers having the same formation ; here the resemblance ceases, and the hickory stands alone in the qualities of its timber. There are six varieties scattered over the different states, and many sub-varieties caused by soil, climate, etc.; only three of these varieties are suitable for carriage work, although all but one are manufactured and sold for spokes and rims.

The first variety is the "Carya Alba," or white shag-bark hickory, found in New-York, Pennsylvania, the Western States, Virginia, and North-Carolina. This tree flowers in May, and the timber when grown on clay soil is good, provided the tree is cut in its prime and properly seasoned ; but when old or of thick forest growth, it is soft and dries brash or brittle.

The second is "Carya Sulcata," or shell-bark hickory ; this tree flowers in April, and is found in the Western States, part of the Southern States, and in the mountainous portions of the Middle States. The timber is red, and very brash or brittle when dried, and is subject to rapid decay.

Third, the "Carya Amara," or bitter nut hickory, is a fine large variety, which flowers in May, and arrives at its best in rich, heavy soils ; it is principally found in Pennsylvania, New-Jersey, and New-York. The timber is of large growth, very fine grained, and elastic, and retains its density or heft after seasoning ; on poor soil, or soil charged with surface iron, it becomes brittle and streaked with iron streaks and not fit for the best work.

Fourth, the "Carya Porcina," or pig-nut hickory ; this is the monarch of the hickories, not from its size but from its superior qualities ; it flowers in May, and is found in southwestern New-Jersey, and parts of Pennsylvania and Delaware. There are several sub-varieties, but all are good timber, being heavy, elastic, and very durable, nearly all white in color, and generally very straight grained.

Fifth, the "Carya Aquatica," or water hickory, grows on wet, marshy soil, and is very deceptive in the appearance of the timber, being fine grained and white, but not durable or strong ; it grows principally in a warm climate.

Sixth, the "Carya Olivaforma," or pecan hickory ; this is a Southern tree, and flourishes in Louisiana and Texas ; it is never used for timber or for building purposes.

Hickory lying on the ground, after being felled for any great length of time, becomes subject to the insect *scolytus 4-spinosus*, described so fully in the last volume of *The Hub*. These pests are mostly bark grubs, and will only enter the hard wood when undisturbed for some length of time.

The oxygen of the air will also cause spokes or rims to become dead or old when they are long exposed in a finished state without paint ; the effect seems to be of the same nature as that of oxygen on metals. It is not advisable to have many small spokes in stock, as there is no remedy for this oxidizing except to cover the wood with some "filler." Rims are effected even to a greater degree than spokes, owing to the action of the steam upon the pores or structure.

Hickory must be piled for seasoning where there is complete circulation of air free from dampness ; where this precaution is not regarded the sap decomposes or ferments, giving the white portions that mottled appearance of maple, shortening the grain, and eventually ending in "doting" or dry rot.

Spokes or rims should never be stored near a horse stable, as the ammonia is sure to kill even the most durable of timber. I have experimented with both oak and hickory, and find that small timber of either kind is made worthless in an atmosphere impregnated with ammonia, and in less than sixty days ; while heavy pieces of oak were affected to the depth of one inch all around in three months, hickory was completely rotted in ninety days, more or less, according to the strength of the ammonia. Split or rived hickory timber can be seasoned in the weather, without fear of staining, if not piled beneath the drippings of trees, or the eaves of a roof, where water impregnated with iron would have access to it. So delicate is this hard timber that the microscopical atoms of steel deposited in the wood by the action of sawing will cause the timber to turn blue on exposure to the weather for a short time ; therefore sawed stock must be housed immediately after sawing.

So great has been the demand for spokes and rims in this country, that some manufacturers cut the timber without regard to time or season ; and to hurry the goods to market, they have drying rooms or ovens where these are subjected to violent heat, the sap evaporated, and an unnatural stiffness given to the articles, which are then sold as seasoned. This is not seasoning, but drying, and it is every mechanic's duty to keep these two terms separate and distinct ; to *season* timber is to subject it to the slow but certain action of the air, which allows the timber to close and harden each pore, converting the sugar and gluten of the living portions of the tree into an impermeable waterproof filling, and admitting of violent vibration without disarranging the fibres, or splintering the wood. Violent heat on the other hand drives off this substance from the outside, while the interior is filled with moisture which eventually seperates each fibre, as it takes the place of that evaporated ; the timber then becomes hard and stiff, is easily split, and having little or no spring, causing the hub and rim to take all the jar or concussion, and sooner or later working loose in the hub. We do not advocate the idea that timber must be hard as ivory, and heavy as lead to make durable work ; good workmanship and proper proportions have more to do with durability than being over particular about each piece used ; but in our long experience we have never met with a wheel-maker who did not lay every mistake or failure of his work to defective material, no matter how good the latter might be ; that this is not always the case is proved by the very many good wheels running under very common work. It is a well-known fact that the durability of carriage wheels depends more on the manner of building than on the material used, always taking it for granted that the material is dry.

After many experiments with hickory the conclusion has been reached that there are only two kinds, good and bad. Any timber that posesses vitality, strength and proper seasoning is good ; any that is dead, brash, or dried out of season is bad. The grading between these points is more or less "fancy," and as this valuable material becomes more scarce, as it must in a short time, this fact will become more apparent. Let us, in constructing our wheel, use no timber but that fully matured, and cut at a time when we can depend on its keeping free from the worm that never dieth. Let us have it seasoned in the good old way by nature, believing, that those who "doctor" wood will meet with as little success as those who have "doctored" our iron ; and we should also take care to have our workmanship as near the mechanical idea of perfection as possible, after which we need have no fear of the prematurely decrepid wheels so fashionable at the present time,

continued next page

Wheelmaking

Chapter II.—Material for and Proportions of Hubs.

THERE are many kinds of timber used in the different states and other localities for hubs, all having some points which deserve recognition, but the majority of the consumers of hubs use either gum (*nyssa multiflora*) or the white rock-elm (*ulmus Americana*), both of which are strongly advocated by those using them, and both seem well adapted by nature for this particular purpose. We will describe the peculiar properties of each, together with the different manipulation required to bring about the best results.

The best elm is found on rocky and rough soil, in all the Northern and Western States of this country, and it should be carefully selected, as there are never many good, tough trees grouped together. A stick five feet long is about all that can be worked from each tree. Great care must be taken in seasoning to prevent checking and splitting, which will occur if the bark is removed from the ends of the stick.

the varnish to deaden and the paint to peel off. Elm hubs in fancy patterns, where the bands are turned small, are not durable, as the amount of end grain exposed makes them liable to shelling, from the same causes ; it is therefore necessary, if this timber is selected by the wheel-maker, to be certain that it is thoroughly seasoned, of hard, firm texture, and of such a pattern that the full strength of the wood is not sacrificed to a false standard of beauty. One cause of the popularity of elm is, that the spokes may be driven tighter sidewise than can be done with any other wood used, but should the material be only dried on the outside, the shrinkage will cause the hub to split open after running a short time. But make the spokes fit perfectly, and drive them down tightly on the shoulders, and you will have no further trouble with good elm, properly seasoned.

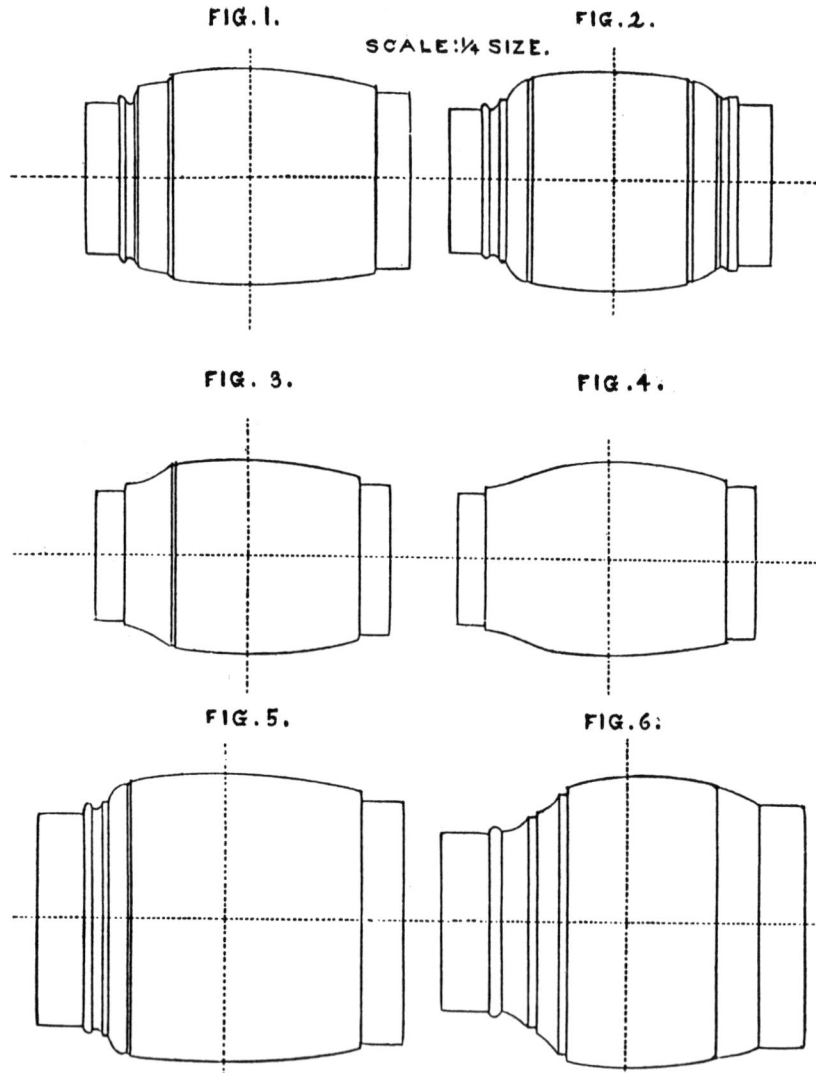

FIG. 1.   FIG. 2.

SCALE: ¼ SIZE.

FIG. 3.   FIG. 4.

FIG. 5.   FIG. 6.

WHEEL PATTERNS.—SCALE, ONE-QUARTER FULL SIZE.—(See descriptions.)

Some, to prevent this danger, steam the blocks for several hours, but this is injurious to the fibre or grain, and it also gives a dead or colored appearance to the wood, and makes it more ready to absorb the grease from the axle boxes after being manufactured—this being the strongest argument urged against elm by those preferring gum.

This liability on the part of elm to absorb grease or oil is mainly due to the rapid growth of the tree, making the annular rings far apart, the porous nature of the substance between acting like a sponge. This also makes it difficult to cover and finish in painting, as the expansion and contraction of these rings under different temperatures soon cause

"Gum" (*nyssa multiflora*), or sour gum, must not be confounded with the sweet gum or pepperidge of the North and West, as it is of an entirely different nature, and is found in but two or three of the Middle States. The trade is principally supplied from the seacoast marshes of New Jersey, where it reaches its greatest perfection. It requires a long time to season thoroughly, and a great amount of care to prepare it for the seasoning. The outer bark must be removed, leaving the "liber" or inner bark on to prevent checking, and a free circulation of air must be allowed to keep the sheds ventilated where it is piled, or "doting" soon sets in ; but get it perfectly dried, and it will last for

continued next page

years, without losing it qualities. It is of very fine grain, and so interlocked in growth that it is impossible to split it open with an axe. The atmosphere has no effect upon it after it is covered with paint, and the pores are so fine and filled with gum that it does not absorb grease as readily as the other woods. One serious fault has been charged to this material, and that is, that the mortises check in the corners; that this is the case can be accounted for by the foolish custom of having hubs oiled to show the wood to better advantage, which is the worst thing that can be done to any surface of wood, as either animal or vegetable oils destroy the tissues of the wood, and the vegetable oil soon vitrifies and draws the wood open in very fine checks or cracks imperceptible to the eye. All turned timber should be painted, and gum in particular, to which two coats of good lead should be applied; also, it should be made a rule to mortise hubs not more than one day before driving the spokes. In this manner gum may be kept in stock for years without losing its vitality, and all fear of checking will be avoided. We have seen gum hubs eighteen or twenty years old, perfectly sound and as good as when first used. The spokes should be make to fit perfectly to the mortises, and not filed, as is often the case when the spoke is large in the tenon. To sum up the whole matter, we can find nothing better to make hubs of than either of the woods above described, but failing to secure these, we might select the hard white birch, of New-York, the locust, the black walnut, or the white oak, all of which are used to some extent in heavy and light work in different parts of this country; but the first two mentioned are decidedly the best of known woods for hubs.

There is another peculiarity about hub material which often helps to defeat the efforts of the mechanic, and it is not confined to any one variety, as all timber used in the round stick or log, on the appearance of the budding season in early spring, feels its influence, and a very perceptible swelling occurs, of the pores leading to the outer growth-rings of the tree, making the timber less able to stand hard driving or boxing, or any operation which gives the expanded inner growths a chance to increase the diameter of the outside circles. This sympathetic action is continued from year to year as long as the wood keeps the shape nature gave it, and when this spring life is lost the wood rapidly decays, as it has then lived out its time.

Having given some of the natural laws governing the subject, we will now proceed to the mechanical proportioning of the hub, to secure the greatest amount of strength and beauty of outline possible, within the limits given.

A great diversity of opinion prevails among wheel-makers in regard to the size or proportion of the "bands" upon hubs, many having no standard rules for governing this vital point in a wheel, and many cut the front bands small to give a neater appearance to the work, regardless of either rules or reason. There is a fixed limit to the sizes of these bands beyond which it is not safe to venture, particularly in roughly paved cities or on broken turnpikes. The following table of proportions has been in practical use for many years, with results justifying its adoption as a standard for this purpose.

We have selected a few of the many patterns and dimensions of hubs now used, to give some idea of the points to be considered under the head of a selection of the best pattern for making durable work. See six illustrations on the opposite page, showing hubs in one-quarter their full size, which may be briefly described as follows:

Fig. 1 shows a very popular style, size 4½ x 7 in., bands turned to the table given above; it is a good, strong pattern, and has at least ½ inch wood from face of forward spoke to beginning of first bead; not less than this amount should be given in making any design for beading.

Fig. 2 shows a hub of same dimensions, with bands turned smaller than the measures given above; the back band was reduced to allow for fancy work, the whole making a very weak foundation upon which to build, the amount of end grain exposed, together with the leverage of the spokes, making a pattern very liable to shell and crack when subjected to undue strain or rapid thumping on rough pavements. We would not advise having a fancy bead at the back, even on larger hubs, as the back band must be turned down to admit the pattern; the little gained in appearance is thus lost in utility, and we should recognize the fundamental fact that *nothing can be truly beautiful unless founded on the laws of mechanics and symmetry.*

Fig. 3 shows another popular style, and has, in the smaller sizes, all the elements of strength necessary; size 4 x 6½ inches; bands according to table.

Fig. 4 shows the plain hub, much used for light sizes; the dimensions shown are as large as this style can be made to look proportionable. It is not a strong pattern, on account of the slope of the curve having to be carried to the front spoke, to ease the abruptness of the lines otherwise obtained; in very small sizes this objection decreases in proportion to the size.

TABLE No. 1.

PROPORTIONS OF BANDS.

Diam. of Hub.	Diameter of Front Band.	Diameter of Back Band.
Inches.	*Inches.*	*Inches.*
3	2	2⅜
3¼	2¼	2⅝
3½	2½	2⅞
3¾	2⅝	3
4	2¾	3¼
4¼	3	3½
4½	3⅛	3¾
4¾	3¼	4
5	3½	4¼
5¼	3¾	4½
5½	4	4¾
5¾	4¼	5
6	4⅜	5⅛
6½	4¾	5½
7	5¼	6
7½	5½	6¼
8	6	6¾

WIDTH OF BANDS.

Length of Hub.	Front Band.	Back Band.
Inches.	*Inches.*	*Inches.*
6	⅝	⅝
6½	⅝	¾
7	¾	¾
7½	¾	¾
8	1	1
9	1	1
10	1¼	1
12	2	1¼

LENGTH OF HUB IN PROPORTION TO DIAMETER.

Diameter.	Length.	Diameter.	Length.
Inches.	*Inches.*	*Inches.*	*Inches.*
3	6	5	7–7½
3¼	6½	5½	7–7½–8
3½	6½	5½	7–7½–8
3¾	6½–7	5¾	7–7½–8
4	6½–7	6	8–8½–9
4¼	7	6½	8½–9
4½	7	7	9–10
4¾	7	7½	10–12
		8	10–12

Figs. 5 and 6 show the American and English patterns in contrast, No. 5 being the regular 6 x 8 inch American hub, the other the same size from the factory of a London coach-builder. Very little comment is necessary, as it will be evident to those posted on wheel-making that the hub marked Fig. 6 would not last in this climate, with so much end exposure and so little solid support in front for the spokes. No. 5, on the other hand, will support any strain that can be thrown upon the wheel, and if of proper material should rival the hubs under the celebrated "One-hoss Shay" of the poet. The notch shown on the back of Fig. 6 is intended as the limit for a chamfered iron band which holds the wood at that point very securely, but the front part of the pattern is decidedly weak. These few samples are given that those wishing to design for themselves, can see what to avoid and what to adopt, being drawn from patterns in every day use. The fashion of some of our builders to copy after the English designs seems to us to be a step backward in this direction, and it should not be allowed to influence our judgment in making wheels to suit the climate nature has given us, but we should always strive to reach a greater perfection of design and greater skill in construction, taking *Originality* for our motto, and *Thoroughness* for our record. H. M. DuBois.

Chapter III.—Principles Involved in Mortising Hubs.

THERE are few more difficult things to do mechanically than the proper mortising of a hub, and it is in this particular that machinery rivals the best efforts of the skilled mechanic, and so general is the appreciation of this that it is not necessary to give directions about the way in which to proceed, but we will confine our attention to the priciples governing the question, leaving it optional whether the hub shall be worked by hand or done by the aid of machinery. At this important point it is necessary to determine the following conditions: how many mortises the hubs shall contain; their relative size compared with the diameter; the manner in which the wheel is to be driven as regards "dish," etc.; and the amount of "brace," "dodge" or "stagger" the spokes are to have. These are all details of the utmost importance, upon each of which there exist the most conflicting opinions, each wheel-maker claiming that his is the best and only correct method, which he can prove by his local experience.

continued next page

Wheelmaking

After investigation we would say, in regard to the first question, "How many mortises should the hub contain," that, this being the foundation of the wheel, it is not wise to reduce its strength any more than absolutely necessary for the requisite durability of the other parts of the work ; therefore the fewer mortises, the stronger the foundation obtained. The following table is prepared on the basis of securing the best results without any regard to English, French or German styles, and the diameter of the hub is supposed to correspond with the dimensions of the spokes to be given hereafter.

TABLE No. 2.

HEIGHTS OF WHEELS.	SIZE OF SPOKES.	NUMBER OF MORTISES.
4 ft. 8 in. and 5 ft.	$\frac{7}{8}$ in. to $1\frac{1}{4}$ in.	14 and 16
4 ft. 8 in. and 5 ft.	$1\frac{3}{8}$ in. to $1\frac{7}{8}$ in.	14 and 14
3 ft. 10 in. and 4 ft. 2 in.		
3 ft. 8 in. and 4 ft. 1 in.	$\frac{7}{8}$ in. to $1\frac{1}{2}$ in.	14 and 14
3 ft. 6 in. and 4 ft.		
3 ft. 10 in. and 4 ft. 2 in.		
3 ft. 8 in. and 4 ft. 1 in.	$1\frac{5}{8}$ in. to $1\frac{7}{8}$ in.	12 and 14
3 ft. 6 in. and 4 ft.		
3 ft. 10 in. and 4 ft. 1 in.		
3 ft. 8 in. and 4 ft.	2 in. to $2\frac{3}{8}$ in.	10 and 12
3 ft. 6 in. and 3 ft. 11 in.		
3 ft. and 3 ft. 3 in.	$1\frac{1}{4}$ in. to $1\frac{1}{2}$ in.	12 and 14
3 ft. and 3 ft. 8 in.	$1\frac{5}{8}$ in. to $2\frac{3}{8}$ in.	10 and 12

The objection that will be urged against the above figures is, that there is too much space between the spokes at the rim. This objection will be answered in a chapter devoted to rims. In regard to the advantages secured by using fourteen mortises instead of the sixteen so long used, we claim first, and most important, more strength in the body of the hub ; second, less danger of the hub becoming cracked and shelled during service ; third, a tenon more in proportion with the width of spoke used ; fourth, more surface to hold the tenon before meeting at the center or box ; fifth, a better proportioned and neater wheel. To secure these advantages requires that the size of the mortise relative to the diameter of the hub should be thoroughly understood, as there is nothing gained if we cut the same amount of wood from the hub in making fourteen holes as we do in making sixteen. Fig. 1 shows a mortising gauge giving the greatest amount that can be with safety taken from the hub by mortising.

The gauge shown on the next column can be relied upon as being correct ; it has been in constant use for a long time. The exact fractions have not been written, but the terms "scant" and "full" have been used, as most rules in shop use are not marked above sixteenths ; those who wish to go into mathematical measurement can do so by using their dividers upon the angular lines shown. The circular lines show the radius of the hub ; the diameter is marked above each arc, and at the base is a diagram of a three-inch hub, showing the proportion of wood left in all the sizes marked, allowing one-eighth more than the square of the widest part of the spoke used, which is more than enough to stand any strain to which the wheel may be subjected.

The subject of bracing, by dodging or staggering the mortises, comes next in order, and needs careful consideration. Many think that if a little is good, more must be better, but this is not the case in this instance, as we soon reach the practical boundary, beyond which we meet so many obstacles in construction that it is safe to say that in *no* instance is it any advantage to give the mortise full stagger, that is, making one mortise the width of the spoke back of the other. The greatest amount, to accomplish the object, has been found to be $\frac{1}{2}$ inch, and this should be reduced to $\frac{1}{4}$ inch on wheels whose spokes measure more than $1\frac{5}{8}$ inch, or where very low front wheels are used, in this case all above $1\frac{5}{8}$ inch spokes should be $1\frac{1}{4}$ inch stagger at most.

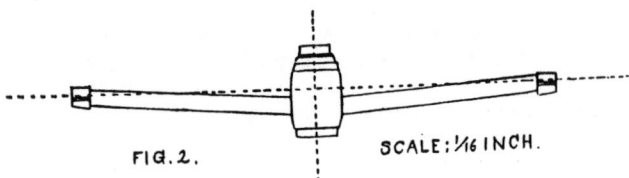

FIG. 2.   SCALE: 1/16 INCH.

MORTISING GUAGE. FULL SIZE.

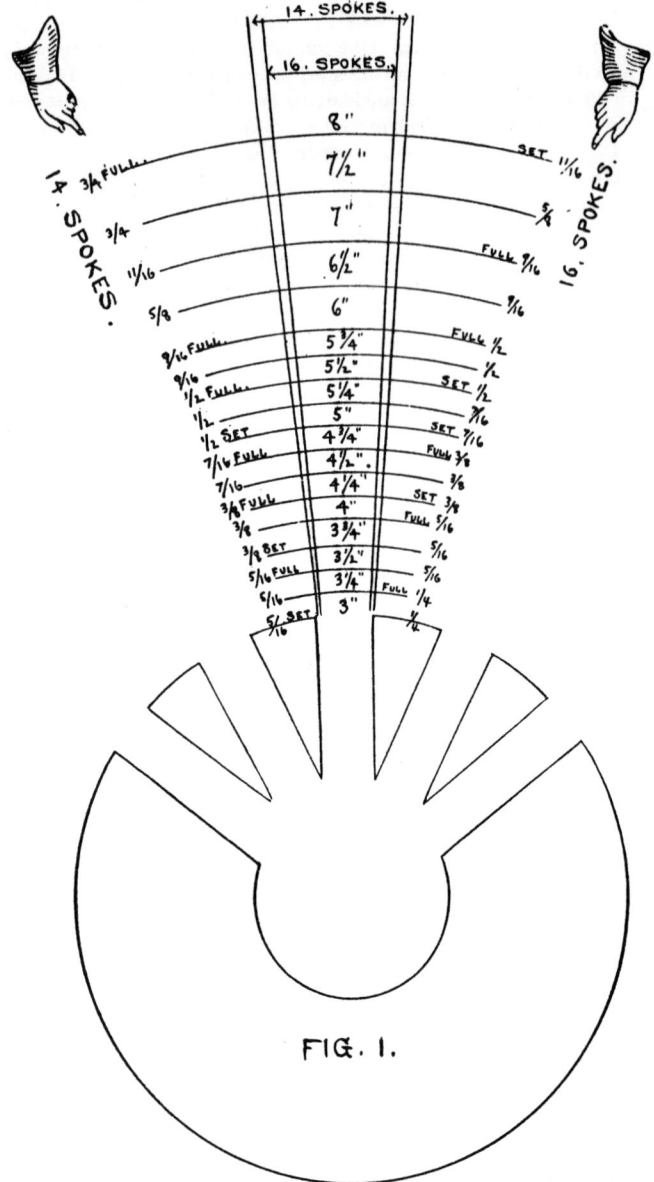

FIG. 1.

In order to understand more fully the disadvantages of too much stagger, attention is called to cut No. 2, showing a section of wheel with deflection in the line of tenon of rear spoke at the rim, which, in case of cross-grain at the upper part of spoke, would soon break off. The line of weight is shown, by the vertical line, to be one inch from the face of the rear spoke, which would make all the strain upon the front spokes, the rear ones only serving as a brace to prevent dishing backward. The wheels from which this diagram was made are now in use in one of the West India islands, and were made to order.

The question then stands in this light : For wheels of light sizes it *is* advantageous to use a limited stagger. Heavy wheels gain nothing by it except appearance, which makes it in this instance a matter of taste independent of mechanics.

We have said in a former chapter that it should be made a rule never to mortise hubs more than one day before using. In explanation of this, it will be found, upon examination, that a hub that has been mortised for some time has become lighter in weight, from the evaporation of the substance upon contact with the air ; but worse than this, the mortises are not in the same shape as when first made ; they will be found wider at the bottom and smaller at the top than they were formerly. These may seem trifles, but it is to these trifles that we owe many failures of otherwise good work.

There is another point to be examined, and that is, when the chisel that cuts the mortises passes through the growth fibres, the edge bends

continued next page

Wheelmaking

FIG .3.

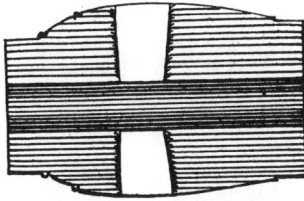

down the grain of the wood, as shown in Fig. 3. This to the eye is not apparent, but on examining the surface with a strong magnifying glass the small parts of the grain are found to be bent as shown, and it is plain that if the spoke be driven before these particles can re-arrange themselves, it will be held more firmly and be less affected by jarring. If, however, these fibres again resume their position, the spoke has not the same tendency to force them down as the unyielding edge of the chisel. We are aware that hubs are seldom mortised to the full size at the factory, but the "opening out" is all done at the back of the mortise, and the front is pared very little. The main object should be to keep the atmosphere from penetrating to the inner part of the wood, and making the shrinkage uneven. Mortises should be examined carefully at the corners to see if there is any fullness at this point. Pressure in driving, if all brought on corners, causes hubs to split, often long after driving. The settling of the rim will, in most cases, cause the hub to open when the pressure is all on the corners of the tenon.

In regard to the amount of "mash," or difference between the taper of the spoke and the mortise, there seems to be no rule, each mechanic making his calculations from the hardness of the hub and relative quality of the spokes. We can only say, beware of *too* much, as more spokes are *mashed* and broken in driving than are worn out by service. Every spoke factory that has a retail department could form a museum of *mashed* spokes which were worn out before they were rimmed. Use, therefore, just "mash" enough to make the spoke drive tightly without *mashing* the fibres and destroying the grain of the tenon. If the bevel on the spoke is too abrupt, it must work loose, no matter how hard it is driven.

In recapitulation, *remember* to leave as much wood as possible in the hub ; understand the nature and peculiarities of the wood used ; avoid all incongruities of construction ; and wooden hubs will not need the aid of patents to hold their own—at all times, and under all circumstances.

Chapter IV.—Spokes, and their Proportions.

In the countries having the highest civilization the old solid wheel was displaced by one containing spokes, at a very early period of the world's history, as we find the Assyrians possessed carts having wheels containing six spokes as early as the year 1680 B. C. The shape of these spokes can not be given, but about 1124 B. C., war chariots were made whose wheels had eight spokes, which were made round and swelled out heavily in the center, and the tenon in the naves or hubs was round also. As remarked in the first chapter we probably owe to the Egyptians the first improvements in spokes, they having flattened the end entering the hub and enlarged the base of the spokes, so that, taking into consideration the height of their wheels, they were well adapted to the requirements of the times. This superiority of manufacture seems to have been retained by this ancient people, as the chariots used by the Romans were nearly all built at Alexandria as late as the year A. D. 66.

The modifications from these ancient styles have been very gradual, being affected somewhat by local tendencies, each country having a peculiar manner of shaping the spokes, but there appears to be no standard of sizes, adapting certain measurements for sustaining a given load, based on the resistance to strain of the material used. In all other branches of mechanics this is considered absolutely essential to the intelligent designing of any kind of work, and it is remarkable that this branch of the trade should have been left to the individual experience of those following the avocation, who, no matter how intelligent they may be, have but a limited opportunity to work out a standard sufficiently correct to be received and adopted as such by those seeking enlightenment on this subject.

To consider the subject from the beginning, we must first analyze a spoke and find its component parts ; and then, from a given size, proportion all others to this standard. The analysis will show that a spoke consists of (A) "Tenon," (B) "Shoulders," (C) "Face," (D) "Back," (E) "Throat," (F) "Body," and (G) "Point." Each of these parts must

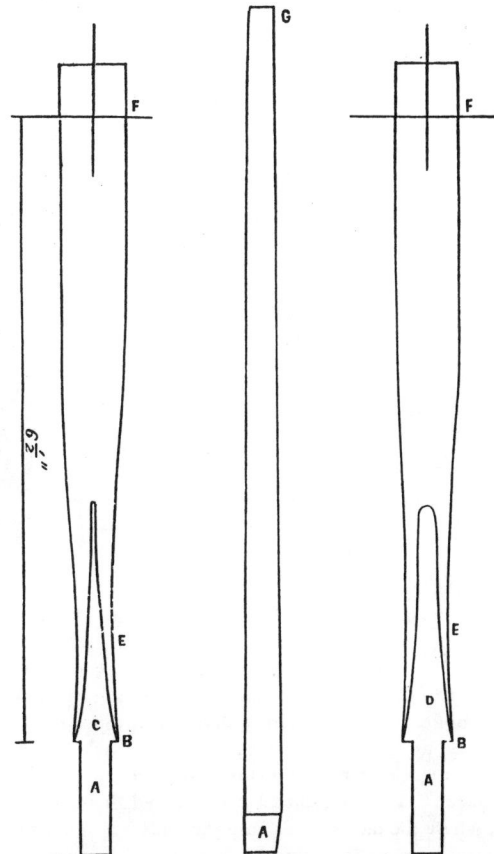

Fig. 1. ½ size. Fig. 2. ⅙ size. Fig. 3. ½ size.

be of a size to correspond with the others, the whole making a spoke capable of performing all that its size demands. Figs. 1, 2, 3, show the front or face, side, and back, of a one-inch spoke, being the size we have selected for testing. The full measures will be found in the accompanying table. See Table No. 3.

Made with these proportions we find, after many and accurate tests, that this size will stand, without deflecting or bending sidewise, a weight of 750 pounds.

This is the mean result of the experiments which embraced from the best to the medium grades of timber. One would suppose that this amount of resistance to pressure would not be required when the actual weight of a suitable carriage is distributed between four wheels, each containing fourteen supports such as we present, but when we consider that the wheel is subject to the strain of the shrinkage of the tires (which often deflects the spokes when improperly throated), and that even metals soon crystallize and break under continuous vibration, we can readily see that wood-fibre is subject to the same laws, and therefore any lightening of these proportions must present less surface to this destroying agency.

We will next consider the "Tenon." It will be seen, by referring to Figs. 1 and 3, that, for the size shown, this is $\frac{5}{16}$ inch. This is enough to accomplish all that can be required at this point for any amount of resistance compatible with the size of the wheel. By the rule of proportion we would naturally suppose that we could find the size of tenon required for any width of spoke, thus : as $\frac{16}{16} : \frac{5}{16} : :$ the size of spoke in 16ths to the tenon required. This would give us a proportionate result, which would be sufficient to stand a proportionate strain, but in driving the tenon into the hub we displace so much fibre-surface that there must be an allowance made, which will be found laid down in Fig. 3, wherein each size is marked, the angular lines showing the size of tenon.

continued next page

Second, we come to the consideration of the "Shoulders." These are for the purpose of limiting the extent of the "driving," and to prevent any settling of the spoke after being driven and in use; they also act as a lateral support to a limited extent, and present a more graceful outline to the center of the wheel. They should be made smaller at the back of the spoke than at the front, as considerable space is taken

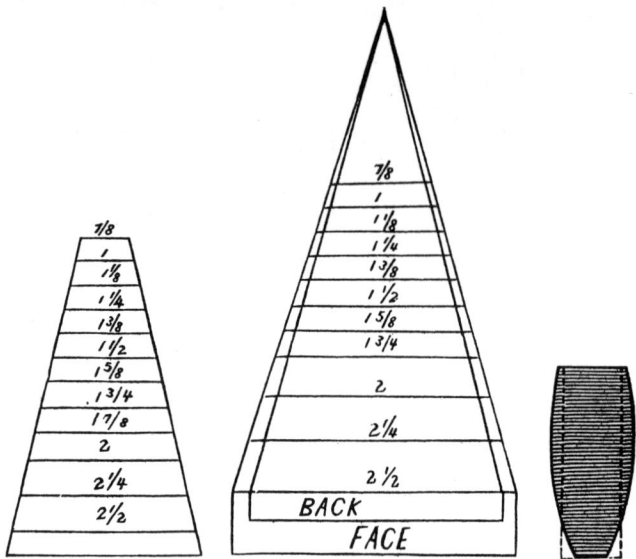

Fig. 4, Full size. Fig. 5, Full size. Fig. 6, Full size.

upon the hub lengthwise, which carries the rear of the spokes to a smaller diameter of the hub than that occupied by the face of the spokes. There should be enough difference between the face and back shoulders to give the same (uniform) appearance to the spacing between the spokes. Diagram No. 5 shows the proper relative sizes of these shoulders, measuring across the face and back, which are governed somewhat by the diameter of the hubs used.

Third, "Face." This is formed by two curved lines, almost meeting at the point where the face rounds into the body, as indicated in Fig. 1 (C), and in this size it is $3\frac{1}{2}$ inches from the shoulders to the limit marked.

Fourth, "Back." This part should invariably be made to contain $\frac{1}{8}$ inch more than the thickness of the tenon, as all the end thrust of the axle is borne by this part of the spoke, which can not be made with any safety to contain less than the measure given. This will be found sufficient for all sizes.

Fifth, "Throat." The lightening of the spoke at this point accomplishes two important objects: it saves the rims from "settling down" upon the shoulders of the spokes, and saves the hub and axle from all violent or sudden concussions, dissipating by vibration the shocks which, directly delivered, would soon destroy the wheel. Properly executed, the throat adds vitality to the wheel,—improperly, a sure cause of dissolution. Throating should also bear a proper ratio to the thickness of the tenon. Diagram No. 6, a cross-section through the throat, shows that the spoke, in the smallest part, should contain a trifle more wood than the tenon. The amount taken from the front should be counterbalanced by leaving that much more at the back. The dotted lines in this diagram show the size of the tenon. "Body" and point will be found in Table No. 3. As shown in Fig. 2, the spoke is tapered from the line of shoulders to the point from the back, while the face line is perpendicular; the reasons for this are obvious.

The figures contained in the Table below (No. 3) are an actual standard, both in theory and practice, being the result of a vast number of experiments during many years. There are many who make their spokes tapered sidewise from the center of the body to the point; in fact this is the general custom, but for what reason we could never learn, as the position of a spoke in the wheel, by the widening of the angle of radiation, gives a lighter appearance to this part when the spokes are of a uniform thickness from body to point, while the tapering mentioned gives less shoulder for the rim to rest upon,—an important and undesirable fact, when we consider the many complaints of "rim-bound" wheels, which are caused mainly by these shoulders being inadequate to sustain the pressure of the loads placed upon them.

TABLE NO. 3.

Size of Spoke.	Thickness of Body.		Width at Point.		Reaches, full body thickness from shoulders.
Inches.	Inches.		Inches.		Inches.
$\frac{7}{8}$	$\frac{5}{8}$	full.	11-16		$6\frac{1}{4}$
1	11-16		$\frac{3}{4}$		$6\frac{1}{4}$
1 1-16	23-32		13-16		$6\frac{1}{4}$
$1\frac{1}{8}$	13-16		$\frac{7}{8}$		$6\frac{1}{4}$
1 3-16	13-16	full.	$\frac{7}{8}$	full.	$6\frac{1}{4}$
$1\frac{1}{4}$	$\frac{7}{8}$		15-16		$6\frac{1}{4}$
$1\frac{3}{8}$	15-16		1		$6\frac{1}{4}$
$1\frac{1}{2}$	1	sc't.	$1\frac{1}{8}$		$6\frac{1}{4}$
$1\frac{5}{8}$	1	full.	$1\frac{1}{4}$		$6\frac{1}{4}$
$1\frac{3}{4}$	1 3-32		$1\frac{3}{8}$		$6\frac{1}{4}$
2	1 3-16		1 9-16		$7\frac{1}{2}$
$2\frac{1}{4}$	1 5-16		$1\frac{3}{4}$		$7\frac{1}{2}$
$2\frac{1}{2}$	$1\frac{1}{2}$		1 13-16		$7\frac{1}{2}$

It may be urged, and with some reason, that all true art should be individual, and not the reproduction of certain set formulas. That this is true to a great extent admits of no argument, and every one interested in the building of carriages should endeavor to build each one as an individual work of art, and not by the foot or square yard; but underlying all art there *must* be the foundation of mathematical mechanics, and the few figures given above will, I trust, prove of service in giving a proper understanding of some of the natural laws governing this branch of the business.

Chapter V.—American vs. Foreign Spokes.

In the proportions of spokes given in the previous chapter, we made no reference to the contour which may be given these proportions without changing anything but the appearance; the figures given had reference only to the American or ovate style of spoke, as this is now used almost exclusively by the builders of this country, and may be termed "American Style" to distinguish it from the sharp-faced or European style, exclusively used in trans-atlantic countries. Many years ago we are told this style was in vogue here, but with the advancement made in the art, it was soon detected to be unsuitable for this country, and the result has been the adoption of the present more durable shape. In this we seem to have reversed the order of nature, as the natives of England and Germany are physically rotund and full bodied, while the representative Yankee is as sharp as the foreign spokes mentioned.

Before going into a comparison of these two styles of spokes, we will review the rule which has governed the designers in all other arts, and see upon what grounds they build the useful and ornamental. It has been laid down, that the object of all design should be to produce as nearly perfection as possible; first, in adaptation to purposes required for strength, compactness, durability, and *lastly*, beauty of outline or contour. These rules may be best studied in architecture, where the designing hand of man has had the greatest opportunity to display its power. Hugh Miller, the geologist, tells us that there is not a column or pedestal, pinnacle or dome, which the great Creator has not already shown in the "testimony of the rocks" long before the age when man appeared upon the scene; showing that design in this direction has been led by the faultless hand of nature, in accordance with the eternal fitness of things. So it should be in everything; the natural laws governing the material used should be made the foundation for design. Even the outline of a spoke should have some reference to the form given the material by nature to resist the stress brought upon the living fibre under corresponding circumstances. When we view the question in this light the foreign spoke is an oddity, nothing in nature showing a sharp frontal elevation backed by a half-oval form, such as shown in Figs. 1 and 2, which illustrate the face and back of an English spoke. Fig. 3 gives a cross section, showing the amount of timber in the body of this style. For comparison we give Figs. 4, 5, and 6, which are the

continued next page

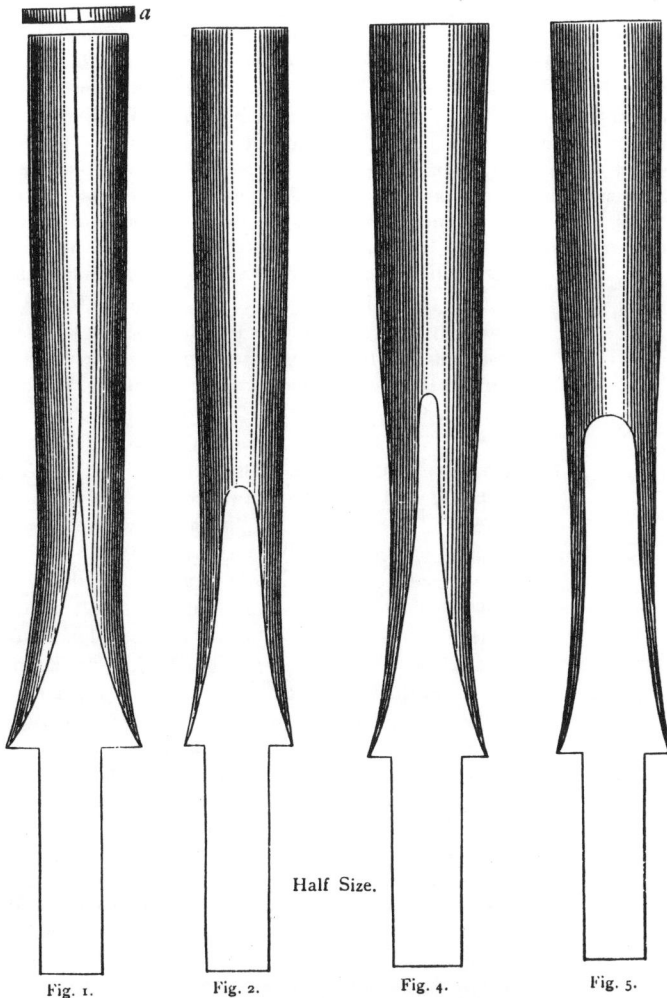

Half Size.

Fig. 1.   Fig. 2.   Fig. 4.   Fig. 5.

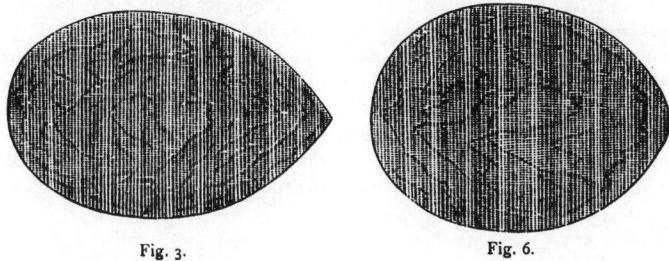

Fig. 3.   Fig. 6.

corresponding parts of the American style. The question is one of adaptability entirely, as after numerous tests we have found that the strength to resist pressure, such as received from the end thrusts of the axles, is about evenly balanced between the two on these larger sizes, each spoke sustaining a weight of four hundred and twenty pounds for hickory, and three hundred and sixty pounds for oak. The tests were made upon each spoke separately, as shown in Fig. 7, the test samples

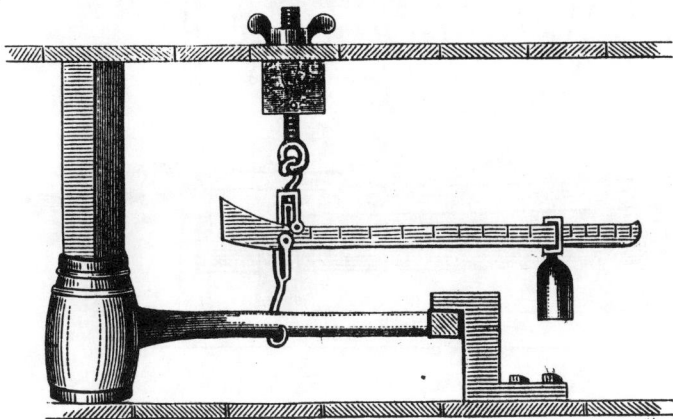

Fig. 7.

measuring one and three-quarters inch across the shoulders, and corresponding measures according to the two styles. The samples were as near a uniform quality as careful selection from the same tree could make them.

If we compare the cross sections 3 and 6 respectively, we can perceive the reason why the spoke with the sharp face could bear as much pressure as the full rounded one. The English spoke has very little taper towards the point, owing to the heavy felloes used, while the American is tapered to suit the narrower rims of this country. We will state here that the weights given above are not the breaking tests, but merely such as would flex the spoke one-half inch from a straight line. In the tests for end pressures the American style gave the same proportionate results obtained in the tests made with the standard size given in the previous chapter; any size contained in Table No. 3 may be worked out by the rule of proportion. The English style gave a greater capacity for bearing weight endwise than the former, owing to the system of having the spoke thicker at the felloe end than in the center of the body, see Fig. 1 a (a design that could be adopted in this country to advantage). This advantage at the felloe end, or point, is counterbalanced by the short, sharp face, cutting the fibres, and exposing the pores of the wood to the action of the atmosphere, which will affect them in a climate subject to such variations of temperature as our own, no matter how well covered by paint and varnish. This abrupt cutting also tends to weaken the resisting power of the shoulders, which in this style are large and not well adapted to "hold up," particularly when the wheels are dished to such an extent as we have seen them. As there are many who do not use this extreme English style, even in that country, it will be well to remember that so far as strength is concerned the English and American are nearly equal, and we must seek for our superiority in some other direction. We can claim superiority, first, in having a design more in accordance with the rules laid down by nature in the structural arrangement to resist similar stress and strain during growth and maturity; second, we have in this form the most durable surface for the reception of the paint and varnish of the finishing process, which surface is less liable to be scarfed or hacked by collisions, which is quite an object in this fast-driving, democratic country, where the rough members of society often try their free and enlightened equality by scratching the paint and varnish from coach and carriage by reckless driving. But the strongest argument in favor of the American style is its suitableness for all sizes, while the sharp spoke is limited to the heavier kinds, any below one and five-eighths inch being much weaker in proportion, than the corresponding size of the former style, while below one and a half inch they are entirely inadmissible. We admit that, individually, the sharp spoke presents a more graceful appearance, but when assembled in a wheel the effect is lessened, and at a distance of fifteen feet, with the wheel at rest, it is lost entirely. Every effect supposed to be gained by this design can be produced by ornamentation at the hands of the painter, without sacrificing the durability of this part of the work from causes mentioned. There might have been something in the idea that the intrinsic value was increased by the amount of time taken to produce this unseen effect, like the value that attaches to old lace valuable from the amount of labor wasted in its production, but this satisfaction is lost when we know that the machinery of to-day turns out this style as well as the more durable, and at about equal cost. The sum of all this is, that the English spoke is admirably proportioned to gain the greatest advantage under the circumstances, but we fully believe that the American style, like the younger races of men, will eventually prevail. It is already preferred by South American and West Indian carriage buyers, and used in Australia to a great extent.

The German spoke possesses all the defects of the English and none of its advantages, being tapered more than the English; to make up for this, some have made the sharp edge or feather quite obtuse, making a kind of compromise between the two styles, neither ornamental nor useful. The French spokes we have examined vary slightly from the German, having ponderosity without corresponding strength. What the more ordinary spokes of the latter countries may be, we have had no opportunity of knowing, as our experience has been with the finer wheels made in the respective countries mentioned. Unfortunately for this criticism, America is the nursery and home of more incongruities and peculiarities in the spoke and wheel line than can be produced the world over. Some of our builders have been obliged to go backward in this direction to suit a foreign taste manifested in our eastern cities,—the effect of easy communication with those countries, but there

continued next page

is no need of our mechanics looking in that direction for instruction in this branch of business, as the verdict is in our favor; and the time will come when this country must lead in the manufacture of the complete carriage and coach, as much as we now lead in the proportion and strength of our wheels, when made upon appropriate mechanical principles. It is generally supposed that the heavy foreign wheels are more durable than the American, and that they will run a greater length of time. On examining into this we find that there is no way of making a comparison by giving the length of time a wheel will last; the comparison should be on the distanced traveled, and the state of the roads, speed at which the distance was accomplished, etc.; with all these conditions balanced, the advantage will be found with the lighter and more elastic wheel, not counting a climate where the thermometer registers from 125 degrees, Fah., in the sun to 10 below zero. Nothing is needed by our mechanics but faithful adherence to the principle of thoroughly mastering the cause and effect of everything connected with the trade, and applying the knowledge obtained, which will be found to constitute the chief element of success in all undertakings, great or small.

Chapter VI.—Bent Timber for Rims.

It is a singular fact that the wheel-makers of this country owe a life-long debt to the warlike occupation of the gunsmiths, as the following preliminary sketch will demonstrate.

In the year 1838, Edward K. Reynolds, a gifted mechanic, native of New-Jersey, resident of St. Georges, Del., and gunsmith by trade, after some years' experimenting, was granted a patent for a machine and process for bending timber ("preferably hickory") into felloes or rims for carriage-wheels. Mr. Reynolds, like all other inventors, met with very little encouragement in his first efforts to introduce his invention, and became disgusted with the indifference of those who should have been most interested in this great improvement. After many disappointments, he found in Geo. W. Watson, of Philadelphia, at that time becoming prominent as a builder of carriages, a patron and friend, who after repeated tests adopted the bent rim, and Reynolds was soon forced by increase of business to remove to Salem, N. J., where he secured increased facilities for obtaining timber and shipping the articles when manufactured, and to this beginning may be traced the era of the *American* wheel made possible by this invention. Unfortunately there was one difficulty: Reynolds' machine was not economical in its working, as nothing but the best timber would bend, the strain upon the fibres causing the inferior qualities to break. This, though limiting the profits of the business, was the means of introducing these rims more rapidly, as, bent in this manner, they possessed more elasticity and were free from any sinking at the joints.

The difficulty of breakage was overcome in the year 1849 by another inventor, the well-known Thos. Blanchard, also a gunsmith, who had already conferred an inestimable benefit upon the carriage trade through the invention of the "spoke turning lathe." Blanchard's improvement consisted in placing a "screw" upon the binding band of a bending machine, which was a perfect mechanical remedy for this wasting of good material, which amounted under the Reynolds system to about thirty per cent. As this improvement caused a revolution in the principles governing the working of the rim in making a wheel, we will consider it in detail, as it is of the greatest importance that it be understood.

If we examine a rim strip before bending, say of one and a half inch in size, the lines of the top and bottom are parallel or of equal length.

Now, if we bend this (the rim strip) around a half circle, or "form," 42 inches diameter, the inside of the rim would measure in decimals 65,9736 inches for the one-half circle, while for the outside of the rim it would require 70,636 inches to the same point, leaving more than four and a half inches for the fibres to stretch in this old style of bending it; it is therefore plain that nothing but the most elastic material would endure this operation. With the "screw" used by Blanchard, the inside fibres are crushed together, or, to use the technical term, "upset," so that the outer fibres retain their normal condition, with the exception of a slight elongation, caused by relaxing the screw during the revolution of the "form" or bender; then instead of having the back of the rim stretched to make up this four and a half inches, we have the inside of the rim driven together to that extent. This can be better understood by referring to Fig. 1, which shows a front view of the "Reynolds-Blanchard machine." The "screw" is seen at A; B is the rim as it appears after the revolution of the form D; C is a sliding bed, which supports the wood during the process of bending. It might naturally be supposed by those not informed, that a rim, after bending, would have a tendency to spring out or away from the circle to which it is bent; but this is the case only so long as the rim retains the moisture and sap left in the pores of the wood after steaming. This, in drying out, closes together the circle, caused by the inner circle of the rim containing more fibres in a given space than the outer; as shrinkage takes place, the inside exerts a force greater than the outside, drawing the rim in the direction mentioned, which accounts for the "flattening" of the joints, so prevalent in wheels made with these rims, as the ends will retain this tendency to curve in. No matter how much the mechanic may open them out while preparing them for the wheel, the fibres will eventually return to the position received in the operation of bending, or a little beyond this point, if the "upsetting" has been severe. Fig. 3 gives an idea of the closing together mentioned. This, under the Reynolds-Blanchard system, and in the hands of a skillful operator, is rendered very regular and uniform. All that is necessary, then, to secure the best results and render these rims durable, is to remember that the manufacturer of these articles makes the "forms" upon which they are bent much smaller in diameter than the heights marked upon the bundles in which they are packed. This is done to facilitate packing, as when first bent they *have* the tendency to force outward; but the wheelmaker must wait till the rim dries before using, at which time they will return to the exact circle first bent, irrespective of the heights marked

Fig. 1.

continued next page

Wheelmaking

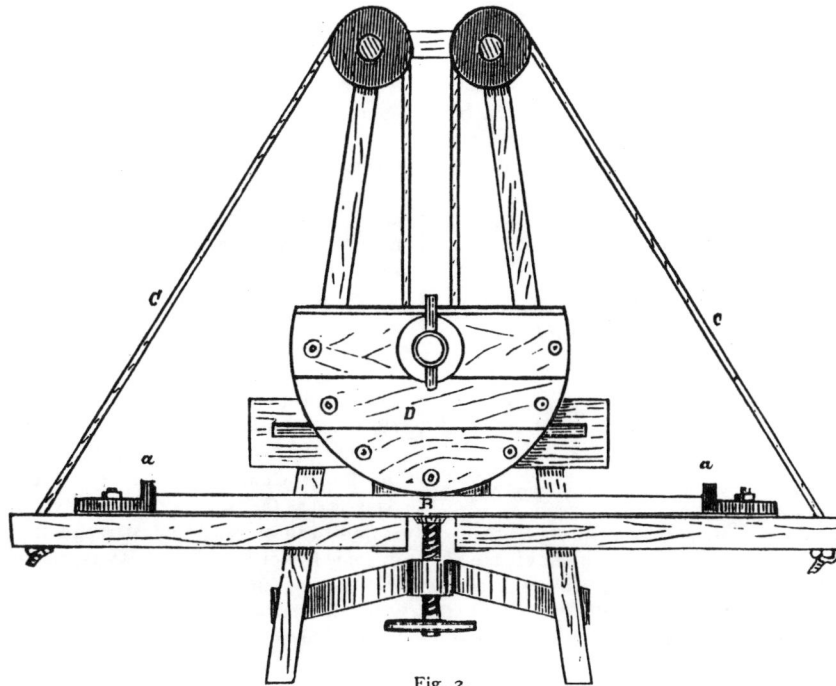

Fig. 2.

upon the bundles in which they were confined. We have often heard wheel-makers remark that it is more essential to have the spokes thoroughly dried or seasoned than the rims. There never was a more egregious error, as there is *no* such thing as durable work if any material used in the work is not *thoroughly* seasoned, and when the rim has reached this state, the distance between the spokes that support it upon the wheel is of little importance. In fact the wider the space allowed, to a reasonable extent, the easier the motion conveyed to the axle, always taking it for granted that the rims used are bent upon a "form" approximating to the circle of the wheel when finished. This brings the strain of the fibres against the tires, which prevents all flattening between the spokes or at the joints, which is inevitable if this natural law be disregarded.

There is another system of bending, in which the rim is confined in the center from which it is bent, each way (see Fig. 2). A A show the stops which are fastened rigidly against the rims. B is the unbent rim. The differential windlass at the rear operates the ropes C C, which draw the levers around the form D, as shown. These stops A A being fastened rigidly, the "upsetting" takes place at the extreme ends, as there is no pressure when the levers first move; then for the same reasons mentioned in the first instance, the shrinkage will affect these ends to a greater extent than the center of the arch, which, drawing naturally, must flatten the rim at this point. It is an impossibility to make a rim of this kind stand the ordinary hard usage of city streets. The shape of the rim may be all that could be desired when first worked, but the draw of the tires and the jar of the pavement soon force the fibres to the position described.

It has become the custom among wheel-makers to have the rims bent with the growth rings perpendicular (see Fig. 4), and for no other reason

than to prevent the splitting in boring the holes for the spokes. This is ingenious, but not correct; as the fault of the splitting lies in the auger-bits used, the "worm" or thread forcing the fibres of the wood apart like a wedge. It would be better to make this tool to overcome the difficulty rather than use the timber edgewise. It is well-known in practice that the timber is less able to stand vibration when the edges of the growth rings are turned in this position; nature has placed them layer upon layer, and it is in this direction they exert the most resistance to strains and weight. The dowels generally used have also a tendency to split the rim when the growth is in this direction shown, and we contend that what is termed "bending the rim the right way," may be *called* right in the work-shop, but is radically wrong in service and point of durability.

There is some diversity of opinion respecting the amount of wood which can be displaced in boring for the spokes, and the rules of mechanics must be modified to suit the circumstances. We are taught in mechanical construction that not more than one-third of the substance should be removed in making a mortise, but to follow the rule in this particular would make the tenons upon the spokes too small for strength. Then of two evils chose the lesser, which we can do by making the holes displace one-half the amount contained in the width of the tread. This, in the cities where the tramway brings an additional strain upon this part of the spoke, will be none too much; but on this subject local experience is the best teacher, and it is our aim in these articles to present the general features of the art, leaving the many minor points untouched, as the local wants of each district must vary with climate, roads, and taste. An old, broken-down wheel may teach more to an intelligent mechanic than much that can be written. Dissection and analysis are the means used in gaining knowledge in every art and science.—Why not in wheel-making?   H. M. DuBois.

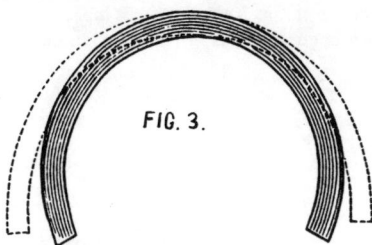

FIG. 3.

FIG. 4.

continued next page

Wheelmaking

APPLIED MECHANICS IN WHEEL-MAKING.

WE have already had the pleasure of announcing that our valued correspondent, Mr. H. M. DuBois, of Philadelphia, has been honored by the Carriage-Builders' National Association by the award of the "Jones-Lewis Prize" (fifty dollars), for his essay on "Carriage Wheels." We present below the first half of this valuable prize essay, and to make room for it, we defer publishing Chapter VII. of his serial, "Applied Mechanics in Wheelmaking," until our February number.

THE JONES-LEWIS PRIZE ESSAY ON CARRIAGE WHEELS.

THE BEST MATERIAL FOR, AND BEST MODE OF MANUFACTURING.

BY HOWARD M. DUBOIS OF PHILADELPHIA.

(CAREFULLY REVISED AND CORRECTED IN PROOF BY THE AUTHOR.)

To intelligently understand and apply the principles of the mechanical arts and sciences in wheel-making requires a thorough knowledge of the materials used, in all their details, including (1) the sources of supply, (2) the differences of qualities and varieties, (3) different modes of seasoning and other preparation required previous to manipulation, and (4) the natural enemies to the materials, encountered both before and after being made up. We begin with a description of the geographical distribution of the timber best adapted by nature for the wheelmaker's use.

HUBS.

As the hub is the foundation of the wheel, we will make it the starting-point in this essay, and begin by describing the material most suitable for its manufacture. Many kinds of timber have been used for this purpose, but the majority of manufacturers prefer either Rock Elm (*ulmus Americana*) or Sour Gum (*nyssa multiflora*); these two seem to possess every requirement, and they are decidedly the best yet tried for the purpose.

ROCK ELM.—Rock Elm is found as far south as North Carolina; it extends northward into the Canadas, westward as far as Illinois, and is scattered to a limited extent beyond the last-named state, being found on the Pacific coast, in Oregon. Covering this wide range, it varies with each locality, the best quality being confined to a limited region, including New-York and the more eastern states, which have the most suitable soil and climate for its perfection, while a fine variety is also grown on the rough, rocky soil of some portions of Pennsylvania. Ohio and Indiana produce the largest quantity; the timber, however, is not so heavy after drying, though it is of fair quality; and the Southern and South-western States produce a quality which becomes quite light when thoroughly dried. In Rock Elm, about five feet of the butt is all that can be used to advantage for hubs, as its virtues are not carried very far upward in its growth. In selecting this timber "standing," it should be remembered that the best are the young trees that grow in open spaces, where the elements have had full chance to render their growth tough and elastic; these are infinitely preferable.

CUTTING AND SEASONING.—December and January seem to be the best months for felling the trees, as vegetation is at rest in the Elm at this season. To prepare the timber for seasoning, the outer bark should first be removed (leaving the "liber" or inner bark), excepting at the ends of the logs or "sticks," where the rough bark also should be left on, to prevent the ends from checking, which would occur if this were removed. The timber should not be allowed to remain long on the ground, as, if so exposed, the seeds of decay will manifest themselves in a very short time. After remaining in the "sticks," in cross-piles, at least two months, the Elm may then be cut into lengths or blocks for further curing. The ends of the blocks should now be dipped into a cement, composed of lard and rosin, which, being elastic, shrinks with the wood, keeps the air from drying and cracking the ends, and also prevents the gluten contained in the pores from being decomposed by the action of the atmosphere. Dried in this manner, without the process of steaming (the practice with some manufacturers), Elm will

be found durable and suitable for the purpose under consideration. The following defects will still remain. Owing to its rapid growth, the concentric growth-rings are quite far apart, and the intermediate spaces more or less cellular, which allows the temperature to affect the outer layers to such an extent that, if the patterns used are not carefully designed, shelling of the surface will occur, and, from the same cause, it will be difficult to keep the hub covered with paint. A still worse defect in this connection is the conveying of the grease, used to lubricate the axle, through the straight pores directly to the mortises, which, by destroying the retaining fibres and the glue, will result in loosening the spokes. These defects are more or less common to all qualities of Elm, but especially to the softer varieties (or to blocks that have been steamed); for this reason it is important, in selecting this fine timber, to remember that there is but one kind or quality fit for good work, viz.: *the very best*, and specific gravity or density is an infallible test of this.

SOUR GUM.—The Gum *nyssa multiflora* must not be confounded with the Sweet Gum or Pepperidge scattered so plentifully over North America, as the two are of an entirely different nature. The true Gum is limited in its geographical range, that fit for hubs being found only in some few of the Middle States and in the swamps of Maryland, and the trade is supplied principally from the seacoast of New-Jersey, where it is found in perfection.

The season best suited for cutting this timber is December or early in January. Being of a peculiar nature, great care is required in preparing it for drying, as, if carelessly handled, or worked before it is thoroughly seasoned, it will be rendered worthless. Unlike the Elm, the outer bark should be removed the entire length of the "stick." Not more than five or ten feet from each trunk will be found tough enough for hubs. After removing the outer bark, the logs or "sticks" are cross-piled under dry, light sheds, where they are allowed to remain for a period of three years or longer, varying with the diameter of the "sticks." At the expiration of this time, the timber is ready to be manufactured into hubs, and it is nearly indestructible, being fine-grained, heavy, and so interlocked in growth as to make it difficult to split with wedge or axe.

TREATMENT OF HUBS AFTER TURNING.—Unfortunately it has become customary, at the factories, to give the turned hubs a coating of boiled linseed oil, in order to show to better advantage the grain and quality of the wood. This treatment is the worst to which turned timber could be subjected, as, in every instance, it has a tendency to separate the medullary rays—those fine lines of pores which separate the growth-rings laterally,—in consequence of the oil vitrifying and drawing the grain open during this chemical change. After repeated tests and experiments, the writer has found that the best way to secure uniformly good results is to have *all* turned hubs painted as soon as they are finished from the lathe; and in this case of Gum hubs, two coats of lead priming or wood filler, with good body, should be applied. If this course is followed, such timber will not be liable to check when driving the spokes, and it may be kept in stock for years without losing its peculiar qualities. Elm, treated in this manner, may also be depended upon to retain its weight and substance, while, if left exposed to the action of the atmosphere, it is sure to lose these qualities.

THE BUDDING SEASON.—The sensitiveness of timber during the budding season, March and April, is a fact which should not be overlooked in the working of round timber, as this sensitiveness renders it difficult to operate upon timber during that season without danger of splitting. There is no kind of timber which is free from this sympathetic action. For this reason the wheel-maker should avoid, if possible, driving spokes during the two months named, or otherwise the hub will often be rendered worthless when the carriage has been used a short time, owing to the unequal strain caused by swelling, particularly where the heart is not in the center of the hub. This sympathy with budding nature continues to be evinced until, from age, the timber has become worthless.

continued next page

Wheelmaking

SPOKES AND RIMS.

We will now proceed to consider the kind of timber best adapted for spokes and rims.

HICKORY.—*Carya Juglandae*, known to the trade as Hickory, is classed among the Walnut family, on account of the leaves and flowers having the same formation, but Hickory stands alone for the peculiar virtues of its timber. There are six distinct botanical varieties, widely scattered over the United States and Canadas. While Hickory may be said to grow in all the timbered states, the boundaries given below show where the varieties attain the most desirable qualifications.

The first variety is the *Carya Alba*, or White Shag-bark Hickory, which is found in New-York, Pennsylvania, the Western States, and part of the Southern States—notably Virginia and North Carolina. This tree flowers in May. The timber, when grown on good clay soil, is excellent, provided the tree is cut in its prime and properly seasoned, but, when old or of thick forest growth, it is too soft and brash.

The second, *Carya Sulcata*, or Shell-bark Hickory, flowers in April, and is found in the Western States, part of the Southern States, and in the mountainous portions of the Middle States. The timber is red, and very brash and brittle when dried, and it is also subject to rapid decay.

Third, the *Carya Amara*, or Bitter-nut Hickory, is a fine, large variety, which flowers in May, and is produced in perfection in rich, heavy soils. It is found in New-York, Pennsylvania, New-Jersey, Delaware, Maryland, and to some extent in North Carolina. The timber is sound and elastic, but when grown on soil charged with "surface iron," it is very apt to contain black streaks, and, in seasoning, to become soft and brash. There is no other variety so sensitive as this to the influences of soil and climate, or, consequently, so changeable,—good and bad qualities being frequently found within a few rods of one another.

Fourth, we come to the *Carya Porcina*, or Pig-nut Hickory, which is the monarch of the Hickories, not from its size, but from its superior qualities. It flowers in May. It is found in the Eastern States, parts of New-York, Pennsylvania, New-Jersey and Delaware. There are several sub-varieties, all of which produce good timber, it being heavy, elastic, nearly all white in color, generally straight-grained, and surpassing all the others in durability.

Fifth, the *Carya Aquatica*, or Water Hickory, grows on wet, marshy soils. The appearance of its timber is very deceptive, being fine-grained and white, but neither strong nor durable. It seems to prefer the climate of the Virginia Valley.

The sixth variety is the *Carya Olivaforma*, but as its timber is not employed we will not describe it further than to say it is esteemed for its fruit, the well-known pecan nut of the south and southwest.

The above-named six varieties comprise the family of Hickories, and, as before stated, the influences of climate, soil, etc., may give rise to many sub-varieties, differing to some extent from the above, and acquiring local names, but all belonging to the botanical divisions named; combined, they cover a geographical range extending from thirty-two degrees to forty-five degrees, north latitude, and from ten degrees west to five degrees east longitude from Washington.

FELLING HICKORY.—There is a diversity of opinions in regard to the season most suitable for cutting this timber, but those who have made this a business and had the longest experience, give preference to the month of August, when the timber is most solid, as vegetation is then at rest. When cut during this month it will retain its weight after seasoning, and will also be free from the attack of insects which so often destroy this valuable timber. The writer has made a special study of these insects for years past, and, as the result of his experience, he would state, as a general rule, to be adhered to by those keeping Hickory in stock, that, whenever possible, it should be stored in some place which has never before been used for that purpose. If this suggestion is followed, and the stock examined in May or June, thoroughly dusted, and then repiled, there will then be absolutely no danger of injury from worms,—provided, always, that the timber was felled in August. These pests are generally first introduced by spokes or rims, in whose pores the eggs have been laid while they lay in the factories where they were given their first shape. August cutting is the only safeguard against them, and if the above rules are strictly observed, timber felled at that time may be stored for years without fear of damage.

Hickory should never be baked or kiln-dried, as the elasticity of the fibre is thereby destroyed, and the timber rendered stiff and rigid,—qualities looked upon with favor by some wheel-makers, but which render the wheel less durable, and much harder upon the boxes and axles.

After collecting the experience of some of the oldest and most successful wheel-makers, we can safely assert that the best quality of Hickory is that which grows on a heavy clay soil, in open groves or hedge-rows, which is notably found in New-Jersey, Pennsylvania, Delaware, and some parts of Maryland. The reasons for the preferences named are, that the trees are mature before they are felled and worked into spokes and rims, while this is not the case with the younger timber found in the West. The latter possesses weight, but requires cooking to stiffen it; it rots from undue dampness or exposure, and its sweet, young substance makes it subject to the attack of worms.

THE DECREASING SUPPLY OF HICKORY.—The writer has personally examined with great care all the sources of supply, and finds that we are fast reducing the available stock of this valuable timber, in view of which he would suggest the propriety of some action being taken by the Carriage-Builders' National Association regarding its wholesale destruction, which now prevails.

There are made annually, in this country alone, at least seventy-five million (75,000,000) hickory spokes. The amount of timber required to produce these would be at least seven hundred and fifty thousand (750,000) cords, and an equal amount is annually cut into rims and gearings. This one and a half million (1,500,000) cords of selected timber would exhaust five hundred thousand (500,000) acres of forest. When we remember the scattered growth of Hickory, which never presents a solid forest growth, but is intermingled with groves of oak, etc., these figures become appalling. Proceeding at this rate, the entire available supply in America is likely to be exhausted in less than ten years. It is time that someone should cry "Halt!"

That this subject demands attention is manifest to all connected with the business. The Government has tried to do what it could, but in a very general way. The remedy must come from those more directly interested, and it is to such organizations as yours that we look for the initiatory steps toward reform.

Having given the natural sources of supply of timber best adapted for carriage wheel making, we now proceed to the second topic suggested in the circular issued by your Committee on Prizes.

IS MACHINERY AVAILABLE?

"Can machinery be adapted to wheel-making; and to what extent?" The modern art of wheel-making is singularly susceptible of the application of machinery, as each of the successive processes required to produce the wheel consists of operations, the regularity and accuracy of which determine the success or failure of the finished product. These operations are emphatically mechanical, so peculiarly so that we have used the term "modern art" to distinguish the present era of manufacture from that in which the wheel was made from the rough timber, split and sawed from the log by the hands of one individual, and with the most primitive tools. The "modern art" has developed into what might be called "the assembling together of the different parts of a wheel." Machinery now turns and mortises the hubs, doing this work more accurately than can be done by the most skilled mechanic; and machinery turns, tenons, throats and polishes the spokes far more accurately than was formerly done by hand. Instead of the old segmental rims, we now have those in half-circles, bent by machinery. All of these detail processes have now grown into immense special industries, and the wheel-maker of to-day begins his work at that point where his grandfather was wont to congratulate himself upon having completed the hardest part of his task. Machinery does not stop half-way in its work, but nearly all carriage-shops are now supplied with improved machines, devised to accurately cut the tenon upon the ends of spokes, and to hold in place the rim, the holes in which are bored upon the same machine, thus still further reducing the labor.

DECREASE OF SKILLED LABOR.—As wheel-making is strictly a mechanical art, the sequence would seem to be, that modern wheels must be better than those made previous to the introduction of machinery, and just in proportion to the extent and mathematical exactness of the machines employed. But it is unnecessary to state that this is not so.

continued next page

Wheelmaking

In seeking the cause why the modern machine-made wheels do not generally possess the superiority that would be expected, some may be led to the conclusion that this is owing to degeneration in the quality of timber now employed, or to a defective mode of seasoning timber; while others may argue that hand-wrought material is necessarily better than machine-wrought. Those skilled in mechanical laws can controvert the latter argument, and point to the more rational theory of the decreased skill of the mechanics now following this vocation. That this decreased skill in the art is the natural result of the abolition of those elementary and secondary stages of what formerly constituted the trade of wheel-making, depriving the artisan of knowledge of the natural laws governing the materials employed, and frequently of the qualities requisite to secure the results required, is obvious; and when we further consider that the proportions of the spokes and hubs are in most cases fixed unalterably in the factories where made, we will readily recognize some of the reasons why the introduction of machinery has not produced results as superior as economical. The growing scarcity of intelligent and skilled workmen at the present time is not imaginary; there are, without doubt, many who can rank creditably with those of any other mechanical trade, but this number grows less with each decade, and must, necessarily, for the reasons above stated.

The remedy for this evil is apparent. It does not consist in going back to what some term "the good old way," but in advancing. One step in the right direction would be the establishment of technical schools, where the apprentice may become familiarized with everything calculated to be of use in his calling,—a movement that has already been successfully inaugurated in France and England.

As an illustration of what can be done in this direction, we respectfully call attention to the Artisans' Schools of Holland, not on account of their advancement in our special trade, but for the admirable system adopted in them, which may be applied to any technical school. In that country trouble was experienced in obtaining skilled workmen. Owing to the decline in remuneration, trades were only half learned, until the places requiring skill passed into the hands of foreign mechanics, of course to the detriment of national style and art. As a remedy the schools mentioned were instituted, and the result was most gratifying, Holland to-day possessing some of the most gifted mechanics in the world.

In this connection allow me to say that too much support can not be given to the periodicals representing our trade. Properly seconded, these have the power to become educators of the mechanics of the future, and every means should be used by those interested in the growth and development of the business in this country, to bring these trade journals before the younger mechanics, in order to stimulate in them the desire to become more thoroughly conversant with their adopted trades.

Nature has favored our own country, above all others, with natural and mechanical means to make us *the* carriage-builders of the world, provided only that native talent is developed, by technical education, to utilize the advantages that we possess.

In concluding this part of the subject, we can only add that there is no limit to the successful application of machinery to wheel-making. The only condition necessary is the *intelligent* application of the same.

MACHINE-MADE *vs.* HAND-MADE WHEELS.—To reach a solution of the question: "Which is the better, a hand-made wheel, or one made entirely by the aid of machinery?" requires that we should first distinctly understand what may properly be termed a hand-made wheel, and just where we are to draw a line between the two. As we have already stated, every wheel of to-day is more or less the product of machinery, and the question can better be understood if put thus: "Which is the better, one made in a wheel-factory as a specialty, or one put together in a carriage-shop by a single operator?"

The gradual decline, owing to the causes previously mentioned, in the skill and the number of mechanics who followed wheel-making as a vocation, in time created a demand for a wheel industry separate and distinct from the carriage trade, where a nearer approach to completing the wheel by machinery would be possible. The fact can not be disputed that wheels made by machinery, under the supervision of persons educated in all that pertains to a wheel and its duties, must be superior to those made in the less accurate manner of handwork, whenever the intelligence of the two corps of mechanics is equal. It may be said in general, to the prejudice of the machine-made wheel, that it is the product of a few skilled artisans, assisted by many unskilled, the latter being more interested in the quantity than in the quality of the work they produce, and educated only in the technicalities of the parts to which they are assigned. If wheels are made under these circumstances by men poorly paid, whose aim is to become mere automatons, rivaling in this respect the machinery they superintend, then the machine-made wheel will be a failure; or, if the aim be simply to produce something cheaper in price, irrespective of quality, then the days of machine-made wheels are numbered, and deservedly so; for, as a country advances in civilization, so each mechanical art approaches an era when individuality is given to each production, and it consequently becomes impossible to make many articles from one set model. This will always be a check upon the overgrowth of the wheel industry, the amount of skill required being the same, in either the carriage-shop or the wheel-factory, when the manufacturer is required to execute each order individually, and not by the hundred sets, as at present. As the question stands to-day, under like conditions of material and intelligence in manufacturing, the answer must be: *The machine-made wheel is the better one*, the proof of which statement we will give further on, in answer to the question: "What is the best manner of constructing a wheel?"

WHAT IS THE BEST MANNER OF CONSTRUCTING A WHEEL?

By Howard M. DuBois of Philadelphia.

(CAREFULLY REVISED AND CORRECTED IN PROOF BY THE AUTHOR.)

To construct a carriage wheel upon scientific principles requires a thorough knowledge of several complex natural laws and principles of mechanics, a proper understanding of which will enable the designer and the maker to correctly adjust the proportions of the different parts to suit them to their several duties. As a first condition, then, it is essential to know what those duties are, and, in the following remarks upon this subject, it should be remembered that this essay treats solely of wheels intended for pleasure vehicles, as distinguished from the heavier, burden-bearing wheels, which latter differ in some points.

LAWS GOVERNING THE CONSTRUCTION OF WHEELS.—The wheels of pleasure vehicles are subject to the following laws.

First, the wheel must support the normal weight of the gearing and

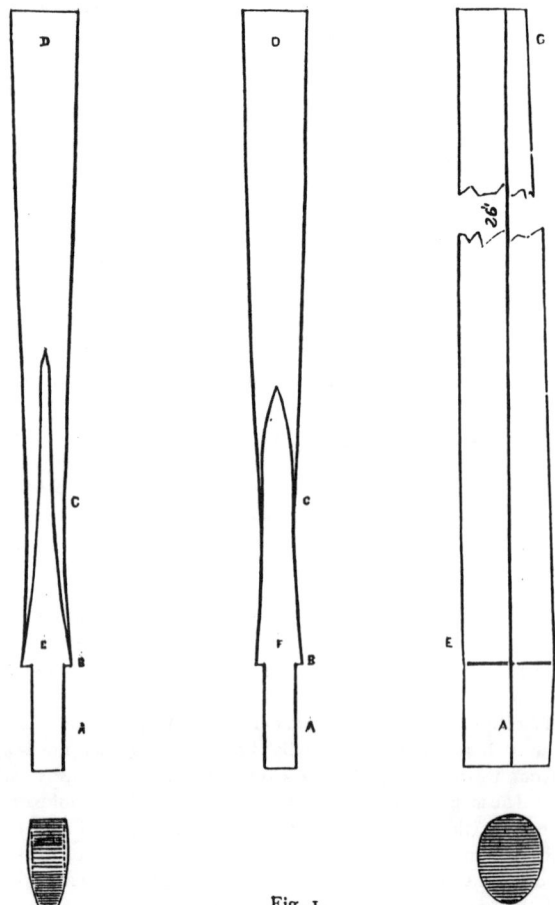

Fig. 1.

continued next page

body, together with passengers and luggage; this normal weight is represented when the vehicle is at rest.

Second, it will be understood that, when motion is imparted to the wheel, the spokes act as levers assisted by the rim to carry the load over a certain space, until the weight is taken by the next spoke in the line of rotation. This intermittent conveying of the weight to the hub and axle, resembles in its effects the striking of so many blows by a hammer, the vibrations equaling at each revolution the number of spokes in the wheel. These vibrations are felt through every fibre of the material, though scarcely perceptible to the touch or eye, but their effects become manifest when, by the heavy blows upon the tire caused by rough roads, stones, etc., they soon lead to the detection of the slightest error in construction, their irregularity constituting the most active agent in destroying the wheel. How to reduce these vibrations and render them harmless, before reaching the hub and axle, constitutes one of the most difficult problems in this art.

Third, the lateral swaying of the load produces a severe strain or end-thrust upon the collars of the axles, and, through them, upon the spokes and sections of the rim which happen to be in contact with the road or pavement at the moment the end-thrust is received. This difficulty can be successfully met and overcome by properly bracing the spokes in driving, and by properly shrinking on the tires.

Fourth, the internal pressure exerted by the tires binding the wheel together, also bind together to some extent the fibres of the wood, but, if weakness appear in any one part of the wheel, this internal pressure acts as a destroying agent.

The above principles point to the four most serious obstacles to be guarded against when proportioning the component parts.

PROPORTIONING THE DIFFERENT PARTS.

We will now proceed to give some idea of the proportions that have been found most effective, under the conditions above named, in accomplishing the most desirable results. We will take for an example a single wheel, and carefully give its proportions throughout; this will be all-sufficient, as, by observing its proportions, those of any other size of wheel may be calculated.

THE SPOKE.—As the size of a wheel is written by giving the size of the spoke, we will give, as our example, a spoke that measures one inch across the shoulders; see Fig. I.

For ease in reference we have divided this spoke into the following parts: A, tenon; B, shoulder; C, throat; D, body; E, face; F, back; and G, end. This drawing is made carefully to the scale of one-half full size, and the written measurements are consequently not necessary. A spoke of these dimensions will stand, without deflecting, an average of seven hundred and fifty pounds direct end pressure, and about two hundred pounds in the direction of end-thrust or lateral

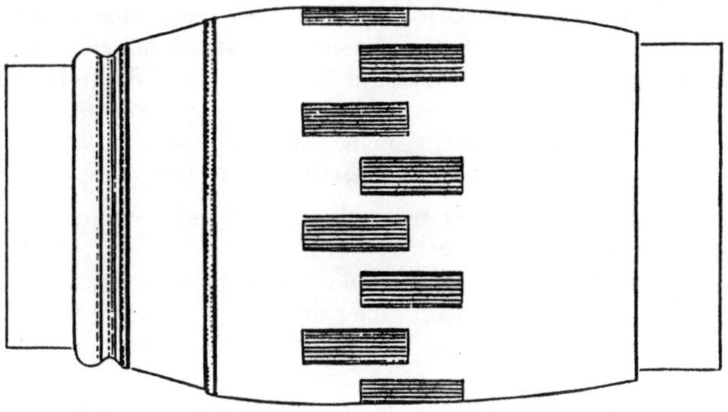

A

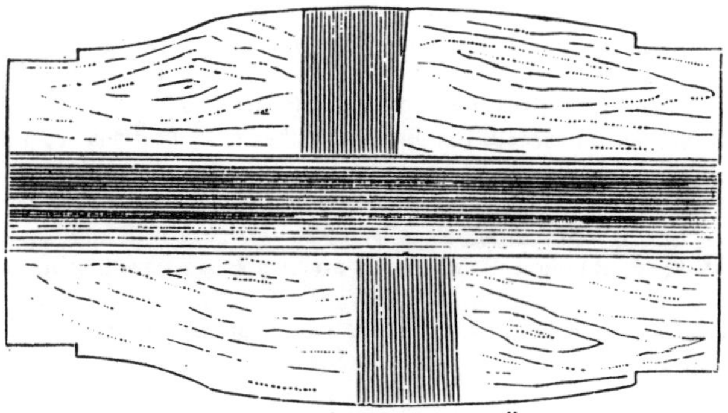

Fig. 2. B

strain. These dimensions have been determined upon as the result of long experience.

THE HUB.—The hub may be of either elm or gum, as previously described in detail, and it should have a contour which will allow as much strength as possible consistent with symmetry. We herewith present a sample hub shown in two ways and patterns; see Fig. II., in which A shows the full hub, and B a sectional view of it cut longitudinally through the center. These cuts show a hub of proper proportion to suit the spoke shown in Fig. 1; they represent the hub in one-half its full length, and its dimensions can consequently be easily reckoned. These cuts plainly indicate the size of the mortise, in width, length, and "mash" or taper.

PROPORTIONS OF HUB-BANDS.—We would here call attention to the bands, which form an important part of the hub, not to be overlooked. A list of the proportions best adapted for these is given in the table, Fig. 3.

PROPORTIONS OF HUB BANDS.

DIAMETER.			WIDTH.		
Diam. of Hub.	*Front Band.*	*Back Band.*	*Length of Hub.*	*Front Band.*	*Back Band.*
3 in.	2 in.	2⅜ in.	6 in.	⅝ in.	⅝ in.
3¼	2¼	2⅝	6½	⅝	¾
3½	2½	2⅞	7	¾	¾
3¾	2⅝	3	7½	¾	¾
4	2¾	3¼	8	1	1
4¼	3	3½	9	1	1
4½	3⅛	3¾	10	1¼	1
4¾	3¼	4	12	2	1¼
5	3½	4¼			
5¼	3¾	4½			
5½	4	4¾			
5¾	4¼	5			
6	4⅜	5⅛			
6½	4¾	5½			
7	5¼	6			
7½	5½	6¼			
8	6	6¾			

Length of Hub in Proportion to Diameter.

Diam.	*Length.*	*Diam.*	*Length.*
3 in.	6 in.	5 in.	7–7½ in.
3¼	6½	5½	7–7½–8
3½	6½	5¾	7–7½–8
3¾	6½–7	6	8–8½–9
4	6½–7	6½	8½–9
4¼	7	7	9–10
4½	7	7½	10–12
4¾	7	8	10–12

Fig. 3.

The sizes and proportions shown in the forgoing table have been in constant use for years.

continued next page

MORTISES.—The number of mortises required should be the next subject for consideration. It has been found that the best results are obtained by giving fourteen to a hub of this size. As a rule, the fewer the number of mortises, the better (in any size), so far as the durability of the hub is concerned, for the reason that the vibrations are then reduced to a minimum, and the "spring" of the wheel is increased, thereby tending more readily to dissipate those shocks which so often injure wheels containing more spokes. For the ordinary height of wheels it would be well, wherever practicable, to use this number of spokes.

These mortises in the hub should not be made until the operator is ready to drive the spokes, as the interior surfaces thereby exposed will

PROPORTIONS OF SPOKES.			
Size of Spoke.	Thickness of Body.	Width at Point.	Full body thickness distant fr. shoulder.
$\frac{7}{8}$ in.	$\frac{5}{8}$ full	$1\frac{1}{16}$ in.	$6\frac{1}{4}$ in.
1	$1\frac{1}{16}$	$\frac{3}{4}$	$6\frac{1}{2}$
$1\frac{1}{16}$	$\frac{23}{32}$	$1\frac{3}{16}$	$6\frac{1}{2}$
$1\frac{1}{8}$	$1\frac{3}{16}$	$\frac{7}{8}$	$6\frac{1}{2}$
$1\frac{3}{16}$	$1\frac{13}{16}$ full	$\frac{7}{8}$ full	$6\frac{1}{2}$
$1\frac{1}{4}$	$\frac{7}{8}$	$\frac{15}{16}$	$6\frac{1}{2}$
$1\frac{3}{8}$	$\frac{15}{16}$	1	$6\frac{1}{2}$
$1\frac{1}{2}$	1	$1\frac{1}{8}$	$6\frac{1}{2}$
$1\frac{5}{8}$	1 full	$1\frac{1}{4}$	$6\frac{1}{2}$
$1\frac{3}{4}$	$1\frac{3}{32}$	$1\frac{3}{8}$	$6\frac{1}{2}$
2	$1\frac{3}{16}$	$1\frac{9}{16}$	$7\frac{1}{2}$
$2\frac{1}{4}$	$1\frac{5}{16}$	$1\frac{3}{4}$	$7\frac{1}{2}$
$2\frac{1}{2}$	$1\frac{1}{2}$	$1\frac{13}{16}$	$7\frac{1}{2}$

Fig. 4.

warp or shrink "out of true," and there will also be a tendency for them to check in the corners, further preventing the accurate fitting of the tenons of the spokes. Long exposure to the atmosphere of the interior parts of the hub will also occasion loss of vitality.

DODGE OF THE SPOKE.—The amount of dodge or "stagger" given the spoke calculated to secure the best results, is found to be one-half inch. Anything beyond this leads to a strain upon the rim tenon, and to a result which outbalances the slight advantages otherwise gained by increasing the amount of dodge above this point.

PROPORTIONS OF SPOKES.—In Fig. 4 we give a table which shows the sizes and proportions of spokes founded upon the principles already laid down.

RIMS.—There are several important details in the working of rims which need a thorough explanation. The most important of these is, the tendency of the fibres to assume the artificial position given them during the operation of bending, as a result of the following causes. In bending a rim around the circular form, the end of the timber is pressed upon by a screw, which forces together the fibres on the inside of the circle, or, as it is technically termed, "upsets" them. This is done in preference to allowing the outer fibres to stretch, as would necessarily occur if the piece were simply bent around the form with the mechanical restraint mentioned. Rims are bent upon forms whose circle is much smaller than that of the bundles in which they are packed. This is owing to a certain amount of spring or reaction in steam-wet timber. On becoming thoroughly seasoned, the rim draws in, as the inner fibres are crushed into shorter space, and their tendency is to resume the circle to which they were originally bent, and—if the "upsetting" has been very severe—sometimes beyond that point. Fig. 5 illustrates this tendency, the dotted lines showing the height or circle the rim has been packed, and the full lines showing the

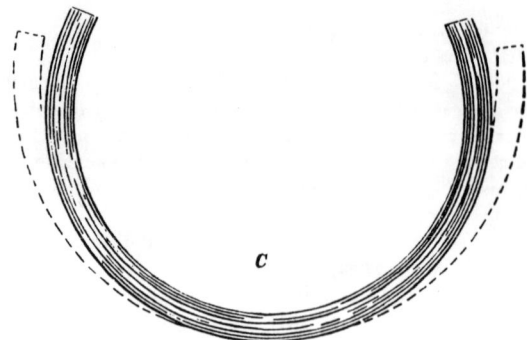

Fig. 5.

springing together mentioned. If, therefore, a rim bent upon a circle or form whose diameter measures 3 ft. 6 in., should be used to make a wheel of, say, 4 ft. 2 in. (the rims being marked the correct height in the bundle or package), the result would be that the joints would flatten, regardless of the "opening out" while fitting them upon the spokes. When we add, that it is the general custom in rim factories to bend them upon circles less in diameter than the wheels upon which they are intended to be used, we can understand the cause of the many complaints of "rim bound" and "low jointed" wheels. On the other hand, if a rim is used which has been bent upon a form as near the height of the wheel to be made as possible, then the resistance of the fibres will be against the tires, and the joints will always retain their proper position.

Much has been said by wheel-makers about the proper manner of bending rims, as regards the direction of the growth-rings. Many argue that, when they are perpendicular, as shown in Fig. 6, cut A, the rim is less liable to split during the operation of boring. This may be true, but nature teaches that wood is better able to withstand the effects

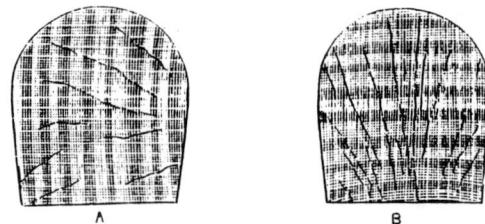

Fig. 6.

of vibration and jarring when these layers are in the opposite position, as shown in Fig. 6, cut B, and also not so liable to split when a dowel is used, as is still the general custom with wheel-makers. If rims are used which have the foregoing requirements, the old proportion of having them measure the same in depth that the spoke measures in width, will be found serviceable; but the more modern style is to have the rim greater in depth, by one-sixteenth or one-eighth inch, than it is in width. This change has been made to overcome the trouble previously mentioned, but without success. These proportions may vary with the wants of each locality, as they are not arbitrary. Many wheels are rendered unserviceable by reason of the amount of wood removed from the rim in boring for the spokes. The amount thus taken out should never exceed one-half the width of the "tread" or width, and this is even more than is compatible with the durability of the rim, but, if the holes were made smaller, the spokes would not be able to resist the extra strain brought to bear upon them by the modern street-railways.

We have now described the more important of the mechanical principles which apply to the component parts of a wheel: the hub, spokes and rims, and we next come to the manual part of the work, which consists in assembling these parts together and forming a wheel.

continued next page

Wheelmaking

PART III.

PUTTING A WHEEL TOGETHER.

The doctrine advanced by a system of theology that "the end justifies the means," is eminently applicable to mechanics.

DIFFERENT MODES OF USING TOOLS.—It was the writer's privilege, during the preparations for the Centennial Exhibition, to see and compare the differences in mechanical operation exhibited by the artisans of nearly every nationality of the world. The mechanics of each country represented showed a different method of attaining the same general results. This was particularly noticeable in the case of the Japanese, who entirely reverse our manner of using tools; the teeth of their saws were made to cut upwardly, and they sawed in this direction; in planing, they drew the board over the plane. When the work was finished, however, no one could detect whether it had been executed by an American or Japanese mechanic. It would be useless, therefore, to lay down any positive rules as to how the wheel-maker's work should be done, but we will give the laws governing the material employed, and we will endeavor to explain them in such terms that any mechanic, so desiring, will be able to understand and apply them.

CONDITIONS TO BE OBSERVED.—We can not lose this opportunity of impressing upon the wheel-maker the importance of observing the following primary requirements of success in his work.

First, be absolutely certain that the timber you are to employ is thoroughly seasoned and perfectly dry before using it in any part of a wheel.

Second, see that the proportions of all the parts of the wheel are relatively correct.

Third, in the important operation of driving the spokes, see that the fibres of grain of the wood are not crushed by being "mashed" into the hub, but see that they are perfectly fitted, with just enough draw given to assure the contact of every part of the surface of the tenons and mortises.

Fourth, remember that fresh, good, tough glue is necessary, to cement together the fibres, the quality of elasticity in glue being also important.

Fifth, remember that a dry atmosphere, during the process of driving the spokes, is requisite to durability; it is therefore advisable to avoid driving when the weather is rainy or humid, and for these reasons special precaution should be taken during the months of March and April.

Sixth, see that the wheel when finished and ready for the tire is free from all strain, as that caused by forcing a spoke into line while rimming; the wheel should be a perfect mechanical whole.

Seventh, artificial heat should be applied to the tenons of the spokes both before and at the time of driving, in order to reduce all swelling caused by dampness taken up from the atmosphere.

To give detailed instructions by which all these conditions may be observed and these ends accomplished, would require more space than can be covered by a single essay. The writer is describing and illustrating this subject more in detail in a series of articles appearing in the current volume of *The Hub*, a magazine well known to your Association and to the trade in general, to which the reader is referred.

RELATIVE MERIT OF WOOD-HUB AND PATENT WHEELS.

If all the conditions mentioned above had been observed, and all the operations been mechanically and faithfully performed, this last topic would have no place in an essay; but through want of skill in proportioning, together with bad workmanship and worse material, a desire was created for a wheel possessing more strength and durability than characterized the ordinary wood-hub wheel. The latter when properly made had stood the test of centuries satisfactorily, but this new western world now came forward and produced something novel, if not ornamental or useful, in the shape of various patent wheels, each of which, if the advertisements could be relied upon, was superior, not only to wood-hub wheels, but to all others. The idea took. Patent wheels became popular. This country seems to be peculiarly subject to contagions of this kind. No trade, no calling, no profession is exempt from them. In this instance the carriage trade was widely affected by the new mania, but it may be said to have passed the crisis, as the "patent wheel fever" seems now to be abating.

MERITS OF EACH COMPARED.—We propose, in the following, to briefly examine and compare the principal varieties of wheels now in the market, noting their relative mechanical merits as compared with the wood-hub wheel. We will endeavor to do this unbiassed by any question of self interest, seeking only to impart general information upon a subject of great interest to the carriage trade. Indeed, this may be spoken of as one of the questions of the day.

CLASSIFICATION OF PATENT WHEELS.—Patent wheels may be divided into two general classes, and one sub-class of the second.

First, those having the flanges of iron cast separately, and forced upon the hub after the spokes (which have short tenons) have been driven; the spokes form a solid arch around the hub, to accomplish which requires that each wheel should have two more spokes than the plain wheel.

Second, those which are made with a band of iron around the hub, through which are "cored" the mortises; this band is forced upon the hub before the latter is mortised, the spokes being driven through the mortises in the band into the wood of the hub.

There is an offshoot or sub-class of this latter patent which is characterized by a club-footed spoke; this is forced into a groove in the hub, but this possesses the defect that the hub, by this groove, is rendered liable to be broken asunder, as after "boxing" there remain only a few thin layers of wood.

The first two classes above named involve the principles upon which all the patents now under consideration hinge. The various modifications of these are surpassing in number if not in merit.

DEMERITS OF PATENT WHEELS BELONGING TO CLASS FIRST.—In analyzing the wheels which belong to the first classification, one must be impressed by the rigidity or stiffness which its principles involve. With wheels made by this system, there is as little real spring about them as in the most primitive of all wheels: those made from the solid block. This rigidity is not imaginary, as a ride over them upon the rough pavements of our city streets will plainly demonstrate. It is by reason of this peculiar quality of stiffness that patent wheels generally have declined in popularity, and it will result, we think, in their extinction. The causes of this stiffness or rigidity lie in the high iron flange used, and in the mass of end fibre accumulated and braced by the arch, which latter occurs just at the point where the ordinary wood-hub presents a natural cushion, counteracting the concussion and jarring of the spokes, produced by the wheel coming in contact with obstructions in the road. This cushion of fibres laid horizontally, as in the wood-hub, forms a wonderfully sensitive spring, firm enough to be durable, and sufficiently flexible to fully answer all the requirements of the case. That this peculiar flexibility exists wherever wood fibre is used in this position, is proved by a trial of the difference between ascending a stairway of iron, and one where the steps are made of wood. Wood is one of nature's most available cushions for the dissipation of constant vibratory motions, and as yet the mechanical world seeks in vain for a substitute possessing all its qualities. Want of elasticity is the great demerit in patent wheels belonging to the classes referred to. No further proof of this is needed than the frequent breaking of axles by reason of the crystalization of the steel or iron, primarily induced by the rigidity of the wheels. This breakage is fully twenty per cent. more common than it was before patent wheels came into use,—and this, too, notwithstanding the fact that, as a rule, axles one size heavier are employed than with wood-hub wheels of the same dimensions.

The question may now properly be asked: "In what respects are they stronger than plain wood-hub wheels?" This is an unsettled point. The iron band certainly prevents the wheel from breaking when subjected to lateral strains or end-thrusts of the axle.

When repairs are necessary (and all wheels need repairing sooner or later), then "comes the tug of war" in the use of patent wheels, and the mechanic who can undergo this test of character without breaking the third commandment, deserves to receive a prize as a model of patience. The additional cost of repairing patent wheels plainly indicates the additional labor involved.

continued next page

Wheelmaking

DEMERITS OF PATENT WHEELS BELONGING TO THE SECOND CLASS. —Wheels belonging to the second class mentioned possess most of the defects and few of the virtues of those of the first class. In comparison with well-made wood-hub wheels they are not so durable, as the chafing of the tenons of the spokes during the operation of driving renders them liable to splinter or break across the back when the tires are put on tightly. Under these circumstances dampness will soon penetrate the injured wood, and its destruction will be further assisted by the oxidation of the iron in contact. Where this injury does not occur, wheels of this kind may, with care, be made to last as long as a plain wheel ; but even then, the "stilty" motion imparted by them to the carriage, and the increased drumming noises which they produce, are features which render them undesirable.

The sub-class before mentioned, allied to class second, is, when considered from a mechanical standpoint, a barbarism. There is not a truly scientific principle involved in the patented portion, and it is a wonder that wheels of this kind should have been used by carriage-builders. In these, a groove is cut into the hub, leaving only a few of the growth-rings of the heart-wood between the box and the "clubbed" end of the spokes, the latter being driven down upon an inverted wedge, which has a tendency to force the hub asunder laterally, but this the spoke is prevented from accomplishing by its resting upon a slight incline in the sides of the iron band, producing this novelty in the way of a wheel: its strength depending upon the arch of the iron band, which prevents the spokes from destroying the hub, while the wedged ends of the spokes do not perform any useful office whatsoever. With wheels of this kind, the writer has in two instances seen the band and spokes loosened from the hub and moving around it, while the hub was immovably fastened to the axle by the binding of the nut on the axle-arm.

There are other defects in these classes of patent wheels, but like the six reasons once offered to explain the non-appearance of a witness at a trial, the first of which was, *that the witness was dead*, so we think we have already given a sufficient number of reasons, and of a sufficiently grave nature, to show why, in our opinion, the patent wheels now known to the trade can never, in any true sense, replace the wood-hub wheel. We utterly leave out of the question, the matter of symmetry and beauty of outline, in which respect patent wheels are so conspicuously defective that no argument is required. Moreover, this is a secondary consideration, as utility is the one grand end aimed at, and time alone the demonstrator of merit.

This reference to time reminds us that it is now time to bring this essay to a close, which we do in the words of Byron, who has said :

> "As forests shed their foliage by degrees,
> So fade expressions which in season please ;
> And we, and ours alas ! are due to fate,
> And works and words but dwindle to a date."

HOWARD M. DUBOIS.

APPLIED MECHANICS IN WHEEL-MAKING.

Chapter VII.—Appliances used in Driving the Spokes.

THE consideration in the previous chapters, of the natural laws and principles to be observed in designing the component parts of the wheel, leads us now to the subject of the tools necessary to successfully accomplish the mechanical portion of the work. This is a subject far too extensive to be treated of in detail, owing to the multitude of special tools invented within the past few years, demonstrating clearly that this division of the business is pre-eminently mechanical,—by which, we mean, that there is no limit to the adaptation of machinery to its different stages or manipulations.

It may not be out of place to add here, that this gradual growth of the use of special tools has certainly lowered the standard of mechanical skill on the part of those who follow this branch or calling. The mechanic of old constructed his wheels from the rough material and with the most primitive tools, depending upon his skill to fashion these elementary substances into the proper proportions. Now the material is formed ready for the finishing process, with infinitely less expenditure of skill and time, thus rendering unnecessary part of the work which formerly gave prominence to this department, and consequently demanding less skill of the artisan, and lowering the standard as before stated. No better proof of this is needed than a careful survey of the bench and tools of the modern wheel-maker ; we mean the average condition, for there are many individual cases where the operators can rank with the mechanics of any other profession ; but the majority will be found, on inspection, to be below the other departments of wood-working in the condition and care of the common tools of the trade, such as saws, planes, chisels, etc., together with the general appliances of the work-shop.

While it is not within the reach of every mechanic to possess *all* the newest tools, it is possible for him to keep those which he does have in the best working order. The cost of a neat bench, "driving horse," etc., is so slight in comparison with the convenience, that any one, with the proper timber and metal work, may construct these without any great trouble. We will proceed to give a description of a "driving horse" having all the requisite qualities, and possessing the advantages of being easily made and operated.

The illustrations, Figs. 1, 2, 3, represent the horse or bench, drawn to the scale of one-sixteenth. Fig. 1 shows the front elevation ; Fig. 2, the top view of the same ; and Fig. 3, the side view, showing the hub in position, with the driving staff, gauge, or sett, ready for use. The letters refer to the same parts in all. It will be seen that an ordinary work-bench forms the foundation. This is made preferably of ash, and, as shown at A, should be four inches in thickness. Into the face of this bench are made two mortises, into which are fitted the lateral projecting bars, B, at the distance of six feet from bar to bar. These bars are fitted neatly, so that they can be removed by withdrawing the dowel pins, P, when the horse is not in use. The uprights or standards, C, are mortised into the bench, as shown in Fig. 2, and can be driven down flush. When the other parts of the horse are removed, the standards belonging to the front bed, E, are fastened securely. An angle iron, shown at G, is let into the bench, for holding the back band of the hub ; also into the bed, E, for the front band. The object of the keys or wedges, F, will be readily seen, as the bed, E, can be adjusted to suit any length of hub, and be secured rigidly by driving these keys slightly. The pressure bars, D, do not need any explanation. H represents a steel spring with a projecting center point, which presses upon the driving "sett" or "gauge," as shown in Fig. 3. This spring is bent, as shown, to admit the introduction of a wrench or lever (not shown) long enough to force the spring out, allowing the gauge or sett to be slipped down till the center point enters a countersink in the gauge-plate. The screw-bolt, I, releases the spring so that it may be raised or lowered to suit any diameter of hub. With this spring the hubs can be used without plugging the ends, a great saving of time, as it can be applied to any size, and holds the "sett" K truly and with uniform pressure. It is entirely out of the way of the wheel-maker when paring out the mortises of the hub, during the operation of fitting the spokes.

Fig. 4 shows a side-view of the "sett" or gauge, K, showing the the steel plate, N, with countersink before mentioned. The screw, L, answers the same purpose as any other sett, to give the inclination or "dish" to the spokes in driving. Figs. 5 and 6 are full-size views of the index slide, which moves upon the "sett," K. The pointer, O, shown in side view 6, marks the height of the sett screw, L, from the center of the hub, and can be adjusted to any desired height by the binding screw, M. The appliances shown are simple and inexpensive, and occupy comparatively little room, the work-bench being left in its normal condition upon the removal of the attachments.

One would scarcely imagine that there could be anything written about so ancient and common-place a tool as the mallet, but from the time of Sampson these have been made of varied forms and substances, and the wheel-maker finds a great difference in the driving

continued next page

Wheelmaking

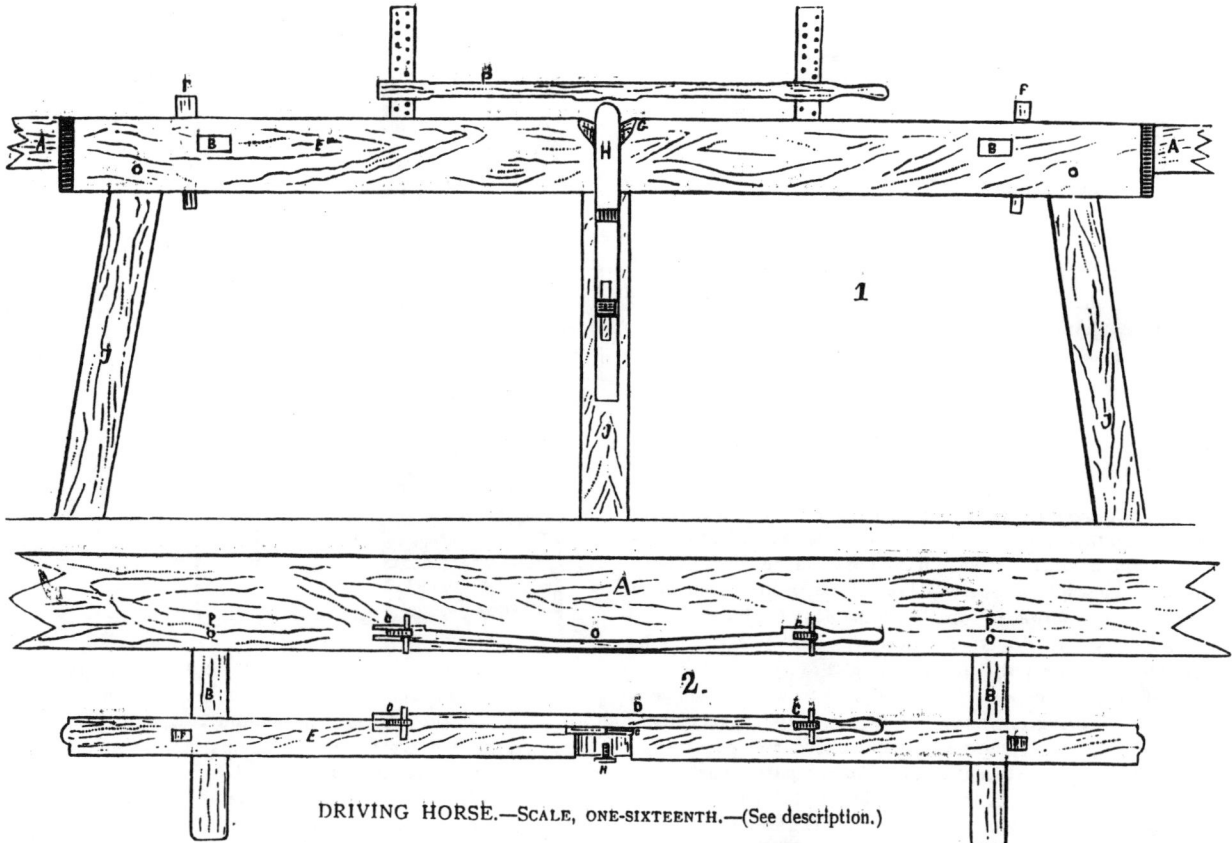

DRIVING HORSE.—SCALE, ONE-SIXTEENTH.—(See description.)

qualities of the various shapes. The best results are obtained by the use of long mallets like the one shown, side and face view, in Fig. 7. This is supplied with a handle, which is shaved quite thin at the neck, while the width is maintained sidewise, as shown in the face view. This insures full spring to the reaction of the blow delivered, while it prevents the tendency to splinter and break, so generally the case when the handle is made thin in both directions.

It has been demonstrated by experience, that a spoke should be driven with as little spring or vibration as possible; therefore, any concussion greater than is absolutely necessary to force the spoke "home" is not only a waste of muscle and energy, but a positive injury. To obviate this, we should have mallets proportioned to the spokes we drive. The dimensions of the one shown are 2¼ inches by 2 inches, 6¼ inches long, length of handle 18 inches, and when made of dogwood, live oak, or hickory, it would weigh 1¼ pound. This size is suitable to drive spokes up to 1⅛ inch, while one made 3 by 2½ inches, 8 inches long, handle 21 inches long, would weigh 2½ pounds, and be suitable for a variety of sizes. These weights and figures are not arbitrary, but the facts are, for there *is* a difference between *driving* a spoke and *crushing* it into the hub.

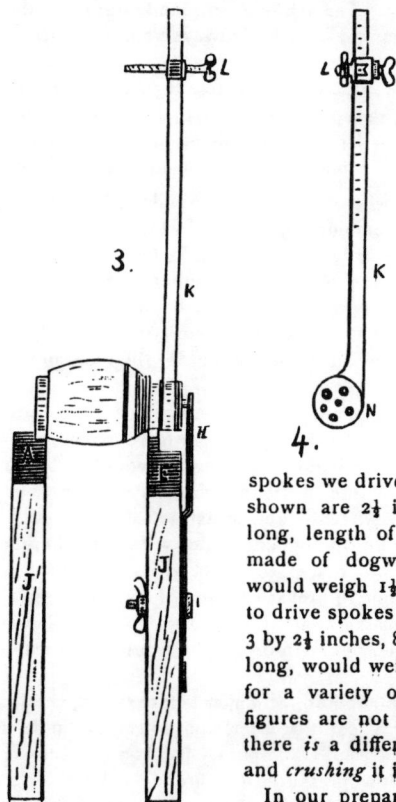

In our preparation for driving spokes the question of glue is an important one. Every wheel-maker has had some difficulty in procuring the proper quality for the purpose. Low grades of glue are of no use whatever in this work. In almost all kinds of joinery the parts fastened by gelatine, or glue, are not subjected to the violent jarring and straining experienced by the spokes in a wheel. The joints of the body of a vehicle receive these con-

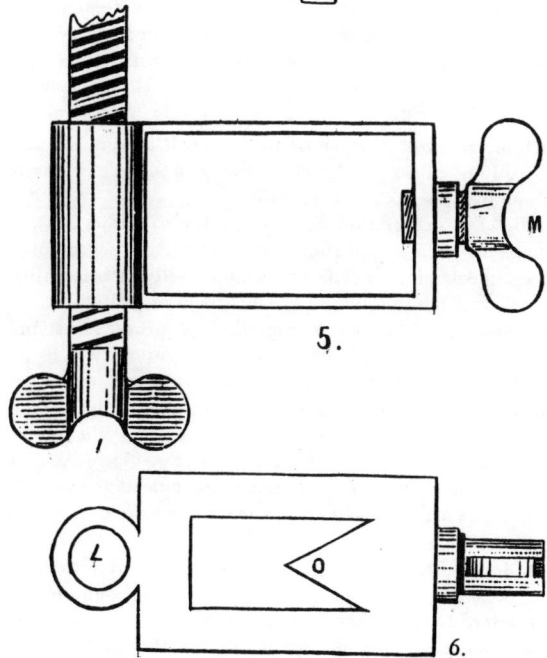

cussions only after they have passed through the springs; therefore, a glue to hold under such conditions must possess *elasticity* as well as adhesiveness.

The very best pure glue becomes in drying very hard, and in some grades brittle. To remedy this, a small proportion of "bonnet" or "millinery" glue should be added. The latter is a glue made from calves' sinews, instead of hides, very white and flexible, that will bend without breaking, and this should be added in the proportion of one ounce of "bonnet" glue to seven ounces of pure transparent glue.

When glue has a dark, muddy appearance, it is owing to impurities and foreign substances other than pure gelatine, which latter is transparent even in thick flakes. Some manufacturers flake their glue very thin to give this transparent appearance. In this case, whether it be French or American, any glue that will not absorb ten or twelve times its own weight of water without "thinning out," is not first quality. For wheel-makers' use the flake or sheet glue is preferable, as there is more difficulty in getting a pure powdered glue, or at least in judging

continued next page

21

Wheelmaking

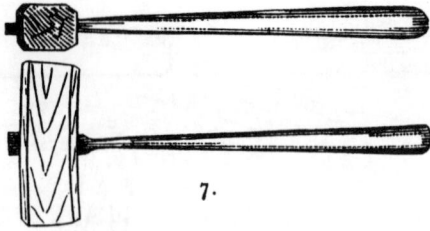

7.

of the quality. The question of cost is of little consequence compared with the advantages gained by using the very best. Good material, intelligently applied, must produce good and lasting results.

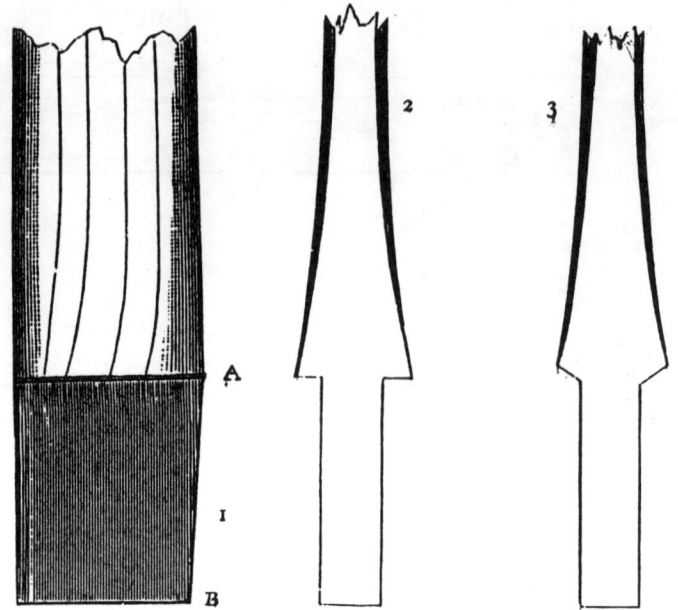

Chapter VIII.—"Driving" the Spokes.

THE march of improvement in the mechanic arts has failed to lessen the barriers between theory and practice. The mere theoretical knowledge of an art or trade will never enable the possessor to appreciate all the difficulties in the way of successful execution ; while, on the other hand, the unchanging routine of purely mechanical motion gives no chance for any advancement in style or design ; each therefore has its sphere of usefulness, and should go hand in hand to this end.

One great difficulty in the way of converting theory into practice is the impossibility of making any fixed law to govern the movements of the artisan in working out a given design. Each mechanic posesses an individuality in the accomplishment of the different parts constituting his trade. This individuality is influenced by his local training to a certain degree, but no two use *exactly* the same means. This dissimilarity in reaching the same results is so wide that in some parts of the world the tools and modes of use are entirely reversed, yet the end accomplished is the same, and no one can distinguish by which set of mechanical motions the work was produced. It will be seen by this, that any attempt to lay down arbitrary laws or rules to govern the *manner* in which mechanical work *must* be done will end in failure, every-one being ready to admit that it is far easier to tell when anything is properly done than to teach how to do it.

In the following description nothing is arbitrarily laid down except-ing the object to be accomplished. The manner of accomplishment rests with the mechanic himself, who should cultivate originality, after mastering thoroughly the principles of his trade or calling.

Having given, in the previous chapters, the proportions for hubs, spokes, rims, etc., also the bench or horse upon which to hold the work while being driven, we will next suppose we have a set of the different parts ready to frame together ; and at the risk of repe-tition we will add that the wheel-maker at this stage of the proceedings should be morally certain that this material is *perfectly seasoned*. No risk should be taken on a point of such vital importance, for what the keystone is to the arch, so is seasoned material to the durability of the wheel.

Our attention is drawn first to the spokes, which we further prepare by bringing the tenons to the proper lengths, which should not be more than one-quarter, nor less than one-eighth, of one inch longer than the width of the spoke across the tenon, at the shoulder. These (the tenons) will also be found too full in thickness for the mortises, as they come from the factory. Great care should be exercised in reduc-ing them to the proper thickness, to keep them perfectly parallel or of the same size from shoulder to end, as any fullness at the shoulder will be sure to check or crack the mortise in the corners when the spoke is driven. The plane generally used in the shop for this purpose is too well known to need any description. When kept in good order it answers very well, but there is room for the inventor in the production of an inexpensive tool for this purpose, easier to manage, and less sub-ject to derangement.

There is great difference of opinion existing in regard to the amount of fullness to give the tenons sidewise, but the majority favor making them to fit so that the pressure of the hand can force the tenon down when the end is applied to the mortise. Reason seems to be on this side also, as nothing can be gained by heavy compression sidewise

except cracking the hub, which may not ensue at the time of driving, but after it has been used in the weather for a time. The difference in size between the spokes at the shoulder, A, and at the end, B, (see Fig. 1,) should be one-sixteeeth inch ; this may be increased on the larger sizes to one-sixteenth full.

There is also a difference of views in regard to the proper shape of the shoulders as they rest upon the hub. It is quite common to have them driven just as they come from the factory (see Fig. 2), while others use them as shown in Fig. 3. This latter is decidedly the bet-ter plan, as the shoulders are not so liable to "shell" off in driving, nor the face of the spoke to split during service, as when used as shown in Fig. 2. One object in using them in the latter style is the trouble of cutting them to the proper bevel, but if wheel-makers would uni-versally adopt the beveled shoulders, they could be made at the factory as easily as the square shoulder. The spoke as originally put upon the market was tenoned in that manner, and has always kept the original shape.

There is some difference of practice, even among the best hand-workers, in reference to sawing the spokes nearly to their proper lengths before driving ; others prefer sawing them after being driven. There is very little choice in this matter. It may be said, in the first case, that the spokes drive with less spring, when cut to the proper lengths ; on the other hand, it is urged that the tendency to "mash" while being driven often spoils the spoke, but the danger of mashing is lessened when a square surface is presented to the blow of the mallet. Whether the spokes are cut before or after driving, they should be selected so that the stiffer ones are reserved for the longest, or hind wheels. These should be again separated, leaving the stiffer timber for the rear spokes of each wheel.

After preparation in the above manner, place them tenon end toward the oven of drying stove. No spoke is dry enough to be driven with-out this precaution ; and if the day is rainy or damp, it would be better to avoid driving until the atmosphere is dry enough to insure the wood against that worst enemy of the wheel-maker, moisture.

This may sound absurd to one who depends upon a certain number of wheels for his weekly pay, but disregard of these seemingly slight points often compels the mechanic to prevaricate and blame the material worked, when the fault lies nearer home. The builder's reputation is often injured when a little patience and forethought would have made all secure.

Our attention is now called to the hub, which, as before stated, should be well protected by two coats of paint or filler, and, to secure the best results, it should be freshly mortised. Temporary bands are necessary to prevent the unevenness of the pressure from forcing the hub open during the process of "driving." These bands, as a general thing, are roughly made by the smith, and used just as they come from the anvil. Some have them wrapped with cloth to prevent bruising the hub.

continued next page

Wheelmaking

This accomplishes the object to a certain degree, but the cushion of cloth allows the hub to expand, and in some cases it cracks out from under them. The best plan is to have the bands turned out smooth in the lathe; this is expensive, but it pays in the long run. After they are properly prepared, care should be used in forcing them on and off. In many cases the bead work suffers from carelessness in this operation. Having adjusted the bands properly, the face end of the hub should be examined, to find if it is perfectly true and square. If there is any fullness in the center, it should be planed or filed until the driving "sett" or "gauge" strikes upon the surface all around.

The hub is now ready to place in the horse for further manipulation, where it is confined by the pressure bars or other device. The driving "gauge" or "sett" is applied as shown in a previous chapter. The slide of this gauge is then adjusted to the height that the wheel is to be driven. For wheels whose spokes measure under one and three-eighths inch, the set-screw of the slide should project so that the forward or face spokes (after being driven) stand back from a perpendicular line, enough to bring the wheel straight when the rims are put on; this would be about one-sixteenth inch full, as an average. For the heavier coach wheels the gauge can be set to any "dish" required. Having adjusted the gauge, we next apply a straight-edge to the hub, by passing it down into and against the face of the mortises. These will be found to have far too much dish for ordinary work, as the straight-edge will show upon the screw of the "sett" or "guage." To remedy this we pare out the bottom of the mortise until the straight-edge stands upon a true line with the end of the screw on the "sett." The paring should be done with a chisel the exact width of the mortise, to insure square corners, which are necessary to a perfect fitting of the spokes.

We will remark here that the straight-edge should be what its name implies, and not a crooked stick, which often goes by this name. We have seen a neat design for a straight-edge, which has a small brass slide resting upon the hub in the same position as the shoulders of the

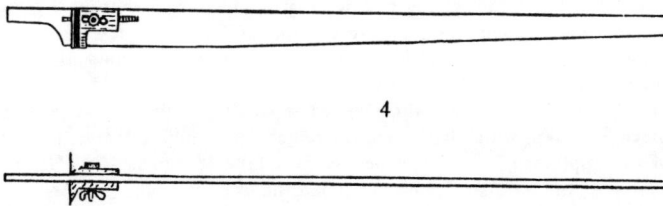

4

spokes. This ensures the line being perpendicular, and avoids running the instrument too deep into the mortise. Fig. 4 gives a side and a face view of this design.

Mortising chisels should be carefully selected from the best steel, as they require to be ground with a long bevel, and not left to wear short and stumpy. To do good work the chisel should pare, not *break*, the wood from the bottom of the mortises. Any chisel which will not bear grinding to an angle of twenty degrees from the horizontal is not fit for this work.

In ordinary hubs mortised at the factory, many are so made that they (the mortises) have to be opened out at the back to fit the spokes. This fitting is accomplished by making a templet representing the tenon of the spoke, minus one-sixteenth inch scant allowance for mash; the mortises are then made to correspond to this templet. Others rest a spoke upon the hub, and by striking lightly with the mallet a mark is made upon the hub as the limit for mortising, the bottom of the mortise being gauged by calipers; this latter plan is very uncertain, as it requires the calipers to be held always in one position to secure uniformity.

The writer's advice would be to try and secure a hand-mortising machine, and with this mortise the hub to the exact size required. If this can not be done, then have the hubs mortised as nearly right as possible at the factory, and fit each mortise with a templet as described. It is very important that these tenons and mortises should fit the entire length, and not be crushed out of shape by forcing far more wood into the hub than has been taken out in mortising. This part of the subject requires the closest attention on the part of the operator, as each different grade and kind of material requires peculiar treatment, and fullest acquaintance of the artisan, to secure the best results. No amount of written instructions can convey the various slight changes necessary to make each part correspond in this particular. Experience can only be earned by practice.

As the hub is of varied diameter in its length, a sufficient amount of wood must be removed to allow the shoulders of the spokes to bear evenly their whole width. To do this correctly, the angle the spoke is to be driven should guide the amount to be taken out of the hub. This is best accomplished by cutting across the grain at the corners of the mortises, and removing the wood so that the shoulder of the spoke will fit to the angle. Fig. 5 shows a section cut through the center of the hub, exposing the mortises; the chisel is shown paring the wood to the angle described. If the hub to be operated upon is of small diameter, there is some danger of forcing down the "walls," or material separating the mortises, during the entrance of the spoke in driving. To

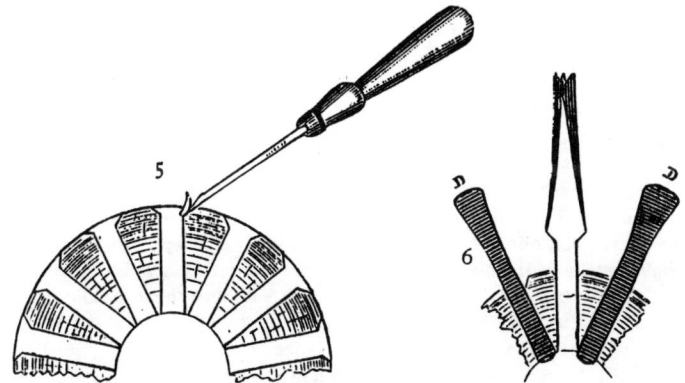

avoid this, it is a good plan to have two pieces of wood made the neat size of the mortises, which are applied one on each side of the mortise operated upon. See D D, Fig. 6, which is a sectional view of fragment of the hub. By removing one of these plugs to each alternate mortise, they may be kept in perfect shape for the driving of the rear spokes, which completes the wheel.

We have supposed the spokes to be driven to be all exactly of one size. To make sure of this, each one should be examined, and all made to match the templet by which the mortises were gauged.

The glue should be made of the consistency of a thin syrup, and used very hot, a small brush being necessary to apply the glue to the mortises, as the application to the spoke alone is not enough, owing to the glue being scraped off by the pressure of the parts in driving.

Everything being prepared in the manner described, the operation of driving begins. The forward or face spokes being driven first, care should be taken to have the tenons of a temperature of not less than one hundred degrees. In driving, the spoke should be held firmly by the left hand, as nearly to the angle required as possible. The blows from the mallet should be delivered squarely upon the end of the spoke, which should be driven home tight upon the shoulders, each one being in perfect line and touching the end of the screw upon the gauge or "sett." Some, in driving, are not particular to bring the spoke to this line until nearly down upon the shoulders. This makes poor work in most cases, as it generally ends in cracking the spoke across the tenon at the back or face, as the case may be, owing to the spoke being forward or back of the proper line. When the hub is properly mortised, and the spoke started at the proper angle, there should be no trouble experienced in making a wheel upon which the rims may be worked without straining the spokes to bring them to the rim, as these spokes invariably "gnaw off" when so forced, proving that the wheel must be free from internal strain to be durable. By deliberation, and doing the proper thing at the right time and place, wheels can be driven without fear of accident or failure, which latter are generally the children of hurry or carelessness.

continued next page

Wheelmaking

Chapter IX.—Rimming.

WE do not know why, but this part of wheel-making seems by general consent consigned to a secondary place in importance compared with the operations first considered in these articles. While we admit that "driving the spokes" may be termed laying the foundation for durability in the wheel, yet we are not willing to second the idea that less skill is required in any one of the different stages of manufacture than in another; and it is beyond dispute, that if this part be left to the unskillful or inexperienced, no amount of good material or time spent on the first process will ever produce a wheel as durable as one in which mechanical skill has been exerted equally in all parts of its development. We must also consider, that the working of circles has always been one of the difficult problems, not only to the mathematicians, but also to the practical mechanics. While the first named have never been able accurately to square the circle, the latter find the greatest difficulty in preventing it from becoming a polygon. This is particularly the case with some of our modern wheel-makers. Instead of circles, the wheels of many of our carriages on the streets, upon examination, will be found to consist of many angles, and justify us in calling them polygons. We all know it is an impossibility to make a wheel *perfectly* true to the circle, but it should approach this shape so nearly that the eye can not detect the deviation. Anything short of this standard should be rejected by the carriage-builder, as such wheels do not wear well and are a source of annoyance to the purchaser, from the incessant "drumming" which is communicated to the body of the vehicle while they are in motion.

There are few of those who work the rims upon wheels that are able to give the exact measurement of their work by rule. This is not an absolute necessity in wheel-making, but it is so considered by all other workers in wood who use circles, and the intelligent wheel-maker will lose nothing by devoting some of his spare time to the geometrical propositions referring to this subject. He will soon be convinced that to work a rim by exact measurement is quite another thing from guess work, and it will also enable him to realize why the fraction of an inch increase, or decrease, in the diameter of a circle alters so rapidly the sum of the circumference. Before leaving this, which may be termed the theoretical part of the subject, we will illustrate a simple manner of obtaining the exact length of a quadrant or quarter circumference; or say the quarter of a rim. Fig. I. represents the circumference of a plain circle, any given number of times smaller than the circle to be measured. Across this draw the line representing the diameter, as shown. Then draw the tangent, A B, then with the dividers set to the same points, as when describing the circle, mark the points C D. A secant, or line intersecting these two points, would meet the line A B, as shown, which line, from the semi-diameter A to the intersection at B, gives the exact measurement of the quarter circumference. By multiplying the number of times the diagram is smaller than the circle to be measured, we have the exact length as given on the common rule. Whereas, if we proceed to figure it out, the diameter of the circle must be multiplied by the formula 3.1416, the quarter of this result, leaving decimals that can not be measured on the ordinary rule, though perfectly intelligible when written, but as we have stated it is not *absolutely* necessary to know the exact measurement of rims. We will not go further in this direction, but resume the manual or practical part of the subject, by taking the wheel as left in the last chapter, viz., with the spokes properly driven into the hub, for the further working of which we require a "bench" or "horse" to hold it during the operation of "rimming." These benches or horses are made in many ways and of many shapes, all approaching the same model. From the many we select one for illustration, as it possesses some new features. Fig. II. shows a

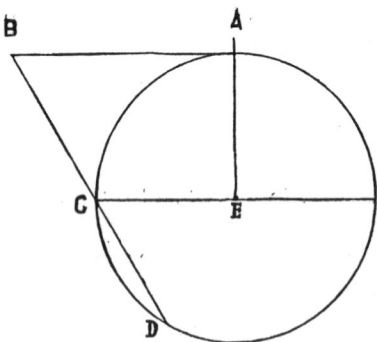

Fig II Side   scale 1/6

sectional view, side elevation, the timbers divided not quite through the center, but nearly so. A A A represent the timbers of the frame; B B is a circle of bent wood, which supports the spokes midway when the wheel is fastened down upon it, keeping the wheel very stiff, quite an object in making sulky wheels; C C are the angular brackets which support the bent circle; E E are stud braces, which pass through the frame or legs and screw firmly into the central iron ring, as shown; F is a piece of iron shafting, supported upon its outer end by the brace G. Upon the shaft, F, is bolted the movable fulcrum of the cam lever, H, by the raising or lowering of which lever motion is given to the ring, I, upon the cam, which communicates the motion to the press or slide, J, the whole being used to settle the rims upon the shoulders of the spokes, a cavity being drilled in the slide, J, so that the pressure is all upon the rim. The casting, L, holds the shaft in place by a key, which is shown, the removal of which releases this part of the appliance, when it is necessary to take it down. Fig. III. is a top view, the letters of reference being the same in both.

Supposing we are possessed of some such "horse" or bench, the skeleton of the wheel is fastened upon it, and the mechanic then gauges the exact height intended, by using a straight gauge or stick, being careful to remove any superfluous glue that may remain upon the hub. After the operation of driving, the gauge used should be pressed upon the hub at the base of each spoke, as the roundness of the wheel depends in many cases upon the exactness of this gauging. The spokes should be sawed squarely through the gauge marks, which leaves the wheel ready for the important operation of cutting the tenons. These should be accurately made, true with the line of weight upon the wheel when finished. The shoulders should be even all around, and present as much surface as each size will allow. To cut them in this manner with the old style of hollow auger worked by a hand brace is an impossibility, as these tools require the aid of a "spoke trimmer," which tool

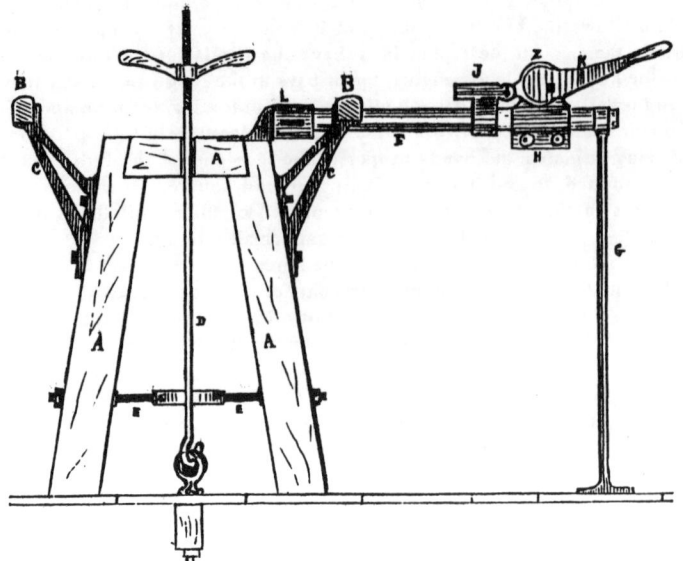

FIG. III.—TOP.

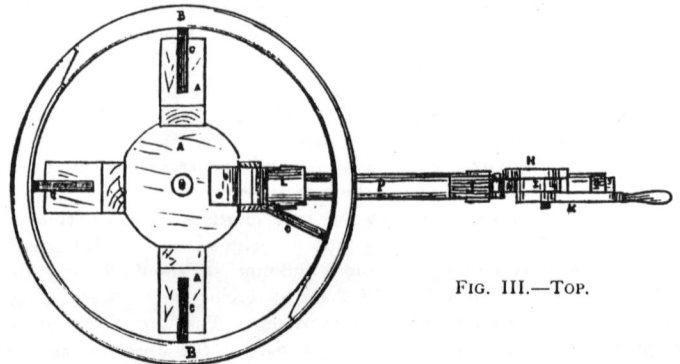

Fig I

continued next page

is made upon the principle of a leadpencil sharpener, and possessing but one cutting edge, will work off to one side or the other, the hollow auger or tenoner follows after, and the consequence is a tenon nearer one edge of the rim than the other. These tools also have a tendency to work out of line laterally. There is no excuse for this now-a-days, as there are several good machines in the market for this purpose, sold at such reasonable rates that the cost is more than repaid in the saving of time alone, leaving out the question of accuracy, which is after all the most important item.

We will suppose, then, our wheel to have been accurately tenoned, and again placed upon the horse. The rims selected for the wheels should now be carefully examined, to see if any are sprung sidewise. If this is the case, they should be faced out of wind by planing with the fore-plane. A rim that can not be brought perfectly true in this manner should be rejected, as no amount of humoring will ever make a good job of a "side kinked" rim. After squaring one side of the rim it is laid upon the tenons of the spokes, when the mechanic can judge of the trueness of the semicircle. If the ends of the rims are depressed, as mentioned in a previous chapter, opening them out will not prevent the joints from flattening during service. Having the rim bent properly is the only way to secure the benefit of this really great invention, and a rim properly bent in this sense means one which is a perfect semicircle, true sidewise, and as nearly square upon the inner side as possible. It is also necessary for the mechanic to know, that as long as manufacturers use timber green or freshly sawn from the log, these requirements will never be met, as the natural arrangement of the fibres under the seasoning process must draw the work from the shapes in which it is first forced; but if the plank from which the rims are made becomes seasoned before working, there is no reason why they should not be satisfactory in all these respects. In working our rims, therefore, we will suppose all these conditions have been fulfilled. At this point it becomes necessary to open the rim enough to pass over the ends of the spokes, as soon as the operation of boring the holes is completed. The rim being bent somewhat past the center of the semicircle, a favorable point is selected for making the joint, which is marked by dividing the distance between two of the spokes, and marking such division on the rim before removing, while the point of junction of each spoke belonging to this half of the wheel is marked upon the inside of the rim with a lead pencil. This is certainly not a very mechanical proceeding, but the writer has seen the experiment tried of mathematically spacing the distances of the holes for the spokes, and then applying the rim to the wheel, but no amount of care in mortising the hub and driving the spokes will bring them to their exact radial angle at the rim, and as it is absolutely necessary to construct the wheel free from all strain, there seems to be no remedy for marking with the pencil as mentioned.

It is just as imperative to have the rim bored truly, as mentioned in the case of the tenons, to fit the holes. The machines mentioned for tenoning also have an arrangement for boring, and by the use of this the rim may be made to fit the wheel accurately; the holes should be made small enough to require driving upon the tenons. It is at this point that the mechanic often gets into trouble through the splitting of the rim. There is little danger of splitting if the holes bored are round. This is the secret of the whole subject, as the rims that split are nearly all started by the "worm" of the auger bit, which by its extreme angle, or taper and double thread, separates the fibres instead of cutting them. After this separation, the bit bores a hole in the shape of an ellipse or oval, the tenon upon entering which wedges the fibres still farther apart, eventually spoiling the rim. Sometimes these cracks do not appear until the jarring of the running wheel opens them to sight. To overcome this difficulty requires an auger made specially, having the "worm" quite slender and single-threaded; the "pitch" of this thread does not force the cutting edges of the bit, but allows the shaving to be cut thin, and prevents all cracking, even in the most flinty wood, also boring a hole as nearly round as possible. The spurs or cutters of these bits should be made somewhat longer than the ordinary length, to prevent breaking the wood from the back of the rim, when forcing the auger through. This breaking out of the wood often gives handmade wheels an untidy look, to say the least of it.

As the inner sides of the rim are rarely square, the mechanic finds some difficulty in making the shoulders of the spokes show a tight joint at the junction with the rim, and to cause them to do so the wood around the holes is removed, as shown in Fig. IV., which shows a section through the rim. This practice can not be recommended, as we have shown that the spoke has little enough bearing, even when every atom of surface comes in contact with the rim; therefore this practice is extremely faulty, as the spoke, supported in the manner shown in the cut, will allow the rim to settle when subjected to actual service, such wheels needing repairs in a short time. A better plan is

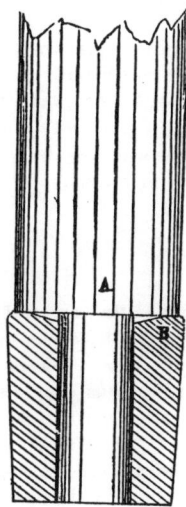

FIG. IV.—FULL-SIZE.

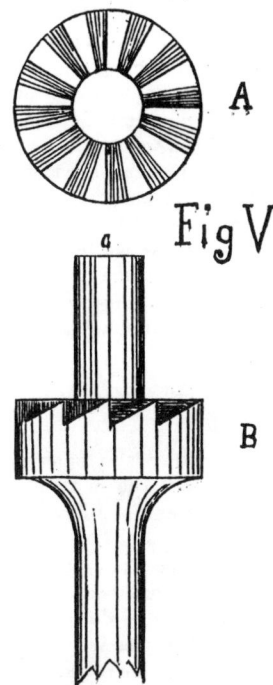

to have a counter boring tool, as shown, Fig. V., A and B. This can be made at trifling expense and by having a number of removable plugs, C, made to fit neatly the different sized holes bored for the tenons. This tool placed in a brace, and a plug inserted in the hole in the rim, by a single revolution will clear the surface around the edges of the hole perfectly square and true, so that the shoulder of the spoke will bear in every part, and if a pressure is applied to the rim, such as can be exerted by the cam lever, shown in Figs. II., III., the joints will be so closely fitted that the wheel will stand a great amount of jarring and strain without altering the relation of the length of the rims to the wheels, or, in other words, becoming "rim bound."

Having settled this, the most important item in rimming, we are confronted with the much argued question in relation to the use of "dowels" in the ends of the rims, and it is a debatable question whether the harm done by their use is not greater than the good accomplished. Many rims split where the cause can be traced directly to the dowels used. Still something is needed to keep the joints together, particularly where the wheels stand some time, which they should do before tiring. Many devices have been invented to do away with them, but in most cases the remedy is worse than the disease. Dowels are still generally used, and are likely to be, as they are more easily applied and are better than any of the substitutes. If we use dowels, let us have them of annealed wire, and as light as practicable, and carefully avoid driving them tightly into the rims. It is also appropriate at this time to mention the practice of shaving the tenon to one side or the other, when the spoke may have been driven out of line, or tenoned to one side. This is miserable practice, and is due generally to the hurry and carelessness shown by many workmen in order to accomplish a great amount of work in a given time; but it is better to be deliberate enough to make each part as it should be, when such botch work would be wholly unnecessary. These tenons can have nothing taken from their substance without detriment to the wheel as a whole. The wheel we have supposed ourselves working is free from any such tinkering. We come next to the rounding of the rim. This varies so much, and is governed by local tastes to such an extent, that we will only say that the more we sink the fanciful and encourage the practical, the nearer we will approach a universal style.

We now come to the last operations in light wheel-making, and this is the part where many who have constructed a good wheel proceed deliberately to spoil what they have been at such pains to make. We

continued next page

Wheelmaking

have reference to the almost universal custom of splitting the tenons on the ends of the spokes, and driving wedges to force the rim upon the shoulders and prevent it from rising. To any one, on reflection, this will be seen to be faulty, and yet the wheel-maker may say: "What are we to do with rims that can not be kept to their place?" We answer there may be a time when wedges are absolutely necessary, but until that time comes do not use them. Make the tenons and holes to fit tightly, and if there is a place here and there where the rim rises off the joint, the tire will bring everything right if it is properly fitted. We can imagine the great strain upon the spokes in the hub, when the wheel becomes loosened and the rims bound, which occurs to all wheels at *some time;* these spokes, then, which are wedged fast on the outer ends must pull from the hub, or draw through the rim, and we leave it to the intelligent mechanic which is the more probable.

There are many minor matters that may suggest themselves to the wheel-maker that we have left untouched in our consideration of the subject; these we leave to the solution of the workbench, remarking finally, that in the foregoing chapters we have written of wheels made exclusively of hickory timber, and we propose now to devote the remaining chapters to oak timber, and wheels made therefrom.

Chapter X.—Oaks.

WHAT iron is to the mineral kingdom, so oak is to the vegetable; in fact, so well known are the virtues of this family, that it becomes difficult to determine which to describe first. For the wheel-makers use, oak may be said to be invaluable, for when in previous chapters we have written that hickory was the timber *par excellence* for carriage wheels, we reserved or emphasized the proviso: "light wheels, whose spokes and rims do not exceed 1⅞ in. in size." Above these dimensions, hickory becomes secondary to oak, owing to the more durable qualities of the latter under exposure to atmospheric changes, and particularly dampness, the penetration of which, into the larger surfaces exposed of the heavier wheels, it is impossible to prevent, the result being, that heavy hickory spokes soon become tender and rotten, and break at the tenons, while those made from oak are not in the least affected by dampness, and even when paint and varnish are entirely neglected, will last an indefinite time; therefore, in heavy coach work, also in wagon and cart building, oak is the only timber universally applicable.

Many of our factories engaged in the preparation of wheel material, receive orders from foreign countries for very heavy hickory spokes, the parties ordering having, no doubt, heard so much of the durability of hickory, that they imagine it rivals the oak in this respect. That this is not the case, except in small or compact pieces, is well known to all who handle carriage wood-work in quantities. Hickory preserved from moisture by painting and varnishing, will last for ages, but exposed in any manner to the weather, it becomes the most tender of woods.

It is incumbent on us, therefore, to become thoroughly acquainted with the oak family, in so far as the species interest the trade, to accomplish which we will illustrate some of the most noted varieties, both for reference and comparison, leaving many of the family out of the catalogue as being of interest only to the botanist, and confining our remarks to the qualities of the timber of those shown, and kindred subjects. Alphonse de Candolle, the celebrated botanist, gives us more than two hundred and fifty (250) accepted varieties, while the Messrs. Michaux, father and son, have carried the division still farther, making the oak question botanically considered about as interesting, to the general reader, as an unabridged dictionary.

The oak, as a family, grows over nearly the whole of the northern hemisphere, excepting the extreme north, while in the tropics and along the Andes the evergreens and deciduous or regular varieties grow side by side, presenting a wonderful difference in leaves and appearance, but easily recognized as relatives by possessing the one common fruit, an acorn and cup, which latter never completely covers the nut.

To bring our subject nearer home, we find about twenty accepted varieties in our Atlantic and neighboring states; these are divided again into as many more sub-varieties, climate and soil making great differences in oaks, as in other trees, which has led to confusion in nomenclature. The practical woodsman has great difficulty in naming the exact variety of the trees on a given tract, as a certain variety is known by one name in one part of the country, while the same variety in another section is known by one entirely different; therefore, in our description we will give the botanical names, and illustrations of the leaves and fruits of those quoted, which may be considered approximately correct.

Nature has divided the *Quercus* into two great divisions, known as the annual fruited, and biennial fruited. These great divisions are easily recognized by the botanist, and are of still more importance to the wood-worker, as nature has placed a greater division between the two in the virtues of their timber; for the former, or annual fruited, in nearly every case, have timber impervious to the weather, and therefore nearly indestructible; while the latter, or biennial fruited, have timber whose structure is loose or porous, and contain the elements of decay, which rapidly develop on exposure to the changes of temperature and weather.

With this long preamble, we begin our list with the *Quercus alba,* or American White Oak, annual fruited. This tree was at one time profusely spread over America, extending as far north as Lake Winnipeg, and south into Florida and the Gulf States. Fig. 1 shows the general shape and arrangement of its leaves and fruit, and as it is the best known of our forest trees it is almost useless to add that its timber has more general uses than any other, excepting the pine.

This variety is not so fruitful as some of the others, which led to the popular opinion that it fruited but once in seven years, which mistake was owing to its tendency on some soils, to abort or throw off the fruit soon after formation. When this tree has room, it attains a noble height, ranging from 70 to 120 feet, and well deserves its common title of "monarch of the American forest;" its slow growth, coupled with its extreme tenderness while young, has led to its gradual thinning out, and under the destructive influences now at work in our forests, from increasing consumption, insects, etc., will eventually end in its scarcity. The time taken to bring it to its proper maturity for timber is eighty or ninety years (though some favored localities, where everything is propitious, may produce good timber in sixty years, but these are the exceptions); oaks may be considered at their prime at the age of one hundred, though they continue to increase in size until they reach two or three hundred years, but like humanity, after the prime is reached, they retrograde. To cut White Oak when young, or as some dealers call it, "second growth," is a mistake, as such trees contain more than a third of their entire substance of alburnum or sap-wood, which is really false wood, containing acids (gallic and tannic), making it subject to chemical changes, which result in decay from "sap rot" and other well known causes; but when the timber is fully grown or ripe, these acids are confined to the liber or inner bark and the slight circle of live wood lying next to it; this allows the seasoning of the true wood without danger from the causes mentioned, showing us that trees come under the universal law of the vegetable kingdom, that the ripening of each is fixed at certain limits, before or after which it is unwise to reap or cut.

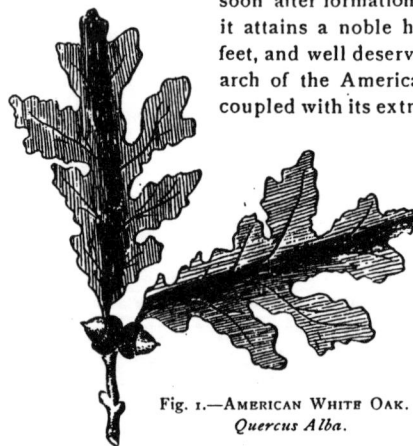

Fig. 1.—AMERICAN WHITE OAK.
Quercus Alba.

Fig. 2.—POST OR ROUGH WHITE OAK.
Quercus Obtusiloba.

We present, as the next variety, the *Quercus obtusiloba* or Post Oak, annual fruited, the leaves and fruit of which are shown in Fig. 2. Its timber resembles the White Oak, but is more elastic, and more diffi-

continued next page

Wheelmaking

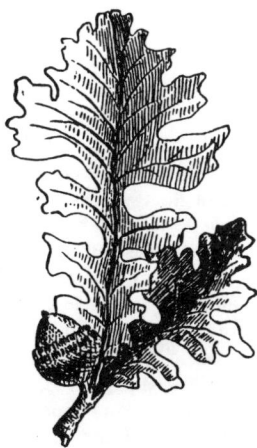

Fig. 3.—Burr or Over-cup Oak.
Quercus Macrocarpa.

Fig. 4.—Swamp White Oak.
Quercus Bi-color.

cult to season without springing. It is admirable for bending purposes. Its durability may be said to exceed the White Oak. The writer examined a beam of this timber a short time ago which squared about fourteen inches, and which was worked into shape more than one hundred and fifteen years ago, and though time-stained and rusty upon the surface, at the depth of an eighth of an inch, looked as fresh and sound as when first worked, though it had been exposed in an open belfry most of the time, uncovered by paint or anything of that kind. This variety is found from New England southward, and prefers poor dry soils, and seldom attains a diameter over eighteen inches; it is ranked next to the Live Oak for density.

The third variety, the *Quercus macrocarpa,* or Burr Over-cup Oak, annual fruited, grows more abundantly in the Western than in the

Fig 5.—Rock Chestnut Oak.
Quercus Montana.

Fig. 6.—Live Oak.
Quercus Virens.

Atlantic States. Its timber is often used for spokes, as it resembles the White Oak closely in appearance, but is not so elastic or durable; and most of the timber of this variety, which has come under the writers personal experience, has not retained its weight in drying, and cannot be recommended for this important use when other kinds are to be obtained.

Fig. 3 shows the leaves and fruit, though these are subject to slight change, owing to soil, climate, etc.

Fig. 4 shows the leaves and fruit of the Swamp White Oak, *Quercus bicolor,* annual fruited, which, as its common name implies, prefers in its growth the low lands. Its timber can not be recommended for spokes, as it is more porous than the other varieties mentioned, though it is durable and makes excellent bent rims for heavy work, drying rigidly, and standing exposure well.

Fig. 7.—Willow Oak.
Quercus Phellos.

Fig. 5 is the Rock Chestnut Oak, *Quercus montana,* annual fruited, and, with the exception of the evergreens, the least durable of the annual species. This timber is not used for spokes, but makes tolerable rims, and is better adapted to other wagon work than any of the biennials.

Fig. 6 shows leaves and fruit of *Quercus virens,* or Live Oak, annual fruited, and is merely shown for comparison with the others, as this, the most durable of the annual species, is too scarce for the wheel-makers' acquaintance in a business point of view.

In Fig. 7 we show *Quercus phellos,* the Willow Oak, biennial fruited, which has leaves somewhat like the live oak, but shows the wonderful change that nature has marked in the production of its fruit, for, as a biennial, its timber is worthless, being red, coarse, and very perishable.

Fig. 8.
Scarlet Oak.
Quercus Coccinea.

Fig. 8 shows the leaf and fruit-stalk of *Quercus coccinea,* or Scarlet Oak, biennial fruited, whose timber is often made into heavy wagon rims, but it is neither strong nor durable, and rots on slight exposure to moisture.

There are two other varieties of biennial fruited oaks, whose leaves and fruit very much resemble the ones shown; these are the *Quercus rubra,* or Red Oak, and the *Quercus tinctoria,* or Black Oak. The first of these, on some soils, produces timber of good quality for some purposes, being the most durable of the biennial species, but the latter is, as a general thing, worthless for any purpose of the wheel-maker; in fact, is scarcely worth using in any department of coach or wagon building, yet quantities of this timber are sold every year for durable oak. Its durability is very uncertain, as it rots often without exposure even when painted, particularly when in contact with iron.

Chapter X.—Oaks

Fig. 9 we give as a specimen of ornamental oak, the *Quercus agrifolia,* or "evergreen," a native of California; it is not a timber tree, the picture being given to show one of the wonderful varieties of the foliage of this immense family.

Figs. 10, 1 and 2, we give as specimens of the oak of Europe, *Quercus rober,* two varieties of which are shown; we do not know whether

No. 9.—California Evergreen Oak.
Quercus Agrifolia.

the pictures of these are correct or not, as we have no specimens with which to compare them, as has been done with the others. Our knowledge of the virtues of their respective timbers is also confined to what we have read, from which it seems that one variety is more durable than the other, but which one it is does not seem to be clearly understood, even in the country where they belong, the evidence being about equally divided between the two; from which we would imagine that either is quite good enough for any coach and wagon work whatever, while some of our enthusiastic acquaintances from the mother country assure us that "'eart of hoak" is stronger and more durable than iron; whether this is the case or not, we leave to the reader, only adding that we are satisfied to close our list with so noble a specimen.

According to an European entomologist and author, M. Coutance, there are ninety-eight species of insects which infest the oak family. Of these, eighteen attack the wood or timber, and ten the bark. From all these we have selected two for illustration, which from their apparent insignificance might be thought comparatively harmless. The first of

continued next page

these pests is called by entomologists, *Lymexylon sericeum* (Harris). A member of this group has done much damage in Europe ; we are told it is rare in this country, but the mere fact of any specimens of such a destructive insect being found here at all, is cause for alarm, for we have seen how insects multiply, under certain favorable conditions, as was the case with the Colorado beetle or "potato bug." We need not fear that wood beetles will ever become so numerous, but their combined attacks are to be dreaded, and guarded against as much as possible. The insectivorous birds, our greatest help

No. 10.—EUROPEAN OAK.
1. *Var. Sessiliflora.* 2. *Var. Pedunculata.*

and reliance in this case, are being driven from our country by persistent persecution, though stringent laws have been enacted in nearly all our States for their protection. It is manifestly the duty of all to see that these laws are rigorously enforced, for this and this only will prevent the oak from becoming scarce for mechanical purposes, a condition of things hard to realize when looking upon our many groves of noble trees.

The second enemy is shown in Fig. 11, and goes by the uncommonly hard name of the *Brenthus septemtrionalis* ("Herbst"). This representation is much enlarged, as the insect is small ; it is very wary, feigning death on the slightest alarm, which makes it difficult to distinguish it from the bark upon which it is running. The worst of this pest is that it attacks healthy trees and sound timber, and makes it unfit for any use whatever. The female, in midsummer, punctures the bark with the long sharp snout shown, deposits the egg, and the larva or worm, as soon as hatched, bores directly toward the heart of the tree, and does not confine its depredations to the sap-wood, like other beetles. This insect prefers the White Oak, and is seldom seen on any other variety. That there is cause for alarm in viewing the ravages of these insects, may be shown when we state that in Montgomery County, and parts of Bucks County, Pennsylvania, the White Oak is nearly all

No. 11.—LYMEXYLON SERICEUM.
(*Harris.*)

No. 12.—BRENTHUS SEPTEMTRIONALIS.
(*Herbst.*)

infested, it being difficult to find many trees that have not been bored from trunk to branches by these small worms, which leave black holes known to users of oak as "pin worm." Nor are these pests confined to the localities specially named, as we have timber from nearly every section with these imperfections, but we mention these particularly as a warning of what will be the general fate of the White Oak if some means are not adopted to arrest these insects in their rapid development. Having furnished a theme which well deserves the careful consideration of all the trade, we will resume the more mechanical part of the subject in Chapter XI.

Chapter XI.—Wheels of Oak.

WE have divided our subject according to a really natural division, following upon the peculiarities of the structure of the two timbers in question, the kind written about in previous papers, namely hickory, being rigid and of equal resisting power or density with the timber used for the hubs, while that under consideration in the present chapter is of a more compressible nature ; and as the hubs for both are made from elm or gum, it follows that what would be the proper treatment with the one kind would not produce good results in the case of the other.

This difference is made wider when we consider the philosophy of the subject which presents this remarkable divergence, and remember that while it requires the utmost skill and care of the maker of light wheels to keep the spokes from coming *out* of the hubs during service, it requires an equal amount of skill and science to keep the spokes from working *into* the hubs of heavy wheels ; a difference which has led to the common notion that there are no individual wheel-makers who can make both light and heavy wheels successfully. If this is the case, it should not be so, as the practical mechanic should be conversant with both branches ; even if he does not carry his theoretical knowledge into practice, he should be able to do so if called upon.

Taking this view, we find our subject is subdivided into four distinct classes, which we will proceed to consider somewhat in detail :

First, heavy coach wheels whose spokes measure 1⅞ in. to 2¼ in. respectively.

Second, heavy wheels with square-shouldered straight-tenoned spokes.

Third, heavy wheels with spokes having shaved or curved shoulders and tapered tenons.

Fourth, wheels with "Jersey spokes."

Taking our first class, we find these heavy spokes should be made of oak for the reasons given in the previous chapter, viz.: durability under exposure to dampness. The only objection we have ever heard urged against the use of oak for spokes of these sizes is that it is not so stiff as hickory. This we grant, and claim that it is a decided advantage, as the more elastic fiber, when used in such proportion as made possible in spokes of these dimensions, is just what is needed to insure durability if properly manufactured. We would not recommend oak for any smaller sized spokes than 1¾ in., but above this size it is the better timber of the two for the wheel-maker's use.

The preparation of the hubs for coach wheels is the same as for the lighter carriage work. We would only add the suggestion, that it would be better to have the mortises in them made "straight,"—that is, without "stagger," as we can not see that this adds any real strength to the wheel as a whole, or to the hub in particular, and is a positive disadvantage to the rim tenons in a wheel of this size, besides increasing the tendency to throw water and mud during wet weather.

In driving the spokes in wheels of this description, we must consider what difference to make in the proportion of the "mash," to insure the spokes staying where they are intended, for our remarks in the introductory about the tendency of heavy spokes to sink into the hubs do not apply altogether to the sizes under consideration in this class, as these wheels occupy the middle ground, where the tendency is about equally divided between coming out and sinking in ; so that there are fewer failures of wheels of this class, when the hubs are properly proportioned, than with any other kind. The use of oak requires that the spokes have considerable more "mash" than was recommended in the case of hickory, though there is little to be gained by any great increase in this direction, for though oak is more easily compressed than hickory, still it is very easy to over-do it in this line, as will be illustrated further on.

Fig. 1 shows a half section of hub and spoke of about the proportions found to answer the purpose for this class of work. It will be observed, by referring to the ¼ scale, that the spoke has about ⅛ in. taper from the shoulder to the point, while the mortise shows an allowance for "mash" or compression, ⅛ in. at the top of the mortise, increasing to 3/16 in. at the bottom, while the extreme surface of the mortise is relieved, as shown in the cut, by reducing the "mash" to scant 1/16, thus preventing shelling or cracking during the operation of driving the spokes. In this class of

continued next page

work the spokes should not be compressed sidewise, as brow or permanent bands are not used, and therefore any strain in this direction will encourage the hubs to crack out to the front after finishing. Bent rims are used on these wheels exclusively, in this country at least, and the remarks made in the chapter on bent timber explained why these should be bent of heights to correspond with the wheels for which they are intended, and where such are used no difficulty is experienced; but heavy rims are more subject to sinking at the joints than lighter ones, particularly when the conditions described are neglected.

When bent rims were first used, it was quite a task to get them upon the spokes

Fig. 1.

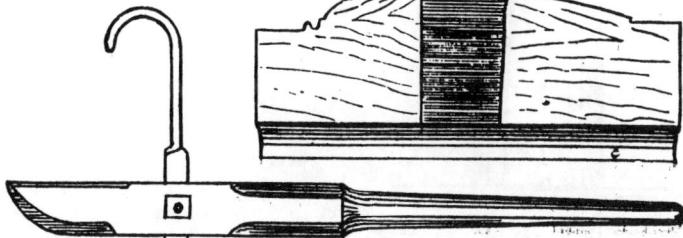

Fig. 2.

of the lower wheels; and even now the mechanic is often puzzled to get the spokes started into the rims, and it requires the aid of some contrivance to accomplish the object. In Fig. 2 we give an illustration of an old but very useful contrivance for this purpose, very appropriately called a lever-hook. We do not give it as a novelty, but there may be some reader who has not seen or used the like. The cut is drawn to the scale of $\frac{1}{4}$, and as its application is manifest, it needs no explanation.

As the rimming does not differ from that described in the chapter on that subject, we come to the wheels of the second class, viz.; heavy wheels with square-shouldered straight-tenoned spokes. See Fig. 3. This class comprises wheels for bearing comparatively heavy loads, and as we have said the trouble to be experienced in these is to prevent the spokes from driving or sinking into the hubs, it would be natural to suppose that all we would have to do to overcome

this difficulty would be to make the shoulders, resting upon the hubs, of larger proportions. This could be done very readily with hubs of these sizes, but unfortunately this plan will not work as well in practice as it looks in theory, for the following reasons: First, it is impossible to secure timber large enough to be completely clear of sap-wood, and this sap is always placed towards the front or face of the spoke; as, being more tender than the heart-side, it would give, if placed to the back, to such an extent that the spoke would not be forced tightly against the face of the mortise. Secondly, it would crack at the shoulder sooner if placed to the back, as oak in drying always bends toward the sapside, and the strain coming from the opposite direction would part the fibers sooner than when the heart is toward the back; and thirdly, because the timber is wider as split from the bolt, allowing the face to be made thus when finished. This preliminary argument shows us that to escape one set of troubles we encounter another, for these wheels whose tires are of heavy proportions are subjected to enormous pressure when these are shrunken upon them; and the spokes are of such dimensions that the wheel can not be dished by the bending of the spokes, as with light wheels, and therefore an undue amount of strain is brought upon the fore shoulder of the spoke, which, assisted by the constant jarring while in motion, causes this part of the shoulder to settle into the hub, which shells off or bruises to such an extent that it eventually ends in cracking the spoke at the back, as shown in Fig. 4. This, of course, does not always follow, for shouldered spokes do good service when not overtasked, but for very heavy wheels, or those intended for heavy loads, they are not so strong as those to be hereafter described. We have now reached a point when the taper and "mash" of the spokes must be made to suit the work intended, for in the hubs used in this class we can give more pressure sidewise than we could where the hubs were driven without permanent or "brow bands," which are generally used in this class of work. The taper too, of these spokes of $2\frac{1}{4}$ in. size and upward, can be increased, following what might be called a sliding scale, making for spokes of $2\frac{1}{4}$ in. about $\frac{1}{4}$ in. taper and the mortise $\frac{1}{4}$ in. "mash" at the top, increasing to $\frac{3}{8}$ at the bottom. This gives some idea of the proportion to be observed, but all such measures must rest to some extent upon the quality of timber used.

Some wheel-makers allow $\frac{1}{8}$ in. "mash" sidewise, opening the mortises at the surface, as shown in Fig. 5, which represents a sectional

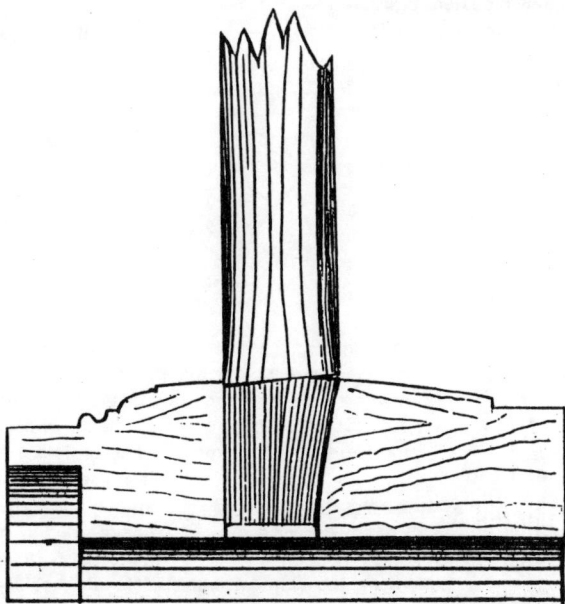

Fig. 3.   Fig. 4.

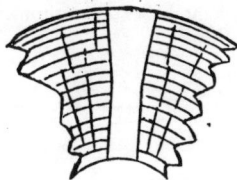

Fig. 5.

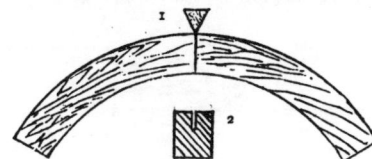

Fig. 6.

view of a fragment of a hub. It will be seen that this takes the extreme pressure from the surface layers of the wood, and also presents an inclined surface which helps to prevent the spokes from settling, which, as stated, is the object in designing these wheels.

Bent rims are used upon wheels of this class up to $2\frac{1}{4}$ in. in depth; above this size sawed felloes are stronger and make the best wheel. Even these sink at the joints to such an extent that every felloe is made higher at the joints than the true circle of the wheel demands. This, of course, makes a wheel of many angles instead of a circle, but when the tires need tightening, the wheel is rejointed, which makes it as nearly round as possible.

There is very little to be said about working sawed felloes, as the manner is the same in all places, and the elements of success depend solely upon the thorough manner in which the work is performed.

Many years ago it was the general plan to make the tenons upon the ends of the spokes square instead of round, as at present. This entailed a great amount of mortising, which greatly increased the cost of making these wheels, which are just as serviceable made with the round tenons. These, besides being more easily made, insure a perfect fitting, which was not always the case with the square tenons. In our large wagon works, where thousands of sets of heavy wheels are made annually, the sections of short felloes are not dowelled as is the general custom, but instead, they use small pieces of iron of triangular shape, average size $1\frac{1}{4}$ in. on each angle. These are driven into each joint, see Fig. 6 (1); a saw kerf is made to receive them. See (2). These seem to answer admirably, and are easily removed when the wheel needs rejointing, after which they can be replaced as readily.

continued next page

Wheelmaking

The question is often asked: Is glueing any advantage to heavy wheels? We think it is safe to say that glue does little or no good when we reach the sized wheel whose spokes measure 2¼ in., though we know many makers who do use glue on all sized spokes, while many more use white-lead and Japan dryer; others, still, use nothing adhesive, but dip the tenons of the spokes into water, to make them drive easily. All of these plans we have known to produce good work, but we prefer, in

FIG. 7.

working oak, to heat the tenons as hot as possible without scorching; these are then dipped into hot water to furnish a lubricant for driving. We have also known those who boil the spokes to make them "mash" easily into the hubs. This is a very poor plan, and does not work well, as the boiling water takes the substance from the timber, also making it more liable to absorb the grease, which soon wrecks the wheel.

We now come to the third class, whose spokes have shaved shoulders and tapered tenons; see cut 7. These we can recommend as making the strongest wheel known to the trade, as will be seen by examining the principles involved in their construction. These spokes are made as shown, and the mortises in the hub are opened sidewise to fit the side taper, the "mash" allowed this way being ⅛ in. scant at the top, increasing to ³⁄₁₆ in. at the bottom. The corners are not removed at the surface of the mortises to any great extent, as the curved shoulders are forced down upon them tightly.

The hub is mortised for the width of the spokes, as shown in Fig. 8. An allowance is made for "mash" of ½ in. at the bottom (c), decreasing to ⅜ on reaching the point (A); from this point the "mash" decreases until

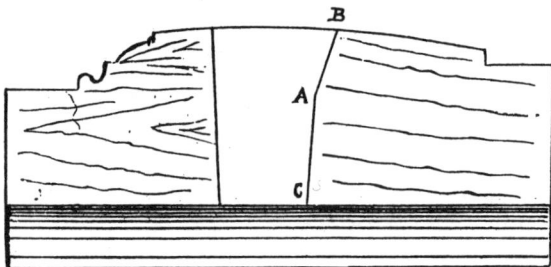

FIG. 8.

at the surface (B) there is left but ¹⁄₁₆ in. In mortising this way the face of the mortise has more dish than the spoke is to have when the wheel is finished, as the hip or projection in the mortise throws the spoke backward in driving. The spokes should be tapered ⅜ in. full for large sizes. With this extended description we will leave the subject for conclusion in the following chapter.

I NOW continue the subject of heavy wheels, with spokes having tapered tenons and shaved shoulders. I have described the manner of mortising the hubs to secure the benefit of the entire width of the spoke. In opposition to this plan many makers on this class of wheels make the "mash" ½ in. or more, and carry this up to the surface of the hub;

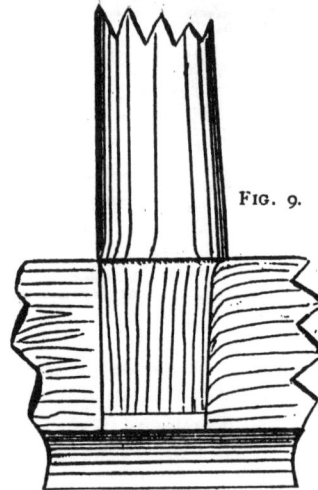

FIG. 9.

the consequence is, that the spoke compresses nearly the whole of this amount, forming as it were a kind of shoulder, which presents the appearance shown in Fig. 9, which represents the spoke reduced by this kind of driving fully ⅜ inch, and it is in reality no stronger than one made of the size to which the spoke has been "mashed ;" so we see that beside the labor required to spoil the spoke in this manner, it only secures the same results as if a spoke two sizes smaller had been used. But when the spokes are driven into mortises prepared as shown in the previous chapter, the whole strength of the spoke is utilized, and the resisting powers of the wheel are proportionate to its size.

The wheels of this class are subjected to the same strain from the tires as mentioned in the description of the class whose spokes have square shoulders, but, unlike them, this force imbeds the curved sides of the shaved shoulders until they form an arch, as shown in Fig. 10. This gives them great resisting power, in which they are assisted by the taper or angle of the tenons, and it is a fact worthy of notice that spokes made in this manner seldom break at the back, even under repeated tightening of the tires, while it is a well known fact that those of the second class mentioned are very liable to this fault.

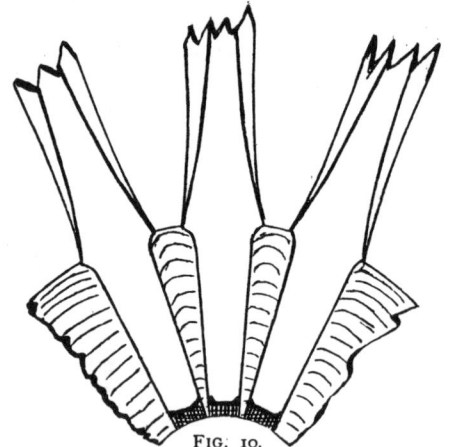

FIG. 10.

In driving the spokes of such heavy wheels it requires something more solid than any "horse" that could be made, and the old style wheel-pit made into the floor of

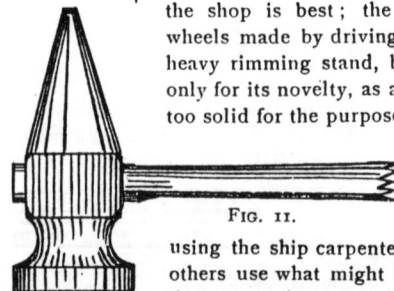

FIG. 11.

the shop is best; the writer has seen very heavy wheels made by driving the spokes horizontally on a heavy rimming stand, but this can be recommended only for its novelty, as a well made wheel-pit is none too solid for the purpose. The tools used for driving heavy work differ very much among the craft, some using heavy wooden mallets with handles 36 in. long, others using the ship carpenter's "maul," see Fig. 11, while others use what might be termed a pole-axe. Any of these seem to accomplish the purpose, but the ship carpenter's "maul" is certainly one of the most effective instruments for driving that can be used by hand. With the heavier spokes it is often necessary to mark the distance the spoke is to be driven into the hub; to do this quickly some of our large works have a block with

continued next page

two uprights, between which is stretched a cord saturated with red lead or other pigment, and as the spoke rests upon the bottom board the line is marked upon the shoulders at the proper point. These are trifles, but where many sets of wheels are to be made, everything that tends to decrease the time in manufacture without slighting the work, should be carefully followed; the individual wheelwright is often nonplussed to know how the work can be made so cheaply in the large factories, even allowing for the use of steam power, but the secret is, system and an eye to small things. We will close our remarks on this class of wheels by describing a large set made by one of the Philadelphia works, to be used in the Mexican mining district; they may be considered among the largest wheels ever made upon the ordinary principles of wheel-working. A view is shown of driving the spokes, see Fig. 12,

FIG. 12.

which gives some idea of their proportions, while the written size we give below. In our sketch we have not shown the shavings and litter upon the floor, therefore the picture may not be familiar to many who have waded through the usual accompanying rubbish of the wheel-shops, which are seldom thoroughly cleaned, except by burning out. The height of the wheels shown were : for the hind wheels 12 ft.; front wheels, 7 ft.; spokes, 7 in. across the shoulders at the hub ; thickness of spokes at the hub, 2¼ in. Felloes, 6 in. deep, 10 in. on the tread. Hubs, 24 in. diameter. The boxes for these hubs weighed 120 lbs. each, while the tires for the hind wheels weighed 1,250 lbs. each; the huge crank axles weighed 2,600 lbs. each ; the whole constituting a monster which looked more suitable for a mastodon or an elephant team than to be drawn by horse flesh.

FIG. 13.

We have now reached the fourth class, which we have designated by the name of "wheels having Jersey spokes." This may be an obscure term to some, but it is not a misnomer, as the wagon-makers of New Jersey can testify, and for transportation of loads over loose or sandy soils there can be no better designed ; the peculiarities of these consist in great strength with light weight, presenting a broad surface at the rims, while the hubs are sloping at the front and large at the back, the spoke being proportioned somewhat as shown in Fig. 13. We give below the dimensions of the different sized spokes which seem to be most in favor.

SIZE.	THICKNESS AT HUB.		THICKNESS.	WIDTH.
	FACE.	BACK.	POINT.	POINT.
2	$1\frac{1}{16}$	$1\frac{1}{8}$	$1\frac{1}{8}$	$1\frac{5}{16}$
2¼	$1\frac{1}{8}$	1	$1\frac{3}{16}$	$1\frac{1}{4}$
2¼	$1\frac{1}{8}$	1	$1\frac{1}{4}$	$1\frac{1}{2}$
2½	$1\frac{3}{16}$	$1\frac{1}{16}$	$1\frac{3}{8}$	$1\frac{5}{8}$
2¾	$1\frac{1}{4}$	$1\frac{3}{16}$	$1\frac{7}{16}$	$1\frac{3}{4}$

JERSEY SPOKES.

When these spokes were all made by hand, the Jersey mechanic, with an eye to saving work, split the rough material so closely to the dimensions required, that when the spoke was shaved there was a decided difference between the face and back in width ; to accommodate this order of things the hubs had to be mortised wider at the front than at the back, as shown in Fig. 14. This custom is still in vogue, although the spokes are all made by machinery at present. In speaking with an intelligent wheel-maker on this kind of work, and asking if there was any other reason for this seemingly troublesome operation, he remarked that the hub being larger at the point where the face line of the spokes came, the mortise followed the gradual lessening of the diameter, making the removal of the wood by mortising equal the whole length of the cut. We can think of no other reason for this operation, and whether reasonable or not they are all made thus.

The rims for these wheels are always wider on the tread than in depth, and the bent timber is very popular, as the circle can be made much better and stronger with this ; and it may be well to add that

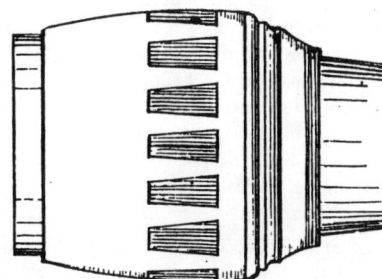

FIG. 14.

the swamp white oak and chestnut oak answer equally as well, on these dry, sandy roads, as the white oak. One of the many advantages possessed by these Jersey wheels is being able to settle the spokes when the tires are tightened, without any increase of "dish" in the wheel ; and when the ends of the spokes are cut from the inside of the hub the wheel seems to be as good as when first made, and many of them are running to-day, the rivals of the old veterans who made them.

We have now reached the conclusion of our series on Applied Mechanics in Wheel-Making. As with all important subjects, we find there are many parts which have not received the degree of attention which rightfully belonged to them, our endeavor being to present the facts in as condensed a form as possible, as being more easy of reference ; and as with all writers who venture upon such debatable ground, there may be many readers who do not coincide with all the views advanced ; this is but natural, as we are all subject to the same conditions, which make it impossible for lawyers, doctors or divines to agree.

Though the subject of mechanics is among the most prosaic and practical, still the differences of opinions are none the less ; some of this opposition of views may be attributable to the policy of blindly following old ideas without recourse to practical experiment to prove or disprove theories, and this leads us to look forward to the establishment of some permanent technical school, an object boldly and ably advocated by the editor of this magazine. While the trade publications have nobly done their part toward the universal spread of information, there remains a great want of some center of scientific information, where experiments may be carried out for the good of the trade, the value of such experiments being in their freedom from cliques or individual interests, and devoted to the enlightenment of *all* sections. The pupils educated in such an institution should be able to push forward the carriage trade of our country until we could occupy the position nature intended. To the credit of one of the departments, it can be said that this is certainly the country of the wheel-makers. H. M. DuBois.

END OF THE SERIES.

The Hub, May 1878-Aug1879

Wheelmaking

DRAFTING DEPARTMENT.

SCALE DRAWING AS APPLIED TO CARRIAGE-BUILDING.

CHAPTER IV.—DRAWING A WHEEL.

To illustrate this subject are given drawings representing the four different views of a carriage wheel as used in making a working drawing. Fig. 16 is a front view of the wheel; Fig. 17 shows the swing; Fig. 18 the wheel as it appears in looking down from the top; and Fig. 19 the same when looking up from the bottom. First draw Fig. 16, then the dotted line A. At right angles with the base line draw dotted line B; at the top, on line A, measure two and a half inches, representing the amount of swing. Draw lines corresponding with lines C, D, and then draw the hub. Let the base line and line E, Figs. 18 and 19, represent line E, Fig. 17. Take the distance from line E to F, Fig. 17, and transfer as shown at F F, F F, Figs. 18 and 19. Take the distance from line E, Fig. 17, on the dotted line A to the face of the rim, and place in Fig. 18, on line I. From points F F, Fig. 18, draw a curve through the point on the line I, giving the outer or face side of the rim as looking down from the top; draw a parallel line representing the inside or back of the rim. Now draw the hub, making the outlines very light. Measure on the base line from line E to the inside of the rim, Fig. 17, and place on line I, Fig. 18; then draw a curve from the back of the rim, at F F, through the point on line I, which gives the rim of the lower half of the wheel; now space off for the spokes and draw them. In drawing Fig. 19 reverse the rims, making the lower half to show on top, and the upper half show on the bottom.

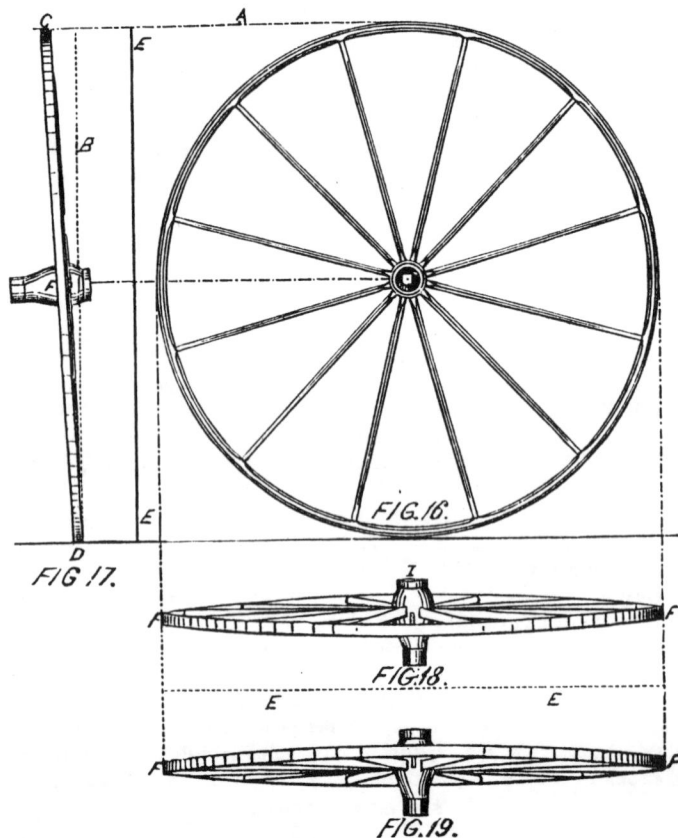

FIG. 16.

FIG. 17.

FIG. 18.

FIG. 19.

Coating of Hubs.

MR. EDITOR:—In reply to your request for my views of the editorial inquiry in the March number of the *Coach-Builders' Art Journal of London*, England, as to the cause of complaints from the patrons of that excellent journal in regard to American "Rock Elm" hubs showing the paint, I give you the following: It may be caused by the hubs being coated on the surface only, while the surface of the hole in the hub, or mortises, if mortised, is left entirely naked, so that the water absorbed in transit cannot escape in reasonable time for drying after their arrival, and consequently the paint is thrown by the water. If the hubs have come in contact with salt water, it might have such an effect.

There can be no advantage in coating mortised hubs with anything to protect the timber, as water from the air would be absorbed in the mortises, and could not escape so well as if the surfaces were not coated. Some good makers immerse the hubs in pure hot linseed oil, coating both the outside of the block and inside surface of the hole. For unmortised hubs this does very well, the oil not being objectionable as a base for paint, but for mortised hubs it is worthless, as the surfaces of the mortises may not be coated.

The best way to treat hubs for export, is to have the block thoroughly seasoned, and box the hubs as soon as finished, while clean and bright. Then the water absorbed will be taken in generally over the entire surface, and when put up to dry it will be thrown off evenly, and there will therefore be less liability of checking, as checks are generally caused by unequal drying. It will also be better for the merchant for the timber to be fully exposed, so that the buyer may see its quality. Otherwise, the buyers in other countries will know the meaning of a coat of color on hubs as we do here.

As to the many patent conglomerations used on hubs, referred to by the editor, I know nothing. JARED MARIS.

Coach-makers' International Journal, May 1881

WHICH SPOKE SHOULD BE PLUMB?

BUFFALO, N. Y.

EDITOR OF THE HUB—DEAR SIR: In setting axles for light work where the spoke is dodged, should the axle be set so that the two *outside* spokes are plumb, or the *inside* and *outside* spokes plumb? Say with ⅜ inch dodge. Please answer through next issue of *The Hub*, and oblige. Yours Respectfully, B.

ANSWER.—Set the outside spoke so that it will be plumb when on the ground.

The Hub, July 1882

Wheelmaking

WHYS AND WHEREFORES IN WHEEL-MAKING.

By Howard M. DuBois.

(Verbatim Report of Lecture delivered before the Class in Carriage Drafting, New-York, May 17th, 1882.)

QUESTION IV: HOW TO PROPORTION THE LENGTH OF HUB.

The length of wearing surface or arm on the size of axle selected, measures 5½ in. from inside of the collar to inside of the nut. In getting the proper measure for the length of our hub, we must allow for countersinking or letting in the swell on the box which covers this collar, and this gives us the actual length of hub to be used as 6 inches, which we proceed to write upon the blackboard.

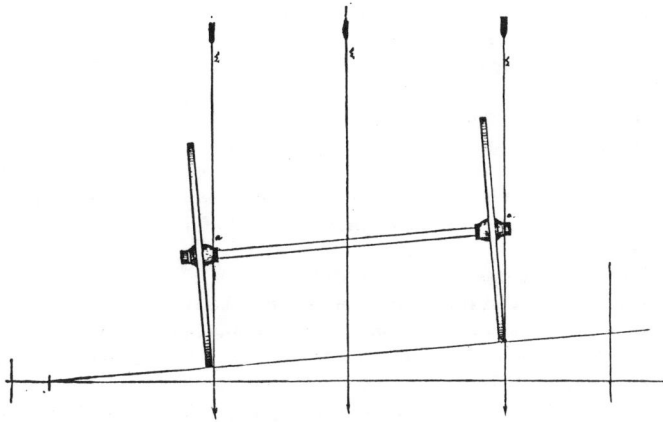

FIG. I. CARRYING OF WEIGHT ON INCLINES.
Scale, about three-eighths inch to one foot.

I will now try to show *why* this is a proper length, or at least to show *why* the axle-arm is not made shorter.

By referring to Fig. I, you will see roughly represented two wheels upon an axle, standing on an inclination of about one in twelve, which is not an unusual one for ordinary vehicles to encounter. Now, gravitation always pulls in a straight or plumb line, and you can see by the lines marked *W W W* that wheels on such an incline do not carry the weight through the spokes, but at an angle, as shown at *a*. Now, it is an engineer's aim and rule not to have what is termed an "overhung bearing;" that is, every journal, pinion, or revolving box, must be made long enough to have the weight to be carried fall inside the wearing surface, no matter what the inclination may be.

By measuring the angle made by the line of weight at *a*, you will perceive that it measures (following our scale) 2½ in. to the center of the box; and as a portion of the weight must be borne by the upper wheel at the same angle, the figures must be doubled, giving as the true length for a durable wearing surface, 5 in.; and to this we add an allowance for possible contingencies of ½ in., making our axle-arm, as before stated, 5½ inches.

I will now call your attention to Fig. 2, which represents a section through the length of our hub, set off by lines marking certain divisions which I will now explain. This line marked No. 1 is the limit for turning on the front band, which, in a hub of this length,

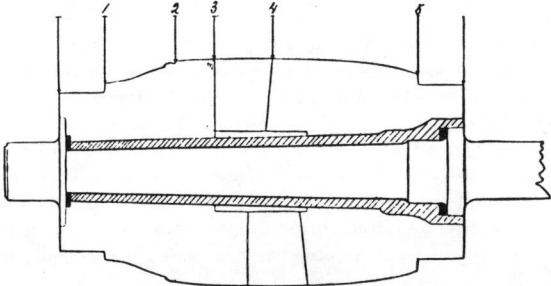

FIG. 2. LONGITUDINAL SECTION THROUGH HUB AND AXLE.
Scale, about one-third full size.

is ⅝ of an inch. This will be found as little a hold for this front band as it is safe to allow. Any less than this would result in the displacement of the band if accidentally struck against any object with the wheel in motion.

Line No. 2 shows the space between the front band and the limit of the bead-work, and within these limits the ornamental work must be placed.

Line No. 3 marks the face line of the spokes or mortises. We have, between the lines 2 and 3, a space of ⅝ inch; and *this is an important point* to be considered, for a less amount than this would not give enough solid wood to hold the strain of the end thrusts to which the spokes in the wheel are subjected. To be on the safe side, therefore, we must confine the ornamental work so that we have not less than the proportion named; and where we use longer hubs, say 6½ in., it would be well to increase this measure to ¾ inch.

Line No. 4 limits the length of the face mortise, which in this case should be 1⅝ inch, full, and we will place this measure in its appropriate place upon the blackboard.

Line No. 5 limits the depth of the back-band, which is generally made ¾ inch deep for this sized hub.

We have thus ventilated the several points in connection with the length of our hub, and now go on to the question of its proper diameter.

QUESTION V: HOW TO PROPORTION THE DIAMETER OF HUB.

We will do this by considering the subject from two points of view, namely:

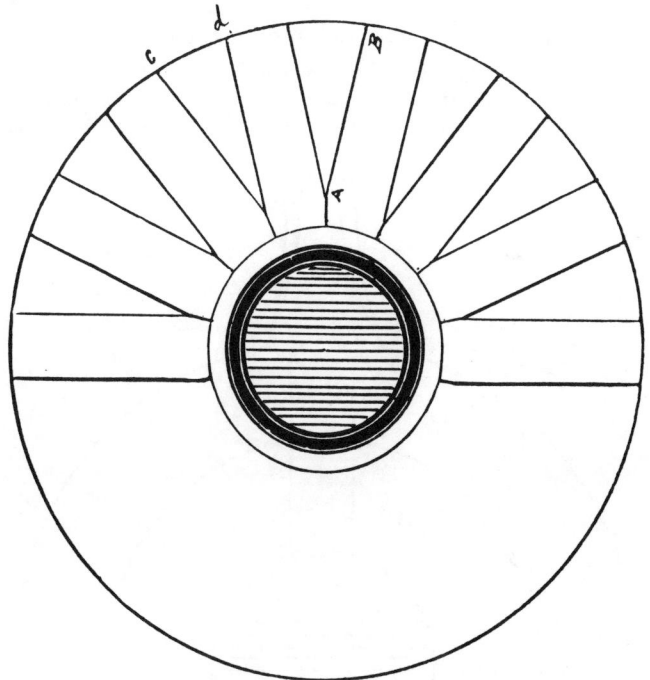

FIG. 3. CROSS SECTION THROUGH MORTISED HUB.
Represented in full size.

As a first condition, it should be borne in mind that, taking the thickness of an ordinary fine axle-box suitable for an axle such as mentioned, we find its average thickness to be ⅛ of an inch, some idea of which we show in the dark band, Fig. 3.

Reason teaches us that the ends of the spokes to be used must not rest upon this box, and so we make an allowance of about 1/16 of an inch.

We have before said that the axle would bear a certain amount of load, and consequently a spoke is required to correspond with this load limit. The one indicated in this case would be 1 in. full, and we must therefore have more than this amount of wood between the box and the surface of our hub, or a distance of 1 3/16 in. from *A* to *B*.

Taking this as a guide, we find that our hub would have to be 3½ in. in diameter. Some idea of this proportion can be seen by looking at the finished hub I now have before me.

We will now write upon the blackboard 3½ in. as a proper diameter for our hub. [This blackboard schedule will appear in full further on.—ED.]

You may say I have told you that there should be a certain amount of wood in the diameter of this hub, but that I have not explained the whys and wherefores of this second condition, and I will now proceed to do so. This brings us to the question of the mortises which are to hold the tenons of the spokes.

continued next page

Wheelmaking

QUESTION VI : HOW TO PROPORTION THE HUB MORTISES.

Wheel-makers do not all agree about the exact sizes of mortises suitable for hubs, but the opinion is unanimous upon one point, namely, that it is not safe to remove more than one-half the amount of wood in the surface of the hub in making them. I think it advisable to leave somewhat more than this amount ; and for a hub of the size we are describing, a mortise $\frac{5}{16}$ in., full, using 14 spokes, gives the best results.

In Fig. 3 we show a cross section through a mortised hub. By making the mortises $\frac{5}{16}$ in., full, you can see that our diameter of 3½ in. gives more wood from C to D than is removed in mortising, while the amount of surface from A to B measures *more* than 1 in., the amount absolutely called for by our proportions.

It will be found a safe plan in designing the proportions of *all* wheels, to begin by laying them off as shown, and when proportions suggested have less distance from the points A and B than the width of the spoke at the shoulder, then it is a question of chance whether such wheels will do service or not.

QUESTION VII : HOW TO PROPORTION HUB-BANDS.

We now come to proportioning the diameter of the front band.

We will take for our hub a band of 2½ inches diameter as the proper size, and give, as a reason, that there should be not less than one-half

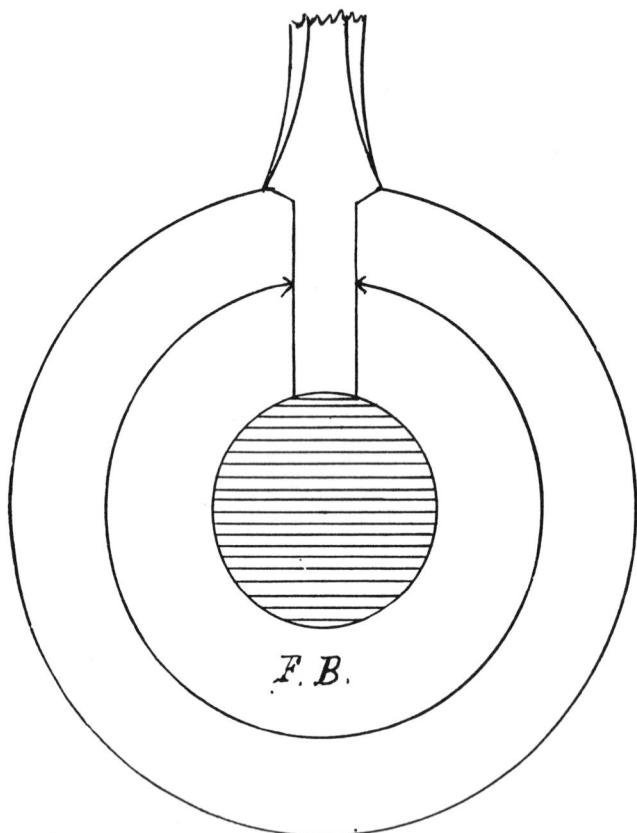

FIG. 4. SHOWING LINE OF FRONT BAND IN RELATION TO SPOKE TENON.
Scale, reduced one-eighth inch from full.

of the length of the spoke tenon in the solid wood comprising the whole length of the hub, to resist the leverage of the spokes. In Fig. 4 the line of the front band is broken, to show the length of tenon in the solid wood as mentioned.

You can see, by this, that when these bands are made small, for appearance sake, we are sacrificing a vital principle for mere beauty of outline, which cannot be done with safety. We will then write on our board 2½ inches as a proper proportion for the front band.

The back band is usually made larger, to accommodate the swell in the box where it covers the collar of the axle, and we will write down for our back band 2¾ inches, which will give a proper appearance to the finished wheel.

QUESTION VIII : HOW TO PROPORTION AND SHAPE HUB MORTISES.

We have before told why the mortises should be $\frac{5}{16}$ inch wide and $1\frac{5}{8}$ inch long. We will now speak about the manner in which these should be made. By reference to this sample, it will be seen that there is a certain amount of stagger or zigzag in these mortises. In practice it has been found that ¼ inch is about the proper amount in a wheel of this size. There may be cases where very light wheels are used, such as sulkies, which have the weight thrown all on one wheel in curving, where more than this amount can be used to advantage ; but in ordinary cases the disadvantage of a stagger, above the ¼ inch named, is so great that it is seldom used in fine work, while for heavy spokes it is best to use less than this or none at all ; but of this we will speak further later on.

You are of course all familiar with the *form* of these mortises ; the face portion or that toward the front of the hub is supposed to be perfectly true or plumb, in so light a hub, while the back is beveled to suit the taper on the back of the spokes, as shown by the line in Fig. 2. It is better, however, to have the face of the mortises made so that by holding a straight-edge against them, they will show a trifling amount of dish or inclination from a straight line ; as the spokes, in being forced down, will settle back to a slight extent in driving.

A very important point in this mortising is to have the corners cut out clean, not having any fullness, or otherwise the spoke tenon will have a tendency to crack the hub. Care should also be used in selecting the quality of material in the hub, as this, being the foundation of the wheel, should be made of suitable timber. Any hub where the grain runs diagonally from the corner of one mortise to another should be rejected, as such will be sure to crack and break out, after being used, and is not suitable for a good wheel.

Having thus gone over the whys and wherefores which most commonly suggest themselves in regard to hubs, we now come to the question of proper-sized spokes.

QUESTION IX : WHAT ARE THE REQUIREMENTS OF SPOKES.

We have already found that for a hub 3½ inches in diameter, we can safely use a mortise $\frac{5}{16}$ inch full, and $1\frac{5}{8}$ inch long, and to suit these mortises we must have a spoke measuring 1 inch plump across the shoulders. This brings us fully to the consideration of these spokes and their proportions.

One of these we have just mentioned as adapted to this size hub, namely, a spoke measuring 1 inch full across the shoulders ; but this does not convey to us any definite idea of the other portions of the spoke or what will be required of it. It will be well, therefore, to go into the points which we are to cover in arriving at the true proportion of these revolving levers for carrying the load, for this is really the office they perform.

Our first requirement, then, for a light wheel, is to select some solid substance which possesses a certain amount of elasticity and is capable of being easily worked into shape ; some substance which can be driven into the mortises described, and which, when driven, will *stay there* under every condition ; and some substance which, while comparatively light in weight, will not break under sudden shock, nor crystalize with constant vibrations. Now, the only substance which completely covers these requirements is properly seasoned hickory timber, and it almost seems as if nature really intended it especially for this purpose.

continued next page

Wheelmaking

QUESTION X: HOW TO GET OUT SPOKE MATERIAL.

As there are many grades of quality of this valuable timber, we must for our purpose select a comparatively stiff, rather than the tough, springy material ; and as our whys and wherefores seem to cover this part of the subject, I will say something about the proper manner of getting the spoke from the rough bolt, for every precaution must be used to secure the strength necessary, and to keep the proportions within the limits of the hub used.

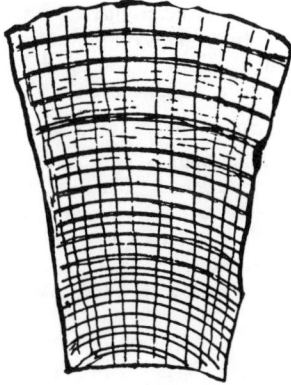

FIG. 5.

We will now refer to Fig. 5, which represents the ends of a rough "spoke bolt," rived or split from a hickory log. The larger lines shown are the "growth rings" of the tree from which it has been taken, and the lighter lines running toward the heart or inside, are the "silver grain" or medullary rays.

To make a spoke from such a bolt as this represents, we must have the tenon which enters the hub made from the part of the bolt nearest the butt of the tree, as this is the toughest and densest portion. We must also be particular to have the growth rings pass *across* the tenon, as shown here in small sketch at the side of the bolt (Fig. 6) ; when the grain passes the other way the spoke is termed a "bastard."

FIG. 6.

One important reason for having the growth *across* the spoke, is that there is less shrinkage in this direction than in the other. The timber is also stiffer, and more able to resist strain by this arrangement.

But it is not enough to have our spokes made with the growth rings across them ; the face or front of the spoke should always be made from the part nearest the sap or outer portion of the tree. The principal reason for this is, that the portion of the wood nearest the heart of the tree is densest, and as our spokes are always tapered from the back, the hardness of this portion of the tenon insures a forcing of the spoke against the face of the mortise. Were the spoke to be made with the sap to the back, it could not be depended on to do this, as the cells of the sap wood are more easily compressed, and the proper amount of wedging strain or force is not gained in driving them when made in this way.

Another point is, that if we were to make spokes indiscriminately with the sap or heart to the front, there could never be any regularity of "dish" drawn in the wheel by the tire. If a spoke is made in the proper way, it will give gradually and evenly forward under this pressure, but if made in the other way they sometimes kink just below the rim, and are always as a rule hard to manage.

These are the main points in getting out the material for spokes.

QUESTION XI: HOW TO PROPORTION SPOKES.

Now we come to the arrangement of the proper details regarding the measurements to be given for sustaining the weight our wheel is to carry.

Were it a question of merely carrying the load which we had to consider, a spoke somewhat like this "skeleton" which I hold in my hand would answer every requirement. This is made the same as the width of our mortise, namely, $\frac{5}{16}$ inch, while it is of the proper depth also, being 1 inch full at the lower end, where it tapers properly to fill the mortise. Now, any force acting against the end of such a lever as the diameter of our wheel would make of the spokes, enables us to reduce the width of the spoke at the *rim end*, and we therefore taper the spoke from the back, carrying this taper up to the rim end. The gain in the leverage allows of making the width at this point $\frac{5}{8}$ inch full or $\frac{11}{16}$ inch at the rim end.

These, then, are the proportions of what I have termed the *skeleton of a spoke*, which you will perceive does not resemble one excepting in its extreme proportions.

FIG. 7.

As I have said before, were the mere carrying of a load all we had to consider, this "skeleton" would be of sufficient strength to do the work, but it would not stand up under the shrinkage of the tire. This developes a new force to be met, and we must provide for this by adding body to our skeleton, and in doing this we can also give beauty of outline to our mathematically proportioned "skeleton."

You may ask, how are we to arrive at the proper amount of material to be added to accomplish the object sought ? I answer, testing machines of any kind are too bulky for transportation, or I would have been pleased to show you some tests made here in your presence, upon the different proportions of spokes. I know these would have been of interest to you, but in all tests made with specimens of timber, the want of uniformity in the material itself, prevents anything but an approximate result. I have, however, made many series of such tests, and will report to you the general result which I have thus obtained. The results were often conflicting, but I have found that a spoke proportioned like the sample I now hold in my hand will sustain a weight of from 700 to 750 lbs., acting in a direct line from point to tenon, without deflecting. When these figures were passed, and the spoke began to bend, the bending was carried evenly throughout the entire length, showing that these proportions are evenly balanced.

The figures, then, for a proper-sized spoke we will now write upon the board, as 1 inch full across the shoulders ; thickness of the body, $\frac{11}{16}$ inch ; and width at the rim point, $\frac{1}{8}$ inch.

continued next page

Wheelmaking

In this other sketch (see Fig. 7) we show a spoke with the shoulders beveled where they are to rest upon the hub. Made in this way, they seldom crack up the spoke, or break off the corners, which the square shoulders will sometimes do. The objection urged to this plan is that the varnish will show a slight cracking around the spoke when made in this way, which it is not so apt to do where the shoulders are square and slightly let into the hub; but I favor the plan here shown, as the spoke can be driven down firmly without danger of cracking or forcing the shoulder off; and durability being the principal object, the cracking of the varnish spoken of is comparatively secondary, providing the spoke remains sound and firm; and that this is accomplished is seen by referring to the specimens which are here before me.

There should also be a difference between the width of the shoulders at the face and at the back of the spokes. This tapering of the shoulders is necessary on account of the difference in the diameter of the hub at the points between where the face and back of the spokes come, and this taper should always follow the amount of such difference in the hub. By this plan we give the wheel a clean look; otherwise it will have a clumsy and filled-up appearance.

QUESTION XII: HOW TO THROAT OUT SPOKES.

Throating out the spokes now comes next in order, and there can scarcely be any set rule given for this portion of the work, as it should be governed by the quality and condition of the material. For this sized wheel I use about the proportion shown in this specimen, the idea being to have as much wood in excess of the width of the tenon here at the back, as has been removed from the face or narrowest portion. With this proportion the spokes will not work loose in the hub, which they are sure to do if not relieved by throating out to a certain extent.

Supposing all our spokes to be prepared upon the measurements given, we are now ready to consider some of the whys and wherefores which present themselves in framing and driving the wheel together.

QUESTION XIII: WHY ARE NOT SPOKES TAPERED BOTH WAYS?

We have incidentally mentioned why spokes were tapered always from the back, and we some time since received a letter asking why spokes were not tapered both ways, which I will now answer.

If we were to taper spokes both ways, there would be no certainty of bringing them to a face line in driving. The "mash" in driving would send some forward, others backward; and the spokes, after being driven, would have nothing to prevent them from dishing backward, as the stiffness of the wheel is in a great measure due to having the face of the tenons at right angles to the greatest strain they are to resist.

While it is absolutely necessary to taper the spokes to wedge them into the hub, still a small amount of taper has been found to give a better result than where greater angles have been used. About $\frac{1}{16}$ inch in a tenon $1\frac{1}{4}$ inch long is considered the extreme limit or right figure.

QUESTION XIV: HOW TO DRIVE SPOKES.

In regard to the manner of driving the spokes into the hubs, I will have nothing to say, as it is the principles underlying the work which particularly interest us at present. Every mechanic has a method of his own to accomplish a given object, and, looking at it in this way, I can only say, drive the spokes into the hub to *stay there;* and to do this, three things are necessary; dry material, good glue, and plenty of patience. "Never get in a hurry," is an excellent motto for the wheelmaker; and, I may add, one which the average wheel-maker often follows to an **extreme**.

QUESTION XV: HOW TO PROPORTION THE DISH

We have before mentioned the amount of stagger or zigzag which should be given a wheel of this size, and as the question of dish is intimately connected with this, we will devote a moment to its consideration.

You are perhaps aware that the nearer the load is carried through the center of the box, the less the amount of draught or power required to move it. Consequently, the least possible dish we can use, the nearer we approach to a perfect mechanical device. There may be some special point which it is necessary to gain, such as an increase of seat-room in the body, which calls for great dish in wheels, but this is a question for the builder and not for the wheel-maker; and one thing is certain, the more dish called for in a wheel, the more the stagger should be reduced; and in no case does dish or stagger add to the strength of a wheel beyond certain fixed limits.

QUESTION XVI: HOW TO PROPORTION RIMS.

It is now time for us to say something about the proportions for the rims, which, of course, must be capable of sustaining their part of the load.

Experiment has demonstrated that a rim slightly heavier in proportion than the spoke measures at the shoulder, is the true proportion for this part of the wheel; and to secure proper room to support the shoulders of the spoke, the rim must have nearly the same measurement on the inside face.

QUESTION XVII: HOW TO PROPORTION THE TREAD.

To secure the greatest ease of draught in a wheel of this kind it must have as narrow a tread as possible under the circumstances. This holds good for all hard roads and city pavements, but on soft and yielding roads the tread should be increased in order to diminish draught.

We will, therefore, mark down for our tread the figure $\frac{7}{8}$ inch, scant, as this is about right for a durable tire for regular work.

QUESTION XVIII: BORING THE RIMS.

Boring the holes for the spokes is about the most particular part in putting the rims on the wheel. These should follow the exact angle to which the spoke has been driven, and should also be perfectly true with the face of the rim laterally.

Also bear in mind that, in boring these holes, we should not remove more than one-half the amount of wood that is shown by the width of the tread.

The conditions named would lead us to determine the size of hole suitable for a $\frac{7}{8}$ inch tread to be $\frac{7}{16}$ of an inch, which gives a sufficient sized tenon on the spokes to prevent them from breaking, and also allows enough of a shoulder to prevent the rim from settling, which is a very important point.

If there is any one of these Whys or Wherefores more important than another, it is, Why do these rims so often split out from the holes made for the spokes? Wheels perfect in every other particular will often show defects in this direction when least looked for. There are several well-known reasons why rims split, but cases often occur where they split without any apparent reason. We will briefly mention some of the most fruitful causes of this trouble.

First, there are some varieties of hickory which, while seemingly possessing every requirement of good, tough wood, yet in reality give evidence, under tests, that the fibers are so bound up in separate layers that the slightest separating force cleaves the grain asunder, this cleavage becoming easier the higher up the tree the cut is taken. The butts of such timber, while eminently suitable for spokes, are consequently totally unfit for rims.

Second, it is necessary, after a proper selection of timber has been made, to have the timber bent properly to get the greatest advantage in this direction.

FIG. 8.

A third cause for rims splitting arises from the shoulders of the spokes being fuller at the inside than on the edge, somewhat as shown here in the small sketch on the blackboard, Fig. 8. Shoulders so made invariably act like wedges, and force the rims to split open as soon as the wheel receives any hard usage. The same weakness may be caused by scooping out a saucer-like excavation around each hole in the rim, as by doing this the support is removed, and all the strain coming upon the outer edge, the rim splits open.

Fourth, a spoke may be used which has been dried until all the moisture is completely taken out of it. A tenon having been cut in this, and the rim driven on, the carriage goes into service, and the carriage washer gives those wheels a taste of how much water can be used to a small quantity of mud, when the consequence is that the tenons absorb moisture, swell, and split the rims open. A hole bored out of the proper line will also cause this trouble.

To avoid most of the troubles just mentioned, we should seek to make square shoulders on the spokes, a rim square on the inside face, a tenon only moderately tight; and then if everything is in proper line, a good hickory rim will then stay there every time.

continued next page

Do not be too particular about whether the wood is white or red, as this has little to do with its fitness. The red or heart wood will be found to hold up at the joints better than the white, if the proper quality is secured. The main point is to secure an *interlocked growth ;* and remember that it is easy to get timber *too glassy and hard* for durability.

COMPLETE MEASUREMENTS OF A LIGHT WHEEL.

STEEL AXLES.	HUBS.							SPOKES.				RIMS.		
	Size.		Bands.		Mortises.			Size.				Size.		
Size.	Diameter.	Length.	Front.	Back.	Length.	Width.	Stagger.	Width at Shoulder.	Width at Rim End.	Thickness of Border.	Number used.	Depth.	Width.	Tread.
⅞	3¼	6	2¼	2⅜	15-16	5-16 full.	⅛	1 in. full.	13-16	11-16	14	1 1-16	1 1-32	⅞ scant.

FIG. 9.

I will now, in concluding, show you a completed wheel which embodies all these points spoken about, and which gives a better idea of how the proportions, thus reached, agree as a whole or in the finished job.

SPECIMEN WHEEL EMBODYING THE PRINCIPLES DESCRIBED.

This specimen has not been made as a mere sample, but is a practical, full-sized wheel, selected as one from among many sets, all made with the proportions I have named, giving a practical support to arguments that might otherwise be thought to border on the *theoretical.* "The proof of the pudding is in the eating," is a homely saying, but conveys an idea distinctly applicable to wheel making ; and I am perfectly willing to have any tests applied to prove the correctness of the proportions advanced as being suitable in the manufacture of light wheels.

After every precaution has been taken by the carriage-builder, to make wheels wear and give satisfaction, there is still a long distance between what is expected by the carriage-user and what could reasonably be demanded from wheels of ordinary proportions. As a rule, the builder tries to secure the lightest proportions possible, and in fact will often sacrifice strength in order to secure beauty of outline ; while the average carriage-buyer is dissatisfied and disappointed if he cannot get away with a fire hydrant or a lamp-post without injury to the wheels of his turn-out.

Trouble is sure to arise from expecting *too much.* There is a limit to what iron and wood will stand ; and when this limit is passed, a speedy collapse must follow. Judgment is required at the hands of the driver of a vehicle, as well as at the hands of the builder. It is a question of judgment all the way through ; and one of the old Welsh "triads," that has come down to us like an echo, quaintly says : " *Three foundations to judgment : bold design, constant practice, frequent failures.*" As I understand it, this school is intended to assist you in gaining these three foundations to good judgment as applied to carriage mechanics. Here you learn the principles of design, hereafter to be subjected to the test of practice, and modified by the lessons often so dearly bought by experience in the factory and repair shop.

I congratulate you upon the progress which you have made during the past winter, and hope that this work of technical instruction will continue from year to year until the perfect wheel and the perfect carriage shall be the rule rather than the exception.

In concluding this paper, I desire to thank you for your considerate attention, and will ask the worthy chairman to make some fitting closing remarks, or suggest points for more general discussion.

WHYS AND WHEREFORES IN WHEEL-MAKING.

BY HOWARD M. DUBOIS.

(VERBATIM REPORT OF LECTURE DELIVERED BEFORE THE CLASS IN CARRIAGE DRAFTING, NEW-YORK, MAY 17TH, 1882.)

The Hub, Aug-Nov 1882

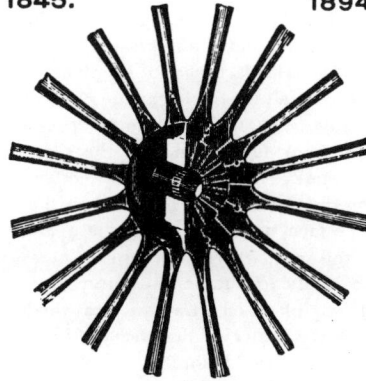

Wheelmaking

COMMUNICATION FROM A CINCINNATI COR-RESPONDENT.

Which Spoke should Stand Plumb—Beveling the Face of Rims—Spoke Tenons generally too large.

62 GEORGE-ST., CINCINNATI, O.

EDITOR OF THE HUB—DEAR SIR: In the July number of *The Hub*, page 223, appeared the question : "Which spoke should stand plumb, with, say, a ⅜ dodged wheel?" and your answer was, to set the outside spoke plumb when on the ground.

Pardon me for disagreeing with the answer. After a practical experience of over thirty years, I find that both spokes, back and front, should stand plumb, that is, from center of the rim to the center of spokes at hub, thus : (see sketch) and that, as for gather, there should

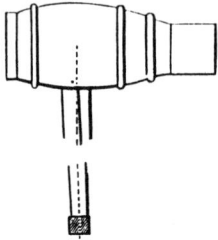

be just enough so as to insure that it is not backward, say to measure 1/16 inch narrower on front than back.

Just one thing more allow me to speak of. There is a practice among wheel-makers and also in wheel factories, in regard to rimming wheels, which I find objectionable. They now, universally, bevel the face of the rim. This is all wrong. The face should be planed to a straight edge, and the bevel should be all on the back side of the rim.

Some claim that it is the only way to get the tenon in the center of the tread. For that reason, I bore my tenons on with the shoulder the largest on the back side of the spoke ; and there is also another reason, namely, the tire will stay on the rim better. That throws the tire a full sixteenth-inch farther to the face, and also gives a larger bearing on the back of the spoke and rim, where the hardest pressure comes. Spokes are then not so likely to buckle and throw the dish out of the wheel, or to crack half-way off at the thin part, near the hub. This way will bring the tenon in the center of the tread. It should not be in the center at the shoulder, as I said before.

Also, nearly all wheel-makers make their tenons too large. They thus weaken the rim, and make it flatten quicker between the spokes. The reason they do this simply is, that a rim is easily set upon the shoulder of the spoke when the shoulder is small all around the spoke, and one blow with a large hammer and set-block will set the spoke into the rim whether the shoulder fits or not ; while if your shoulder on the spoke is of good size and the rim is hard, it must be bored on square and no gaping. A hint to the wise is sufficient. D. C. SHOTTS.

ANSWER.—All makers of first-class wheels will agree with Mr. Shotts that it is not correct to bevel a rim from the face side. There are, however, exceptional cases where it is advisable, as in the case of very heavy work, where the rims are very deep, and the tread required narrow. In this instance the rims are "swelled" or rounded from the face as well as the back, in order to avoid boring the holes so far out of center of the rim.

We would also say, as regards boring large holes in rims, that the wheel-maker has another object in view from the one mentioned by Mr. Shotts, namely : Since the use of street-railways, it has been found that turning a carriage wheel out of the track often results in breaking the spokes at the rim end ; and as it is better to choose the less of the two evils, wheel-makers sometimes increase the size of the rim tenons to avoid this breakage, even at the risk of having the rim flatten between the spokes.   DuBois.

PLUMB SPOKES.

FLINT, MICH.

EDITOR OF THE HUB—DEAR SIR : I see in your July number, page 223, in answer to an inquiry from B, of Buffalo, that you answer, "Set the outside spoke so that it will be plumb when on the ground." Do you mean plumb the face of outside spoke, or center line of spoke ? If either of the above is correct with a ⅜-inch dodge, will you be kind enough to say why, and oblige,   ANOTHER B.

ANSWER.—We are well aware that there is great diversity of opinion as to the proper manner of plumbing the spokes of a wheel, even among the most experienced builders. Some prefer to plumb the front face of the front spoke, and consider this the only proper way. Others prefer to plumb from the center of the rim to the center of the two spokes at the hub, and consider this the only correct way. (See "Communication from a Cincinnati Correspondent," below.) Some prefer plumbing from the back face of the front spoke ; and others, still, plumb the center of the front spoke. It will be noticed that a slight variation must result from these different methods, and the best of these has never been determined beyond the reach of argument. Experience would seem to be the true teacher in such a case ; but, in this instance, equally careful observers have seemed to derive different lessons from equal experience. New-York carriage-builders generally follow the rule of plumbing the front face of the front spoke, the vehicle being unloaded ; while the Philadelphia method, as stated on page 225, is to plumb the back face of the front spoke.

The Hub, Nov 1882

Wheels, Wheel Stock and the Construction of Wheels.

Jared Maris.

WHEELS.

In teaching the novice to make any structure, it is common to make a drawing of the same and give specific dimensions. This method appears well in print, and the cuts and tabulated statements

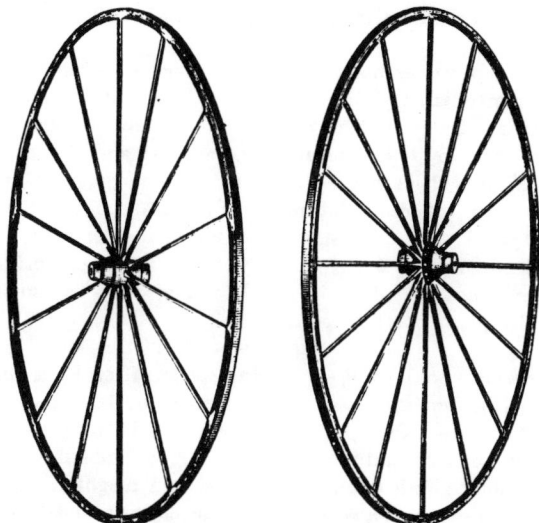

PLAIN WHEEL.   PATENT WHEEL.

give an appearance of exactness that inspires confidence. The method referred to is simply laying out a piece of work that may be made without a thought on the part of the learner, more than to copy; and after executing the work he has no practical knowledge, though he may have acquired more skill in the use of tools, and if the job is a good one he may fancy himself master of the art. We prefer to teach wheel-making on what we will call the "suggestive" method. If asked for the kind of timber for a hub, we would say: What size wheel, and is it to be banded? If asked what size a hub should be, we would say: Tell us the size of your axle, including box, and the width of your spoke; and so on.

LARGE HUBS.

For farm wagons, white oak is generally used, because of cheapness and good qualities for holding the tenon of the spoke. Chestnut oak is sometimes used; with what success, we do not know, but from our knowledge of the timber we can think of no special objection to it. Birch is preferred by some; its closer grain and free-splitting qualities may be objectionable, but it has the advantage of appearing well. Elm large enough—that is, not too soft—is not in supply, and, for work liable to go unpainted, it is too subject to decay. Black locust has several qualities to recommend it. It is not, however, in sufficient supply for general use in farm wagons, and is too costly. It is largely used in hubs from 6 to 8 inches in diameter, for express work by small makers. The advantage over other timber is that its tensile strength is nearly double that of oak, and, therefore, the spoke tenons may be large to the size of hub. The wood absorbs very little water, and changes but little in diameter by changes from dry to wet, and it is not subject to decay as other timber; and on this account, and because of its great tensile strength, it may be refilled a number of times, and will last without painting.

The timbers most favored for these are white or rock elm and gum. The latter is less liable to split than the former, though more

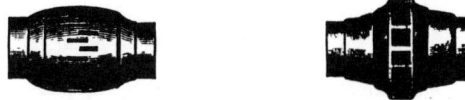

SMALL HUBS.

liable to season checks. The tensile strength of elm is greater; but the advocates of gum say it "hugs the tenon better." Both are much subject to decay, but gum the most.

For banded hubs, we would prefer ash to any other wood. Its tensile strength is about one-fifth greater than elm, and it shrinks and swells less and takes glue better.

DENSITY OF HUBS.

The required density of a hub depends very much upon the density of the spoke. A very hard hub should be used only with an equally dense spoke.

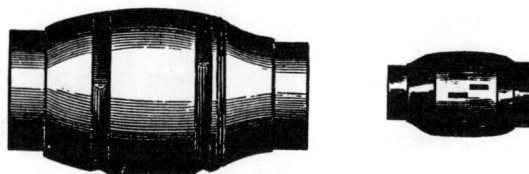

BANDED HUBS.

Hubs secured by bands shrunk into grooves near the spoke, or with band capping the ends of the raised part, coming flush with the wood at the crown, have their advocates, who claim that these bands of whatever shape secure the wood against splitting and swelling from absorbing water; also, that being so secured spokes can be driven much more securely by making the tenons tighter sidewise, and furthermore that the hub will bear refilling, and will not swell and cause wheels to dish.

To this the objector replies, that, admitting security against splitting, the additional cost and weight largely overbalance this advantage, because a good hub is necessary in any case and the cost of bands is a hundred-fold greater than the loss by split hubs properly boxed, and further, that in time the paint will break at the wood and iron joint, even if entirely hidden; that driving spokes so as to compress tenon more than a good hub will stand with temporary bands, injures the tenon; and that compressing the hub and securing it with bands diminish the desirable quality of a wood hub, viz., its elasticity. These are the respective claims, as near as we can give them; and now, in the language of the renowned Israelite, you "pays your money and takes your choice."

PATENT WHEELS.

We shall speak of all wheels made with hubs composed of wood and metal, where the metal forms all or a part of the bearing of the spoke, as "patent wheels." Because nearest in form to the common wheel, we will notice a hub encased with metal over its entire crown. In this metallic covering mortises are made so as to receive the spoke in full size. It is claimed for these that all the advantages against splitting are secured that may be had with bands and iron, that provision is made against increased dishing from swelling of the hub, and that, covering the crown so as to form no end joint with the wood part of the hub, there can be no trouble with paint, as it is claimed that there is with bands. It is also claimed that they have the appearance of a wood-hub wheel with all the advantages of a patent wheel in the matter of appropriating the strength of the large part of the spoke to reinforce the tenon, and are therefore able to make a stronger wheel at that point. They further claim that they secure a large mechanical advantage in placing the weak point below the surface, and gaining leverage for protection of the weak point by the bearing of the spoke against the iron.

continued next page

Wheels, Wheel Stock and the Construction of Wheels,

JARED MARIS.

Objectors admit the claim as to splitting, but interpose objections of unnecessary additional weight and expense, destruction of elasticity of hub, and liability to breaking of tenons when not seen, because of alleged difficulty in getting mortises in the metal true with mortises in the wood, and the depth of the countersink in the iron being insufficient as a reinforcer to support a strained tenon. The objector also claims that the bearing of iron against the spoke is so narrow and sharp that the wood yields and the support is reduced, so that the spoke has, in fact, no acting tenon but that inserted in the wood core. It is further claimed that the water will get into mortise or countersink in hub, and rust it, and render that part of the support to the spoke inoperative; and that the cast metal will be rusted, and destroy the paint where the iron terminates.

PATENT FLANGE WHEELS.

The advocates of this method of making hubs claim that they secure all the advantages of a large hub with the neat appearance of a light one, and are able to use the larger part of the spoke as a tenon, and have that point in the wheel that is weakest in an ordinary plain wheel the strongest laterally; and those using an arch claim the wheel to be stronger vertically. They claim that by having the bearing for the spokes so high they gain greatly in leverage, and the surface being so large they avoid indentation of the wood by the iron. They also claim that with this method they make a wheel more symmetrical by having the spokes in line, and by this means give better bearing for the rim, because each spoke makes the same resistance, while the dodged spokes do not; and that this method provides for making wheels that will stand on country roads when made of cheap timber; also, that they avoid trouble from decay and shrinking, swelling and cracking of hubs; and, finally, they claim economy in that they can make a wheel with such hubs that will stand using cheaper timber than is necessary for other styles.

As per contrary, the advocates of all-wood hub and band-hub claim that the patent hub is not only large, but unsymmetrical and out of proportion to spoke, and that the great strength gained at this point is at the expense of all the other parts of the wheel, and, in fact, that the rigidity is a disadvantage in light work to the entire vehicle. In reply to the claim that the rim is better supported, they say that the support is so rigid that the rim must yield, and therefore cannot stand; and that, the spokes being so arranged that each one performs the same function as the other, the wheel is made an absolute unit as to resistance to tire, bearing of burden and concussion, and therefore, when the tire is on and the spokes sprung at their center, any stroke upon the wheel is received by the whole wheel as nearly equally as if it were a solid plank, and it trembles like a Chinese gong and shudders, in running over a hard, uneven road or paved street, like an ague fit, wearing itself out, together with tire and axle, springs and rider, by means of its excessive so-called strength. This refers to light work. As an addition to the above arrangement, it is claimed that the wood part of the hub is not large enough to supply sufficient cushion for the axle-box, and it is therefore liable to break, as is the axle for same cause.

HOW WHEELS SHOULD BE MADE.

To give any definite sizes or proportions other than general rules, would require more space than can be given to this article; and we do not regard it as the best method of instruction. Knowing the size of axle-box, add to it enough to insure $\frac{1}{16}$-inch play. Add to this the width of spoke tenon, with ⅛-inch added, multiplied by 2; this will give diameter for small hub in fine work, where the best of materials are used. For cheap hubs and spokes, and for heavier work, more should be added to size of hub, because softer, and the shoulders of spoke will be more deeply seated. After boxing, the spoke tenon should always be longer than its width.

SIZE OF SPOKE.

This depends so much upon height of wheel, strength and character of material, and place to be used, that any definite specification cannot be given that will not be as likely to confuse as to instruct. Size of tenon depends on material and size of hub, and material in both as well. Size of round tenon must be in proportion to width of tread and quality of rim and spoke.

SHAPE OF SPOKE.

Spokes should be so shaped in the barrel as to yield forward when the tire forces the wheel into dish, and to yield as uniformly the whole length of barrel as possible.

DEPTH OF RIM.

Depends upon quality of rim and spoke. If spoke is better than the rim, the rim may have more leverage by length of tenon. If rim is better than the spoke, the spoke may be relieved by shorter tenon and consequently shallower rim. Proportions should be given in accordance with quality of material and height of wheel, a low wheel requiring a deeper felloe, because concussion is not relieved so much by yielding of spokes.

When should the timber be cut? From July to February. How seasoned? Oak hubs in closed buildings after roughing off bark and sap; elm and gum same. Locust shrinks but little, and will require less care. For light wheels, hub-blocks should have ends dipped in a composition of resin and tallow, coal tar or pitch, or be heavily painted, to exclude the air.

All blocks should have holes bored through them, roughed off, and placed in well-ventilated rooms, and moved from side to side and rehacked two or three times per month, and all mold should be wiped off. When dry, a condition that cannot be determined by time, they should be turned and exposed, after oiling with pure hot linseed oil, to the sun and wind, to determine whether they will check. They should be mortised immediately before using, if possible.

Blocks after being half seasoned may be put in the dry house to finish. This will save trouble in assorting. The best ones will be known by the cracks. Good blocks require much time and great care in handling.

RIMS.

Sawed so that the holes for tenons will run with the medullary rays, they will have greater power to resist concussion, and shrink and swell more. If they are bent with the heart side in, they have greater density for bearing on shoulder of spoke, and the more porous side will absorb more water than the heart would, and decay more quickly. If bent with sap side in, the open grain is packed by the upset and there is more danger of checking under the tire. If the strips are sawed so that the hole for the tenon runs with the annular rings, the wood is not likely to check, but timber has less resilient power in this way. If cut bias or diagonally, there is less liability to split at the tenons, but they spawl off the edges or corners under the tire.

There are objections to each way, and each has its advocates If we had our choice, we would have them sawed with medullary rays and bent sap side in. In timber of equal weight, heart or dark-

continued next page

colored wood shrinks and swells less, is less liable to decay, has greater resilient power, keeps its place better, and makes a more desirable rim than sap wood. A large per cent. of the failures in rims of otherwise good timber is in parting lengthwise after being subjected to pounding on the road. It is commonly called "windshake." It is not "windshake;" it is not the result of frost. It is a dead or diseased condition of an annular ring or growth. It can be detected always by examining the end of the rim, if it extends to the end; but it often extends no more than one or two feet, and do not extend to the end. In such cases it cannot be detected. Cracks from freezing would be more general. "Windshakes" never close, as a wound in wood never heals.

A "windshake" by reason of constant friction caused by storms is worn on the bearing surfaces. We suppose it to be caused by some wound by worm-seeking birds, or other cause that robbed the growth by bleeding or admitting the air to dry it for a space; but this is conjecture. All hickory, of all colors and in every locality, is alike subject to it. Checks are sometimes caused by not subjecting the rim strip to steam a sufficient time to soften uniformly.

The size of hole for tenon depends upon quality of spoke and rim. What is called the long rim is satisfactory to those who like a rim-bound wheel. Rims are commonly made too deep, and rounded too much. When too deep, there is too much leverage on tenon. When rounded much, the weaker point is in a line with the force of concussion. The rim when ready for tire should rest equally on the shoulders of all spokes, and the ends should form a close joint, or nearly so, on the inside, and be open at back in proportion to dish intended for it. The rim should be dressed so that the tire will fit when the proper amount of dish is drawn.

SPOKES.

All spoke timber, whether oak or hickory, if large or old, should be seasoned wholly in the atmosphere. The use of steam or artificial heat increases its brittleness. All young, second-growth timber that is too limber should be exposed to the atmosphere six to twelve months, according to size, and then put in dry rooms at about $70°$, and gradually increased from $25°$ to $40°$ until it reaches $140°$, according to quality of timber. Those who saw spoke-cuts to center, and open as we do an orange, steam them to loosen the bark. Exhaust steam is used, and only enough to loosen the bark. No appreciable change is wrought in this timber, except in color. Direct steam would not injure the timber, when young and limber.

FRAMING WHEELS TOGETHER.

Mortises in hubs should be $\frac{1}{32}$ narrower lengthwise to each one inch in depth. The front end wall of mortise should be plumb; sidewise the mortise should be straight. The spoke tenon should be exactly the mortise, and fill very tightly lengthwise, and as close sidewise as the hub will bear without splitting, if not banded, or if banded as tight as it will bear without raising the surface of the hub. This, however, will not apply unless the spoke is harder than the hub. Good glue should be used by drivers who apply it thoroughly; for those who don't, a cheap article will do. Spokes should be seated to the amount of dish required.

Amount of dish should be drawn by the tire enough to insure the keeping of the tire tight. If a wheel is to be used on the Pacific Coast, the wood will shrink more, and it will in the hot sand expand more, and it should therefore have more dish than if used in Chicago, or if to go to England it should start with as little as possible.

No doubt, many a lazy reader who will neither reason nor think will say: "Well, just how much for each place?" I reply: "How high is the wheel? What stagger has your spoke? Give us size of spoke? Inform us of weight of tire? Is it a patent wheel? Is the rim hard, or will it yield at bearing on spokes, and if so how much? Is the hub soft?"

Some dish should be given the wheel in setting and driving the spokes. For learners I will explain. Drive a spoke in opposite sides of a hub with, say, one inch dish, and put this beginning for a wheel on its face on the floor, where there is a hole to receive the hub. The spokes will rest on the floor at their ends, but nowhere else. Step up on the top of the hub, and the weight springs them down to the floor, because the spokes receive the burden in the way it has least resisting power. Get down from your perch on the hub, and your spokes spring back into dish, so to speak. Now nail a block at the end of each spoke, and get on the hub and add 500 pounds more, and still there is dish. You call into requisition what is called the crushing power of the timber—*i. e.*, what it will bear endwise—and the blocks on the floor and the hub receive the weight. Now take up the hub and drive all the spokes, and lay a tire over to fit against the ends of all the spokes. It takes the place of the blocks. Now you put a ton's weight on the hub, and the dish remains in the wheel.

Now, what carries this weight? Snap the tire, and you say the tire does. Let the hub crush, and you say the hub does. Let a few spokes spring sidewise, and you say the spokes. It is all together; the tire made them a unit. For reasons given, it is obvious that some dish should be framed in the wheel. Also for another reason. When spokes are staggered, and the front one is straight when drawn forward, the shoulder on back part of front spoke describes a larger circle, and it is farther from this shoulder to the hub, and consequently the weight is unequally distributed, the front ones carrying most of it. When front spokes are driven $\frac{1}{16}$ back and are drawn forward, they will sometimes spring the spokes sidewise and the wheel will play in and out like the bottom of a tin pail, and another trouble is that when drawn forward the lengthening of the front spokes, while the back ones are shortening, draws and makes the bearing on the back spokes unequal—sometimes breaks the tenon or causes "chewing."

After enough dish has been driven in to insure equal bearing of back and front spokes (in staggered wheels,) against the rim, enough must be drawn by the tire to spring the spokes forward enough, so that when the rim settles down on shoulders of spokes, or the tire elongates by heat, the resilience of the timber will follow the tire and keep it tight. If the tire enlarges by heat, or the rim shrinks, or the spoke shoulders press into the rim so that the spokes have room to go back to their condition before the tire was put on, the wheel becomes rim-bound. In this condition the weight flattens or depresses the part next the ground, and forces the rim off at an opposite side, and as it runs the change moves round the circles continually, until spokes are drawn, shoulders mashed, or tenons broken. The wheel is no more a unit, but a naught.

When bent rims are used, hickory is preferred for light work, oak or ash for heavy. Butt cuts of good young trees so sawed as to have the medullary rays run with the spoke tenon, are probably preferable. When so sawed, the bark side should be the outside, says one, because it is softer and upsets better in bending. Says another: "The heart side should be in, as it gives a more solid bearing for spoke shoulder." Weight being equal, the dark wood is best, because it keeps its place better under the tire, and shrinks and swells less when exposed to the elements, and aside from that it is not so liable to decay. It is claimed that rims are less liable to split when sawed so that the tenon goes through "bias" or obliquely across the annular rings. This is true as to liability to check in boring or driving on; but in practical use they shrink and swell unevenly, and throw the tire-bearing face out of true, and too weak points are made, so that concussion causes a spawling off when the weaker way of the grain has to resist the force.

As wheels are now made, the rim is the weak point. The size of hole to receive spoke-tenon depends so much upon quality of timber in both spoke and rim, that any specification is more likely to mislead than instruct. Depth of felloe also depends very much upon quality of timber in both. A deep felloe on a weak tenon gives too much leverage, and it is therefore better to have tenon shorter. When a poor rim is put on a good spoke, the rim may be deeper.

continued next page

Wheelmaking

All rims should be carefully examined at the ends for annular rings that have the appearance of dead wood. Where these are, the rim usually separates. It is called a "wind shake," though probably caused by a wound in the bark of the tree. This condition sometimes extends not more than one foot in length, and may occur in a rim and not show at either end. The length a rim should be when ready for the tire depends on many things. If the wheel is light and high, it requires more dish and therefore a shorter rim. If low, longer. When left long and contracted with the tire, they upset at the weakest points, making them weaker. If not shortened, the wheels are rim bound—*i. e.*, the rims do not rest on the shoulders of the tenons, in which case the loaded wheel is never sound, and its constant changing wears it out. A rim-bound wheel is worthless.

Hickory is generally used for light rims, and uniformly two pieces to the rim. Ash, if young and of good quality, answers a good purpose. For larger wheels oak is better than hickory, for various reasons, among which are the following, *viz :* The white oak tree lives to more than twice the age of the hickory, and does not begin to decline till after the size at which the hickory has attained its full age, and white oak lumber can therefore be obtained heavy enough for this purpose that is of good quality. For section rims, ash is better, for the reason that its crushing strength is greater and it keeps its place well.

What are called wind shakes, give more trouble in rims than any other cause and there is no means known to avoid them. The cause is a dead growth, probably caused by a wound in the tree from a birdpeck bleeding it at the season when the sap is in motion. This loss of sap and presence of air keeps the annular ring or layer of the season from fully uniting with that of the former for some distance from the wound. The influence of the wound may not extend more than a few inches, and therefore may not extend to the end of the rim piece and cannot be noticed. It is well, however, to examine the ends, and a skilled or experienced observer can detect the dead growth in an end view of the piece. Much has been said on the question of the best way to cut rim strips. Various methods have their advocates. If cut as in *Fig. 1*, they are not so likely to split in boring and driving the round tenon. They, however, are more likely to spawl off at the corners, the weak way of the timber being required to resist the concussion from the tire. All things considered, it is best to saw the strip as shown in *Fig. 2*, and bend with heart to the tire. Weight being equal, the dark wood makes the most satisfactory rim, either heavy or light.

There is much discussion as to whether wheels with 1-inch tread or even ⅞ endure more with what are called second-growth rims, or a good quality of what is called black hickory. All rims should, as well as the balance of the wheel, have a coat as soon as made, either of filler or hot oil, *i. e.*, as hot as water at boiling temperature. Oil heated to the point of ebulition injures wood instantly. The spoke shoulder where the rim rests should be well saturated and a heavy coat of lead, preferably red, should go on under the tire. Every avenue for water should be closed both as to spoke and rim. The ends of rims should be coated with lead and dowell hole filled.

Wheels are in better condition for the painter and smith as soon as they come from the finisher, than they can ever be again. It is presumed that the joints all fit, and any change, especially if it be shrinkage, injures them. It is best, therefore, to keep them as nearly in the condition they were when made, as possible. If the object is to test the wheels to determine whether they are seasoned and fit for use or not, a purpose may be served by keeping them in a dry place, but this might better be done with the stock before framing.

 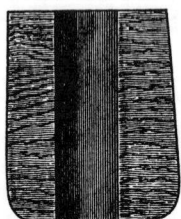

Fig. 1.   **Fig. 2.**

When tenons are so made as to exactly and tightly fill the mortise, they are likely to keep their place if good glue entirely envelops them.

DISH IN WHEELS.

Spokes should be driven so that a line from the front side will strike the hub forward of the spoke at the tenon, in order that the diameter be less than if the spokes were in a straight line. The tire should be put on so as to draw the rim as much forward of the point to which the spokes stand (which should be, in a dodge spoke wheel, about ⅛ inch to 4 feet, height of wheel) as will allow for the expansion of the tire at the greatest heat it is likely to be subjected to in use. This bends the spokes forward, and their tendency to straighten will keep the tire tight. The additional dish which is "driven" in the wheel, will bring the crushing power of the entire timber in the structure to resist the strain most frequently put upon it. If the spokes are driven straight, and the rim dressed on tread at right angles, when brought forward by the tire, the diameter is increased as it is brought forward, the front spokes bear most of the pressure of the tire, and the back ones make greater resistance to the forward movement, which throws their tenons out of line and the front spokes are liable to spring sidewise. This being the case, the wheel is liable to spring backward from the center like the bottom of a tin pan, and the wheel is worthless. A wheel is never thrown backward and forward in this way at the hub without the spokes springing sidewise. Wheels driven with spokes in line, require more dish than is given front spoke in staggered work, on this account.

Size and weight of wheels vary these suggestions, much. It would require a large volume to contain all the technics and measurements necessary to be known by a skilled wheelwright.

BOXING WHEELS.

The simple fitting of an axle box in the hub would not seem to be difficult, and yet there are more disasters from failures to do it well than any other work of like amount on a vehicle. It can only be avoided by care in one who has a knowledge of the powers of resistance in the materials used. Care and mechanical common sense are absolutely required in boxing wheels. If there is a fool in the shop, better put him to setting a top than a box. The top is seen before shipping.

There are several reasons why the boxes are often out of center of the hub. If the fault lies in the boxing machine, as is sometimes the case, the box must be wedged toward that side of the hub, in order to have the wheel true, although we will say here that wedging boxes is a bad practice, and should not be encouraged.

If the wheel is properly made in the wheel shop, and then it is found that the boxes have to be set out of center of the hubs, the trouble will usually be found to lie with the blacksmith ; the tires are not heated equally around its circumference, and then, when they placed on the wheels, are cooled off too suddenly. The part cooled off first will throw more dish into the wheel than that part cooled off last. You will notice that when a wheel is twisted, the lighter the wheel and the tighter the tire, the more noticeable the twist will be. It is sometimes almost impossible to pound the rim level with the tire, it slipping from one side to the other. A wheel hooped under these conditions will have the hub out of center, and it can be remedied by the blacksmith's heating the tires uniformly and cooling them off gradually. Nothing is more annoying to the wheel-maker than to find the hub to be out of center, either front or back.

Prior to putting on the tire the wheel is an agregation of parts, —hubs, spokes and rim—and each performs its office individually. When bound together by the tire, the wheel is a unit, and receives and resists all concussions as a whole. It is as true of a wheel as of an arch, that it cannot be stronger than its weakest point. Therefore, *proportion* is the sesame to successful wheel making.

Many questions arise as to detail that cannot be considered here. Such as the size of round tenon to the width of tread. This depends so much upon comparative quality of spoke and rim, that no rule in measurement can apply. Depth of rim depends so much on height of wheel, quality of timber and weight, and kind of tire, that measurements would only darken counsel. So long as wheels are round, there will be no end to them, or to methods of their construction.

The Carriage Monthly, May 1883-May 1884

PROPORTIONS OF WHEELS IN RELATION TO THEIR "TIRES."

THERE is no distinct mechanical trade so dependent upon another for success or failure, as that of the carriage wheel-maker upon the blacksmith to whom is entrusted the completion of the wheel by putting on the tires.

The making of carriage wheels has become such a distinct industry, and so widely separated from the shop in which the wheels receive their finishing touches, that a failure is often the result, which, under other conditions would not be likely to occur were every operation conducted under one roof and one supervision; this being the case, we propose to include the wood-work and smith-work in one article, and view both branches in one and the same subject, as their mutual interests are so intimately combined.

It is a fact well known to the manufacturer of carriages, that there is a kind of lottery about the purchase of wheels; and a very great irregularity in the wearing qualities of this, the most important part of the carriage. There does not seem to exist any *standard of quality* by which the length of time and amount of service a wheel of certain price might be expected to give, and the fact still remains that wheels of the cheapest grade, and supposed to be made from material not good enough for fine work, often outwear those of the highest grades made from choice selected stock; and many builders can bear witness that it frequently happens that the most particular customer will, by some strange fatality, get the worthless wheels, if any such are encountered.

There is no reason why there should not be a standard of quality, in proportion to price; nor is there any reason why *chance* should enter into the calculation. The foundation and key-note of this standard can be reached only by a uniformity of *proportion* and a definite idea of how to properly put on the tires. It is evident that the most skillful smith can never make a durable wheel from one which does not contain the proper elements of proportion, and which has not been properly constructed. It is equally evident that the wheel-maker can never be assured of the durability of *his* work, no matter how good the stock, or how closely graded and proportioned, if the smith is ignorant of the proper manner of handling, and correct understanding of the peculiarities of the metal with which he binds the whole together. It is safe to say that more ignorance exists in regard to the true mechanical principles which form the foundation of this subject, than any other connected with the carriage industry.

The conditions under which this question of durable wheels must be considered are manifold, but we will try and touch on each point, and give the general laws which govern them, leaving the many individual cases to be worked out from these general principles; and we see no reason why the carriage-smith and wheel-maker, though seemingly widely separated in their respective callings, should not be more closely identified, at least so far as their mutual knowledge is concerned; to which end we will separate our paper into two parts, entitling the first:

"HOW WHEELS SHOULD GO TO THE SMITH."

It is necessary to state at the beginning that our figures relate to wheels in which the tires "draw" the wheels into "dish," and do not apply to wheels the spokes of which measure 1 5/8 in. or over, as this is about the point where other conditions arise, which require a separate treatment.

Every wheel-maker is supposed to know how to proportion the stock of which the wheel is to be constructed; but, as most of our wheels are made in large factories, this proportioning depends upon the amount of skill and experience which the manufacturers possess. This knowledge has gradually left, or is leaving, the carriage factory, and the apprentice of to-day stands small chance of becoming familiar with this portion of the work, unless we begin our subject at the very beginning, and touch the elementary principles of making wheels, which we have, from time to time, presented in these pages. The beginning, then, of all calculations for durability must be made upon the basis of the plain wood-hub wheel, which gives a working limit of strength for all other kinds. We begin by asking how shall we proportion a wheel, and where shall we make the starting point. The first and fundamental rule to be determined, is the diameter of the hub to be used in a given sized wheel, and this should be governed in all cases by the diameter of the *outside* of the *axle-box*.

When the amount of load which the wheels are intended to carry has been settled upon, a suitable sized axle is selected, and the boxes of these should be carefully measured. The diameter of the hubs for these axles should be enough to allow for using a spoke whose length of tenon in the hub, clear of the box, is at least 1/8 in. more than the spoke measures in its width at the shoulders where they rest upon the surface of the hub.

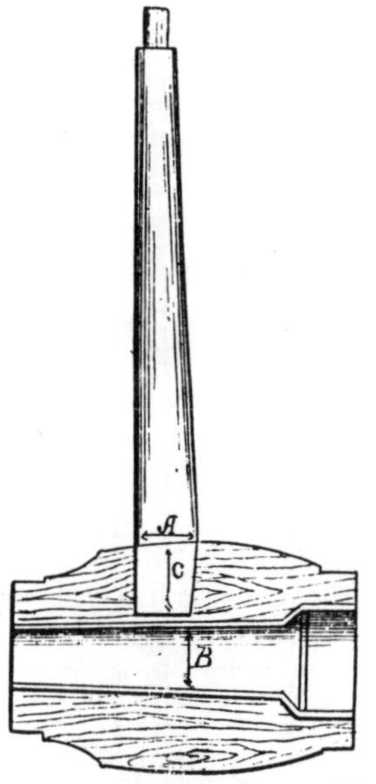

Fig. 1.

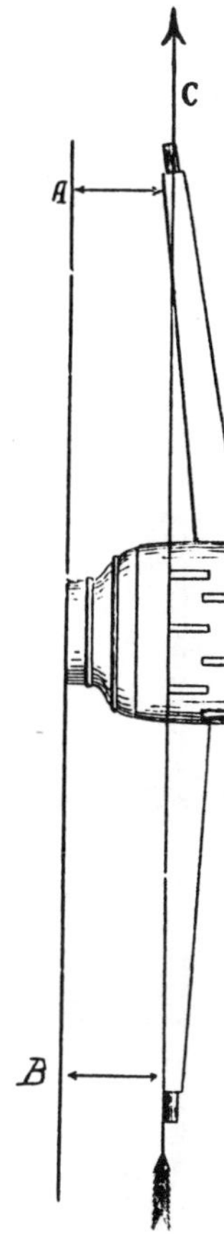

Fig. 2.

By referring to our sketch, Fig. 1, this point, which is really the danger line in wheel-making, is more plainly shown. The axle-box B is let into the hub, leaving enough wood to allow the spoke from the line C to be, at least, 1/8 in. longer than the width of the spoke at A. This important point in mechanics has much to do with our after subject; for, if this simple rule of proportion be neglected or over-looked, no addition of strength in other portions of the wheel will compensate for the oversight.

With this rule as a key from which to begin our calculations, it is not necessary to give any regular table of sizes of hubs to be used with given sized spokes; but, to more fully illustrate, we give a comparative table showing what these proportions figure out in average practice; these figures being the limit for lightness, taking into our calculation the ordinary axle-boxes now generally used.

PROPORTIONS OF WHEELS IN RELATION TO AXLES AND CURVES IN DISH.

Style of Job.	Approximate Weight. lbs.	Diam. of Hub.	Size of Spoke.	Thickness through Body.	Width at Rim end	Depth of Rim.	Tread of Rim.	Size of Axles.	Will carry. lbs.
Buggy, with top.	240	3 1/4	15/16	5/8	11/16	1	13/16	13/16	400
Buggy, with top.	300	3 3/4	1	11/16	3/4	1 1/16	7/8	7/8	500
Phaeton	300	3 7/8	1 1/16	23/32	13/16	1 1/8	15/16	15/16	600
No-top Surrey...	400	4	1 1/8	13/16	7/8	1 3/16	15/16	1 1/16	700
Top Surrey......	500	4 1/4	1 3/16	13/16 full	7/8 full	1 1/4	1	1 1/8	900
Extension Top...	600	4 7/8	1 1/4	7/8	15/16	1 5/16	1 1/16	1 3/16	1200
Rockaway	700	5	1 3/8	15/16	1	1 9/16	1 1/8	1 1/4	1500
Heavy Rockaway	1100	5 3/4	1 5/8	1	1 1/4	1 3/4	1 1/4	1 3/8	1800

As we have intimated, no table can arbitrarily represent an infallible set proportion, for this rests entirely, as before said, upon the thickness of the axle-boxes to be used, and these vary to quite a considerable extent in the various kinds now in the market. Having gone thus far, we have found how to calculate the proportions of wheels in the *diameter of the hubs* and the *sizes* of the spokes; we must now consider how many of these spokes to put in our wheels, and what position they should occupy in the hub to give the best results.

To keep as much solid wood in the hub as possible, and still have enough spokes to support the rims properly, are the main points to be gained; and, for wheels of the ordinary heights, 14 spokes all round will give better results than any other number. On the lower heights of

continued next page

wheels, 38 in. diameter and under, 12 will give equally good results. We are now speaking of the lighter spokes, after attaining a size of 1¼ in. and above. On extremely low wheels, 10 spokes will give the requisite amount of carrying capacity, but the numbers here given cover what

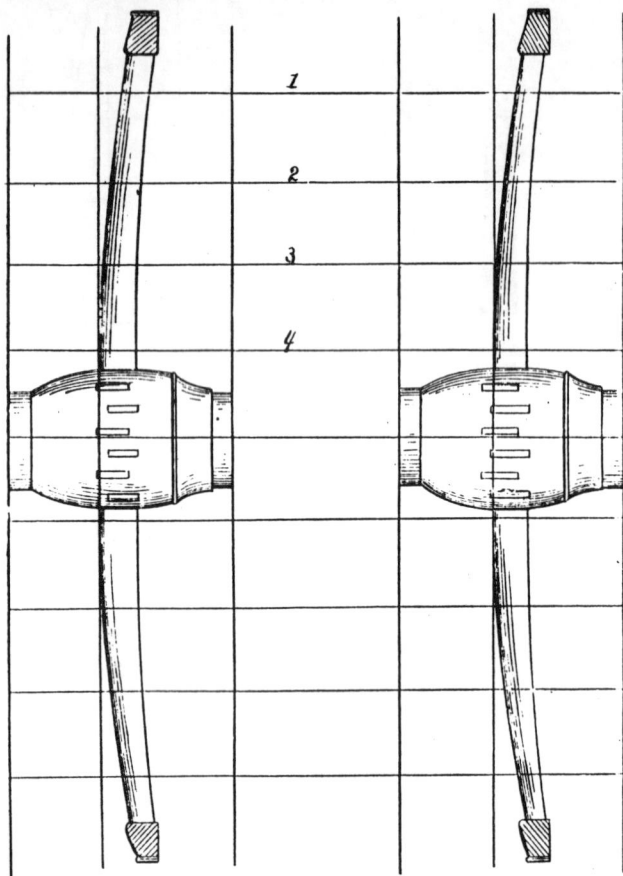

Fig. 3.   Fig. 4.

is necessary for *strength* and *durability*. There may be reasons why these numbers may not suit under altered conditions, but any changes should be made in the direction of a greater number, as we have given the minimum quantity that should be used. Having settled on the number to be used, we next consider what position they should occupy in the hub—by position we mean the manner in which they shall be driven into the hub. Shall we place them in a straight line around the hub, or shall we place each alternate one back a certain distance, making what is called "dodge" or "stagger"? After an exhaustive series of experiments, we can safely say that light wheels should be invariably staggered, and the amount of this stagger was found to give best results when ranging from ⅜ in. to ½ in. This gives a resisting power to the wheel, when properly done, that is of incalculable advantage, but otherwise is a source of weakness. For example, when the spokes are above the sizes we have given as our limit, say 1¾ in., or when the spokes are improperly "framed" in the wheel, the disadvantage becomes apparent at once. To get all the advantages that this "dodge" or "stagger" can give, the spokes should be driven into the hubs in reference to the amount of "dish" the wheel is to have drawn in it by the shrinkage of the tire. It has been said before that a wheel, when entirely finished, should be a unit as regards strength. To secure this result the *face* spokes of a light wheel should be driven straight, while the rear or staggered spokes should be made to stand forward just the amount the wheel is to have dish when the tires are on. By referring to our sketch, Fig. 2, it will be seen that the staggered spoke has been mortised to stand forward as mentioned. It must be borne in mind that our sketches are not made to any scale, and are intended merely to convey the idea, to do which they are exaggerated to a certain extent. The reason for having the rear spoke assume the position as shown, will be apparent when we consider that these spokes are

Fig. 5.

to act as a species of "strut-brace," and should they stand on the same line as the face spokes, the tires, in drawing the wheel forward into "dish," would bring all the strain upon the fore shoulders of the spokes, which would have a tendency, more or less, to crack them; and also, the spoke under these conditions, would only act as a brace after it was relieved of all *forward* pressure. In this sketch the rear spoke shows, at A, the amount of "dish" the tire is to draw in the wheel, when done, while the face spokes as shown at B are straight, as first mentioned; and further, the wheels having such spokes, should they need re-tiring, would eventually have their rear spokes cracked across the backs, by the increase of forward pressure, while the position as shown in our sketch secures a perfect "brace," and causes the rear spokes to resist lateral strains from their first inception, which a wheel made in this manner can be depended upon to perform.

PROPORTIONS OF WHEELS IN RELATION TO THEIR "TIRES."

THE proper shapes to be given to the spokes are all highly important, as their proportions in thickness of "body," "taper" and thickness of "throat" all play an important part in making the wheel what it should be when it goes to the smith-shop. For illustration we take a spoke

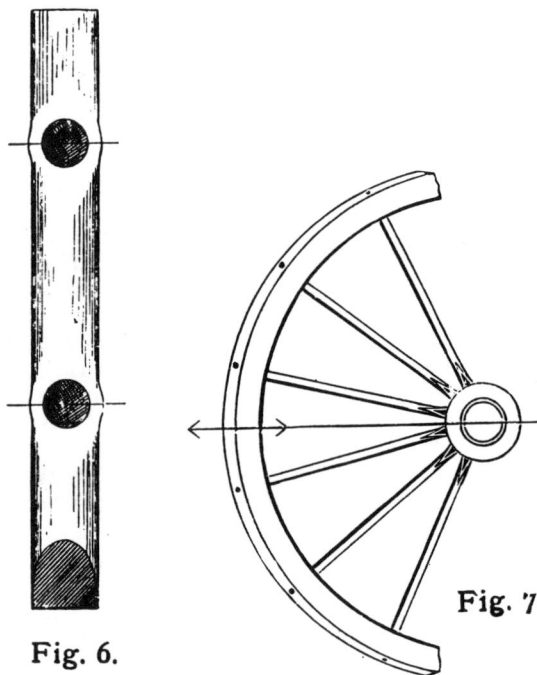

Fig. 6.   Fig. 7.

suitable for a vehicle to carry 500 lbs.; not only must we provide a spoke capable of sustaining its proportion of this load, but also one capable of doing it for an indefinite time without failure.

By experiment and actual practice, we find, for example, the diameter of the hub being 3¾ in., that a truly proportioned spoke would be of the following dimensions: depth or size of spoke, 1 in.; width across shoulders at hub, ⅝ in.; thickness of body, 1⅛ in.; width of point at rim, ¾ in.; thickness through throat, 7/16 in.; reaches, full body thickness, 6 in. from hub. A spoke of these general dimensions will bear, individually, about 750 lbs. in direct line, without deflecting or bending sidewise, while the "taper," as above mentioned, allows of a regular curve being given to the wheel in the process of tiring. To give some idea of these curves, and the important part they occupy in our subject, we must call attention to the fact that, when the shrinkage of the tire draws the wheel into "dish," the spokes, in yielding to this strain, should do so in regular curves, beginning slightly at the hub and gradually increasing as they near the rim. In Fig. 3 we show a gradual curve, as given when the spoke is properly tapered. The wheels which are shown in Figs. 3 and 4 are supposed to have an equal amount of "dish."

continued next page

Fig. 3 should show a curve extending from the cross-line marked 4, to the rim, or line marked 1, while Fig. 4 is intended to represent this curve confined to the cross-lines 1 and 2, extending partly into the space between 2 and 3. Spokes are frequently proportioned in such a manner that these curves are confined to, or between, the lines 1 and 2, thus lessening the chances of wearing in a proportionate degree.

Fig. 4 also shows the result of having the spoke too heavy near the hub and too light at the rim end; a very serious fault, which soon causes the tenons of the spokes in the rim to break off, from the shortening or quickening of the vibrations in service. This also accounts, to a certain extent, for the "hardness" perceptible in the riding of some vehicles, which peculiarity can be plainly felt beyond any ease of motion to be obtained by the use of springs; also, in this connection, we ask attention to Fig. 5, which illustrates a heavily proportioned spoke. Should the spoke be left extremely heavy at A, also proportionately heavy at B, and possess very little taper, the tires will bring a heavy strain upon the point D, while the shoulders at C have really all the load to carry and must be firmly bedded in the hub. If they do not crack up the spokes with the slightest strain, resulting finally in the breaking of the spokes across the back as shown; our example being somewhat exaggerated, to impress the importance of true proportion. Still, this trouble can frequently be met with in practice, in various degrees or stages, and many cases of failure can be traced to this species of faulty construction. The sizes or proportions

TABLE OF PROPORTIONS OF SPOKES TO GIVE PROPER CURVES IN TIRING.

Size of Spoke	Length of Tenon in Hub	Width of Tenon	Width at Shoulders	Thinnest part or Throat	Thickness at Body	Width at Point
15/16	1 1/8	5/16 full	1/2 full	3/8	5/8	11/16
1	1 3/16	3/8	5/8	7/16	11/16	3/4
1 1/16	1 1/4	3/8 full	5/8 full	7/16 full	23/32	13/16
1 1/8	1 5/16	7/32	11/16	1/2	13/16	7/8
1 3/16	1 3/8	7/16	3/4	9/16	13/16 full	7/8 full
1 1/4	1 7/16	1/2	13/16	5/8	7/8 15/16	15/16
1 3/8	1 9/16	1/2 full	7/8	11/16		1
1 1/2	1 11/16	9/16	1 5/16	3/4	1 set	1 1/16
1 5/8	1 13/16	9/16 full	1	13/16	1	1 1/4

given in the accompanying table will insure the graduated curves which we have described, and which should be found in all properly constructed wheels. This table of proportions corresponds with what we have tried to make clear, and these proportions have invariably been found to give excellent results in practice.

We have now reached the important subject of the rims, which should be selected from wood possessing natural stiffness, and, to secure the best results, the "growth rings" of the timber should be in the direction of the depth of the rim and not the reverse way. This prevents splitting, or at least gets the strength of the timber in position of greatest resistance to splitting. These points having been attended to we cannot insure absolute trueness in boring the holes in the rims, unless the preliminary process of "facing" the rims out of "wind," has been accomplished, as it is from this "face" that we must look for a proper fitting and "straight" wheel; and we may add here that the smith cannot be held responsible for the rim "creeping" out from under the tires, as this trouble is caused to a great extent by the wheel being "crooked" before tiring. The smith may draw wheels straight laterally with the tires, but the constant jarring over pavements gives them a tendency to come back to their position as first made.

These preliminaries settled, the rims are supposed to be firmly driven down upon the shoulders of the spokes. If we are to send the wheels ready finished to the smith-shop, we must have them properly "jointed." This settling the rims solidly down upon the spokes may have made the joints of the rims too tight, or, as it is termed in the shop, they become "rim bound;" and, just here, it is appropriate to speak of the practice of putting tires upon wheels with what is termed a "long joint," which means, in plain English, that the shrinkage of the tires will "upset" the end grain of the rims and cause the rims to come down solidly, thus securing a more durable job than would otherwise be the case.

We have examined many rims which have been, in service, tired with

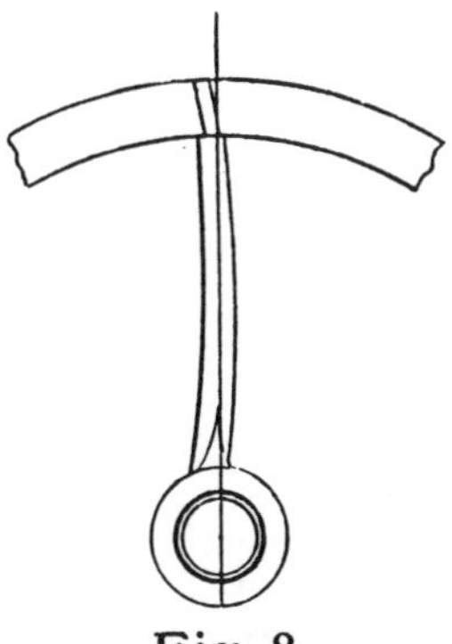

Fig. 8.

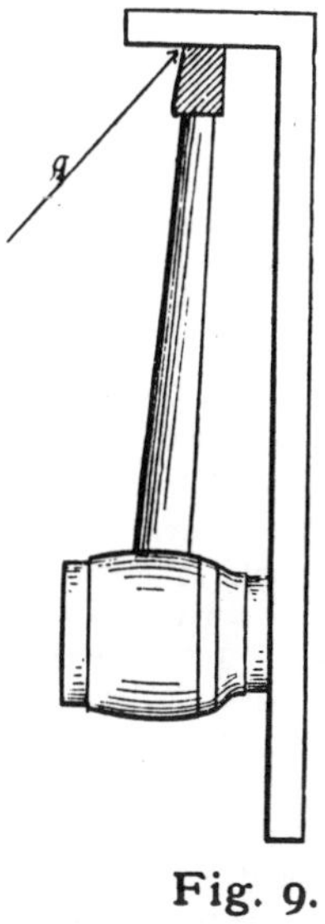

Fig. 9.

this "long joint," and have found that the "upsetting," in most cases, has been made at the expense of the wood left at the sides of the holes nearest the joint, causing them to swell as shown in our Fig. 6; this being the weakest part of the rim. We cannot recommend the "long joint," over the more mechanical method of having the rims just the proper length to secure the required "dish," and drawing of the wheel tightly together. It is also absolutely necessary to have the joint stand square with the tread of the wheel, and not at an angle. The joint should also stand true to the center of the wheel; see Fig. 7. It is well to have this joint somewhat tighter at the bottom, nearest the inside of the wheel, as, made in this manner, the tire will draw it square.

We show in Fig. 10 a kind of miter clamp used by a well-known wheel-maker. The sketch shows the idea clearly, and insures getting the joints to stand correctly with the center of revolution.

Fig. 10.

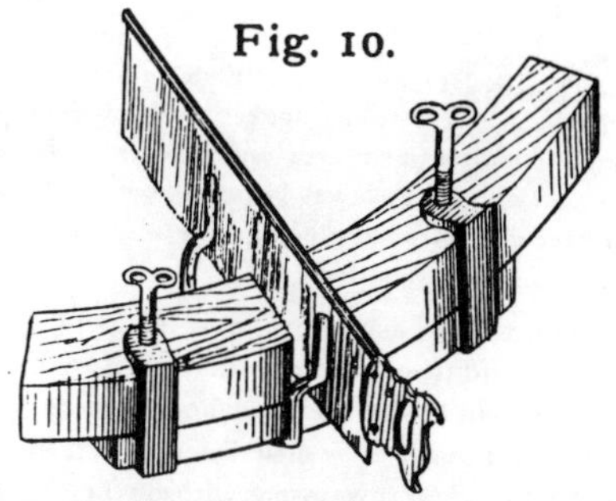

Dowels being used to a great extent in factory-made wheels, we would advise cutting through these when made of wood, or, if iron ones have been used, the rims should be knocked back and these removed entirely. A failure to do this may result in the rims splitting laterally from pressure in service, which puts severe strain upon the joints. Before proceeding further, we must define mechanically just what "dish" really is.

To fully understand this, we take, for example, a wheel which has been prepared for rimming, and, when finished, to measure 44 in. The exact circumference of this wheel would be, in close fractions, 138.2304 in. in "jointing." If we remove an amount of wood equal to twice the thickness of an ordinary saw, we reduce this circumference to 138 inches, and the diameter of the wheel, when the tires are shrunken and finished, $\frac{8}{100}$ of one inch. The shoulders of the spokes having been cut to make the wheel exactly 44 in., this reduction in circumference of the rim, and consequent reduction in diameter, curves the spokes forward until they accommodate themselves to this new diameter. These curves we have roughly shown in our sketch, Fig. 3, and the result we call "dish."

From this it will be seen that the question of jointing is a delicate one, and should always be done with a regard to the points named, as the slightest amount taken from the rim draws the wheel into "dish" in increased ratio, and the strain of putting on and removing tires to get them set properly is a severe operation, which often results in serious injury to the wood-work. It sometimes occurs that one wheel will not assume the same "dish" as the rest of the set. The cause of this may be found in some defect in jointing or some fault in dressing the tread of the rim.

continued next page

Wheelmaking

It sometimes happens that, even with the greatest care on the part of the smith, the spokes of a wheel will bend laterally, and this is where faulty proportions show plainly, for, should the "throats" of the spokes have too much wood removed, the tires will kink them sidewise instead of bending them forward as they should do. Another cause for spokes bending laterally lies in the holes through the rims not being true with the center of the hubs. This is shown in Fig. 8. The hole in the rim, being bored toward the side of the hub, curves the spoke as shown, also, the ends of the spokes, if left too long or projecting through the rim, will kink the spokes; these should be cut off exactly even with the tread of the wheel when the pressure of the tires is put on. All these things are to be met with in practice, and, no matter how slight the deviation from true lines, the effect is sure to make itself known in time to cause annoyance and trouble.

We now come to the question of dressing the tread of the wheel. This point is also one which should receive great attention. The plan giving the best results is to have the wheel so nearly correct that dressing the tread *square* with the *face* of the hub will give an even draft to the tire, as shown in Fig. 9. It often happens that the back of the tread, as marked at A, is left higher than the face, thus allowing more pressure back than on front edge of the tire, when set; this will prevent the wheel from going into "dish," but at the same time it does so at the expense of the rim, which, under these conditions, has a tendency to split at the holes where the spoke-tenons go through.

It is better practice to have the tread of the wheel perfectly square with the hub, as shown, and depend upon a proper "jointing" of the rim. The tires are supposed to be "true" on their inner surfaces, consequently, the shrinkage is in a direct line; and this change in diameter, by the contraction of their particles, should be solidly met by the entire width of the rim. This prevents the tendency of the tires to slip off, and keeps the rims from "creeping" out from under portions of the tire, which they are subject to doing when the tire does not fit solidly all round. Where durable work is the rule, the wheels, as made by the wheel-maker, are kept in stock to season or settle before tiring, and in the majority of cases the wheels will be found to have changed from the lines originally given them, the rims invariably requiring to be jointed over, and the amount of shrinkage shown by the rim is surprising, even after stock has been on hand long enough to be thoroughly dry. The mechanic must never forget that wood tissue will *always* shrink on the exposure of a *fresh* surface, no matter to what degree it may have been prepared primarily.

We have now gone over the main portions of our subject, but to give it all the attention that each detail should receive would require a full volume. We have shown enough, however, to prove to the young mechanic the importance of thoroughness and attention to details, seemingly trifling when viewed separately, but which, combined, make success or failure in their observance or neglect. We hope shortly to present another treatise on this subject entitled : "How wheels should go into service from the smith-shop." HOWARD M. DuBois.

The Hub, Feb-Mar 1887

QUESTION ABOUT DRIVING SPOKES.

PINEVILLE, ——, Nov. 28th.

EDITOR OF THE HUB—DEAR SIR : Which, in your opinion, is the best way to make a heavy wheel : to drive the spokes in a hub dry, or to put it in boiling water long enough to soften to some extent ? My plan is to dip my hubs in hot water, and give my spokes ⅛ in. mash in width, and ¹⁄₁₆ in. sidewise ; and I have discontinued glueing. Is this right ?

Yours truly, R. T. W.

ANSWER.—The question asked by R. T. W. is of general interest to the trade, and we therefore answer it in detail, and illustrate the manner in which heavy wheels are put together in an establishment having a well-earned reputation for the durability of their trucks and wagons, whose wheels have been tested for many years under every possible condition.

In using the term "heavy wheels," we refer to wheels with spokes whose dimensions are large enough to resist the shrinkage of the tires without springing into dish. Such wheels have the hubs mortised as shown in the accompanying sketch marked Fig. 1.

The face of the mortise, when tried with a straight-edge, shows the mortise line forward of the position the face of the spoke is to occupy when driven.

Comparing the mortise shown in Fig. 1 with the dimensions of the spoke illustrated in Fig. 2, it will be found that the bottom of the mortise is ³⁄₁₆ in. less than the corresponding part of the spoke tenon, while at the part marked B this mash is reduced to ⅛ in., and further diminished at the surface of the hub to ¹⁄₁₆ in. See A. This reduction of mash is made to prevent surface checking, and also to secure the full width of the spoke across the shoulders.

A spoke of the dimensions given in our sketch, if the shoulders are cut on the proper angle, will drive *back* about in proportion to that shown in our drawing.

The tendency of the spokes in heavy wheels is to sink into the hubs under service, while the reverse is the case with light wheels, whose spokes are inclined to work out. The use of glue under these conditions before the spokes are driven home. An equally effective plan is to heat the spoke tenons in a hot-oven, and then drive them while hot. An extremely hard hub can be sufficiently softened by pouring a very small quantity of boiling hot water through the mortises while fastened in the wheel-pit.

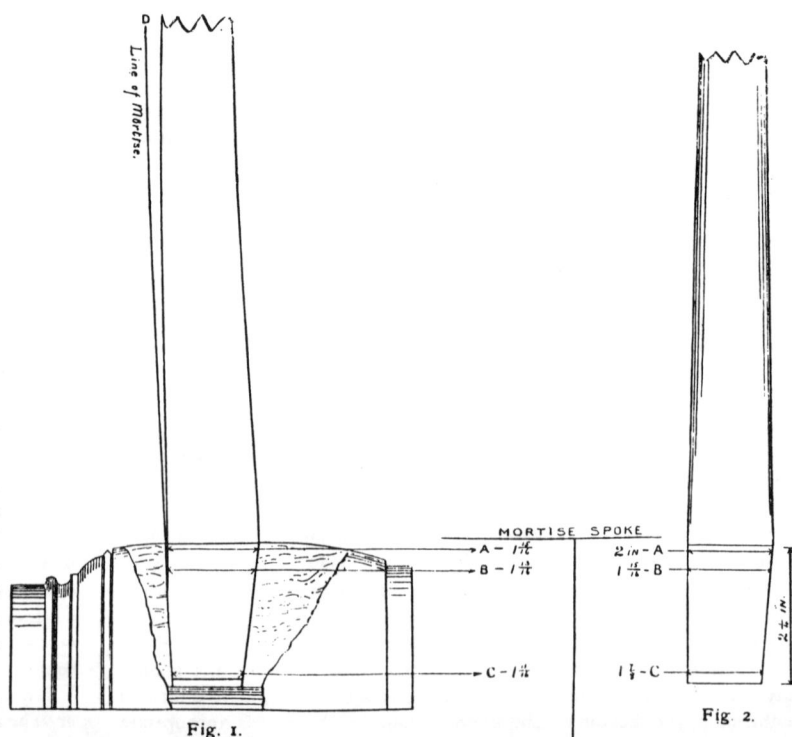

Fig. 1.

Fig. 2.

The above simple rules, mechanically followed by *The Hub's* correspondent, should insure durable heavy wheels if applied to good dry timber. H. M. D.

The Hub, Jan 1888

Wheelmaking

THE "DANGER LINE" FOR WHEEL-MAKERS.

(See four Illustrations accompanying.)

As by far the larger number of carriage wheels are now made by machinery and by special manufacturers, it is natural that these should often be made the subject of comparison with those formerly made by hand and under the personal supervision of the carriage-builder; and it

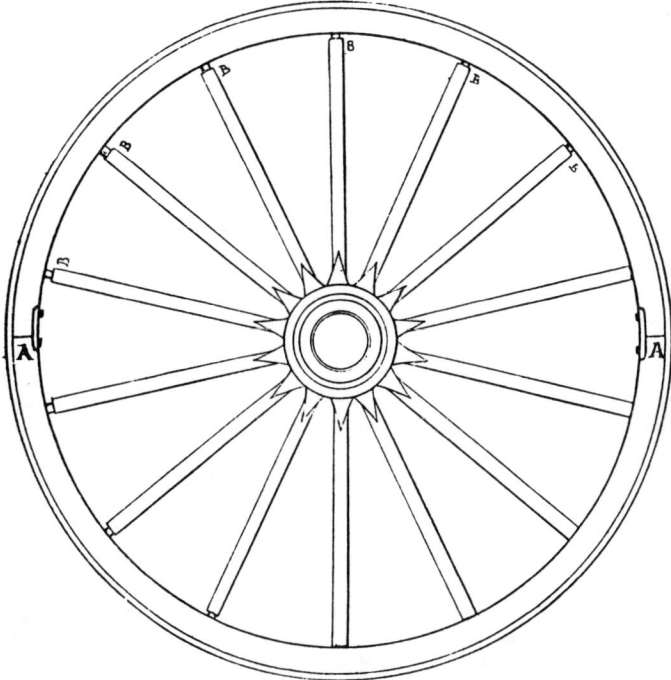

Fig. 1.

must be admitted that the result of such comparisons is generally unfavorable to machine-made wheels. Close investigation of the facts proves that there certainly does exist a difference, and shows that there is a greater average of machine-made wheels that go wrong than of those made by hand in private carriage-shops.

When we remember that success in wheel-making depends mainly upon dry material and the mechanical exactness with which the work is performed, it would seem reasonable to expect that the reverse would be the case; the fact that it is not, leads us now to consider what we will term the "danger line" in wheel-making, as offering an explanation of the problem.

When wheels are made by hand in the carriage-shop, the wheel-maker not only produces the wheels completely, but, when these are taken to the smith-shop to have the tires put upon them, he again goes over them carefully in preparation for this operation, and particularly so if they have been made up for some time. In this case the joints receive special

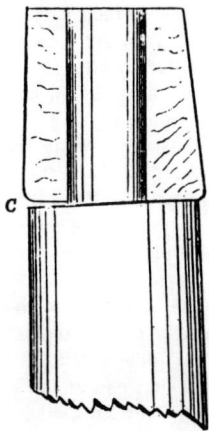
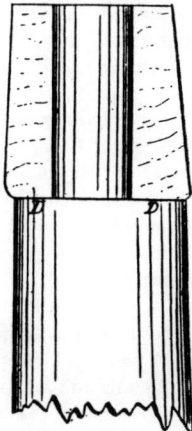

Fig. 2.   Fig. 3.

attention and are re-cut, while the rims are settled down firmly on the spokes, thus insuring a solid wheel, free from all liability of being rim-bound. This question of being rim-bound is the vital one; and, in the writer's opinion, it marks what we have called the "danger line."

On the other hand, where carriage wheels are bought from the wholesale factory, the services of a regular wheel-maker are not required by the carriage-builder, but the duty of attending to the preparation of the wheels for the tires is given to what may be termed a general woodworker or jobber. No matter how skillful this person may be, he does not feel the same responsibility, nor does he work from the same standpoint as the person who really makes the wheels. Under these conditions it is not too much to say that fully fifty per cent. of machine-made wheels are rim-bound to a certain extent before they go into service. That wheels can be durable in this condition is an impossibility.

To explain more fully just what we mean, we give in Fig. 1 an exaggerated sketch of a wheel whose rim is tightly bound by the tire, as shown at A, A., while the shoulders of the spokes have been settled down until they work in the rims as shown at B, B, B. In practice it would be impossible for a wheel to be in such an extreme condition as

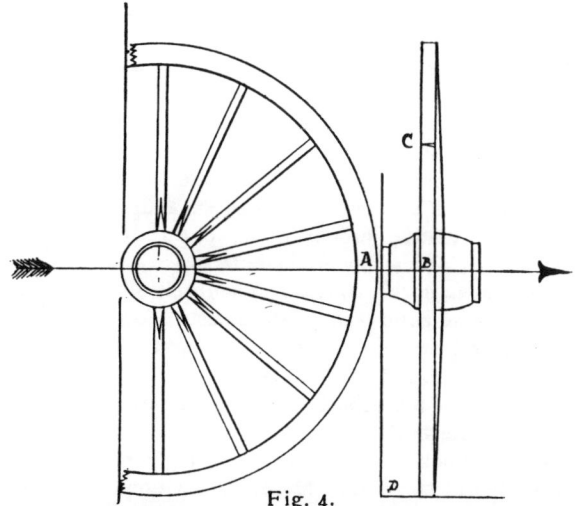

Fig. 4.

here shown; but, so destructive is this dangerous fault, that, if the rims be bound sufficiently to even break the varnish and paint at the shoulders of the spokes, the wheel has already passed the "danger line," and may be irretrievably ruined, unless the tires are at once cut and the rims jointed and again brought tight upon the spoke shoulders.

It sometimes occurs that the outer ends of the spokes are wedged in the rims, and in this case the spokes are inevitably drawn from the hub, and no form of construction has ever been found proof against this leverage. In this way many wheels have been rendered worthless, which, under proper treatment, would have proved extremely durable.

One condition that leads to this danger lies in the impossibility of keeping woodwork in the shape in which it originally leaves the workmen's hands. No matter how dry the material may be before being worked, it will inevitably change its position more or less when allowed to stand and settle: and, in the case of wheels, it is no uncommon occurrence to find the shoulders of the spokes and the inside faces of the rims assuming the position shown at C in Fig. 2. In this case the draw of the tire cannot bring the rim down solid, as it should be, or as shown at D, D in Fig. 3. Where this fault in construction is found, it should be remedied before the tire is applied; for, if not, the wheel will soon work the spokes loose, and the rim will become bound in a short time.

To prepare a wheel properly for the tire, it should first be placed upon the rimming-horse, the rims should be knocked back, the dowels removed or sawed off, and the rims set down upon the shoulders of the spokes by using a hollow block and heavy hammer or sledge. The joints should then be sawed squarely, as shown by A, B in Fig. 4. In many cases the joints will be found to stand as shown at C. These should be brought square, even at the expense of cutting down the spoke shoulders to get wood enough to saw through. It is also highly important to have the tread of the wheel dressed perfectly square with the face of the rims, which should be on a line, as shown at D.

With the above points carefully attended to, no reason remains why factory-made wheels should not be quite as durable as those made by hand, always providing that thoroughly dry material is used.

We think we have pointed out the actual "danger line"; and, with the better understanding of this important point which experience must give, we may hope to see this line gradually disappear, and scientific tiring take the place of present unsatisfactory methods.   H. M. DuBois.

The Hub, June 1888

Wheelmaking

A TREATISE ON WHEELS.

By James Small, Plough and Cartwright, of Mid-Lothian, Scotland.

(Originally published in 1790.)

INTRODUCTION.

By H. M. DuBois.

THE following treatise on wheels and axles is highly interesting from the fact that it shows that the principles underlying the successful making of wheels and the proper proportioning of axles were understood and written about at a very early date. A careful perusal of Mr. Small's treatise will convince the American reader of the fact that, while great improvements have been made in lightening the relative proportions of axles and wheels, still the mechanical rules upon which they are constructed have never varied. The remarks here published relate to cart or heavy wheels only, but the geometrical problems are all correct and apply equally as well to light wheels and axles.

It has been a matter long held by the older wagon and coach-builders, that the large wooden axles of ancient times were of considerably lighter draught than the modern ones of iron or steel. This question of difference in draught rests altogether on the correctness and scientific proportioning of the parts, and it is a lamentable fact that few modern manufacturers follow the scientific rules as laid down in this treatise. The result of this is increased friction. But where the taper of the axles conforms strictly to the dish of the wheels, as here taught, the draught must be the same under all conditions and in all cases.

As a means of comparison of the teachings of the present and the past, this little work is of particular value.

Mr. Small's book has become exceedingly rare. The writer found one copy in the British Museum, also a reprint made by the Agricultural Society of Dublin, dated 1802. These together with an original copy belonging to a friend in London are the only ones the writer has ever been able to find. The drawings are fac-similes of the originals, slightly reduced. The phraseology of some of the passages has been changed in order to make them more intelligible to American readers.

* * *

I. OF WHEELS IN GENERAL.

To attempt to prove that a carriage is more easily drawn upon wheels than upon sledges would be an affront to the understanding of the reader.

But whether high or low wheels are fittest for the purpose has been a subject of dispute even among persons of skill. Reason and experience, however, seem perfectly to agree in this, that wheels whose centers are on a level with the moving power will be easier drawn along a level plane, and that the higher a wheel is, it will more easily get over any obstacle if the moving power is not below its center. It seems to follow, therefore, that carriages drawn by horses or oxen should have wheels whose centers have the height of the draught line, that is, of the shoulders of the horses or the yokes of the oxen. This is true, however, only in case of a horizontal road; in going up hill the distance of the line of draught from the road is somewhat less, because, when a man or any other animal is standing upon the side of a slope his height is inclined to the slope although it is perpendicular to the horizon; as this is the situation in which it is of the greatest importance to diminish the labor of the cattle it follows that the height of a wheel's center should be somewhat less than the line of draught. But what this difference should be depends upon a great number of circumstances which those who make wheel carriages should be best acquainted with. Four feet eight inches is reckoned a very good height for a cart wheel, and has been thought preferable to any other. But the great loads which are drawn in the coal carts at Glasgow, which go upon wheels above six feet high, and other instances of a like kind, show that great advantages are to be gained by using the high wheels, and that the disadvantage arising from the greater weight of the wheels may be disregarded.

The great inconvenience which attends the high wheels is the greater pressure upon the cattle going down hill, because the support which the cattle can give in this case is very low placed, but even in this case the instances above mentioned show that the height of wheels may be increased beyond the ordinary practice.

II. OF THE "DISH" OR CAVITY OF WHEELS.

Suppose that a person totally unacquainted with the practice, were employed to make a cart. He would naturally make the axle straight and the arms of it without taper. He would set the spokes of the wheels at right angles to the axle, by which means they would be perpendicular to a level road and thereby give the most perfect support to the carriage and its load. Such wheels would be flat, and their sole or tread would be parallel to the axle. But experience has shown that such wheels are liable to great inconveniences: when going in wet and muddy roads all the dirt taken up by the wheel comes down upon the hub, gets between it and the axle and soon wears it prodigiously. Such a form of wheels would also require a great distance between them, and consequently a great breadth of road in order to give room for the body, but a greater trouble is found in such wheels bearing hard upon the inner portion of the axle. The swaying of the load also causes a greater strain upon the spokes backward, and the wheels in time would become dished or hollowed on the side next the body.

The reason of these two last inconveniences is this: it is seldom that two wheels are exactly on the same level. When one of them is lower than the other, either by getting in a rut or being on the lower side of the road, the greatest part of the load is upon that wheel. In this situation the tread of the wheel is not perpendicularly below the end of the hub, which is supported by the spoke. The spoke itself leans over and if not of sufficient strength would break inwards, it therefore bears hard upon the inner side of the mortise and causes the hub to bear hard upon the inner end of the axle arm.

Experience has shown that these inconveniences may be greatly removed by setting the spokes into the hub so as to point outward a little; by this means the wheel gets what is called a "dish," and the spokes no longer form a flat surface. Let us now suppose that two such wheels are put upon an axle whose arms have no taper, the wheels would be as wide below as above, and the spokes would never be plumb or perpendicular to the ground. Suppose now the two axle arms are bent downwards, this would make the wheels closer below and wider above. The bend of the axle may be such as that the spoke which rests upon the ground will be perpendicular. This form of wheels will remove some of the above-mentioned inconveniences.

The wheels being wide above, the mud which is now carried up by them will not fall so much upon the hub, and there will be greater room for the body of the carriage and its loading. But one of the great inconveniences still remains. When one of the wheels is in a rut the spoke which is then supporting the carriage is leaning outwards from the point of the hub and bears hard upon the inner end of the axle arm; it is therefore found more expedient not to bend the axle arm so much as to bring the two spokes whose felloe ends are on the road to be perpendicular to the horizon or parallel to each other; they are made to be a little wider below than at the center, by which means when one wheel is in a rut the spoke which is then supporting the load may be perpendicular. It is even found necessary that when the wheel is in this situation the spoke should be a little without the perpendicular below, for the chief jolts which the wheels receive are from the outside, and when the wheel is set a little against these jolts it is better able to resist them.

Experience shows that notwithstanding all the "dish" that has been given wheels, they still go from the face and lose their cavity when they fail. It might be supposed that it would be an advantage to give the wheels a still greater "dish," but this is limited by other considerations. The spokes must be driven very hard into the hubs, in order to have sufficient strength; they must not be very much tapered at the tenon, otherwise they would be apt to work loose. This being the case the spokes cannot be set very obliquely into the hub, because the wood of the hub would then start up at the heel of the spoke. When it is properly driven a dish or cavity of 3½ inches may be given with safety to wheels 4 ft. 8 in. diameter, and in general ¾ inch of dish may be given for every foot of diameter.

The spokes should be set so far from the outer end of the hub that a perpendicular from the tread to the under side of the axle may fall between an inch and two inches without the middle of the hub or bearing of the boxes; by this the bearing upon the outer end of the axle will be somewhat greater than at the inner shoulder when the wheels are running on the level. This should be so, for the inner part of the axle arm being much bigger than the outer it has more friction and therefore should have less pressure; also every sinking of one wheel below the level of the other causes it to bear harder upon the inner shoulder. This position of the spokes remedies both these inconveniences.

continued next page

There are still some persons of experience who have proposed that the wheel should be made without any "dish," and they propose to remedy the inconvenience which arises from unequal pressure upon the outer and inner ends of the axle arm by placing the spokes not in the middle of the hubs but toward the outer extremity.

This will only remove the inconveniency to a certain degree, but still leaves the other defects which have been already mentioned without any remedy. The experiment has already been tried, and the axle arm has been found much worse at the inner shoulder, and the wheels proved much weaker, failing much sooner by getting a "dish" inward; in fact the "dishing" of wheels which was not a natural thought has either been found by accident or has been invented by some person of great ingenuity, and the course of uniform experience has fully confirmed its advantages.

III. OF THE LENGTH OF HUBS.

Too short an axle arm has too little power to guide the wheel in a steady direction unless very tightly fitted into the bushes or boxes, and too long a hub is inconvenient both on account of its bulk and of the dirt which it is apt to carry; 12 inches or thereabouts seems a very convenient length for a hub exclusive of the cupping at the ends.

IV. OF THE TAPER OF BOXES.

It is found necessary to line the wearing surface of the inside of the hubs with iron. This is generally done by driving two iron rings called bushes or boxes into the opposite ends of the holes made for the axles. The axle arms are made to fit exactly the inside of these bushes or boxes. Owing to jolting the greater strain comes upon the inner portion of the axle, and wooden axles generally require a size 5¼ or 5½ inches at this point for ordinary wagons or carts.

When speaking of the dish of wheels in the second section, we supposed the axle arm to be without any taper; in order therefore that the supporting spokes may stand perpendicular to the ground, or nearly so, it is necessary to bend the axle arms downward, but this would be attended with very great inconvenience; when the wagon is going upon a level road, the wheel would be continually sliding in upon the axle and would press hard upon the shoulder and friction be prodigiously increased, this would be much more sensibly felt in the draught when the load is tilted by the wheel sinking in a rut or natural inequality of the road. This trouble is remedied by a very simple contrivance, the tapering of the axle arm, the intention therefore of this contrivance being to hinder the dished wheels from pressing in upon the inner shoulder of the axle. The manner of executing it must be taken from these *principles*.

In the *first* place then, the under surface of the axle arm which rests in the boxes must be level; if it be higher at the inner ends the wheel will slide in and press hard upon the shoulders; if higher at the outer end, the wheel will press upon the lynch-pin or nut. In the *second* place, the spoke which touches the ground or which is supporting the load, must either be perpendicular or must lean a little *inward* above and *outward* below. These two conditions will determine the quantity of taper which must be given to the axle and boxes.

In the first place let us suppose that the supporting spoke must stand perpendicular. In plate, Fig. 1, let L, K A, B, represent the taper hole through the hub; the lower side A, B must be level. Let E be the point of the tread which touches the ground, draw E, D perpendicular to A, B, let Q, O be the center line of the hole. This will be the line around which the wheel really turns, E, D will cut this line in C., let G be the

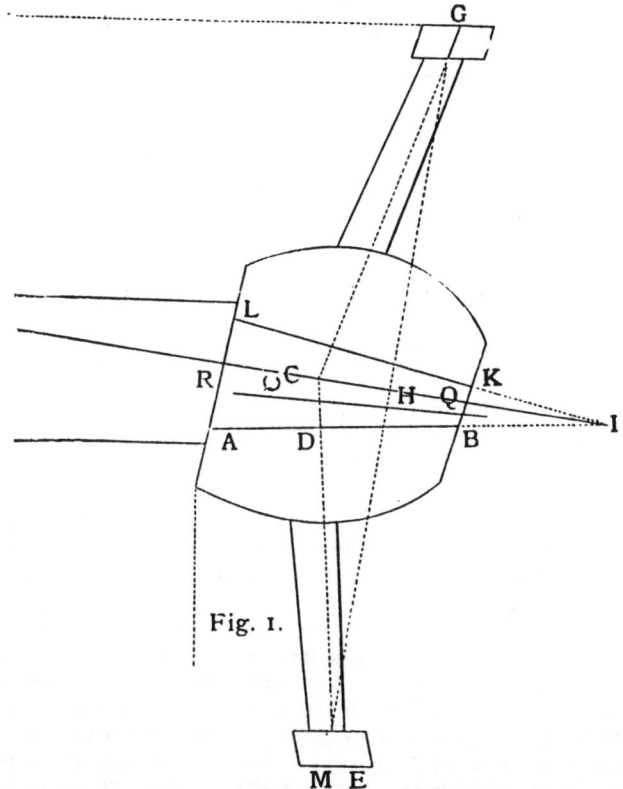

Fig. 1.

uppermost point of the wheel exactly opposite to E, draw G, C. For the same reason that E, D is perpendicular to A B, G E will be perpendicular to L, K, the upper side of the hole in the hub, draw G, E, cutting the line of the center in H, then C, H is the true dish of the wheel; let the lines L, K meet in I, then because the two triangles E, C, H and I, C, D have a common angle at C, and equal angles at H and D, the angles at E and I are equal, therefore the half taper of the axle and boxes is just the same per foot as the real dish of the wheel, and we may use the following proportion. As the half diameter of the wheel is to its *real dish* so is the length of the hub to half the taper of the boxes or to half the difference of the diameter of the inner and outer ends of the boxes or bushes.

On the other hand, the "real dish" of the wheels may be gathered from the taper of the axles and boxes, for if we draw B, R parallel to the center line Q, O, then Q, R will be equal to O, B, the half width of the outer end of the box, and R, A will be half the difference of the diameter of the inner and outer ends of the box, or will be half the taper of the boxes corresponding to the length of the hub which must be counted on the center line Q, O. Now, B, R is equal to O, Q, also B, R is to R, A, as E, H is to H, C, that is the length of the hub is to the half taper of the boxes as the half diameter of the wheel is to the true dish. Thus from the dish of a wheel already made we can choose the proper boxes which will answer for it and taper the axle accordingly, and from the boxes already cast we can make wheels of proper dish for carts. A pair of boxes of which the inner is 5¼ inches diameter and the outer 3¼ are found to answer very well and require a real dish of 2¼ inches for a wheel 4 feet 8 inches diameter.

In the next place let it be supposed that the standing spokes should spread out below, which is thought most proper. This may be done by increasing the dish of the wheel or diminishing the taper of the axles.

It is most convenient to have both ends of the axle arm made to one taper. This, however, is not necessary, providing that the taper of each part of the axle be made to fit the bush or box that is to run on it.

continued next page

Wheelmaking

V. OF FITTING THE BUSHES OR BOXES.

This must be very exactly done, otherwise the wheels can never run well. The first circumstance to be attended to is the setting them fair, that is, so the center line of the two boxes may make one straight line which should be the center line of the whole eye of the wheel or hole of the hub.

In Fig. 2, let A, B be the center line of the eye of the wheel. It is plain that if the inner box C, D, E, F be set more down at C than at E, it will be nearer the center line at D than at E, and will grate upon the axle at D which will immensely increase the draught, and if the outer box be more set down at K than at H it will be nearer the center line

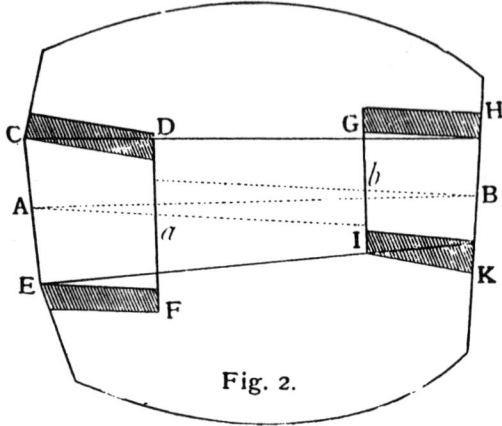

Fig. 2.

at I than at G, which will have the same effect. Their center lines will not in this case fall in with A, B, but will stand at A, *a* and B, *b*. Great care must therefore be taken to avoid this and make them stand "fair." If both the boxes have one taper their inside should run straight with the lines C, H and E, K.

The *second* thing to be taken care of is that the center line of the bushes or boxes may be perpendicular to the face of the wheel, without which the wheel cannot turn true on the axle arm. This may be discovered by turning the wheel round an axle arm which fits it, and holding a steady mark to some place on the face of the ring or felloes; this will touch the face all round as it turns, if the center line is perpendicular to the face. If it does not, there is an error which if not helped will cause a false motion of the wheel and make crooked ruts in the road, equal to the distance of the different parts of the face from the steady mark above mentioned.

VI. OF THE TREAD OF A WHEEL.

This should be made to press fair on the road, especially if the tread be a broad one. When the axle is straight or level on its under side, the tread should make the same angle with the face which the taper of the boxes make with their center line excepting what is to be allowed for the declivity of the road. Therefore, use this proportion: as the length of the hub is to half the taper of the boxes so is the breadth of the tread to the quantity which it should be above the square. Thus, if the length of the hub be 12 in., and the taper of the boxes be 2 in., and the breadth of the tread 3 in., then as 12 is to 2 :: 3 : $\frac{1}{4}$, and the inner edge of the tread should be $\frac{1}{4}$ in. above the square with the face of the wheel.

VII. OF THE FITTING OF THE AXLES.

In order to do this properly we must know: 1st, what is called by the workman the "dish" of the wheels, which is very different from what I have called the real dish, and the methods for finding it; 2d, the way of finding the length of axle beds which will answer to any given wideness on the road, the axle being straight below, in order to find what length will answer for any given hollow, and what hollow will answer to any given length of axle bed. When everything needful has been calculated the wood for the axle must be prepared and marked out or set out (as the workmen call it) with great exactness; that is, the lines must be truly drawn, which directs us for working the axle so as to make the arms fit the boxes, and the wheels run right on the road, and the work must be truly executed according to these lines. When nearly done it will be proper for the workman to try whether the wheels gather inwards equally on both sides both *below* and *before*. This he may do by applying a line or ruler diagonalwise across from the under part of the right wheel to the upper part of the left wheel, and then from the under part of the left wheel to the upper part of the right one. If the wheels be properly set these two trials should give the same distance on the line or ruler. Thus, sup-

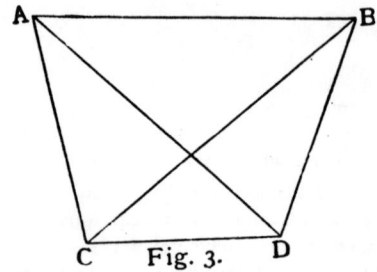

Fig. 3.

posing A, C, Fig. 3, and B, D, two wheels gathering below between C and D. If B, C is found shorter than A, D, then either B is too far in or A is too far out; then the distance A, B and C, D should be measured to know if the gather is as much as required. It must be borne in mind that this gather will always be equal to twice the "*real dish*" of the wheels. If it is too little, A and B must be brought a little in, if too much, the contrary, but in all cases the wheels must have equal inclinations.

VIII. OF FINDING THE "DISH" OF WHEELS.

What the workmen find most convenient to consider as the "dish" of the wheel is considerably different from what I have termed the "real dish." Fig. 4 represents a wheel as seen edgeways and standing on level ground with its tread O, A. The line H, I, which is the lower side of the axle arm, is supposed to be level or parallel with O, A. The line

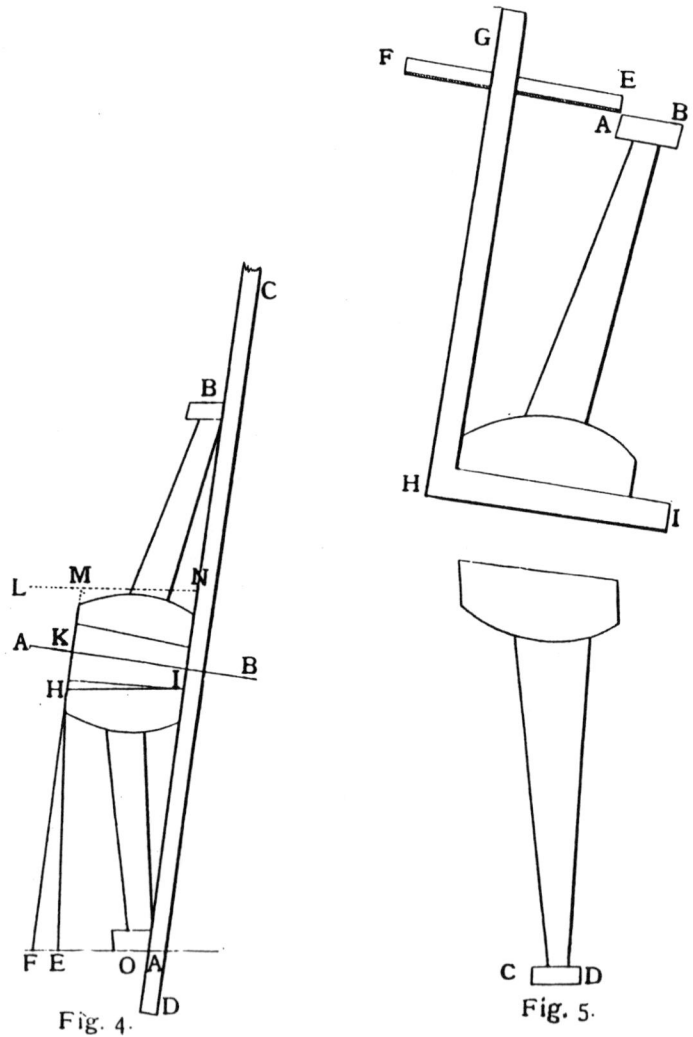

Fig. 4.   Fig. 5.

M, H, F is parallel to the face of the wheel C, D; I, K is parallel to the center line A, B; H, E is perpendicular to the ground. Now, E, O, the distance of this perpendicular from the inner side of the tread O, is called the "dish" of the wheel by workmen.[6] It is plain that the distance of the point E from the like point corresponding to the other wheel is the distance between the shoulders of the axle or the length of the axle bed. This being attended to, the two following methods may be taken for finding the dish of a wheel. *First*, by a square and rule.

Let A, B, C, D, Fig. 5, represent the ring of a wheel, G, H, I a square, the stock H, I applied through the eye of the wheel to the boxes.

continued next page

The blade is applied to the inner end of the hub and reaches as far up as the ring of the wheel, then a rule, E, F, being held square over it will show the distance between the tread of the wheel and the inside of the square, which is the dish sought, allowance being made for the distance between the square and the point of bearing of the inside box. *Second*, by calculation.

Let a straight ruler, C, D, Fig. 4, be applied to the face of the wheel as near the hub as possible. Let a straight rod, L, M, N, be put through between the spokes square on C, D; this will show the distance M, N between the face of the wheel and the inner end of the hub; then say, as this distance is to half the taper of the boxes so is the height of the lower side of the axle arm (supposed to be level) to a fourth quantity. This must be added to the breadth of the tread, and the sum must be taken from M, N. The remainder is the dish required, for it is plain that F, A is very nearly equal to I, K, which is equal to M, N, and that I, K : K, H :: H, F : E, F, and also that E, O is the difference between F, A and the sum of F, E and O, A. Thus, if the distance of the lower side of the axle arm from the ground be two feet or 24 inches, and the taper of boxes two inches, and the distance of the inner end of the hub from the face of the wheel be 12 inches, and the breadth of the tread be 2½ inches, we have 12 : 1 :: 24 : 2, which is F, E, then F, E and O, A make 4½ inches. This taken from 12 inches leaves 7½ for the dish of the wheel or one foot 3 inches for two wheels. This calculation may be done by the sliding rule which is used by cartwrights; set this rule 12 to 24 and then I will stand opposite to "2" for F, E, and proceed as before.[6] [This measure gives the actual allowance for width of body over the track.]

IX. TO FIND THE LENGTH OF THE AXLE BED.

This is the distance between the two inner ends of the two hubs. 1st. where the axle arms are straight below, from the proposed wideness between the wheels within on the road, take the dish of both wheels, the remainder is the length of the axle bed.

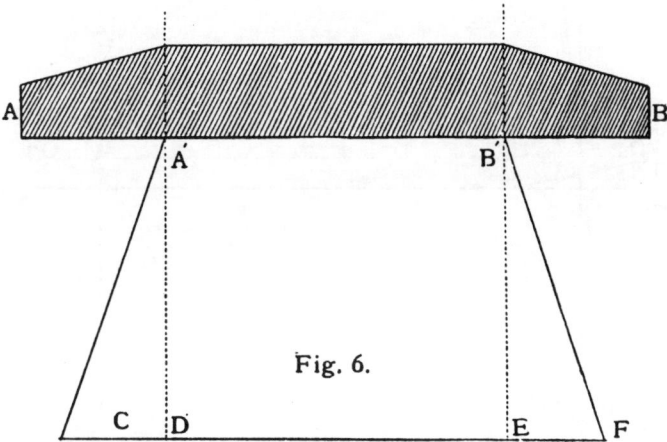

Fig. 6.

EXAMPLE.—Fig. 6, wideness on the road within C, E =

	4' 6"
Dish of both wheels C, D+E, F,	1' 3"
Length of axle bed,	3' 3"

The truth of this rule is plain, by inspecting Fig. 6, where A, B is the axle, C, F the wideness within on the road C, D, and E, F the dish of the wheels, also, A', B' the length of the axle bed.

2d. When the axle is "hollow" below. This is often the case, because the bed of an axle is sometimes confined to a certain length, especially such as have iron arms, whether they are fixed to wagons or carts. In such cases they must be made hollow or rounding, or the axle arm must be laid into the bed with a bend downward, in order to bring the wheels to a right gather or make them run a straight track on the road. This is also necessary in order to make the tread of the wheel press fair on the road. These circumstances give rise to two questions or problems.

1st. Having given a certain length of axle bed and a certain wideness of track, to find the degree of hollow, that is to say, how far the under side of the axle arm should be below the level at the out end of the hubs.

RULE.—Find the length of the axle bed, which will give the wheels the required wideness or track on the road. When the arms are straight below, take *half* the difference between the given length of bed and the length then found; then say, as the height of the under side of the straight axle from the ground is to this *half* difference so is the length of arm to the hollow required.

EXAMPLE.—Suppose the given length of bed is 3 ft. 6 in., and the wheels to have the same dimensions of the last example, namely : 1 ft. 3 in. of dish, etc., and the height of the under side of the axle from the ground 2 ft. or 24 in., and the length of the arm 12 in., then the length of the bed will be 3 ft. 3 in. by the last rule; the difference is 3 in., half of which is 1½ in.; then 24 is to 1½ as 12 is to ¾ of an inch, which is the hollow required. The reason of this operation is very easily understood.

Let A, B (Fig. 7) be the given length of the bed, F, K, the given wideness on the road, and let C, D be the length of the bed, which would give the wheels that wideness on the road when the axles are straight below. Let C, E, be the length of the arms. Draw C, L perpendicular to the horizon, then L, F is the dish of the wheels. Now, suppose the axle bent down to the position C, M, the line C, F will take the position C, G in a manner that the angle E, C, F will be equal to the angle M, C, G, and therefore the angle E, C, M will be equal to the angle E, C, G. Draw E, M perpendicular to E, C, and F, G perpendicular to C, F. Now, as F, G and F, H may be esteemed as nearly equal, we may say as C, E : E, M :: C, F : F, G or F, H. Now, F, H is half the quantity by which the distance between the wheels is diminished by bending down the axle from C, E to C, M. If, therefore, A, C be made equal to F, H, and A, O be made parallel to C, M, the wheel placed on the bent arm P, B, A, O will have the same wideness on the road as when on the straight axle Q, E. Therefore, A, C is really the half difference between the two axle beds when E, M is the hollow or bend of the axle, therefore, the rule given is just. It is not mathematically just, because C, G is not equal to C, F, but the difference of A, C or E, M from the truth does not amount to the 500th part of one inch.

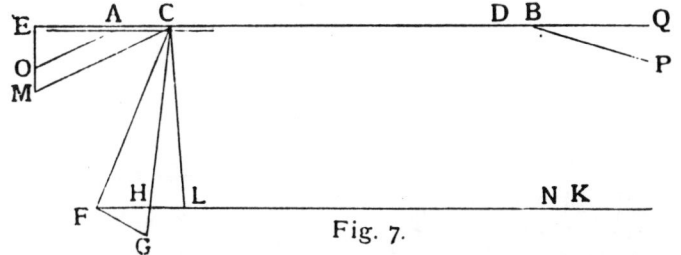

Fig. 7.

QUESTION 2.—Given a certain hollow, E, M, and a certain track, F, K, to find A, B, the length of the axle bed.

RULE.—Find as before the length of the bed C, D for an axle straight below; then say, as the length of the arm C, E is to the given hollow E, M, so is the height of the lower side of the straight axle from the ground C, L to the fourth number, which will be A, C. Twice this added to the straight axle bed will give the hollow axle bed.

EXAMPLE.—Suppose ¾ of hollow, and 4 ft. 6 in. track are given for the wheels which we have so often mentioned, then 12 : ¾ :: 24 : 1½ = A, C. Now, the double of this is 3 in., and C, D is 3 ft. 3 in., therefore, A, B is 3 ft. 6 in.

The reason for this is evident from what has been said already.

In the foregoing calculation, the height of the under side of a straight axle from the ground was always one of the number to be worked with. In order to find this, we must refer again to Fig. 4. By examining this, you will find that H, F is equal to I, A, which is half the breadth of the face of the wheel, deducting half the width of the box at I, even with the face of the wheel. Measure I, E along the box I, K was found in the way formerly directed. Then we have I, H : I, K :: H, F : H, E, the height from the ground. I may here observe that H, E will hardly ever differ 1/10th of an inch from H, F or I, A, so that it is hardly necessary to make a calculation for it.

I conclude this part of the subject by observing that if the taper of one axle be less than that of another, the axle whose taper is smallest will require to be farther bent down in order to give the same pair of wheels the same gather below, or to put the wheels in the same position for running. The bend to be given in this case is just equal to one-half of the difference between the two tapers of the axles.

continued next page

In Fig. 8, let A, B, D, C be straight below, and have two inches of taper. Let L, I be the center line and K, I a horizontal line. Let A, P be the face of the wheel, and L, Q a perpendicular from the center of the outer bush. It is plain that K, L is equal to one-half the taper or difference between the diameter of the boxes. It is also plain that L, I

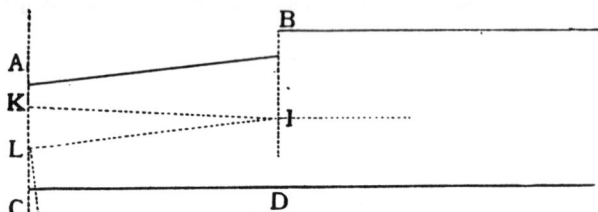

Fig. 8.

is to L, K as L, Q is to Q P, which is one-half the gather of that wheel. Again, let E, F, H, G, Fig. 9, be another axle straight below, with only one inch of taper. M, O will in like manner be half the taper, and half of K, L, Fig. 8; S, R, Fig. 9, will therefore be half of P, Q, Fig. 8. If, now, we make S, T, Fig. 9, equal to P, Q, Fig. 8, we must bend the

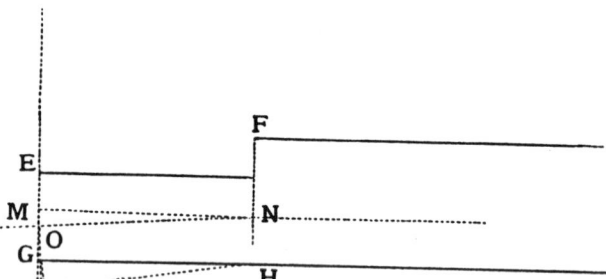

Fig. 9.

arm to the position H, V, Fig. 9, with its lower side so that G, V may be equal to M, O. This will bring down the center O as much further as it is already below the level M, N, and then the face O, T will be the same as that of the face L, P, in Fig. 8, and consequently both wheels will have the same gather below.

X. OF THE GATHERING OF WHEELS FORWARD.

It is found by experience that wheels run better when they are gathered a little forward, or incline towards the foreside of the axles.

The taper of the axle arm is the cause of this. The wheel being pressed forward by the axle arm, it must be forced outward by the taper, unless it be balanced by an inclination inward. The more, therefore, the arm is tapered, and the less the wheels are gathered below, the more gather forward will be required. About 1½ inches gather forward is found by experience to answer for wheels 4 ft. 8 in. high, with boxes of 2 in. taper and an axle straight below.

The rule for bringing wheels to their gather is this: The axle being made straight on the under side, a line must be drawn along the middle of the under side, then, the length of the arm and of the bed being all marked out, set off $\frac{1}{18}$ of an inch or a little less towards the foreside of the axle at the out ends or heads, and take the points thus set off for the centers from which the out ends are to be squared.

But that every person may be satisfied as to the method of finding these distances, the following proportions will answer to every case. As the half height of the wheel is to the fourth part of the intended gather, so make the arm to the distance sought. This proportion may be thus demonstrated from Fig. 10.

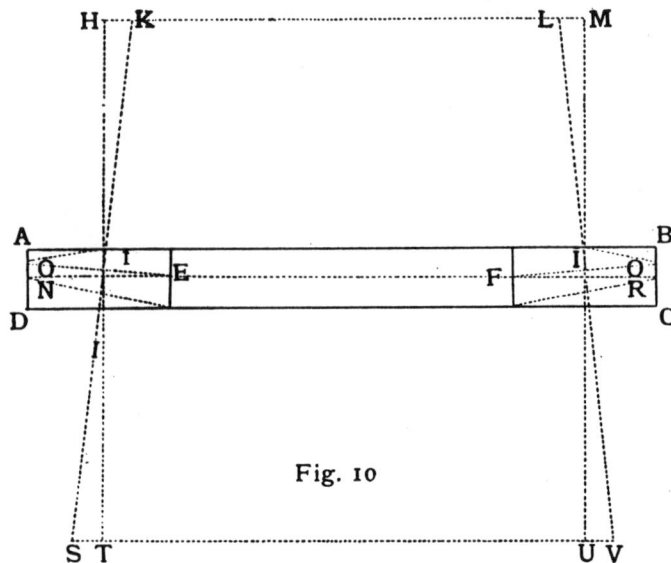

Fig. 10

Let A, B, C, D represent the under side of an axle; N, E, F, R a line drawn along the middle; N, O the distance set forward at one out head, and R, O the distance set forward at the other. Then, since the center line of the axles must correspond with the center line of the boxes, and the face of the wheel cuts these lines at right angles, if the centers of the out heads were continued in the straight lines N and R, the face of the wheels would cut that line at right angles at the lines H, T and M, U, but when the centers of the out heads are placed in O, O, the faces of the wheels will cut the lines E, O and F, O at right angles, and therefore the angles O, E, N and H, I, K are equal, so also the angles S, I, T and U, I, V are equal, and O, F, R and L, I, M are equal. Therefore, as I, H, the half height of the wheel, is to H, K, a fourth of the gather (that is a fourth of the difference between K, L and S, V), so is E, N the length of the arm to N, O, the distance sought, and so of the other end because I, L : L, M : : F, R : R, O.

continued next page

XI. OF THE MANNER OF PLACING THE CARRIAGE ON WHEELS.

Let A, B, C, D (Fig. 11) be the body of a cart, and suppose it placed on a pair of wheels which require no packing, the half height of the wheel being N, P. Suppose it again placed on a pair of wheels whose half height is O, P, so as to require a packing to raise it to the same height with the other. The height of this packing will then be equal to N, O on a level road. This will make no difference of weight on the horse's back, but if the cart should be going down a hill, making an angle with the horizon of about 15°, it will make a considerable difference, as may be seen by the diagram; for suppose I, K to represent such a declining road, then if a perpendicular to the horizon be raised to the center of the high wheel it will cut the body at E, G, whereas a perpendicular passing through the center of the low wheel will cut it at F, H. From this it is evident that the higher any carriage is raised above the center of the wheels, the greater weight in such cases will be thrown on the back of the horse, and in going down the horse will have too much weight, and in going up too little. The higher any carriage is thus raised and the more any road declines the greater will this inconvenience be shown. There have been many attempts to alleviate this, the most effective being to lower the shafts below the shoulders of the horse in going up hill, and in raising them a corresponding distance going down hill by adjustment of the harness.

LAST. OF FIXING THE AXLE TO THE CARRIAGE OR CART.

After an axle has been properly fitted into the wheels, special care must be taken that it be not turned out of its right position by the manner in which it is fixed to the carriage. Axles are sometimes fixed directly to the body of the carriage, and sometimes to the shafts or timbers, the body lying loose above these. In whatever manner the axle is fixed, it must be kept while working in the position for which it was designed or drawn.

Therefore, when the axle is fixed to the body of the carriage and that body is so placed as to be level when going upon level ground, the axle must be fixed with the plane of its upper side parallel to the body of the carriage.

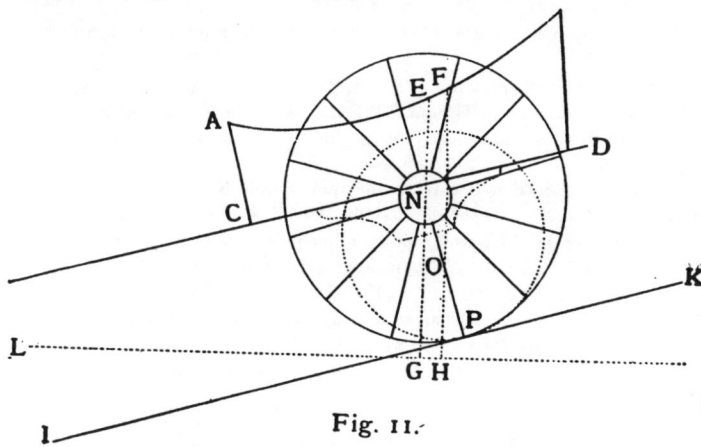

Fig. 11.

When the axle is fitted to the shafts it must be considered how much higher the shafts will be at the back band than at the axle. When the carriage is going upon level ground, this may be known by the height of the wheels and the height of the horse. (See Fig. 12.) Let A, C, E represent one of the shafts, A, F, the height at the back band, and B, C, D, a level line passing through the upper side of the axle. It is plain that to keep the axle in the right position, its upper face must be kept in the line B, C, D. If it be merely applied to the outer side of the shaft A, C, E, the wheels will be made to have too much gather forward.

The right slope to be given to the axle or shaft where they are fitted together may be found by the following proportion. As the length of the shaft from the foreside of the axle to the back band is to the difference between the height at the axle and at the back band, so is the breadth of the axle on its upper side to the depth to which it must be let into the shaft at the hind side more than the fore side, for A, C is to A, B as C, D is to D, E.

From this it is also plain that the gather forward may easily be either increased or decreased by the manner in which the axle is fixed on, and that a proportionable difference will at the same time be made on the swing or gather below or width of track.

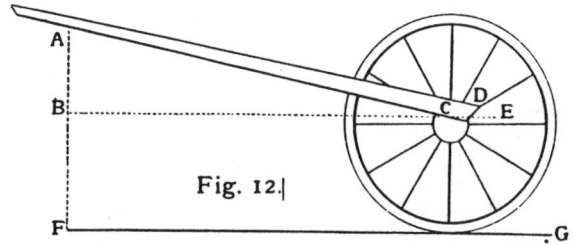

Fig. 12.

Those having charge of two-wheelers should be careful to keep the shafts always about the same height by the manner of yoking the horses, because some alterations will be made by every change. The buyer of a cart should get from the builder a note of the difference between the height of the wheels and the height of the back band to which the axle has been adjusted, or better still, the cart should be built to suit the height of the horse to be worked. Any alteration in the height of the horses must be corrected so that the axle will work at the exact angle to which it was designed.[7]

THE END.

The Hub, Jan-Apr 1889

Wheelmaking

WHEEL MAKING.

As now Performed in Wheel Factories in this Country.

(See Illustrations accompanying.)

In no department of carriage building has machinery wrought so complete a revolution as in that of wheel making. There are carriage workmen now actively engaged in the business who can recall the days when they begun the wheel at the chopping block, hewing the spokes with a broad axe, then squaring up with a draw knife and plane, then tenoning and rounding up, all by hand. The hubs were turned but the boring and morticing was all hand-work. So, too, all work upon the felloes, except the heavy sawing, and in many shops even that, was performed by hand labor. Now all is changed, machinery everywhere, from the rough riven spoke blank and the log as it is cut from the tree and sawed up into rim or felloe blocks or strips, the hub block, to the finished morticed hub, and from the whole, grouped together as a wheel ready for the carriage builder's use.

It is interesting to trace the course of each division, from its rough state until it is made up into the perfect wheel. Beginning with the hub, which is either of elm, oak, birch or locust, cut from logs of a size to permit the heart to form the center, otherwise it will lose its circular form and become more or less elliptical while shrinking.

The log is first cut up into blocks of the required length and a hole is bored

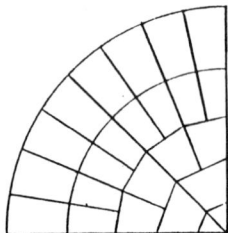

Fig. 1.

RIVED SECTION.

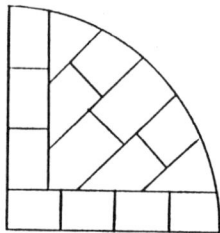

Fig. 3.

SQUARE SAWED SECTION.

through the heart of each block. It is then put into a lathe and turned to the required size and of uniform diameter. It is then seasoned and turned to the pattern required for the finished hub; before being turned, however, the hole is bored out to insure a perfect center. The hub is now ready for the wheel maker. If it is for any one of the patent wheels, the iron bands, which have previously cleaned and trued by being forced over a mandril, are forced on by hydraulic pressure, after which the hubs are ready to be bored and morticed. If the plain wood hub is to be used, the back end band is forced on before it goes to the boring and morticing machine.

There is a difference of opinion as to which is the better course, to fully season before boring and morticing or to bore and mortice and then set the hub away for further seasoning. It, however, appears to be settled that the sooner the spokes are driven into the mortices after they are finished the better, and that if any shrinkage takes place thereafter, it tightens the grip on the tenon.

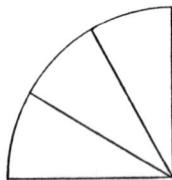

Fig. 2.

SAWED SECTION.

The spoke undergoes a number of manipulations before being ready for the wheeler. In its rough state it is put upon the market in three classes.

The rived, the triangle split from the log, as shown by Fig. 1, which represents a quarter section laid off in the general direction of the lines assumed by the edge of the split pieces. The rule governing is to split from the heart to the bark retaining the heart as a common center. The second is the sawed, as shown by Fig. 2, these are sawed by the log being centered and the saw carfs running toward the heart as a common center, varying the direction only where the log is sufficiently large to allow for more than one spoke in depth. This plan was introduced by those wheel makers who were compelled to cut stock from small trees. The third form is the sawed blocks with square sides, these may be cut by first being sawed up as for plank, and then sawed to strips, but since the introduction of quarter sawing machinery, it is just as easy to saw up, as shown by Fig. 3, thus retaining the growth rings in much the same position as when rived.

The present custom by wheel manufacturers who get out their own stock is to thoroughly season the spoke, by air and artificial means, before turning. When seasoned it is taken to the lathe in the rough form, square, or as shown by Fig. 4, and at the rate of two thousand to two thousand six hundred they are

rough turned, as shown by Fig. 5. The round portion to within a few inches of the end is turned to the required shape and size, ready to be smoothed, the tenon being squared. The next step is heading and tenoning by which the square end is shaped, as shown by Fig. 6, this perfects the tenon as to thickness and the width of the spoke above the tenon. The spoke then goes to another machine where it is throated and given the form shown by Fig. 7. It is then faced and tapered by which process the front edge is straightened and the back edge of the tenon given the required bevel. Sanding the face and back follows, after which the spoke is trued up and smoothed by being worked over five sand and emery belts of varied de-

Fig. 4.

THE SPOKE PLANK.

grees of coarseness. Fitting up the tenons is the last operation before going into the hands of the driver, making in all eleven operations through which the piece of wood passes on its way from the riven block to the finished spoke,

The third division is the rim or felloe, technically these names represent the wood band which form the outer or bearing portion of the wheel; "rim," indicating the bent pieces, and "felloe," the sawed pieces.

In preparing the rims, the log is first steamed to remove the bark, it is then sawed up into planks of required thickness, then cut up into strips, and by some piled under sheds to air season for a time, by others the strips are immediately steamed and bent and then seasoned. Bending with the

Fig. 5.

FIRST FORM AFTER BEING TURNED.

perfected machinery of to-day is a simple operation and one in which the loss is reduced to a minimum. When bent, the felloes are seasoned, then they are dressed up, when, like the hubs and spokes, they are ready for putting up.

The first operation is driving the spokes, this is performed by machine or hand labor. The perfected spoke driver is a machine the hammer of which is operated by power, and is so hung that the blow is nearly as elastic as that given by hand. With such a machine one man will drive about as many spokes as two men can in the old way. In driving by hand, heavy spokes are driven by one man who swings the hammer aided by another who places the spoke in the mortice and guides it so as to assume the correct position when driven. Light spokes are driven by one man who drives as well

Fig. 6.

WHEN TURNED AND HEADED.

as guides the spoke. The popular hammer for driving is an iron head similar to that shown by Fig. 8, about four inches long, the face being one to one and one-eighth inch in diameter. The handle is of fine hickory from eighteen to twenty-four inches in length, and for about half its length from the head it is five-eighths to three-quarters of an inch wide and scant one-eighth inch thick at the thinnest point, the bilge being of a size to suit the grip of the driver. With a hammer of this kind, an expert never fails to send the lightest spoke "home." Almost without exception spokes are driven with hot glue.

From the driver the wheel goes to a machine where it is placed upon a mandril and the spokes are cut off; technically this operation is "cutting to a

continued next page

Fig. 7.

FINISHED SPOKE.

height." From this machine the wheel passes to another which forms the tenon on the felloe end of the spoke. The next step is "marking off," that is, marking the felloes or rims for the spokes, for no matter how careful the workman, nor how true the machinery, the spokes will not radiate from the center on the same line, and the spaces between the ends of the spokes will vary, not much perhaps, but enough in case the holes were all bored at equal distances from each other in the rims or felloes, some of the spokes would be sprung by driving on the felloes. After marking, the holes are bored through the felloes by a machine, the table itself and the guards which hold the pieces to be bored, being adjustable to any desired position, and the auger boring from a center on a line representing that of the spoke in the hub. Next comes the cutting off of the felloes or rims and doweling, followed by rounding and polishing. All then goes to a man who drives on the rims. This, though an apparently simple operation, requires care to insure correct length of felloe and a level tread, the latter being perfected by the wheel being placed upon a mandril which passes through the hub and by which the tread is held against a revolving head wheel with emery face which trues and smooths the tread.

As a last operation, unless screws are run into the felloes to prevent their splitting, the finisher cleans up every part and turns the wheel over to an-

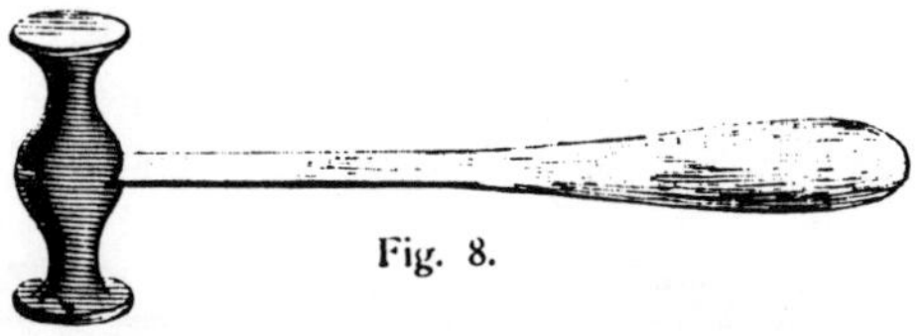

Fig. 8.

A SPOKE HAMMER FOR DRIVING.

other operator for inspection. And no manufacturer who has any regard for his reputation will allow a wheel to leave the factory, no matter what its quality, without a thorough examination.

The perfection to which machinery has been brought and the skill of the operators has made it possible to construct wheels more thorough in all details than was ever possible by hand-work, and where failure occurs it is the result of carelessness.

In no country are wheels of any kind better made, while in none other do the light wheels bear any comparison either in mechanical perfection, symmetry or strength to those made in this. And so rapid is the operation that a plant employing two hundred workmen can turn out fifty thousand sets, or two hundred thousand wheels a year, a wheel every three hours per man for every working hour in the year. Fifty years ago he was a smart workman who could make a set of light or medium carriage wheels complete in one week for which labor he received fifteen to eighteen dollars. Originally there was a strong prejudice against machine-made wheels, there being a vague idea that from some cause or other the work performed by the machine was inferior to that performed by hand. If this ever was true it is not to-day, as better wheels cannot be made than are now made in properly conducted wheel factories.

The Hub, Aug 1891

PROPORTIONS OF SPOKES.

(See Illustrations Accompanying.)

THE majority of builders do not stop to consider how much the proportions of the spokes have to do with the appearance and durability of the wheel. "A spoke is a spoke," no matter what the proportions, regardless as to whether it is to be used for a light or heavy wheel. To assist those who care to use spokes that are correct in form, we give illustrations of four sizes. Fig. 1 shows a spoke with a 1⅜ in. face, ⅞ in. tenon, and fully 1¾ in. deep. The triangular face is 3 in. long; thickness at the thinnest part of the throat, ⅞ in. The diagram A shows the form full size, at a point 10½ in. above the shoulders; at the

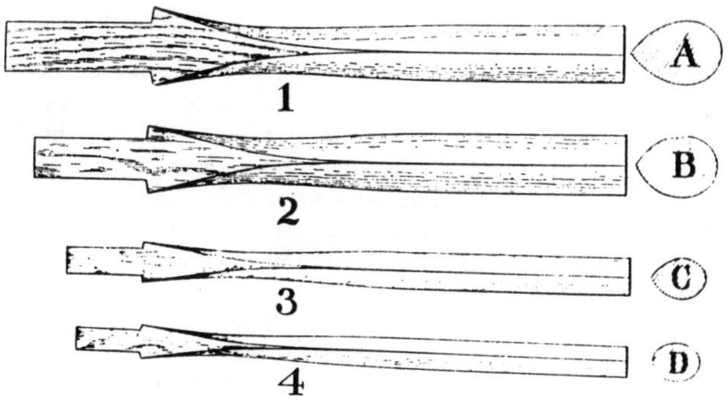

PROPORTIONS OF SPOKES.

pillar it is scant 1¼ in. thick and 1½ in. deep. Fig. 2 shows a spoke 1¼ in. face, ¾ in. tenon, 1¾ in. deep. The triangle, 2¾ in. long. Diagram B shows the form full size, 10½ in. from the shoulder; thickness of the throat, ¾ in. At the felloe, the spoke is 1 x 1¼ in. Fig. 3 shows a lighter spoke, with a full ¾ in. face, scant ½ tenon, 1⅛ in. deep. Triangular face, 2¼ in. long. Thickness at the throat, ½ in. Diagram C shows the form full size, 10½ in. from the tenon; size at the felloe, full ⅝ x ⅞ in. Fig. 4 shows a buggy spoke, ⅝ in. face, ⅜ in. tenon, 1 in. deep. Length of triangle, 2¼ in. Diagram D shows the form full size, 10½ in. from the shoulders; size at the felloe, ⅝ x ¾ in. The spokes shown by Figs. 1, 2 and 3 are finished to a sharp edge on the face, all the way from the triangle to the felloe. Fig. 4 has a round finish.

The Hub, Dec 1894

Early Recollections of Wheel Making.

BY HENRY L. DUBOIS.

About the year 1850 the writer was engaged by Edw. K. Reynolds in Salem, N. J., in the manufacture of carriage rims and shafts. At this time there were only two other manufacturers in this line in the United States, and all in a very small way. One was at St. George's, Del., and the other at Brandywine Springs, Del. Both of these factories were working under the Reynolds patent, which was the first patent ever taken out for bending wheel rims.

At this time there was a great prejudice among carriage manufacturers against using bent rims. On account of being something new the use of them was adopted very slowly. The parties first adopting them in Philadelphia were Geo. W. Watson, Thirteenth and Parrish streets; Wm. D. Rogers, Sixth and Brown streets; William Ogle, Eleventh and Chestnut streets; John Hollohan, Eighth and Filbert streets.

The machinery used at that time, the Reynolds patent, had no upset to the bands, and in bending around a circle the timber was liable to stretch, and on that account about one-third of the stock was lost by breakage. In 1852 the factory at Salem, N. J., was destroyed by fire, and the plant was moved to Laurel street, below front, in Philadelphia, Pa. This plant was abandoned on account of not being able to get timber, as it was thought there was no hickory or oak suitable for this business outside of Delaware and New Jersey. This shows how ignorant the trade was in regard to the growth of this timber in Pennsylvania and other States.

In 1852 Mr. Blanchard, of Massachusetts, invented a bending machine which, by using end pressure on the timber, overcame the breakage. The first machine under this patent was used by Maynard & Hutchinson, Trenton, N. J., and the second by John G. Davis & Son, Philadelphia, although I had been using a similar machine of my own invention at Yardleyville, Pa., not knowing that the Blanchard machine was in existence, as I had never seen or heard of one. After seeing the Blanchard machine I used it and abandoned my own; and although the trade was very slow in adopting this process, the same principle is now in universal use.

When the timber was supposed to have been exhausted in Delaware and New Jersey, the writer made several trips through Pennsylvania, in search of suitable material. One in particular, I recollect, was to Downingtown, Pa., and there I became acquainted with a gentleman by the name of Henry Hoopes. I explained what I was after. He said there was plenty of hickory there and took me along the Brandywine River, where we cut down some large shell-barks which I condemned, and returned to Philadelphia and reported there was no good hickory in Pennsylvania, although Mr. Hoopes insisted that there was plenty of it. I had given him the full explanation of the kind required and for what purpose it was to be used, and I have always thought the result of the information obtained from me was the starting afterwards of the firm of Hoopes Bro. & Darlington, of West Chester, Pa.

To show our ignorance of the supply of carriage material in Pennsylvania at that time, we know now that since then there has been millions of feet cut of the finest hickory in the world in Bucks, Chester and Lancaster counties in Pennsylvania, and the writer has made a number of very successful trips since his first venture, some of which were very thrilling.

At this time all carriage makers made their own wheels, and there was not a single wheel factory in the United States. The first spokes I ever saw turned on a spoke lathe was about 1852, by Eldridge & Fitler, at Hanover street wharf, Philadelphia, Pa. They afterwards moved to Front and Canal streets, and were succeeded by John G. Davis & Son, Mr. Eldridge retiring and Mr. Fitler going into the hub business.

At this time all carriage hubs were made of gum, elm being of comparatively late date, on account of the material checking and causing great loss of labor and stock. Mr. Fitler tried a number of ways to season blocks artificially, none of which were successful, but at one time he almost succeeded in burning his place down. From my observations in this line, seeing from 25 to 30 per cent. of the hub blocks lost by checking, etc., I gave it my study for a long time to find some way by which these blocks could be saved and the timber improved. I had tried a number of ways but did not succeed in my efforts until 1891, when I patented a process which I sold to the Standard Hub Co., limited, of Philadelphia, Pa., now the Standard Hub Block Co., Marion, Ind. Artificially seasoned hub blocks are now, like bent rims, used by leading manufacturers in America and Europe. While the trade was again slow in adopting this improvement, they now realize its advantages.

Knowing there was still room for improvements in the manufacture of wheels, and to reduce the cost and time, I turned my attention to seasoning white oak spoke billets, which require from one to two years, according to the size, to season by ordinary drying. I obtained a patent in 1894 for seasoning spoke billets, by which process a billet can be perfectly seasoned within two to four weeks. Second growth white oak now can be seasoned in five days, keeping it under a temperature of 130 degrees without showing a sign of a check, and the quality will be raised about one grade.

The first person to use a bent rim to my knowledge in South Jersey was Lewis McBride, of Bridgeton, and if I mistake not this is near where the Ware family came from; but there was no CARRIAGE MONTHLY issued at that time, and very few other periodicals, and none in the interest of the carriage trade, as this was before the use of telegraph or electricity. So great was the prejudice in regard to turned spokes, that when I was with George W. Watson in 1858, they were still shaving all the spokes they used by hand. They also mortised the hubs by hand.

The Carriage Monthly, Jan 1895

This letter relates to above article.

WHO MADE THE FIRST BENT RIM?

EDITOR OF THE HUB—SIR : I claim I was the first one who made bent rims. The first ones were put on a vehicle for Nathan S. Thrupp, and were billed May 29th, 1835. The rims were bent all in one piece, and joined. I can to-day show two sets of carriage wheels I built for Peter M. Mower, July 10th, 1840, wherein I put rims bent all in one piece, 1¼ in. width. Those wheels are still all in good order, and the rims are as good to-day as they were at the time when they were put on. I am now eighty-three years old, and I have been in the carriage business most of my life.   JAMES HANSEN.
Box 179, SAUGERTIES, N. Y.

NOTE.—We have no reason to doubt the justice of Mr. Hansen's claim. If there are any other claimants to the honor, let them speak up, and name the facts on which they base their rival claims!

The Hub, Oct 1884

II
DISHING OF WHEELS

57

DISHING OF WHEELS

Among all the aspects of wheelmaking, none provoked more discussion and disagreement than dishing, the curvature in the wheel (with the concave surface normally facing the outside of the vehicle) that provides the wheel with greater strength. In fact, the amount of dish required for certain types of vehicles, and the reasons for dishing, are still debated to this day. The article on p. 63, "Dishing Wheels and Axle Setting," illustrates the intensity of opinions on this topic.

The most important reason for dishing is that it aligns the bottom spoke that contacts the ground perfectly vertical (referred to as a "plumb" spoke) so that no matter how much weight is applied to the wheel (within the limits of the vehicle), the spoke has the tendency to remain vertical and thus less vulnerable to side thrust and subsequent breakage. In addition, a dished wheel is less likely to become rim-bound as the wood shrinks.

The dish of the wheel is acquired by the angle on the front of the hub mortise and the angle on the shoulder of the spoke tenon, which is driven into the hub mortise, kicking the spoke toward the front of the hub. The amount of dish can be controlled by how tightly the iron tire is installed.

One writer expresses the opinion that patent wheels required more dish than banded hubs due to their tendency to "go back." The article on p. 60, "The Dishing of Wheels," warns of the danger of over-dishing the wheel, while another article, on p. 68, "Dish in Wheels," discusses the effect of humidity on the dish in wheels.

DISHED COMPARED WITH STRAIGHT WHEELS.

THE readers of THE NEW YORK COACH-MAKER'S MAGAZINE, on page 145, of volume III., have had their attention called to this subject, by an article credited to *The Carriage-Builders' Art Journal*, in which the writer's mind seems to me to be very much befogged, to say the least of it. I consider the subject to be one of great importance to the community, if not to carriage-builders themselves. The dish to wheels, and the taper to the arms, were originally very ingenious inventions, and strictly in accordance with the laws of mechanics, which will always remain the same, no matter what our wishes may be for a change, so as to agree with the caprice of fashion. The writer of the article alluded to has divided his subject into four parts, two of which I will review, because, by so doing, some common errors can be refuted, and some truths established.

He says, the advantages of dished wheels are as follows: *First*, there is a tendency to keep the tire tight.

Second, the bearing of the wheel and axle-box is against the collar of the axle instead of on the axle-nuts.

In arguing his first proposition, the writer speculates on vibration rather queerly, as follows: " There is always more or less vibration in a carriage wheel, as it passes over the road ; now, with an upright wheel, the vibration would occur on both sides of the wheel. Any practical mechanic is aware that if his tenons are continually moved backward and forward in their mortises, they must inevitably become loose ; and this is what would happen to the spokes and stocks of an upright wheel." Further on, he says : " By slightly coning (dishing) the wheel, it acquires many of the strong points of the arch ; in the first place, the vibration is very much reduced, and, in running, as it were, hard on the axle-collar, the woodwork is pressed tightly into the tire, which will remain tight for a longer time."

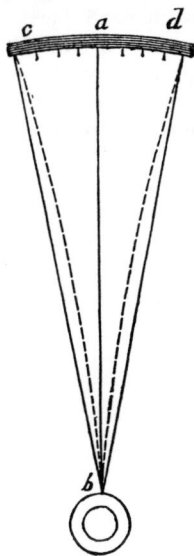
Fig.1

Now let us understand what is meant by vibration on both sides of a wheel. Suppose *a*, Figure 1, is an elastic rod, fastened in a mortise, *b* ; take hold of the end *a* and draw it to *c*, then let it loose, and its elasticity would carry it back towards *a*. When it had got to *a*, the propelling force that it had acquired would force it on to *d*, then the elasticity would carry it back to *a* again, and the propelling force that it would acquire would carry it on towards *c*, perhaps to the first dot, next to *c*, and in that manner it would go backwards and forwards until it would finally rest at *a*. This would be vibration, which, by constantly repeating, would loosen the tenon, *b*. But can the spokes in a wheel have any such vibration ? I think not. Draw the end of the rod *a* to *c*, then fasten a strong iron rod across the curve or angle at *b* and *c*, and then you have destroyed its power of vibration. Why? Because the rod fastened at the ends *b* and *c*, and running

across the angle or curve of the dotted line, is not as long as the curved or dotted line, and when the elasticity undertakes to straighten the curved rod, it makes a strain lengthwise on the rod which is fastened at the ends of the curve. Now this performs the same office as the tire to the dished wheel. If you undertake to bend the rod towards the curved line, you find it has assumed the properties of a brace or arch. We will soon see that the dished wheel was made in accordance with this principle. As for a wheel vibrating on one or both sides, we know it to be an impossibility. The ground the wheel runs on would hold the lower side of the wheel from any sideway movement, if it was inclined to make any such movement. The apparent vibration a wheel has is the vibration of the axle. There is not and probably cannot be any such invention as will keep the axle from vibrating.

The writer's ideas in relation to the properties of the dished wheel run in a singular direction ; for instance, the way he accounts for its keeping the tire tight is in this wise : the wheel, " in running, as it were, hard on the axle-collar, the wood-work is pressed tightly into the tires, which will remain tight for a longer time." I cannot see why a dished wheel should run (or wear) any harder on the collar than a straight one. If we want to make a wheel run hard on the collar of the axle, we incline the point of the arm forward, which holds the forward side of the wheels nearer together than the back, by which operation, when the carriage is put in motion, the wheels are inclined to run together, and are continually crowding on the collar. Reversing the arm, brings the crowding on the nut or point. Either of these positions should particularly be avoided. The weight of the carriage and load pressing down on the arms and axle-box, cannot be avoided ; but the side pressure can, by making the wheels run straight forward, instead of partly sidewise. A carriage that *held the tire on*, in the way the English writer speaks of, would be a poor thing.

The arguments used in the second division of the Journal's subject are the climax of stupidity. He says (2) : " By using an upright wheel on a horizontal axle-arm, there would, in use, be as much tendency for the axle-box to bear against the nuts in front, as against the collar behind. Now, it is of the greatest importance not to throw any unnecessary strain on the nuts, as the point is the weakest part of the axle, and the nuts would not, for any lengthened period, bear the strain, without stripping off the threads of the screws, and dropping off." The writer finally contrives a way whereby this strain on the nuts is *reversed* on the collar—as he thinks. The nut on each arm would bear a strain of at least two tons ; now let this strain be taken off the axle-nuts and put on the axle-collar, which would make eight tons for four wheels, pressing them together. Did the writer think how many horses it would take to draw his carriage?

In this country a style seems almost universally to prevail, of making wheels straight on the face side, and a very little crowning on the back side. This is a great mistake in our mechanics, which I should be glad to see corrected ; but I am afraid our English friend's arguments in favor of dished wheels are not as good as his intentions. I think their superiority for strength and wearing can be accounted for in the following manner. Spokes, when set angling, make the diameter of the wheel smaller than when set straight. For instance, in Figure 2, the diameter from

continued next page

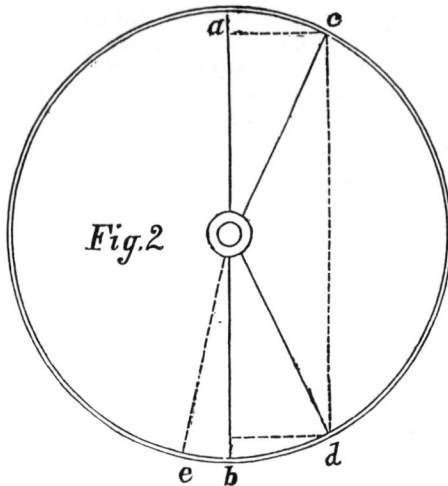

Fig.2

c to *d* is less than from *a* to *b*. A wagon is always so constructed that when a strain comes sideways on a wheel, it is always from the out, or face side, and on the bottom of the wheel, crowding it in. If you raise the left side of a wagon above the right side, you move the centre of gravity from the left nearer to the right, and the right wheel receives all the weight that is taken off from the left. In Figure 2, the strain would come at *d*, crowding it towards *b*, and it would not require much load to carry it to *b*, and even break your best hickory spokes, if it was not for that very ingenious contrivance, the tire. That is a strong iron which will not allow the diameter of the wheel in any way to be increased, It holds it at *c* and *d*, and if the side *d* is bent towards *b*, it must enlarge the diameter of the wheel, consequently the strain comes on the tire, drawing it lengthwise, and as it refuses to part, it, in turn, throws the strain back on the ends of the spokes, and they operate as a brace, instead of a lever. This is a beautiful arrangement of the law of mechanics, by which the utmost strength of the wood and iron in a wheel is obtained.

Now, what is the modern improvement of this good old style? Why a straight wheel! Put the strain on a wheel at *b*, in Figure 2, and bend it towards *e*, and you have diminished the diameter of the wood part of the wheel; loosened the tire from the felloes; almost entirely lost the support you could have got from the tire; thrown the strain sideways on the tenons of the spokes; and, in fact, abandoned every strong point of the wheel for the weakest!

I have had an opportunity, for thirty years, of observing and comparing the difference in the wear and the strength of the two kinds of wheels, and it is my firm belief that the present style of making wheels sacrifices from one-half to nine-tenths the wear of all wheels made. Also, I have a wagon, in the wheels of which the spokes are drove for one-and-a-fourth inch dish on the face side, which has now been in use for six years, and the tire has never been re-set, and I do not expect it will ever be necessary to re-set it.   H. H.

Berlin, Wis., January 20, 1861.

WOOD-SHOP.

GEAR DEPARTMENT.

[Communications, both of inquiry and information, are earnestly invited.]

THE DISHING OF WHEELS.

BY HENRY F. PORTER.

The Vertical Spoke and its Dangers—Why Wheels are dished—How to make Wheels durable—The consequences of insufficient Dish—The Long-Rim Theory exploded—Dish in Germany—Dish of the principal kinds of Carriages—The Track Question.

THE time was when the straight wheel was considered the most stylish, and it may have been so for some bodies. It is capable of a high rate of speed, as is illustrated by railroad-car wheels or velocipede wheels, but only on perfectly smooth grounds. In carriage-making, experience has long since shown that for quick driving over ordinary roads, which always presents more or less obstacles, the straight wheel is utterly impracticable for service, and that a certain amount of dish is required to produce strength. It is laid down by all mechanical law that the arch is capable of the strongest resistance, and this principle is followed out by dishing the wheels and producing an arch by the position of the spokes. How much the amount of dish should be for a particular size of wheels, and for a certain use, is an important question, and in the following I shall give my views on the subject. The amount of dish varies more in different countries than is perhaps known, namely, from ½ inch on the face spoke, as is usually made on a one-passenger wagon in this country, up to 13 and 14 inches in Germany. No doubt there may be good reasons for

continued next page

Dishing

the latter extraordinary amount of dish, and that it is practicable can hardly be questioned, not only because it continues to be made there, but also because the good reputation of German workmanship is acknowledged by all.

In the construction of carriages, the durability of the wheel is a very essential feature; for no matter how perfect the draught and make of the body, if the wheels are bad, they become worthless, and a new set is necessary. It is the dish which contributes toward making wheels more durable. To illustrate this, let us suppose that the spokes are vertically driven into the hub, as our Fig. I. shows, and that the wheels have steel tires, which will not expand. When such a vehicle is run over rough stone pavements, the constant strain and pounding will tend to settle the rims on the spokes, which must invariably cause the loosening of the tire, and make the wheel dish backward. This tendency is produced by the natural law of gravity, namely, the inclination is in the direction where the weight or pressure is acting, a fact which is well established and well understood. A wheel without dish must become what is called "rim-bound," and this creates another and still greater evil, namely, the spokes are driven into the hub, and they either become loose or the wood is chawed off the rim. Precisely the same effect may be produced when the wheels are not dished to a sufficient degree. Let us suppose that a customer is using a carriage for a certain length of time, during which the dish of the wheels becomes less than it was originally. That all wheels have the tendency to lessen the dish, I have stated in a former article in The Hub. Now, if the dish should get so much less, that the above-described action, which causes the wheel to get rim-bound, takes place, the owner of the carriage is not aware of it, but relying upon the reputation of the manufacturer, continues to use it, and the wheels will soon be ruined past repair. A new set is the only remedy. Such a set of wheels may on the start have been made of the best material, and with superior workmanship; but they lacked in the most important principle, a sufficient dishing or the proper arch, which alone could give them the required strength and durability.

After thus showing the effects of the vertical wheel, I will next speak of the dished wheel, a wheel forming an arch, as drawn in Fig. II. A pressure applied to one leg of the angle formed by the arch will react or be transmitted to the center, which in the wheel is the axle, and the strain is thus directly brought to bear on the axle, relieving the wheel itself to a certain degree.

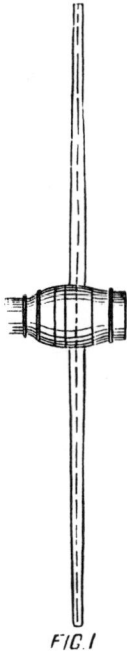

FIG. I

And it is because of this principle also that the face side of the spokes should be the strongest, and that they should be taken from near the bark of the trunk, as stated in the June number of The Hub, page 60. Another illustration of the necessity of dishing wheels is furnished by the fact that, in turning at a great speed around a corner, the strain on wheels with vertical spokes would be in one and the same direction on the wheels on either side, namely, to the outside of the turn, while on dished wheels the strain is equally divided on both sides, the dish of one wheel counteracting the effects of the strain upon the other. Still another and most important advantage of dishing wheels is, that this preserves them from the danger of becoming rim-bound. As the dish tends to decrease, the rim will be subjected to a pressure causing its expansion toward the tire, this being the natural action of the wood when brought in this state. The long-rim theory is disposed of at once by this fact.

I have thus shown the disadvantage and impossibility of the vertical wheel, and the necessity and benefits arising from dishing wheels, and the question next in order is this, What is the proper degree of dish? I do not hesitate to say that it is impossible to lay down a certain rule for all instances, as the dish depends upon many considerations, such as height and size of wheels, condition of roads, weight to be carried, width of track, and so on. For the information of such readers of The Hub as may desire to refer to it, I will state in the following the amount of dish which I give to wheels on the different kinds of carriages:

In conclusion, I would add that an excess of dish not only mars the beauty of a vehicle, but may, under certain circumstances, be very detrimental to the strength of the wheel. Although the dish made in Germany is very considerable, yet the heavy wheels made there can easier stand it, and they are made not only stouter, but also lower. The law prescribes the track in that country, and as it is rather narrow for heavy work, there is no other way to make a wide and comfortable body than to dish the wheel to the utmost; and still it is not infrequent that by the lateral motions of the body the panels are damaged by coming in contact with the rims. On the subject of track I intend to write at an early day, and will only say here that I concur with the view expressed by Mr. Muller in the columns of The Hub, that a uniform track can not and should not be arrived at by compulsion through laws.

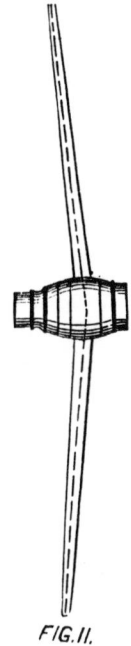

FIG. II.

The Hub, Sept 1872
See Table next page

Dishing

continued from previous page

The Dishing of Wheels

1. *One-Man no-top Box Wagon*, for city use.
   Wheels, 3 feet 11 x 4 feet 2 inches.
   Track, 3 feet 10 inches.
   Hubs, 3¼ x 6 inches.
   Spokes, ⅞ inch ; tire, ¾ inch.
   Dish, ⅛ inch from face spoke.

2. *Top Box Wagon*.
   Wheels, 3 feet 11 x 4 feet 2 inches.
   Track, 4 feet 5 inches.
   Hubs, 3⅝ x 6½ inches.
   Spokes, 1 inch ; tire, ⅞ inch.
   Dish, 3-16 inch from face spoke.

3. *Physician's Phaeton*, with durability for hard service.
   Wheels, 3 feet 4 x 4 feet.
   Track, 4 feet 8 inches.
   Hubs, 4¼ x 7 inches.
   Spokes, 1⅛ inches ; tire, 1 inch.
   Dish, ¼ inch from face spoke.

4. *Four-seat Phaeton*, with half-top or extension-top.
   Wheels, 3 feet 6 x 4 feet 2 inches.
   Track, 4 feet 8 inches.
   Hubs, 4⅝ x 7 inches.
   Spokes, 1¼ inches ; tire, 1⅛ inches.
   Dish, ¼ inch from face spoke.

5. *Six-seat Phaeton*.
   Wheels, 3 feet 6 x 4 feet 2 inches.
   Track, 5 feet.
   Hubs, 5½ x 8½ inches.
   Spokes, 1½ inches ; tire, 1⅜ inches.
   Dish, ⅜ inch from face spoke.

6. *Coach*, for hard service.
   Wheels, 3 feet 6 x 4 feet 2 inches.
   Track, 5 feet.
   Hubs, 6½ x 9 inches.
   Spokes, 1⅝ ; tire, 1½ inches.
   Dish, ½ inch from face spoke.

7. *Coach*, for private use.
   Wheels, 3 feet 6 x 4 feet 2 inches.
   Track, 5 feet.
   Hubs, 5⅜ x 8½ inches.
   Spokes, 1½ inches ; tire, 1⅜ inches.
   Dish, 7-16 inch from face spoke.

8. *Six-seat Rockaway*, for family use in city.
   Wheels, 3 feet 3 inches x 4 feet.
   Track, 4 feet 10 inches.
   Hubs, 5⅜ x 7½ inches.
   Spokes, 1⅜ inches ; tire, 1¼ inches.
   Dish, 5-16 inch from face spoke.

9. *Landau*, for private use.
   Wheels, 3 feet 4 x 4 feet.
   Track, 4 feet 10 inches.
   Hubs, 5⅜ x 7½ inches.
   Spokes, 1½ inches ; tire, 1⅜ inches.
   Dish, ⅜ inch from face spoke.

10. *Three-quarter Landaulet*.
    Wheels, 3 feet 4 x 4 feet.
    Track, 4 feet 10 inches.
    Hubs, 5⅜ x 7½ inches.
    Spokes, 1 7-16 inches ; tire, 1¼ inches.
    Dish, 5-16 inch from face spoke.

11. *Full-size Landaulet*.
    Wheels, 3 feet 4 x 4 feet.
    Track, 4 feet 10 inches.
    Hubs, 5⅝ inches.
    Spokes, 1½ inches ; tire, 1 5-16 inches.
    Dish, ⅜ inch from face spoke.

12. *Clarence*.
    Wheels, 3 feet 6 x 4 feet 2 inches.
    Track, 5 feet.
    Hubs, 6 x 9 inches.
    Spokes, 1⅝ inches ; tire, 1½ inches.
    Depth of felloes, 1¾ inches.
    Dish, ½ inch from face spoke.

The Hub, Sept 1872

Dishing

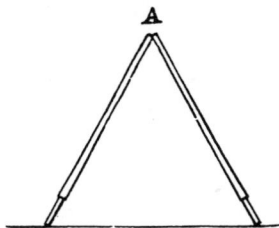

In response to an urgent request, Mr. William H. Stewart, of Orion, Wis., has kindly consented to give us his views on several subjects connected with wheels, as follow :

"Mr. EDITOR : I have been, thirty-five years engaged in the carriage business—not at the desk, with a pen, but in the shop with hammer and shave in my own hand, and I think that, by so long a practice and tolerably close observation, I ought to have learned something, so that I would not have to guess at it, or take other men's theory for it.

"First, in order to give the spoke that advantage over the influence of weight with which it has often to contend, it is absolutely indispensable that we take into consideration all positions in which the spoke is obliged to stand to hold its burden. If standing in the show-room on a level floor, it matters but little whether the axle is exactly straight or a little down, or even up at the point, or whether the spoke is exactly plumb, or out a trifle at the outer lower end. It will bear any reasonable load in either case ; but if we would see the principle properly applied, we should take the wagon out on a side-hill, where one side of the wagon will be much lower than the other, and then load it. You will now see that most of the load is thrown on to the lower wheel on each axle, gravity giving the load an inclination to move in a direction for the center of the earth, and you will see your plumb spoke standing out at an inclination of perhaps thirty or forty degrees, and obliged to hold two thirds of the load in that unpromising position.

"Now, let me come to the point. Here is where spokes get used up—by *side-strain.* If spokes got no other strain than directly endwise, then one half the weight of spokes, as used now, would do just as well, and in that case, what nice little wheels we would make, wouldn't we ? But as soon as a wheel is put to practical use, that moment it commences to carry its burden over uneven ground, and it must be so made, and the axle so shaped and so conformed to the make of the wheel, that it will be able to stand up to its task. You will see at once that, as a wagon passes over rough ground, the wheels drop, often suddenly, into low places or holes, and the load is thrown suddenly sidewise and downward, and the spoke thrown out of plumb, and obliged to receive the weight of the load like a huge sledge-hammer at a side-strain, and bring if up standing. This is what uses up our best wheels, in spite of us.

"Now, let me show it up in another way. If we take two spokes that are able to hold 1000 pounds weight each directly endwise, standing plumb, then let us stand the same spokes thus :

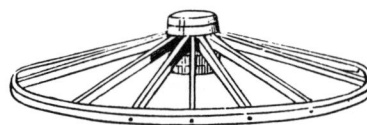

with upper ends in contact and the lower ends fixed so as to prevent sliding ; they are out of plumb several degrees, but will still be able to hold nearly the same load, if applied at *a.* But let either of them remain in that inclined position alone, removing its mate, and it will go down at once with only one half of the load that the two could hold united. The reason is very plain. United, they act mutually to prevent all side-strain ; but alone, the side-strain is what causes it to break. Here is the only reason why we are obliged to put a certain amount of dish in our wheels. In so doing, the wheel becomes an arch, the hub being the keystone, the spokes the side-walls, and the rim the base. Let me here put a dishing wheel, considerably exaggerated, on its face, and let it lay on a floor, so that the rim may

come in contact with the floor, leaving the hub elevated, and resting only on the spokes ; then let us load the wheel by applying all the weight it is able to bear on the hub. In such a case, let me ask the wheel-maker, 'Is the strain all endwise on the spokes, or is it all side-strain?' You can but see that a good wheel, while the tire holds, will sustain many thousands of pounds, because all the strain is *endwise* to the spokes ; but if the tire gives away, then the strain is instantly changed to side-strain, and the spokes go down at once. Now, when by uneven roads the load is thrown sidewise against the lower wheel, the wheel could easily stand the strain, if the upper part of the rim could rest against some support, and then the case would be similar to that of loading the wheel when lying on its face ; but as it is impossible to have that support to the upper part of the rim, then the lower spokes in the lower wheel get much side-strain ; and what sound mechanic does not know that, under these circumstances, the more the point of an axle is bent down the worse the case is made ? My friend, don't drop the points down, and don't teach others to do it, but keep them straight on the bottom ; or if you must take an alternate in regard to deviating from the straight line, take it as you would poison—*the less the better*—and let it be upward instead of downward. We live in an age of progress, and I don't believe it is actually necessary for us to be fixed or stereotyped in styles or forms. We are not obliged to still ride in the old 'one-hoss shay' which I well remember once seeing as the height of style.

"When our great-grandfathers made their wooden axles, common sense taught them that the bottom should be straight, and that it was useless to have the point as large as the shoulder ; that gave the wheel an incline outward at the top, which they were satisfied to put up with, as it served to throw the mud and dust away from the 'one-hoss shay,' and as that inclination of the top of the wheel outward caused the wheel to run sideways —as in leaning a wheelbarrow to one side to turn it around; they were smart enough to counteract that outward tendency by pitching the point of the axle a little forward, to prevent the wrought-iron box from cutting off their fancy old linchpin.

"I hold that, as we now have good iron or steel axles, their size is so reduced that, by keeping them as nearly as is practicable of the same diameter at the point as at the shoulder (which is no objection as to friction, as would be the case with the old wooden axle), and the nearer a uniform size ; and the straighter the wheel, the better it will appear, the nearer plumb the spoke will stand (under circumstances above described), the fewer alternates we will have to take—and I always take alternates as I take pills, when nothing else will possibly do—and the nearer our carriages will run, like best machinery of other kinds, on a ———————— STRAIGHT AXLE."

The Hub, Nov 1875

Dishing

REGULATING THE DISH OF WHEELS.

HAVING received several questions on the subject of equal dish in wheels, we have referred them to three competent mechanics, one a life-long wheel-maker, and two practical smiths, who have had a long experience in tire setting, and their replies are as follow :—ED.

From "Wheel-maker."—Much depends on the selection of the stock. I endeavor as nearly as possible to have the hubs of the same weight, density, and grain, and perfectly seasoned. I next select the spokes, weighing by hand every one, and I try to have the grain alike in all, and also test their elasticity. The same precaution I also take with the rims. In the mechanical part it is my great aim to make my work as nearly perfect as possible. In finishing my wheels, I make them so that the tread is at strictly right angles with the face, which may be proved by placing a straight-edge across the face of the wheel, and by applying to it and the tread of the wheel a trying-square, which at once points out any variation.

To those not accustomed to this method, it may appear to be tedious, causing a loss of time, but I find it pays in the end. In selecting stock I may occupy a half hour of time, but I have become so expert that I can detect a very slight variation in the weight of the material.　　　　　　　　EDWIN.

From "Tire-setter."—Before touching the tires, I give each wheel a thorough inspection, and ascertain that each spoke is properly seated at either tenon. I also ascertain whether or not the spokes are all straight. I also prove the tread of the wheel as to whether it is true and square with the face, and that the dish of the wheel is true at all points, after which I have the wheel so drafted that when the rim is seated on all the spokes, the joints are close, and also see that there is no gaping. The crooks in the tire are removed by hammering on the flat. The twist I remove with a twisting-wrench. After all this, I cut off, bend, and weld the tire, and measure the tire while cold. If I have the proper amount of draft, I heat the tire and place it upon the wheel, and if it dishes too much or too little, I at once remove the tire, apply the proper remedy, and re-set. Practice produces speed, and by this method I do as much as any of my fellows, and have the dish of my wheels so near alike, that $\frac{1}{64}$ of an inch will cover all discrepancies.　　　　　　M.

From "Tubal Cain."—Regarding the equal dish of wheels, it is my opinion that, so far as appearances are considered, they would look better to be all of the same dish. But then a 3 feet 4 in. wheel should not be dished as much as a 4 feet 2 in. wheel, else in reality the lower wheel will have the most in proportion to its diameter. But again, a small amount of dish makes no practical difference so long as the spokes are not sprung. My experience proves beyond a doubt that patent wheels require about ⅜ of an inch dish, as they are inclined to go back ; while the common wood hub wheel should have not more than ⅛ of an inch, ordinarily. The wheels should have sufficient swing to make the whole surface or width of the tire set level on the floor when loaded, or there will be an undue strain on the tenons of the spokes at the rim.　　　　　　TUBAL CAIN.

The Hub, July 1877

THE WEAR OF TIRES.

DOVER, N. J., July 12, 1880.

EDITOR HUB : Supposing a carriage has a 5-inch swing, why is it that the tire wears out on the inside first ? Please answer in the next *Hub* and oblige.　J. G.

ANSWER.—The swing makes no difference. Probably one of two things. Either a want of correspondence between the taper of the axle-arm and the dish of the wheel, or using too light an axle for the work it is to perform. To prevent the tire wearing unevenly, dish your wheel the same fraction of an inch to the foot that your axle-arm tapers, that is, if the axle-arm tapers one-eighth of an inch to the foot, the wheel should have a dish of one-eighth of an inch to the foot. This will bring a horizontal bearing on the arm of the axle which, besides causing the tire to wear evenly because it bears flat on the roadway, will decrease the friction on the arm, reduce the breakage to a minimum, and also reduce the draft. Attention must also be paid to having the axle stout enough to prevent its sagging in the middle, which would cause the tire to spread wider on the bottom than was originally intended, and cause the trouble you describe.

The Hub, Aug 1880

IRREGULAR DISH OF WHEELS.

EDITOR OF THE HUB—DEAR SIR : Can you tell me why some wheels which were true as a die when the tires were first put on, should dish more in one section than another. Sometimes I have wheels which will in a few days after being tired show a difference of from ⅛ to ¾ in. in the dish. "WHEELWRIGHT."

ANSWER.—There are many causes of which we will mention some of the principal ones. We doubt if in a set of spokes a dozen could be found which have nearly the same weight, and, while they may all have been made from the same tree (which is not probable), all will not have the same resisting power. Those which have been on the outer portion of the bundle will have had the advantage of the seasoning process, which will make them more refractory than those from the inner section of the bundle. There may not be a very marked difference in the spokes in weight, seasoning and sustaining power, yet if we get all the weak, light and less-seasoned at one section, and the reverse at the other section, we may look for irregularity of dish.

In handling the spokes, the expert can readily tell the heavy spoke from the light one. A light test proves the stubborn stick or shows the yielding one. The expert can also tell if the stick is well or only partly seasoned. By a careful mixing of all the spokes in driving we get the strong spokes where they will support the weaker ones.

Another cause of irregularity of dish is in the dressing of the spokes. One spoke is reduced more at the throat than the other. If we drive all the over-dressed in one section and the full ones in another section, we may look for a discrepancy in dish. By calipering the spokes at the throat before driving and arranging the spokes so as to have, as nearly as possible, a light and heavy spoke alternately, we may also to a great extent lessen the chance of irregularity.

Again, if the spokes are all of the same caliber throughout, all of the same resisting power and weight, we may, by improper drafting or jointing the rims, throw one section completely astray from the other. Unless the joint is perfect, as soon as the tire is placed on the wheel and ends its contraction, that section which has been improperly jointed will make itself manifest in the dish. Even if we have taken all the precautions possible, the smith in placing the tire on the wheel may undo al our work. If the tire when placed on the wheel should be much hotter at one section than another it will be more expanded than at the other, consequently on being quickly cooled with water its greater contraction at such place or section will produce a greater dish than where there was less heat.

Wheels in which the spokes are " dodged " are less liable to discrepancies in dish than those driven without dodging. If you will season all your spokes alike and follow the other suggestions and heat the tire uniformly you will reduce the irregularity to a minimum.

The Hub, June 1889

Dishing

MECHANICAL LIMIT IN THE "DISH" OF WHEELS.

THE custom of making carriage, coach and wagon wheels with more or less "dish" or concavity, is one that necessarily followed the use of shrunken tires, and is with few exceptions general; but what in the first instance was unavoidable has come to be considered as adding greatly to the strength of the wheel in service, and is consequently over-cultivated. In going into the subject we find that the advantage so clearly seen where used in moderation, soon becomes a positive disadvantage mechanically when carried beyond certain limits. It is a very general idea that the wheel in this shape forms an arch, and is thus enabled to stand great lateral strain. We admit this is so under given conditions, and to a certain extent. When a wheel is placed as shown in Fig. 1, the spokes with the stock or hub do form a regular arch, and the more the spokes are inclined or dished the greater the resistance at the points *a, a*, from weight applied in the direction of the arrow. But now, if the wheel be placed upright as shown in Fig. 2, A, we find that instead of having two points of resistance, as in Fig 1, we have but one, marked *c*. The force of any thrust on the axle in the direction of the arrow would tend to push the spoke in contact with the ground at *c* into a more upright position. The tire resists this pressure

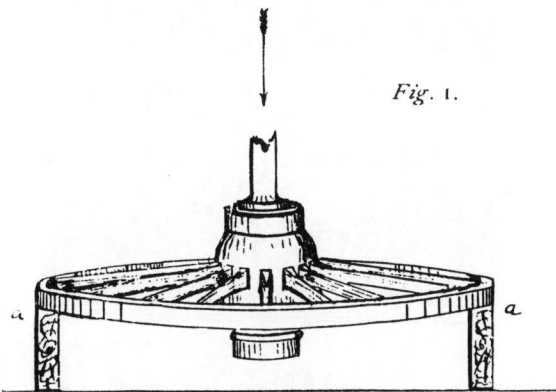

Fig. 1.

and draws downward from the top of the wheel, or the part most directly opposite that receiving the strain. In our sketch we have purposely exaggerated the "dish," to more clearly show the principle. It will be seen from this, that any strain brought to bear upon the point opposite *c* would operate upon the upper spokes as levers; the amount of inclination increasing the leverage upon the axle-arm, and from the rotation of the wheel its tendency is to shell out the back of the hub and work the spokes loose; nor is this all; the tires of necessity are made from bars of steel or iron, and are, when welded, of uniform inside diameter throughout their width. This being the case, the "tread" of the wheel is worked true with the face of the hub, as

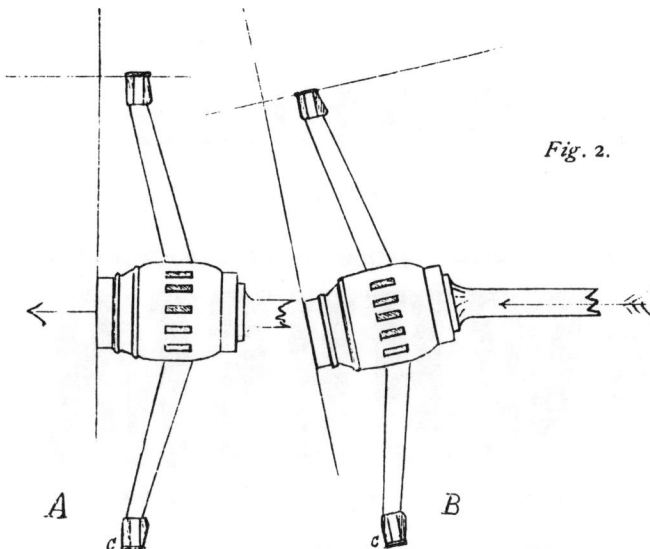

Fig. 2.

shown by the cross-lines in the sketch. The position of the wheel as shown at A is the proper one to secure the benefit, if any, from the "dish," as the "tread" is flat to the road; but practically, to carry the load upon a plumb spoke and through the center of the box, it must be set as shown at B, the exaggeration of our example clearly showing the disadvantages, as the axle-arm must be set to compensate for the "dish." This brings all lateral strains (as indicated by the arrow) above the center of the hub, causing very uneven wear of the boxes, also carrying all sudden shocks to the shoulder where the collar and axle-arm join. An examination of broken axles will show that in nine cases out of ten they break from the top (see Fig. 3). Most of those the writer has examined have revealed the presence of an old crack, as shown in the enlarged section E, T being the top of the axle-arm. We do not say that the "dish" of the wheel is answerable for all of this, but these lateral strains are more severe upon the bearings than the weight carried directly down through the arms, and anything which increases the leverage against the axle-arm must play an important part in this matter.

It may be said that the wheel is bound together by the tire and resists pressure as a unit, but this it does only under the conditions shown in Fig. 1, or in case of concussion on outside of the tire. That the forces operating laterally against the wheel are felt in detail by the spokes in contact with the ground at the time, is clearly proved by the tendency of light wheels to turn "inside out" or "buckle" when turning sharp curves at high speed. If the wheel resisted as a unit, this could not occur, for we know that the tire or rim does not increase in diameter, but by the bending backward of the spokes in contact with the

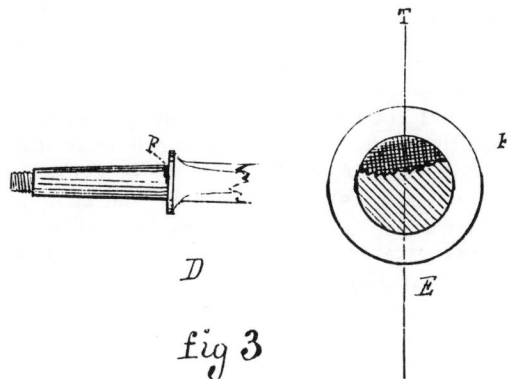

fig 3

ground, and compensating movement of those in upper part of the wheel, one revolution throws the rim over to the back. These forces operate upon all sizes of wheels. The more rigid the spokes, the more pressure on the hubs and leverage on the axles.

An important feature of this subject is the question of "stagger" or "dodge," which is of the greatest advantage to light wheels which have little dish; but as soon as the concavity of the wheel is increased, then it becomes of no real use, and causes weakness instead of adding strength. In Fig. 4 we show

continued next page

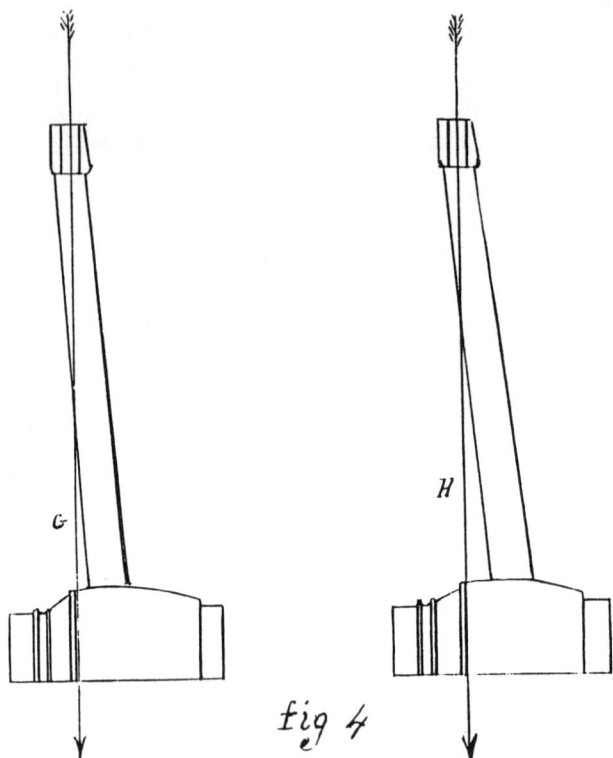

fig 4

sketch, drawn to scale of 1½ in. to 1 ft., representing the half-section of a wheel 4 ft. diameter, 1¾ in. spokes, 1¼ in. dish, G being the face spoke and H the back spoke, with ⅛ in. "dodge" or "stagger." It will be seen from the lines drawn, that we have more than reached the mechanical limit,—the line of strain falling outside of the face of the spoke, while the back spoke H does not support the weight through more than one third its length. That these principles are thoroughly understood by the leading carriage-builders in this country is evidenced by an examination of their work, and a very full and complete list of dimensions for wheels coming within mechanical limits can be found in *The Hub* for April, 1879, where J. D. G. gives valuable statistical data in full detail.   H. M. DuBois.

The Hub, Mar 1881

Dishing

WHEEL-SHOP.

AN ENGLISH OPINION ON DISHING WHEELS.

[The subject of dishing wheels is an ever-live one, and we quote the following intelligent description of the principles involved, from a long serial by Mr. W. T. Casson, on the general topic of carriage wheels, now appearing in the *Coach-Builder's, Harness-Makers and Saddlers' Art Journal*, London.]

THE question of how much dish ought to be given to a wheel is partly an open question; no hard and fast laws can be laid down, as much depends upon the individual taste of the builder or customer. A small wheel requires less dish than a large one, as there is not such a side pressure put upon the spokes.

If too much dish be given to the carriage wheel, as is usually made, the wheel runs all on the outer edge of tire. Many a wheel may be found where, by having excessive dish, the front edge of the tire is worn out before the back is little more than touched. This is owing to the axle having been set so that the under spoke shall stand plumb, and the wheel not having been felloed so that the front diameter shall be less than the back. Had this been done, the tire would require to have been conical to fit, and to have been put on from the face instead of from the back, as is usually done.

To remedy this running on the front edge of the tire, less dish must be given to the wheel; and to give the wheel the strength required, the spokes must be "reeled" in the nave. This term is sometimes called "dodged," "staggered" or "speched," according to locality and usage. By this means every alternate mortise is made a half-inch or more behind the front spoke line, and thus a wheel with 1 inch dish has almost the same strength and advantages of one with 1½ inch dish, and the tire rests upon the ground almost level. If carriage wheels were made according to scientific theory, every dished wheel would have to be made conical on the tread, or beveled in proportion to the amount of dish. The question as to whether this additional labor would benefit the wheel, or make it wear better, is very doubtful, as every time the wheel is newly tired the felloes would require to be beveled more in proportion to the extra dish given, and the tire would have a constant tendency to slip off the front. By giving a moderate amount of dish, and reeling the spokes, the wheel can be made strong and durable, and run with almost as little friction as if perfectly straight up, and rest almost as flat upon the ground as could be desired.

Absolutely conical wheels are therefore unknown in carriage building, or are only met with when they are made to gratify some whim or fancy. It is one of those cases where scientific theory must give way to practical utility, and, instead of making the tread of the wheel at right angles to the spokes, it is made at right angles to the plane of the rim.

The following table gives the amount of dish that may be given upon an average to wheels of different heights. It should, however, be borne in mind that wheels for drags, or any top-heavy carriages, would require more dish than the wheels of a like diameter for a pony cart. In this respect much must be left to judgment, for an experienced carriage-builder will see at a glance if a wheel is strong enough to support its load, without consulting tables. Wheels upon coaches used for carrying mails and passengers in parts where railways have not yet penetrated, have been known to fail under a heavy load, solely through not having enough dish. For such top-heavy vehicles, double the dish may be given with advantage.

HEIGHT OF WHEEL.	AMOUNT OF DISH. Measured from front spoke.
2 feet o inches	1½ inch.
2 " 6 "	⅝ "
3 " o "	¾ "
3 " 6 "	⅞ "
4 " o "	1 "
4 " 6 "	1⅛ "
5 " o "	1¼ "

Double the height of wheel, double the dish.

This table must be taken for entirely new wheels, for ordinary work, and tired. Every time the wheel is repaired or the tire contracted, the dish will be largely increased, so it is advisable to begin with as small an amount as can be safely given.

In four-wheel carriages, the front and back wheels ought to dish in proportion to their height, so that the axles may be set the same and be of the same length. The front and back wheels will then track exactly, and the wheels line evenly across the face.

Thus, a set of wheels of 3 feet and 4 feet, require, according to the above table, to dish ¾ inch and 1 inch respectively, and if the axles be of the same length, the wheels will line truly across the face, and all the under spokes will be plumb spokes.

The accompanying illustrations will show this; in Fig. 1 the wheels are dished according to the preceding table, viz.: for 4 feet 1 inch dish; and for 3 feet ¾ inch dish; the under spokes are plumb, and the rims of both coincide.

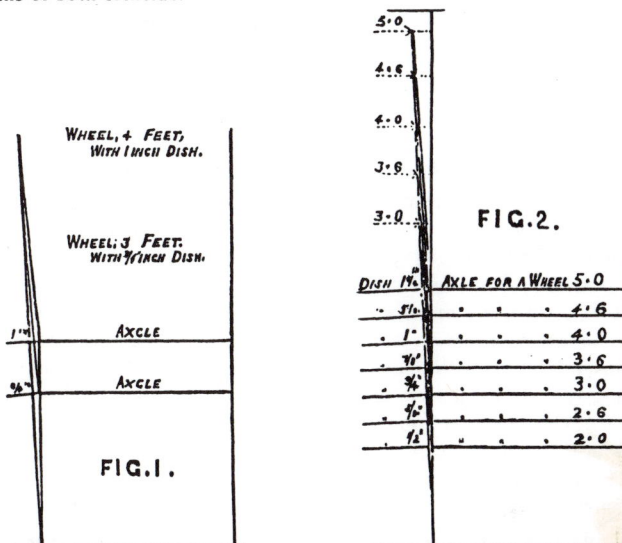

The other, Fig. 2, explains itself, where the under spokes and rims of wheels, from 2 ft. to 5 ft., are shown to occupy the same lines, when dished according to the before-mentioned table.

There are one or two things which prevent the above from being mathematically correct, but were any attempt made to give the correct figures, few if any would care to fully understand them, and fewer still could work to the figures so given. It is also difficult to get two wheels to dish exactly to $\frac{1}{16}$ inch, so it would be useless to give minute measurements. If it should be found that the foregoing table gives the dish too small or too large, it can be altered in proportion to the figures there given. A scale can be made in a few minutes that will give dish required for sets of wheels, without risking the errors of an arithmetical calculation. See Fig. 3, which explains the construction; the method of working is as follows:

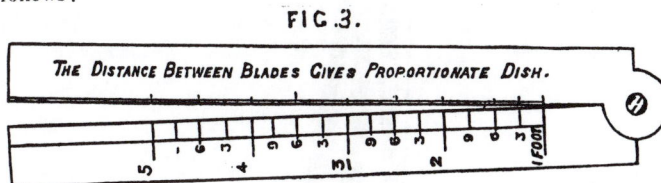

A set of wheels is required of the extraordinary size, say of 2 feet 6 inches and 4 feet 6 inches, the axles to be of equal length, the under spoke plumb, and the faces of the wheels to line. If the front wheel dishes 1 inch, the dish of the hind may be found by setting the blades 1 inch wide at the 2 feet 6 inch mark, and then measuring across at the 4 feet 6 inch mark for that dish.

A pair of hind wheels may be required to match a pair of front ones, or *vice versa*; then, whatever dish the existing wheels may have, set the scale to it, and measure off the dish across the blades at the height that the new wheels require to be.

To measure the plumb spoke of a wheel, it is necessary to put a back spoke of one wheel and the front spoke of another vertical, and measure from the back of one to the front of the other. If the distance at the nave be equal to that at the felloe, the spokes will be plumb, or rather the mean of the two spokes will be plumb. Allowance must always be made for the axle bending under the load, and throwing the wheels out wider at the bottom.

The Hub, May 1884

Dishing

The Meaning of the Word "Stagger."

This is a word generally used in connection with the wheels of a carriage, and has reference to the spokes when they are displaced or

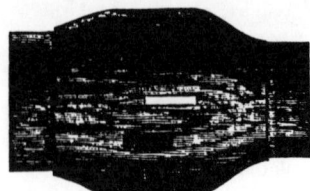

Fig. 1.

Fig. 2.

transposed from each other, or staggered. *Fig. 1* illustrates to good advantage what is meant to be conveyed, showing a hub with the mortises staggered where the spokes are to be inserted. There are two objects for staggering spokes, it giving the wheel more stability and also durability. It is well known that when spokes are staggered in a wheel, it gives it more strength to resist the side motion of the carriage, as is shown in *Fig. 2*. The strain a wheel encounters sidewise is considerable, and the stagger helps to withstand this strain very materially. As an illustration of this, if you stand on the floor with your feet close together, you can be easily pushed over, but if you set one foot forward, you will stand much firmer.

The limit of stagger to be given to a wheel, has never been published for all sizes of wheels ; but too much stagger is impracticable, as we illustrate in *Fig. 3*. Suppose we give 1¼ inches stagger to the wheel, the front spokes driven straight and the back ones to suit the front ones. Put the tire on the wheel and give about ¼ inch dish to the front spokes ; the back spokes have to follow the front ones, and they will be shortened, or, rather, the inclination of the back spokes is so much greater than that of the front spokes ; and in dishing the wheel the front spokes will not shorten at all, while the back spokes will shorten as illustrated in *Fig. 3* ; *A* the front spoke driven straight ; *B* the back spoke, having 1¼ inches stagger ; dotted line *C* the dish

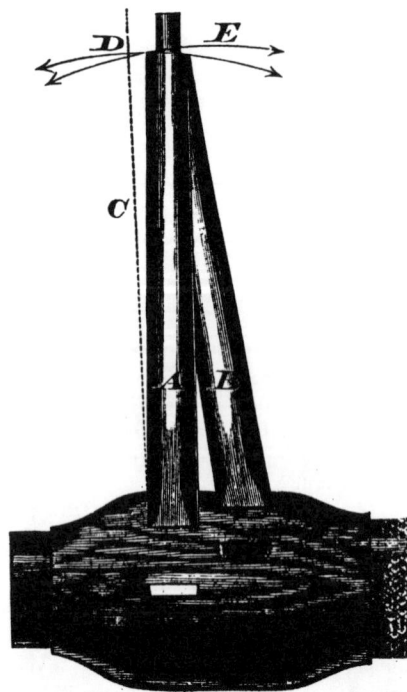

Fig. 3.

given the wheel when it is tired ; *D* the movement of the front spoke ; and *E* the movement of the back spoke after it is tired. The result of too much stagger is to bend all the front spokes on a light wheel, and it will either force the tenons out of the rims on the back spokes or

break off the back spokes in the course of usage ; the back spokes will almost all check at the back of the tenon, just below the shoulder, at the square end. The difference in the length of spokes at *A* and *B* can be seen on dotted line *C*, between curved lines *D* and *E*.

The Carriage Monthly, May 1885

DISH IN WHEELS.

The dish of wheels is a subject that should interest all buyers, as well as builders of vehicles. A large quantity of wheels are made, straight and dished by the tire, from a quarter to half an inch. This dish puts the back bone in the wheel for strength, provided all the other parts are properly constructed. Without the dish, the best wheel ever constructed would soon become comparatively worthless. There are many men who think when the tires are tight, the job is all right, even if the wheels have straightened up. This is right under certain conditions, as in the case of one that has been completed in the fall or winter and placed in a repository where there is little moisture, and where the wheels will strengthen more than if in a repository where subject to atmospheric changes.

When the tires become loose under such conditions they should not always be reset, as it will be found that when run out in the spring, the dampness will cause the wood to swell and soon dish the wheels, even more than they were originally. Wheels have been dished an inch on some jobs under such circumstances, thereby causing many spokes to spring or bend, and oftentimes the tire to burst under the great pressure caused by the expansion of the rim.

The one course to follow is to keep an eye now and again upon the wheels while in use, and when they show straight upon the front, have the tires set immediately, and the wheels will last much longer than if neglected until they begin to show weakness.

One important reason why the above idea should be adhered to, is, that when the tire has drawn the wheel over to the requisite dish, say ½ inch, the axles are set and the track of the job obtained, say 4 ft. 8 in. from outside to outside of the tire, upon the floor, the axles are generally set so that the spokes will stand about plumb from the floor, and when the wheels straighten up, the track is 1 in. less than when the job was run out, and the trouble begins. The "plumb" spoke is not plumb. It stands under, and becomes subjected to twice the strain that it would have were the ½ in. dish in it.

The result of this is, the spokes soon become loose in the hub, and the builder is accused of using poor wheels. Another reason why the dish should be retained in the wheel is that the axles will wear much longer, owing to their being set to correspond with the dish.

When a job has stood in the repository all winter, and has become thoroughly dried out, and the wheels straightened, the tires reset and the wheel dished to the original ½ in., and then run out in the spring, the wheels will go over to one inch, in a week or less, and the result will be a wider track than was at first intended.

Many builders reset the tires upon jobs that have stood in their repositories during the winter and have thus become dried out. These should never have been reset.

In building wheels, and all kinds of woodwork for vehicles, the workman should keep in mind one thing ; that is, the contracting and expanding of timber by heat as well as moisture, and contraction by a cold as well as a hot, dry atmosphere.

In the building of wheels the spokes must be dry, while the hub and rim need not be perfectly so, because the hub contracts to the tenon therein, and the rim to the rim tenon.

If the rim is not dry when the tire is put on, it will soon shrink away, leaving the tenons of the spokes poking through the rims and against the tire, which prevents the tire from setting down upon the rim as it should, and thus furnishes a good reason for contending that the spoke tenons should be cut off a little below the rim.

If the rims contain any moisture when the tires are set and should shrink away afterwards, there is no other alternative but to remove the tires, then cut off the ends of the spoke tenons and reset the tires.

The Hub, April 1894

III
HUBS

Hubs

Hub Bands

Hub Boxing/Setting

Hub Machinery

HUBS

The subject of hub design would probably be a close second to dishing in the race for the variety and intensity of the opinions expressed. A discussion of banded hubs versus the patent wheel (which eliminated the necessity of staggering the spokes) is as apt to spark a lively debate among wheelmakers today as it did in the past.

Many wheelmakers felt that the patent hub (see Chapter VIII), which came into use around 1857, was too rigid and lacked the flexibility of the wooden hub. This was, of course, a major contention of Howard DuBois. Despite this, the strength of the patent hub made it very popular among wheelmakers and owners alike. Prior to the patent hub, the hub mortises were staggered to preserve wheel strength. In banded hubs, an excessive amount of wood removed, with an in-line spoke pattern, results in a weak hub. In response to this problem, wheelmakers decided to stagger hub mortises. This led to the popularity of the patent hub with some type of iron enclosure, which strengthened the hub and eliminated the need for staggering the spokes. Also, the articles in the "Hub Boxing/Setting" subchapter discuss the various methods of securing axle boxes.

Another issue that generated discussion involved the splitting problem inherent in seasoning hub blanks. One writer argues that it is important that hub bands be put on the wheel as soon as possible to avoid splitting, while another argues for banding the hubs when the iron tire is sweated on. A third point of view represented here notes that temporary bands can be used to alleviate the problem. In 1891, a patented process was introduced to season the blanks artificially (kiln drying), although, as usual, wheelmakers were slow to accept this process. Finally, in 1895, a power press was developed to apply hub bands evenly.

Until 1858, hubs were still being mortised by hand with wooden mallet and bruzz (wheelwright's corner chisel). In that year, the Hayes Hub Boring and Mortising Machine was developed. The term "hub boring" in this context means drilling holes which are chased by the square mortise chisel. Hub boring usually involves reaming the tapered hole for the axle box. In the early shops, this was done with a two-handled wheelwright's reamer. (Some of the larger reamers had a hook on the small end to which a chain with a weight was attached, helping to pull the reamer into the hub; the wheelwright had only to turn the handle which, on the larger hubs, was a job in itself.) With the introduction of the Hub Boring Machine, this "boxing" could be done more accurately, eliminating the trial and error method of fitting and wedging to center the box.

Hubs

HUBS.

A CORRESPONDENT, recently, among other questions, asks, "Why is it that some carriage and wagon-builders make or use smaller hubs for the front or forward wheels than they do for the hind or back wheels? Is it for strength? At all events, please explain."

If we were not positive of the sincerity of the writer, we might be prone to give a facetious Yankee answer by asking another question. Why are the front wheels lower than the back ones? Which is the briefest answer to the question.

All carriage and wagon-builders do not make a difference in the size of their hubs for front and back wheels. Only those do it who are or have educated themselves up to the proper standpoint in the excellence of the craft. In the length of the hub for the average vehicle there is no difference; it is with the diameter only. It is the usual custom to invest the hind wheels with two more spokes than are given to the front ones. To facilitate turning and to level up in construction because of the front pivoting the front wheels are, absolutely, necessarily made lower. The educated builder knows full well that if he were to make the front hub as large as the hind one, symmetry and harmony of construction would be lost. The front wheel would be fat and bulky in appearance; and there would be a great waste of daylight and timber between the shoulders of the tenons; the hind wheels would look lean and spooky. Where the spokes converge or set in the hub they would look crowded and out of proportion.

It is as though one were to take a piece or strip of wood 12 in. long and divide it up into twelve spaces, and then divide another piece of the same length into fourteen spaces and place them a few inches apart; an apparent paradox would be the result. It would be necessary for the two to be placed together before you would be convinced that both pieces were the same length. If the second piece were made long enough to put in fourteen spaces nearly as wide as are the twelve spaces, we get another paradox, and the average looker-on would be convinced that both pieces were the same length; so it is with the wheels in question. If both hubs are of equal diameter, the front hub will appear, with its 12 or 14 spokes, greater than will the hind hub with its 14 or 16 spokes; the greater the discrepancy in the height of the wheels the more paradoxical the appearance becomes. The hubs, spokes and rims of the front wheels look fat and squatty, and the hind ones lean and shadowy.

In the great number of kinds of carriages and wagons built there exists the necessity of balancing up appearances and strength. On some vehicles a 1 in. diameter axle would be of ample strength, and yet, the hind one would require to be 1¼ in. or more in diameter. In this case the hub is not only made smaller but shorter, and the spokes correspondingly lighter. But such is the skill of the educated builder that the change or discrepancy is not detected. In conclusion, we can only add that to become convinced of the correctness of our rule, as set forth, it is best to experiment with the strips of wood and then with the hubs.

The Hub, Oct 1889

THE CORRECT SIZE OF HUB AS PER AXLE.

ONE of our young subscribers writes that he wants to learn some standard rule by which to get the diameter of the hub to suit its length of axle-arm.

On the length of the axle spindle depends the length of the hub, not its diameter. On the diameter of the spindle depends the inner diameter of the box, and on the outer diameter of the box, at the point where the spokes are inserted, depends the diameter of the hub at the largest part, or at the bulge. From many tests by experts, certain standards have been arrived at, and give standard and complete satisfaction. From time to time we have given standards and standard gauges of the various modes with which manufacturers have to do. Some little while ago we gave a chapter in the "Smith Department" on standard threads, nuts and sizes. We know of no better comparison than the nut. The nut has to meet a certain amount of resistance; first, it must contain the worm or thread, and enough of it to just balance the thread on the bolt; just enough to stand the maximum standard strain without stripping. With the nut, the metal on the sides must equal the diameter of the bolt to sustain the pressure. But let it be remembered, there are the four corners or angles which lend large aid. Then, again, the nut in thickness must equal the diameter of the bolt, or, on the whole, the nut of the bolt which is ¼ in. diameter must measure ½ in. on each of its sides, and be ¼ in. thick.

Now, we get back to the hub business. Let us suppose that we are using a one-inch axle—that is, one inch at the collar, by which we mean an axle which measures one inch on each of its faces at the back of the collar. The box, if of ordinary grey or cast-iron, the shell ought no to be less than $\frac{3}{16}$ in. in thickness, making a combined thickness of metal of ⅜ in. With a box of malleable cast-iron or malleable cast-steel, we could dispense with $\frac{1}{32}$ in. in thickness, or $\frac{1}{16}$ in. in total diameter, which would make the diameter $\frac{5}{16}$ in. instead of ⅜ in. With the improved wrought-iron box, as made by the Dalzells, another $\frac{1}{16}$ in. is gained in mean outer diameter. The spindle at shoulder or collar is 1 in.; at end of taper, ¾ in., which would make the central point of spindle ⅞ in. in diameter, and with ⅜ in. diameter added for mean thickness of box, we have 1¼ in. We all know that wood is not as tenacious as iron, and that if a tenon is 1 in. wide and 1 in. long, and the tenon is inserted in a mortise less than 1 in. deep, that we can by force draw the tenon out, owing to loss in tenon and mortise, by compression of fiber. To overcome this, we increase the tenon and depth of the mortise; and, for absolute safety, add ⅛ in., which combined, would make 2¼ in. Then, if we make the problem 1⅛ × 1⅛ × 1¼ we get 3½ which gives the minimum diameter of the hub for maximum diameter of axle box.

There still remains a very uncertain factor—shrinkage, for which we must make full and due allowance under all circumstances. Long experience enables us to place it almost anywhere from $\frac{1}{16}$ in. to ⅛ in. and not be many lines out of the way. To be on the safe side of the fence, we would then make the hub not less than 3⅝ in. in diameter, and, for greater safety, would add ⅛ in. more, making the maximum outer diameter 3¾ in. for the forward wheel. For continued information on this point we would call attention to the article on "Hubs," on page 493, of the October issue.

The Hub, Dec 1889

Hubs

SHELLING OF HUBS.

THE shelling of hubs on the front end, where the band score connects with the beginning of the swell of the hub, is common and very objectionable. We are apt to impute this result to many different causes, but in fact there are but two causes, and both of these are the result of one prime one, namely, the front band of the hub has been improperly fitted where it joins with the swelling portion of the hub, or it is driven on too tight. If the back portion of the band be improperly fitted and forced on too far, it severs the fiber, as shown at C, in Fig. 1 (A, band

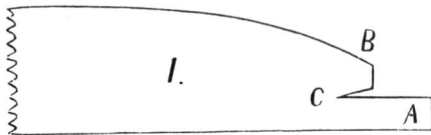

score; B, swell of hub), thus allowing water or the damp atmosphere to enter at that point, which causes a swelling of the wood at that point, which action makes a greater breach of the fiber, and allows of extended shelling up of the hub, from that portion, held secure by the front hub-band.

The second cause is driving the band on too tight, after the hub has been reduced to a shell on its front end, to admit of the rim of the nut. We have noticed cases of this kind where the band score part of the hub has been torn loose from the outer shell, as shown in Fig. 1, and not unfrequently we have noticed that the same bands have split at the back portion, because of their improper fitting. This action produces the same result as that mentioned above.

To overcome the evil, we must fit our bands properly, and also fit the hub to receive the band. Fig. 2 shows a hub as properly fitted to receive the band, D representing the band score, and E its connection with the shell of the hub, sloped back or beveled in such a manner as to admit of counter-sinking the band, which, when on the hub, only serves to bind the hub more closely than before. Fig. 3, section of band, more clearly shows our meaning,

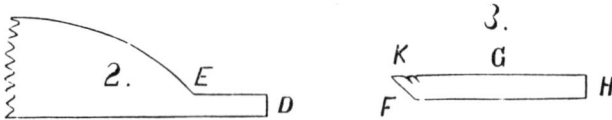

G being the outer diameter of band, H outer or front end, and K the molding on back of band, placed there expressly to allow of the proper amount of counter-sinking, as per the beveled portion F. Upon comparing F, of the band, and E, of the hub together, the reader will at once see that this is the proper and only method by which we may overcome the evil. The score ought always to receive a thin coating of paint or P.W.F. before driving the band on, in order to exclude dampness.

The Hub, Jan 1878

CRACKING OF HUBS.

SAN ANTONIO, TEXAS, January 19, 1880.

TO THE EDITOR: Will you be so kind as to inform me what to do to keep second-growth elm hubs from cracking? We do a great deal of new work, and use second-growth elm hubs for our wheels, and the only trouble we have is about the hubs cracking after a few months' use. If you can give me any information on this matter, I will be very thankful to you.

Your old subscriber,   F. W. LANGE.

ANSWER.—In the absence of data, we would say that the trouble complained of by F. W. L. is often caused by using the hubs too fresh. Hub material of all kinds is seasoned in the rough sticks or blocks; after this, they are turned into the finished shape, and to do this, all the driest part of the timber is removed in the turning. If made into wheels before they have had time to settle or harden, they are nearly certain to crack after slight use. The remedy, in this case, is to carry stock enough to allow a reasonable time for them to harden.

Second: Cracking often ensues from fitting the spokes too tight sidewise. These may be mashed into the hubs without showing any checks at the time, but the action of the atmosphere upon the compressed wood of the spoke tenons, will eventually crack the hubs. In this case the remedy consists in fitting the spokes accurately, so that very little more wood is forced into the hubs than was removed in making the mortises.

Third: If carriages are imperfectly housed or sheltered, the wheels are the first to show the effect of this exposure, and the hubs in particular. We have seen good work placed under a tree, in lieu of a carriage-house. Paint and varnish can not save the hubs where this is the custom.

H. M. DuBois.

The Hub, Feb 1880

Seasoning Elm Hubs.

MR. EDITOR:—Will you please inform me through the columns of the MONTHLY how to properly season elm hubs?   Yours, truly,   T. A. M.

In seasoning hubs, the principal objects to be attained are the shortening of the time of seasoning, and preventing them from splitting. It was a custom, when time was not so valuable as at present, to put the hubs under hay for a whole year, and this produced good results, as they dried very slowly, and did not split, and the quality of the wood grew better rather than worse. From our observation at various wheel factories, the method of drying hubs is much different at the present time. The hub blocks are sawed to the various necessary lengths, and these lengths dipped in soluble resin to keep the air from penetrating too suddenly into the ends of the hub, thus keeping them from splitting. After this they are placed above the floors on racks, lying on top of each other in two rows, and left there until they are supposed to be dry, and turned into shape afterward. We would be pleased to have those possessing better and more practical knowledge of seasoning hubs give their methods in the pages of the MONTHLY.

The Carriage Monthly, Mar 1885

Hubs

WHY DO HUBS "SHELL?"

It is a provoking and annoying fact, that after the most careful attention to details, and the closest inspection of the stock and workmanship, wheels will sometimes be troubled with this "shelling" or "ring cracking" in the end grain of the wood from which the hubs are made. Now, if this were a trouble arising from the use of poor stock, and confined to this class of material exclusively, it could soon be remedied by closer inspection and the use of better stock ; but as it is found to affect all grades of material, both poor and good, it therefore rests with the manufacturer to investigate the *causes* which lie at the root of the trouble, and to intelligently set about seeking a remedy, providing such exists.

There are few carriage-makers who have not had some annoying experience in this line, and many plans have been proposed and tested to overcome the difficulty. One of the foremost promoters of the trouble has been found to consist in having the front and back bands or hoops pounded into the end-grain of the hub, somewhat as shown in our sectional sketch of a hub marked Fig. 1.

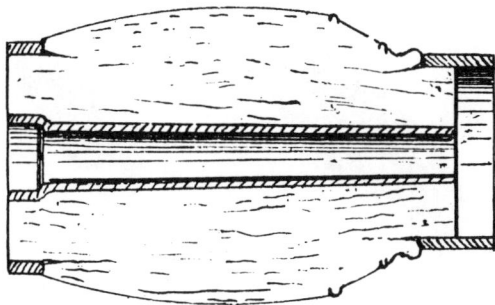

FIG. 1.

This condition is of course liable to cause even the soundest timber to "shell out," for the strain having once separated the fibers, they can never recover, but will spread further, until the wheel is ruined.

This manner of starting or assisting the trouble was soon discovered, and many of our builders of good work, now have the bands of their hubs beveled, as shown in sketch No. 2 ; to suit which, the metal band is correspondingly beveled, thus pressing down, and binding upon the end fibers of the wood. This is a purely mechanical remedy, and, while being a vast improvement, it still does not get down to the real solution of the problem, for hubs protected in the manner shown are yet found to shell in some cases.

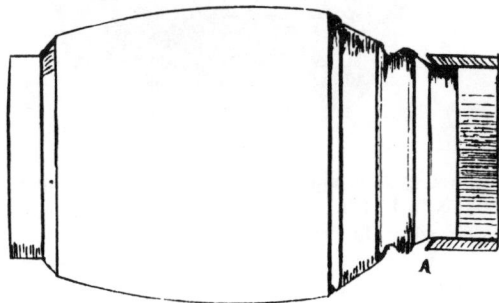

FIG. 2.

We will now examine the question from another standpoint ; and to fully cover it, we must look somewhat at the natural structure of the material from which the hub is made.

It matters little what the wood may be, whether elm, gum, or any other of our hub woods,—all are composed of cells and fibers whose layers are to a certain extent similar in their grouping. The round blocks from which the hubs are manufactured, are as a general rule thoroughly seasoned, and all the "cells" and "ducts" of the timber are flattened and devoid of moisture. This is a necessary condition to prevent the hub from splitting under the operation of driving the spokes. The wood is supposed to have shrunken until there is no more shrinkage to it, and after being made into the finished wheel, it is still kept in the driest state possible ; and after leaving the body of the hub, is painted and varnished, and the bands, or rather the front band, is *driven* upon the end, and, to make a neat fit the wood is often filed and the

band forced *tightly* to its position. In this condition the carriage goes into service ; and then, through the usual changes of temperature and humidity, the flattened "ducts" and "rays" of the wood must absorb more or less of the dampness and become slightly enlarged, and all parts of the wood not rigidly held down by the iron bands must increase somewhat in size. As the layers of fibers are comparatively long, those held down by the bands separate slightly from those more free, and this slight separation of the fibers soon spreads and ends in a complete "shelling" of the entire hub.

You may ask : "How can wood absorb dampness, when well covered with paint and varnish?" Moisture is an enemy very difficult to overcome, and wood is very susceptible to Its influence. The amount of moisture taken up by the wooden platforms on large scales during rainy weather necessitates special adjustment for damp days ; and we can all testify to the swelling and shrinking of doors covered with multitudinous coats of paint, and quite old enough to have abandoned bad habits, if age alone were a sufficient remedy.

Now, in suggesting a practical remedy for this shelling, the writer would state that, in conversation with Mr. Jas. Long, of Long & Silsby, of Albany, N. Y.,—a gentleman well known to the trade for thoroughly practical ideas,—Mr. Long expressed his conviction that hub-bands which were comparatively loose, and which required a "bushing" of canvas or leather to keep them tight, were a sure cure for this trouble of shelling. Some time after the above conversation, the same ideas were expressed to the writer by Mr. Phineas Jones, of Newark, N. J., whose extensive experience as a wheel-maker has given him many opportunities for testing this point. The suggestion, therefore, seems well worthy of the attention of the trade.

This bushing of leather or canvas must be a sure remedy, for it is founded on natural laws that any substance, between the rigid iron and the shrinking and swelling wood, which will compensate even in a slight degree for this difference between wood and metal, will be found effective ; and as the trouble is not an imaginary one, and the remedy thus simple and effective, and indorsed by such authority, I think it well worth discussing as a practical question interesting to every wheelmaker, and of great present interest to the trade.

I earnestly hope that the above statement may result in calling forth further expressions on the subject.

The Hub, June 1883

Boiled Hubs.

MILLVILLE, O., March, 1884.

MR. EDITOR :—Will you please answer the following questions in the April number? What is your opinion of boiling hubs for light wheels so that you can give the spokes more drive in width and thickness? Which will make the best wheel, a 1-inch spoke driven in a dry hub with 1-16 drive in width and the thickness of the tenon to fit the mortise, or the same size spoke with a 1-16 more drive each way in a boiled hub, and will the glue hold as well in the boiled hub? What kind of glue do you think best; white or yellow ?

Yours, truly,   WOOD-WORKER.

We do not believe in boiling the hubs, nor are we in favor of having them too dry. If the hubs are boiled the wheel-maker can give more drive to the spokes, as it makes it softer and more elastic, but after they are driven the hub will dry to its original size and the spokes will cause an overstrain that will result in splitting or checking afterwards and will also crush hub which is one of the principle objections. Our best wheel-makers require the hubs to be perfectly dry before driving the spokes ; just enough drive should be given that the hub can bear in such a state without checking them. To prevent checking the hubs are dipped in hot water for a moment to take the dryness from the outside. Another objectionable feature in boiling hubs is that the glue will not stick as well as by the other method, as the harder the wood, such as gum, elm or hickory, the stiffer the glue has to be, while in a boiled hub in driving the spokes the glue assimulates with the water in the hub, and the consequence is, that what glue is left will be of no use for the purpose intended.

Good glue depends altogether on its quality and adhesiveness. We have seen both white and yellow glue as clear as anyone would wish that was valueless for this work, while others of less clearness was the best that could be obtained. We would advise buying the best glue in the market, boil some in the usual way and glue two pieces of hickory together ; let them stand 24 hours, and then split it at the joint, and you can then observe whether the glue is first-class or not.

The Carriage Monthly, Apr 1884

Hubs

Hub Bands.

Mr. Editor:—What is the proper width to make the front band of a hub? My reasons for wishing to know, are, my employer and myself fail to agree in this one particular and I am in search of authority on this point.

B. And, Ind.

The hub band, that is the front one, never ought to be longer or wider than will cover the axle nut or cap safely, and prevent the same from becoming bruised. Longer than that is superfluous, less than that would cause a general wearing or hammering on the nut or cap, every time the wheels would collide with anything.

Coach-Maker's Int'l. Journal, Apr 1872

Marking Bands.

G. W. H., of Troy, N. Y., says: "Please illustrate how to mark off hub bands so that the holes will be equi-distant." Take a piece of board sufficiently large to lay on your largest band, find the center and draw a number of circles each one-fourth of an inch larger in

diameter than the preceding one, then divide the outer circle into three equal divisions and draw a line from each point—as per our sketch—to the center. Place your band on the board so as to conform to any given circle on the board which best accommodates it, and then you can see how easy it is to mark the band.

G. S.

The Carriage Monthly, Jul 1884

HOW TO MAKE HUB-BANDS.

A Michigan correspondent wishes to be informed how he can make hub-bands, and have them perfectly cylindrical. He writes that he takes the greatest pains possible, and that, after they are thoroughly welded, he swages them with a large flat swage on the beak horn of the anvil, but yet fails in getting them to come out perfectly true.

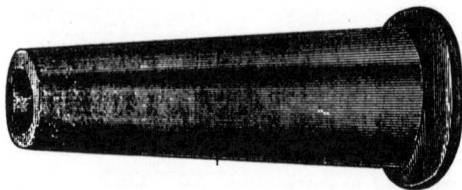

If the writer will examine our advertising pages carefully, he will find in back issues that a number of band mandrils have been illustrated, which are supplied in various sizes and with proper taper to give the band a finish when on the hub. First weld the bands thoroughly; then heat to a moderate red heat, and round up quickly on the mandril, but be careful to remove the band before all the red heat is lost, or you will experience trouble in removing the band from the mandril.

The accompanying sketch shows a plain mandril. Some are fluted, which aids materially in removing the band if it becomes partially cold.

The Hub, May 1886

SMITH-SHOP.

"The smith his iron measures, hammered to the anvil's chime."
—Longfellow.

BEVELING HUB-BANDS.

In a former article on the subject of beveling bands, we promised to illustrate and describe how to bevel hub-bands by means of the ordinary modern drill and hub-borer. As the work properly belongs to the smith-shop, we confine our illustrations to the tools necessary for use with the drill only, from which suggestions will present themselves for fitting up the hub-borer.

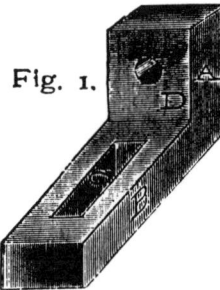

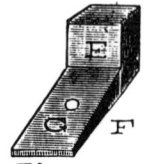

To begin with, we must prepare the clutch or chuck, for which purpose we make three pieces, as per Fig. 1, where A represents the portion coming in contact with the band; B, the portion resting on the surface plate at J J J, (Fig. 3); and C, a slot for the insertion of a set-screw to secure the clutch to the surface plate at J J J (Fig. 3); D is a hole in A, for the insertion of a binding screw to additionally hold the band.

Fig. 2 shows what is known as a "dog," to be secured firmly to the surface plate at L L, L L, L L, and from a slide to hold the clutch, Fig. 1, in position, which in turn holds the band centrally. In this, E rests against the clutch, Fig. 1; and F is the part which is secured to the surface plate by a bolt or rivet at G.

Make the clutch, Fig. 1, of $1\frac{1}{2} \times \frac{5}{8}$ or $1\frac{1}{2} \times \frac{3}{4}$ in. iron or low-grade steel. Make A, 1 in. over all, and B, 4 or 5 in. long, with the slot C, 3 or 4 in. long; and secure by a $\frac{5}{8}$ in. screw, with a $1\frac{1}{4}$ in. square head, by $\frac{3}{4}$ in. thick, and, under the head, a plate $1\frac{1}{2}$ in. square, by $\frac{3}{8}$ in. thick.

Make the "dog," Fig. 2, of $2 \times \frac{1}{2}$ in. iron; E, $\frac{5}{8}$ in. over all, and F, $2\frac{1}{2}$ in. long. Secure by a $\frac{1}{2}$ in. bolt or rivet. Six pieces like Fig. 2 are required.

Make the surface plate 18 in. or more in diameter, and, if of cast-iron [which is best], let it be $\frac{3}{4}$ in. thick. If of wood, make of two pieces of 1 in. hard wood, with the grains crossed so as to prevent warping, and cover the whole with sheet-iron, $\frac{1}{16}$ in. thick. From the center at H, strike a number of circles the same as at K, about $\frac{1}{8}$ in. distant from each other, and have them quite plain. This precaution will give much assistance in placing the band centrally.

We have said, make A 1 in. over all. Where the bands are 1 in., or less than 1 in. high, it will be necessary to block up, so as to raise the edge to be beveled above the part A, so that the cutter may not come in contact with the same.

continued next page

To center the band true, and hold it in proper position, great care must be exercised that the set-screw holes, J J J, are equi-distant from each other on a circle, and equi-distant from the center. The slides LL, LL, LL, must also be so placed, with reference to all the other parts, as to move from the mean center. The set-screws in A must also be the same height from the surface-plate, and strictly parallel with the sides of A, to prevent what might be termed "cocking" or shifting of the band. This brings us to the end of the bottom work.

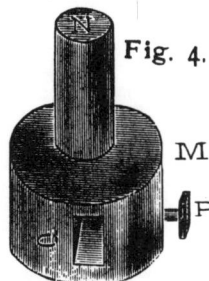

Fig. 4.

Fig. 4 shows the chuck for holding the "tool-post," as iron-turners would term it, N being the portion which is inserted in the drill, or power-shaft of the drill, where it is held by a set-screw P; M, the body of the chuck, having the slot O, for the insertion of the "tool-post," Fig. 5; and P, the set-screw to hold the "tool-post" in the desired position.

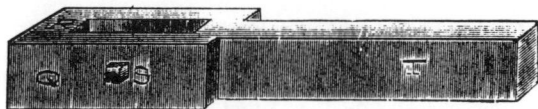

Fig. 5.

Fig. 5 shows the "tool-post," with Q the head, and slot R, for insertion of tools, Figs. 6 and 7, as may be required; S is the set-screw to hold tools secure; and T, the shank, which inserts in slot O, and allows of moving the tools out or in to suit the diameter of the band.

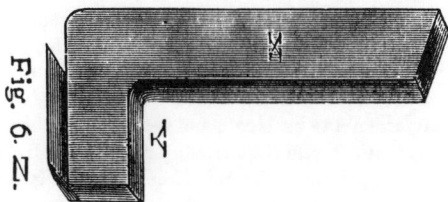

Fig. 6.

Fig. 6 shows a tool for surfacing the band; X being the shank, Y, the turn off, and Z the cutting portion.

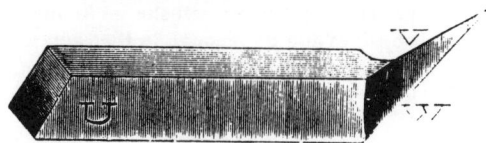

Fig. 7.

Fig. 7 shows the beveling tool, with U, the shank; V, the cutting edge; and W, the back, which is left the full thickness of the tool. The cutting portion, V, must be of the same bevel on the hub. Make these tools of ¾ × ½ in. best cast-steel. Temper the same as you would a drill. Have the cutting edge keen, and sharpen by grinding only. This, we believe, finishes with the tools.

We will now explain about the band turning. First place the band so as to conform with the circle that best suits its diameter, set the clutches up against the band, and secure lightly with the set-screws. Next allow the surface cutter to drop down on the inside of the band, and cause it to move around, which will show you whether your band is centrally placed. If it is not, then, by slacking off one screw, and setting up another, you can soon get it in proper position.

When accurately centered, set the band well down on the surface plate, and set the surface cutter so as to true up the upper portion. When this is done [and be sure and do the front end first] then back up on one screw, remove the band, turn it over, and free off the back end. Next remove the surface cutter, and set the bevel cutter, and bevel the band to suit.

In using both tools, do not attempt to crowd. Proceed with slow speed. Any attempt at rushing will be liable to cause displacement of the band, or injure the tools. After the tools have begun to cut well, the speed of the drill may be accelerated, but feed slow and with great care. Your bands should then prove as good as if done on the best turning lathe. EXPERT.

The Hub, Apr 1886

Marking Bands.

BY UNCLE STEVE.

Without a proper appliance for marking bands for securing holes to get them equi-distant, either with two, three or four holes, would require much lost time and much close measuring.

By the simple devices shown in this chapter the measurements may be readily taken on the spaces accurately marked off.

Fig. 1 takes a board of sufficient size to lay out a circle large enough to suit to the diameter of the largest band you make; then make a series of inner circles ¼ inch between, until the inner circle is small enough to accommodate the smallest band you make. Then if for two holes, make one diametrical line. If for three holes lay out the outer circle in three equal divisions as *A, B, C,* and mark from such points to the center. If for four holes then lay out at *Fig. 2,* with cross diametrical lines *D, E, F, G.* Place the band on the board to suit the circle which it is nearest in size, and mark the band as per diametrical line.

Fig. 3 shows the method of marking bands for plain ragging. Make the circles on the board as per in *Figs. 1* and *2,* and then divide the outer circle as near as possible into spaces of 1 inch,

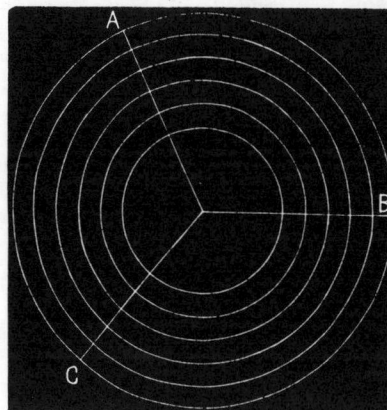

Fig. 1.

and make the diametrical lines as per the sketch. Place the band on the board and mark off as per the diametrical lines. As the band decreases in size, so decrease the cutting point of the

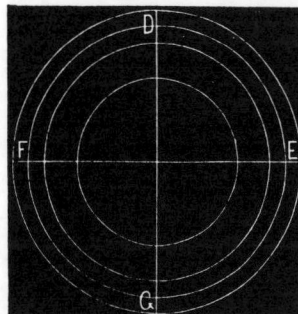

Fig. 2.

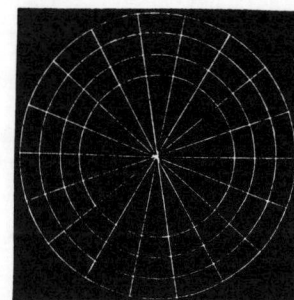

Fig. 3.

rag or rag punch; three or four sizes will be found sufficient for general uses.

It will not take over two hours to make the whole of them, and with them you reach accuracy and save much valuable time.

The Carriage Monthly, Aug 1889

Hub Bands.

BY UNCLE STEVE.

After many chapters we have at last reached the matter of the bands of the hubs.

The tires have been welded, fitted and prepared for heating. If we were to attempt placing the tires on the wheels without first binding the ends or bulge of the hubs, it is more than likely that we would have one or two split hubs in the set. It is not safe to trust entirely to the temporary bands which are placed on the hubs for the wheeler to allow him to drive his spokes with safety. If the same

Fig. 1.

are strong and tightly fitted, they would in ordinary cases answer, but there are so many things against them, it is well to put on at least the permanent back bands, and fit well the temporary front bands. When the front bands are flush with the front face of the hub, it is just as well or better to put on the permanent front band. More or less water or dampness reaches the hole in the hub and causes the hub to swell, which causes too great a strain on the temporary band. When the front band projects beyond the front face of the band, it is in the way of the wheel-boxer, and then, again, when the box is forced in the hub the band would cause the hub to swell, which means, in plain words, the destruction of the hub.

Many smiths in fitting temporary bands, make the same of flat or band iron, which I have ever found a poor plan. They are soon beveled by drawing to fit, and in forcing them on the hub, destroys the flat surface of the same, making it a difficult job to fit and secure the permanent bands thoroughly. Then again they are apt to mar and cut up the outer end of the score, which leaves unpleasant gaps which show. prominently when the permanent band is driven on the hub. The best iron to use for temporary bands is a thin oval iron. It allows of easy fitting and easy forcing on the score. The pressure is chiefly at the central part of the score, and should the hub swell and shrink a little, the score has a nearly full bearing space the whole distance.

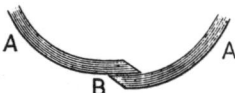

Fig. 2.   **Fig. 6.**   **Fig. 3.**

It is customary with some smiths to make the bulge bands or flat iron. While this practice would answer very well on the hubs of large merchandizing vehicles, it would not harmonize with vehicles of a lighter order, no matter whether for pleasure or light business wagons. Wherever it has been necessary for me to put on bulge bands, I have always used half-oval or half-round iron, as per *Fig. 6*, *A* the flat part fitting on the hub, *B* the half-round or half-oval surface, *C* shows where the sharp edge has been removed from the band before welding, which allows the band to hug close to the spokes without cutting them.

The best sized iron to use for hub bands for ordinary vehicles is for back bands, No. 9 iron, and front bands, No. 8 iron. If the bands are to be turned off in a lathe, then use for front bands No. 6 iron, and No. 8 for the back bands. It is an ordinary easy trick to weld an ordinary band, say from ½ inch wide up to 1½ inches wide, but when we get beyond that size, about one-half of the average smiths will find when done that his band has become bell shaped, nor can he tell what causes it or how to cure it.

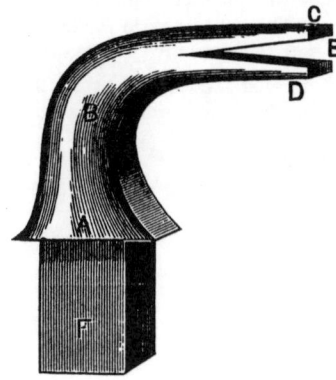

Fig. 4.   **Fig. 5.**

To prevent it, we must begin right. *Fig. 1* presents the band prepared for welding, *A* shows the flat surface with two holes for securing screws, *B,B* are the ends prepared for welding. It will be noticed that the ends are not ragged and full of spaces for the collection of and holding impure matter, slag and other anti-welding agents.

After the band is cut off and prepared for welding, as per *Fig. 1*, heat the band the whole length and place it over a large swage or a wood block hollowed for the purpose and with a ball pene hammer or large fuller, concave the band slightly as shown in *Fig. 2. A* end of section of band, *B* concaved surface as per explanation above. When the band is concaved, heat to a good red heat and bend over the horn and avoid short kinks. Make the lap and fit nicely as at *B*, *Fig. 3; A,A*, section of the band, heat again to a red heat and with an old file remove all the slag or scale and apply the flux. [I would mention right here, to upset ends of band when preparing for welding, to allow for loss in working and oxiding.] Raise a good mellow heat and weld in one heat, which you can do if your heat is clear and clean, and you were not born tired. After the weld is made, clean off the slag, worn flux and scale and swage on the horn of the anvil at the weld to smooth it, and then true up on the mandrel and draw (if necessary) to its proper size. By this time the band is ready to place on the hub; it will, on the sides, be nearly at right angle with the face, which latter feature is absolutely necessary to make a perfect good fitting band, and one which will harmonize with the hub.

The cause of the bands becoming bell-shaped is due to too much manipulation at the edges and not enough in the center. When the band starts to bell do not attempt to help it out by drawing the center. Your efforts will only produce an increased width of metal

Fig. 7.

continued next page

without improving the shape. To bring it back, do it either by shrinking or by upsetting with swages—by swaging about one-third on each edge, leaving the center to take care of itself. Some people will tell you it is impossible to shrink iron by heating and cooling. Tell them to go to pot or Hackensack. If your iron forms a welded circle, you can shrink it, as per *Fig. 9*, showing a band. *A* one end; *B* another; *C* the cylinder. Heat the whole band; then put *E* in water two-thirds up to the upper end, *F*, as per dotted line *D*, and let it remain until *F* begins to show a loss of heat. The trick is this: the part you cool, when cool, only goes back to its normal cold condition, but the heated part is forced to shrink with it in its hot state, and necessarily becomes smaller. Next turn the band over and heat *K* and cool it, and give *I* a chance to draw in its share.

In a former chapter we made mention and gave illustration of a tool or pattern for making holes in bands and have them uniform.

Fig. 8.

Fig. 5 shows a method of preparing a band to prevent shelling of the hubs. *B* flat section of band; *A* the edge; *C* the bevel edge. Where lathes are in use, this band is a good one; without a lathe to form the bevel, they are too costly. *Fig. 4* is a handy tool for bending bands. *F* shank to fit an anvil; *A* shoulder; *B* neck; *C* upper jaw; *D* lower jaw, with rounded upper face; *E* recess in which to insert band for heading.

Band mandrels are a handy tool, and well earn their cost. *Fig. 7* shows a large one, with pedestal, *B*. *A* working part; *C* the end; *C* shows the cylinder; make 3 feet high and taper ½ inch to the foot. This can be rolled around the floor and used anywhere. *Fig. 8* shows a hand mandrel for bands. *E* body; *D* the end; *A* the band piece; *B* swell of same; *C* where *A* connects with *E*. *A* is round wrought-iron, about 1 inch diameter, and is inserted in mold, and the cast iron poured around it; taper ½ inch to the foot. Both mandrels ought to be turned off true in the lathe.

I have no love for the butcher who drives nails in the hub to secure the bands. Screws are much better, and more readily removed. Do not burn the score with hot bands. Make the band a tight fit and heat hot enough to not burn, but to drive on snugly. Keep the water off; more wheels are destroyed by water at the beginning than is of benefit to the trade. Now that the bands are ready and on the hubs,

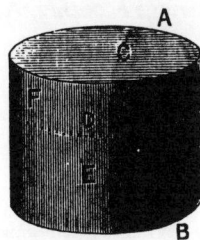

Fig. 9.

if you have been real smart, the tires are also ready to go on, and the wheel is ready for the wheel-boxer.

I do not favor the punching of the holes in the bands for nice work. Wait with the front ones until the same are done and drill them. The back bands may be drilled before welding.

See the chapter on tongs for tongs with which to hold bands while manipulating them. After the band is welded, all the slag and dirt ought to be--while hot--removed with a file, or it will be found hard work to file them.

The Carriage Monthly, Dec 1889

To Find Circumference of Bands.—I noticed in December's number of THE AMERICAN BLACKSMITH an article or a rule for finding the circumference of a band or circle. It is a very good rule, but I think I have a better one to give my brother smiths. It

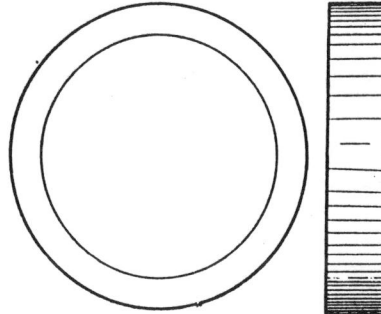

TO FIND CIRCUMFERENCE OF BANDS.

is one that is used by all machine forgers. I have used it for years and find it "Oll Korrect". For example, you get a band to make as shown in the engraving. Add one thickness to the inside diameter and multiply by 3.1416. 8 inches +1 inch =9 inches x 3.1416= 28.2744 or about 28¼ inches, and you allow for your weld, about one-half the thickness of the iron. ISADORE H. DUBE.

The American Blacksmith, Feb 1906

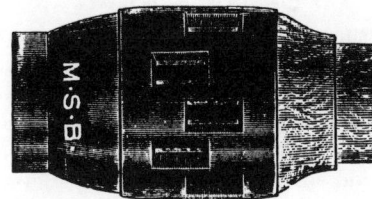

M

M S B

B

OR,

THE MARIS SHELL BAND FOR HUBS,

Wheel-Horse and Mallet.

SPOKE SHAVE, EDITOR.

Having followed our wheels through the hooper's hands, they are again brought to the wheeler for boxing. Much depends upon the same being properly done. Care should be taken not to damage the hub in wedging. Get the centre of hub on the back, and cut out for the box about one-sixteenth less than the back end of box. This may not always hold good, for the reason that some hubs are very hard—we must use our judgment. Our object is to get our box in tight enough to avoid the necessity for wedging on the back. We should leave at least one-sixteenth of an inch, or, in other words, cut out the hub one-eighth inch larger than the body of box to prevent the tar of axle coming in contact with the end of tenon, which is very damaging to the wheel. When ready to insert the box, we should cover every portion of the hub on inside with white lead, also lead the box; this will, in a measure, prevent the oil penetrating the hub; it also prevents the box from leaking, as many have sand holes in them. The lead on the box forces to the back, and that in the hub to the front, closing up every cavity, and preventing oil or grease from entering either back or front.

Coach-makers' Int'l. Journal, May 1869

DOLE'S HUB-BOXING MACHINE,
WITH THE SILVER IMPROVEMENTS.

The annexed engraving illustrates a most useful machine, one which has attained a popularity almost incredible. Ten thousand of them have been sold, and wherever put into service, they have recommended themselves by the precision and speed with which they performed their work. Our cut shows "Dole's Old Standard" Hub-Boxing Machine, with SILVER's Patent Open or Adjustable Feed Nut and Gauge Plate attached. The complete machine, and the parts drawn separately, as seen at *Figs.* 2, 3 and 4, are so well defined that a description

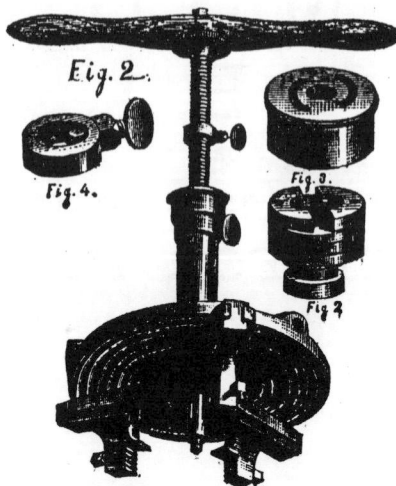

of them is hardly necessary. It will be remembered that the "Old Standard Machine" was considered about as good an one as could be desired. But as one step forward in improvement prepares the way for still further advancement, it is not surprising that the "Old Standard Boxing Machine" should come in for a share of change. The Silver Patent Feed Nut and Gauge Plate formerly applied only to Silver's Hub-Boxing Machine, having met with such universal favor, Messrs. Silver & Deming were induced to apply them to the "Old Standard" also, and they are now making a large proportion of the Dole Machines with the feed nut. This nut is made in two sections, and is so constructed that the two halves are separated from or closed on the mandrel by simply turning the cap to the right or left, as the case may be. By this means the mandrel can be raised or lowered without the trouble or delay of turning it through the nut, thus effecting a great saving of time. The gauge plate can also be raised or lowered on the mandrel without turning.

The manufacturers sell the machines as improved, or where parties have the old machines, and wish the feed nut attached, they can be accommodated by sending them to their shop.

We append the prices, which will be convenient for reference:

Price of Machine with Improved Nut or Gauge Plate.—No. 1, or small size, $23; No. 2, or medium size, $26; No. 3, or large size, $30.

For Attaching the Feed Nut to Old Machines.—No. 1, $5; No. 2, $5; No. 3, $6.

Address all letters to SILVER & DEMING, Salem, Columbiana Co., O.

The Carriage Monthly, Apr 1871

Setting Boxes.
BOONE, IOWA, July, 1874.

MR. EDITOR:—Will you please explain how a box should be set in a buggy wheel? For instance, take a hub 7 inches long and a box 6 inches, would it be proper to set it flush with one end, and if so which end, back or front? By answering the above through the MONTHLY you will confer a favor on a reader of the MONTHLY.   E. J. L.

ANSWER.—In all ordinary carriages the boxes should be let in to within 1/4 of an inch of front of hub, and cupped out on both ends for nut and collar. The hub should not be over 6 3/4 inches long for 6-inch box, allowing 1/4 inch cup front for rim of nut, and 1/2 inch back for collar; but where the hub is 7 inches, the collar must be let in 3/4 inch.

The Carriage Monthly, Aug 1874

Hubs

Oil in Hubs.

If there is any one thing which is liable to reflect discredit upon a carriage builder, it is, after the vehicle has been in use a few months, to have the spokes loosening in the hub in consequence of the oil, with which the axles are lubricated, entering the hub to the mortise, and thereby dissolving the glue at the mortises, and causing the spokes to become loosened. This serious matter is attributable to, in chief, two causes: First the imperfect setting of the axle box, or letting the same into the hub too loosely. Another great fault is the construction of the axle box and axle, the back end of the box being made to fit into a chamber or groove in the "collar washer," and the front end of the box being fitted so as to fit into a groove or chamber in the nut. The axle we refer to is the one invented by the late Mr. Saunders, of Hastings upon the Hudson, which has been for many years considered the best axle in market. In each case, the hub and box are of the same length, which necessitates the formation of a channel in the hub at either end around the box, which same channel becomes a conduit for the oil, which becomes limpid while the vehicle is being propelled at the ordinary rate of speed. The same feature occurs in the employment of the mail patent axle. The leathern washer, and the iron washer or moon plate are let into the hub, in order that there may be no projection at the back of the hub. This space is left in the same condition in which it is cut. There is not a particle of paint applied to protect the end grain of the hub from the oil; the consequence is all the oil coming in contact with the unprotected fibers is readily absorbed, and when once absorbed continues its searching operations until it has destroyed all our handicraft.

The allowing of the back end of the box to run in a chamber in the collar washer is an old feature, and is copied from the Collinge patent axle, Mr. Saunders being the first to apply the same to buggy axles. Although an old feature, and a deteriorating one as applied to buggy axles, it is not without its good effects when properly applied. To apply properly, the box should be made long enough to protect beyond the back portion of the hub sufficiently to allow of its insertion in the collar chamber, and the formation of a groove between the hub and collar chamber, into which the spent oil may locate or lodge, and be thereby prevented from entering the hub. The box ought to be turned off to a true circle on its outer surface, so as to allow of a precise boring of the hub. The box ought to be coated with some quick drying mineral paint, and then pressed into position. These precautions will prevent the oil from entering the hub, and will tend to enhancing our reputation as carriage builders.

The Carriage Monthly, Nov 1874

SECURING AXLE-BOXES WITH SULPHUR.

ONE of our Iowa subscribers has addressed us the following inquiry:

" Some fasten the axle-boxes in the hubs of their wheels with sulphur in lieu of wedges, by first melting the sulphur and then pouring it around the box. Will this make durable work ? or can it be depended on ?"

This is the first time in many years that the matter of securing axle-boxes with sulphur has been presented to us, and we had hoped that this fallacious theory had become obsolete.

One of the great aims of all iron-workers in either cast or wrought-iron has been to discover the best and surest method of removing sulphur and its traces. The purer the iron, either cast or wrought, the less sulphur it contains.

When leather boxes were employed for wagon-axles—now many years out of date—sulphur was employed to harden the leather, and also to retain the box in its position in the hub. The leather box was first cut and then sewed together upon the axle-arm; it was made to revolve upon the arm a few times each hour, for a day, or until it had conformed exactly to the shape of the axle. It was then placed in position in the hub, with the axle-arm inserted, after which the melted sulphur was poured into the space between the box and inner portion of the hub, the sulphur having in this instance a twofold purpose—first, to harden the leather, and, second, to secure the box in position. This is, no doubt, what has given rise to securing axle-boxes with sulphur.

Progress has placed leather boxes, and the above mode of securing them, several decades behind us. The securing of axle-boxes with sulphur has a deleterious effect upon the axle-box; for, by successive heatings, more or less of the penetrating acid impregnates the axle-box, which at a later date is imparted to the axle-arm, and when iron becomes impregnated with sulphur, it disintegrates.

Again, we are loth to believe that any subscriber of The Hub is so unmechanical or so ignorant, in this day of labor-saving machines, more particularly *hub-boring machines,* as to so cut out hubs for the insertion of the axle-box as to require the aid of any substance to secure the box in position.

The proper way to box wheels of any class of vehicles, is to do it with some one of the many improved *hub-borers*—any of them is cheap enough, and there is not one but what will pay for itself, in boring six sets of wheels. By this means, the hub can be bored to the exact shape of the outer circumference of the box, requiring the use of the triangular chisel or gouge only, to remove the wood or to make a channel for the combs or flange of the box. Then, by making the receptacle of the box one-hundredth part of an inch smaller than the outer portion of the box, we can, by means of a screw-press or heavy hammer, force the box into position in such a manner as to require no other fastening. It is better, however, to coat the outer portion of the box with some preparation that is non-absorbent of animal oils, which otherwise might permeate through the box and enter the mortices in the hub, and thereby cause a loosening of the spokes. For this purpose, we would advise boiled linseed-oil and ochre, which acts as a lubricator in pressing the box into the hub, and is also a non-conductor of animal oil.

Leading carriage-builders of this city and vicinity all observe the above rules. They no longer cut out the hub with a gouge, or surcharge the box with sulphur; neither do they use wedges, and yet the boxes remain in position without becoming loose, even until the vehicle is worn out.

The Hub, Nov 1874

Hubs

HOW TO BOX WHEELS WITH RUBBER-CUSHIONED AXLES.

IN order to explain clearly, to a correspondent, the latest and best method of boxing wheels with the rubber-cushions, we recently visited a carriage-shop in this city where rubber cushions are now the rule rather than the exception, and present below the facts there gathered.

The preliminaries, such as centering the hub and adjusting the wheel to the hub-borer, were similar to those employed in preparing the ordinary box. In this instance a Kritch machine was used, and the wheel-maker set the cutter so as to remove sufficient wood from the center of the hub to make the diameter of the hole just the size of the inner diameter of the retaining sleeve, explaining that this was necessary to permit slight oscillation of the box, one of the chief characteristics of this invention. He removed just enough to prevent any portion of the wood of the hub from coming in contact with the metal of the box proper.

He next took the exact depth of the back cushion and retaining sleeve combined, and also the mean outer diameter of the cushion, and bored the back part of the hub to exactly suit the dimensions just obtained. He then fitted the retaining sleeve by making a smaller bore, at the bottom of the large bore, of the same diameter and depth as the flange part on the retaining sleeve.

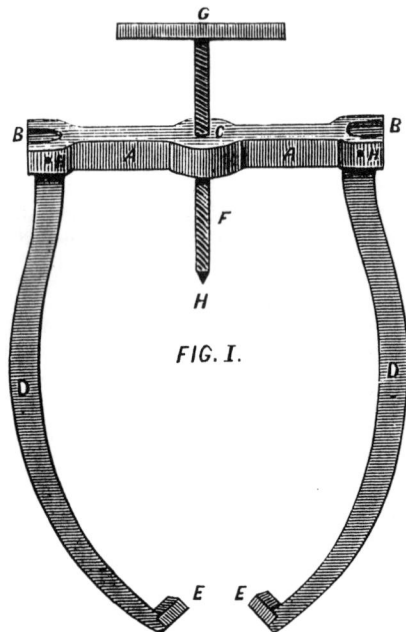

FIG. I.

The next process was ascertaining the mean depth and exact outer diameter of the front cushion, after which he bored the front of the hub so that the front cushion just fitted into the space bored, so that when the box is put in its place permanently in the hub, the compression nut will screw down at least ¼ inch below the end of the box. The wheel was then removed from the Kritch borer, and the retaining sleeve placed in the back bore, when the position of the fins of the retaining sleeve and the raised slot for the insertion of the spur of the box were properly marked. The sleeve was then removed, and with a proper chisel small grooves were formed for the tight fitting of the fins and raised slot of the retaining sleeve, the space for the slot being about $\frac{3}{16}$ or ¼ inch longer than the spur on the box, to allow for lateral movement of said spur, and after this had been done, a thin coating of paint, as a lubricator, was applied to the outer portion of the back cushion and the back bore of the hub, for the easier and better forcing in of the cushion. The retaining sleeve was then set in the hub and forced into position tightly, the back cushion placed upon the box, and the box carefully inserted in the hub, care being taken that the spur on the box was properly inserted in the raised slot of the retaining sleeve.

The front cushion and the front bore of the hub were next painted; the front cushion inserted in the front bore, encircling the front end of the box, and the compress nut placed in position on the box, and screwed down until a proper tension had been reached, when a proper fitting round block of iron, as shown by Fig. 2 (see subsequent explanation), was inserted in the back part of the box. The clamp (see Fig. 1) was then applied, the hooks E E taking the front of the hub, and the screw F at its rounded point, H, inserted in the recess shown by B in Fig. 2, and a gentle strain was applied until the box was in position, when the wrench (see Fig. 7) was applied to the compress nut until the box and cushions were tightly packed in the hub, or the nut screwed down at least ⅛ inch below the end of the box, as above mentioned. We supposed this to be the end of the process, but such was not the case. A thin, sharp knife was next used to remove the rubber protruding from the back and from the compress nut in front—the protrusion of rubber at the ends being evidence that the holes are completely filled. On testing the box it was shown that there was no necessity for further truing up the box, as it was now seated firm and true. The time required for the operations above described was about the same as with the ordinary box.

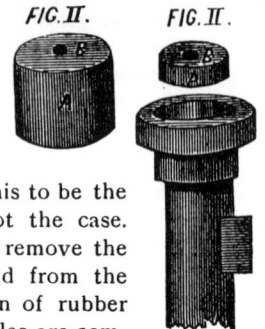

FIG. II.   FIG. II.

The wheeler strongly recommended that the hub-bands proper should under no circumstances be set until after the boxing process is completed, but that close fitting temporary bands should be used while boxing. He also laid stress on this point: In cases where cushions are to be applied to wheels of old vehicles, of which the hubs are of insufficient size at their ends to receive the cushions properly, the cushions may be put upon an arbor and placed in a lathe, while with a sharp cutting tool the outer diameter is slightly reduced, or the same may be done with a sharp, keen cutting file. But in no case should the cushion be reduced in size any more than ⅛ the thickness of its outer edge.

The accompanying woodcuts, illustrating the foregoing remarks, may be explained as follows: Fig. 1 shows the clamp used in boxing, wherein A represents the cross head, with slotted ends,—B B, for receiving and retaining the arms,—D D, where they are secured by bolts H H; C, in the same cut, shows a round enlargement in the center for the insertion of the screw F, of which G is the handle. The point H is rounded and hardened to allow of easy turning on the block.

E E are hooks for taking front of hub.

Fig. 2 represents a round block of metal, A showing its outer surface; B, a shallow recess for the insertion of the end of screw F, at H, Fig. 1. This block must be of sufficient caliber to nearly fill up the outer recess of the back portion of the box, so as to avoid injuring the box at that section.

continued next page

Hubs

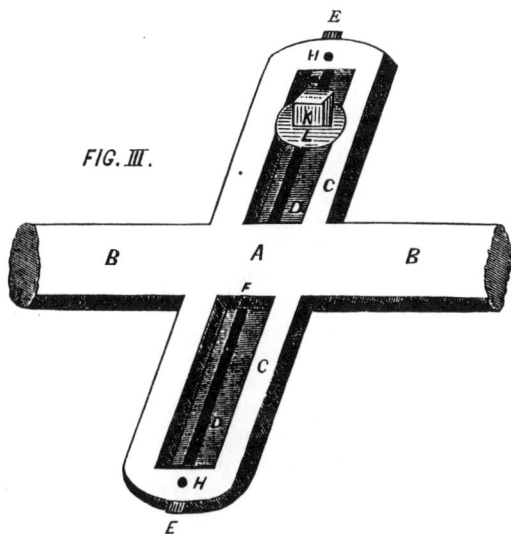

FIG. III.

Fig. 3 is a sectional view of an improved adjustable wrench used in boxing for rubber-cushioned axles, which is adapted to suit any size of compress nut. A is its mean center of body; B B are sections of the handle; C C are projections for securing the movements; D D are slots for the admission of the movable pins; E is the head of the bolt A, shown in Fig. 6, which is made of steel, and threaded so as to fit the threads in hole C of Fig. 5, and pointed as at E, Fig. 6, to fit in the recess at F, Fig. 3; it is held in position by a toggle-pin passing through at H, Fig. 3, and fitting into the recess C, Fig. 6, which holds the bolt in position and allows of the moving of the pin, Fig. 5, so as to adjust the wrench to suit any size of nut; K (Fig. 3) shows the nut applied to end A, Fig. 5, and resting on the washer L, Fig. 3, to prevent wabbling of the pin.

Fig. 4 shows more in detail the bolt described in Fig. 3.

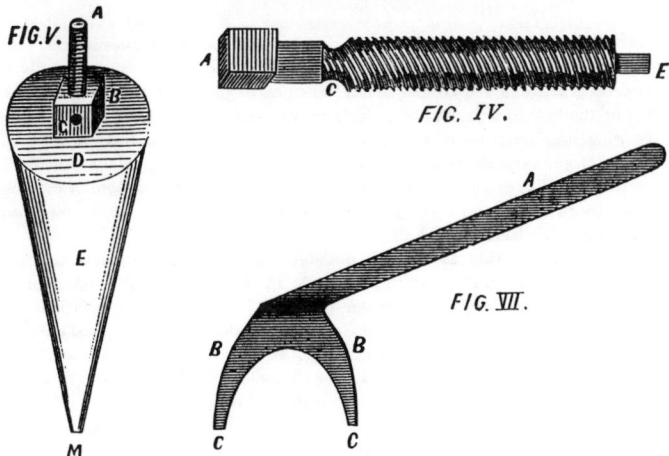

FIG. V.

FIG. IV.

FIG. VII.

Fig. 5 shows one of the two pins forming the wrench; A being the end for nut; B, the square bearing and seat of threads for shifting bolt at C; D, the upper base of the pin; and E, the rounded and working part of the pin, the point of which, M, enters hole in the compression nut.

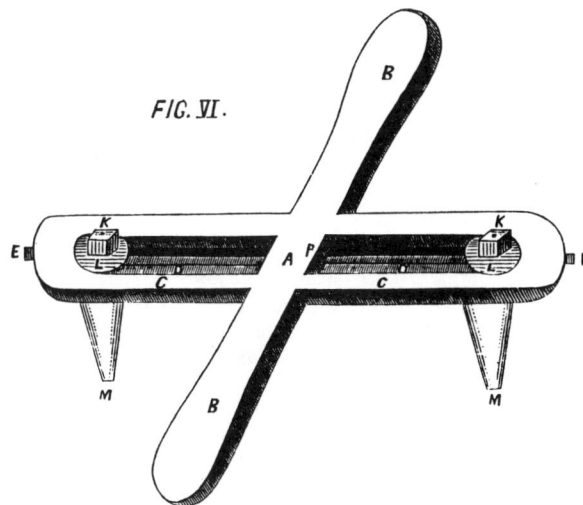

FIG. VI.

Fig. 6 is a simple outline showing the completed tool; A being the body of the tool; B B, sections of the handles; D D, slots; M M, pins or points for inserting in the holes of compress nuts. Make parts shown by Figs. 4 and 5 of good steel. The adjustable wrench above described was the invention of Mr. J. L. H. Mosier, of New-York.

Fig. 7 shows an ordinary wrench used by the majority of builders; A being the handles; B B, the arm, from which are turned and formed the points C C, entering the holes in the compress nut. This wrench ought to be made from a fair quality of steel, and the only objection to it is, that a different sized wrench will be required for each different size of compress nut, while Mr. Mosier's is adjustable to all sizes.

Beside lubricating the cushions as above referred to, it should be observed that the outside of the compression nut should also be lubricated with common brown or yellow soap, moistened, in order to prevent the rubber adhering to said nut, and thereby entirely preventing any cutting or mutilation of the cushion, while screwing the nut into its place.

The Hub, May 1878

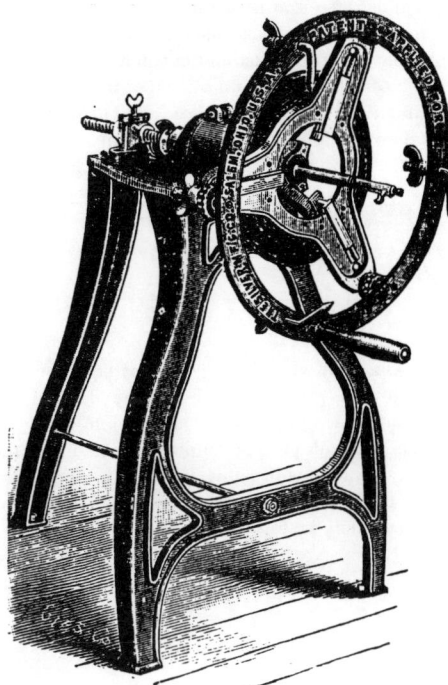

HOW TO BOX CUSHIONED AXLES.

CUSHIONED axles have now come into such general use, that any information that can be given as to the method of applying them, which is of necessity different in some respects than that usual with the ordinary axles, is worthy of a place in the trade journals. We therefore give below, in detail, the precise manner of setting the boxes and cushions, now employed by those having the longest experience in their use.

First.—Measure the length of box and hub, to ascertain if the one is suitable for the other. Then measure the *inner* diameter of the retaining sleeve, which will be the diameter of the *inner* bore of the hub; then measure the central diameter of the hub, to ascertain, if after the center bore is made, the length of spoke tenon will be of sufficient length; if not, then a larger hub is necessary. Now prepare the wheel for boxing, by placing a temporary band on each end of the hub, not so tightly as to compress the wood.

Second.—Center the hub, and bore through it a straight hole, the size of the inner diameter of the slotted retaining sleeve, which fits over the spur upon the box. This will leave sufficient space, as the cushions yield to pressure, for the box to oscillate within them without touching the hub.

Third.—Make the hole at the butt of the hub the *combined depth* of the *back* cushion and of the *ring* portion of the slotted retaining sleeve. The diameter of this hole must be the exact diameter (never larger) of the cushion when off the box; so that when the cushion, fitted in its place on the box, and slightly swelled thereby, is inserted in the hub, it will fit the hole very tightly. It can be easily ascertained whether the bore is of the proper diameter, by trying the cushion after the cutter has made its first mark. Then make a smaller bore at bottom of the large bore, to receive the smaller or flange part of the retaining sleeve, the bore to be the exact depth of the flange; and cut two small grooves in the smaller bore, into which the two small fins on the sleeve are to be forced and to fit tightly. Also, cut another groove to admit the spur on the box, which groove should be one-eighth to one-quarter inch deeper than the length of the spur. The object of this is to allow for a lateral motion of the box. The slotted retaining sleeve, in connection with the spur, prevents the box turning in the hub, and must be tightly fitted in its place, so that the space between it and the end of the hub will be equal to the length of the back cushion.

Fourth.—The smaller or front cushion. Allow space to cut in for the axle-nut, then bore a hole in front of hub, the depth of the exact length of the cushion. The diameter of the front hole of hub, for the front cushion, must be the exact diameter of the cushion when off the box, for the same reason as stated for the back cushion. The front cushion can be readily forced into its position in the hub, by using a piece of wood formed or shaped at its end to fit the end of the cushion.

Fifth.—After the bores and grooves are all made, take first the slotted retaining sleeve and place it in its proper position at the bottom of the back bore; then see that the back cushion is in its true position on the butt of the box, and insert the box and cushion in the hub, being careful to guide the box so that its spur will enter the slot in the retaining sleeve, at the same time forcing both box and cushion *inwardly* with a clamp as far as they will go; then force the front cushion in its place, and screw home the compression-nut, with a wrench made for the pur-

pose (see Fig. 1) so that it will clear the end of box not less than one-eighth inch, which will give the cushions the desired position, and prevent the axle-nut bearing upon the compression-nut.

The practice has been adopted of painting the inside of the holes and the outside of the cushions with paint, to act as a lubricator (white-lead is too thick), just previous to inserting them in the hub, otherwise the cushions can not so easily be forced into position.

Special attention is called to the necessity for using a clamp (Fig. 2), by which, when fixed on the butt end of the box and front of hub, the box can be properly forced into its true position, causing both butt of box and back end of cushion to be even with the face of the hub. Should any rubber protrude beyond the butt of box and hub, by reason of being tightly forced by the clamping, it is further evidence that the cushion occupies all the space allotted for it, and that care has been exercised in making the hole not too large for the cushion. The part protruding should be trimmed off smoothly and neatly.

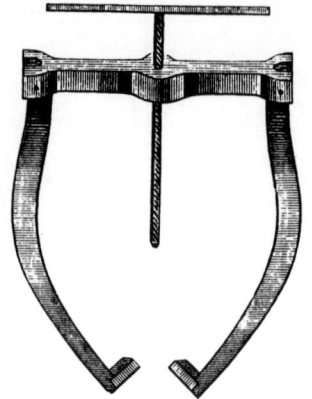

After fitting these boxes once, it will be found as easily done as in the case of other and ordinary boxes, with the advantage of being decidedly better fitted.

To remove the box from the hub when required, it may be readily done by forcing the box inwardly with the clamp placed on the butt, thus freeing the compression-nut from the front cushion, allowing it to be unscrewed; then changing the clamp to front of hub, it can be easily forced out.

When setting the boxes and cushions, the outside of the compression-nut should be lubricated before screwing it into its place within the front cushion. Made as it should be (to fit tightly), this nut will sometimes adhere so tenaciously to the rubber, while being screwed in, that at times it will cut and mutilate the cushion, unless adhesion is prevented by using, as a lubricator, common brown or yellow soap, slightly moistened, or its equivalent.

A tight fit is absolutely necessary. This not only gives the requisite elasticity, but also renders the joints absolutely oil-proof!

The Hub, Dec 1879

BUSHING WHEELS.

——, ONT., Dec. 1880.

To THE EDITOR: Would you kindly reply to the following question in your next issue of *The Hub*, to the signature of Y. Z., and much oblige me: Is it the general practice in the United States, when carriage wheels are being bushed, to cut out the hole in front of hub exactly the diameter of axle bush point, so that no wedging may be required there, while the hole in back of hub is cut out $\frac{1}{4}$ in. all around larger than the bush, this being $\frac{1}{2}$ in. larger in diameter than the back end of bush? This extra cutting out is said to be done for the purpose of making room for the wedging and trueing of the bush in the wheel, and so confining *all* the wedging to the back of the hub.

My own opinion is, that as little as possible ought to be taken out of the hub to place the axle bush there, and that the hole in the back of hub ought not to be more than $\frac{1}{8}$ part larger in diameter than the bush, this being sufficient for wedging and trueing up of the bush in wheel, and this only when bushed with gouges by hand; but if done by a machine, less than $\frac{1}{8}$ part should serve the purpose. The front I would cut out to the diameter of the bush, but I would have no objection to putting a wedge or two there if required. Y. Z.

ANSWER.—With the present appliances for boxing wheels, the holes are made to conform as nearly as possible to the size of the box, excepting where the ends of the spokes are removed to prevent resting upon it. The work can be done so closely that but few wedges are required to true the wheel, and those are driven at the back. The majority of carriage-builders think, as you do, that as little wood as possible should be removed from the hub in boxing. Hand work with gouges requires more room in trueing than machine work. H. M. DuBois.

The Hub, Feb 1881

Hubs

Boxing Wheels.

Lancaster, Pa., March, 1883.

Mr. Editor:—Will you please answer in your next issue the following question in regard to boxing wheels? I often have boxes get out of center ¼ inch in the back of hub when they are wedged, and yet are in the center in front when true on the rims. It usually occurs on light wheels, and is very noticeable. Yours, truly, SUBSCRIBER.

There are several reasons why the boxes are often out of center of the hub to more or less extent. If the fault lies in the boxing machine, as is sometimes the case, the box must be wedged toward that side of the hub, in order to have the wheel true, although we will say here that wedging boxes is a bad practice, and should not be encouraged in the shop.

If the wheel is properly made in the wheel shop, and then it is found that the boxes have to be set out of center of the hubs, the trouble will usually be found to lie with the blacksmith; the tires are not heated equally around its circumference, and then, when they placed on the wheels, are cooled off too suddenly. The part cooled off first will throw more dish into the wheel than that part cooled off last. You will notice that when a wheel is twisted, the lighter the wheel and the tighter the tire, the more noticeable the twist will be. It is sometimes almost impossible to pound the rim level with the tire, it slipping from one side to the other. A wheel hooped under these conditions will have the hub out of center, and it can be remedied by the blacksmith's heating the tires uniformly and cooling them off gradually. Nothing is more annoying to the wheel-maker than to find the hub to be out of center, either front or back, of the hub, or that the rim touches in several places.

The Carriage Monthly, Apr 1883

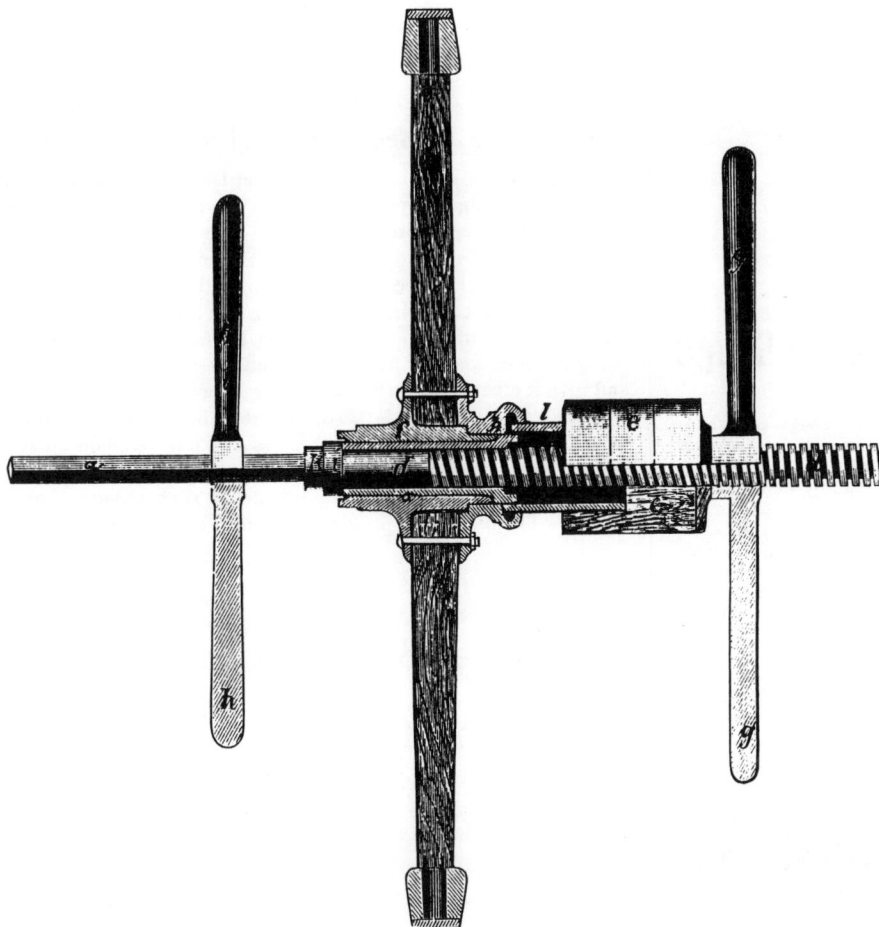

Hub Boxing Apparatus.

The accompanying illustration was taken from *Der Chaisen und Wagenbau*, the only journal published in the interests of the carriage trade in Germany, by George Meitinger, of Munich.

In the majority of the shops the practice is to pound the boxes in with a hammer, that is, a piece of wood is usually placed against the box and struck with a hammer, and if this does not work satisfactory an iron fitting on the surface of the box is pounded upon, almost invariably injuring the box, this being more especially so when the boxes are driven outward. Sometimes the boxes have been so set that they will not move at all, the box is broken to pieces, and every carriage-maker knows the trouble of replacing it.

This device is intended to draw the boxes out by the aid of a lever, moving round a square cut thread, and is worked as follows: The collar i is put on the spindle d against the shoulder of the spindle, and then the spindle is put into the box as illustrated on k, i and d; then band e is put over the spindle d, against the back end of hub h; e, e is put over the spindle d, and band l, the lever nut, is forced against e, e, and the box drawn out. The lever h, h prevents the turning of the spindle from the application of force by holding it. In *Fig. 1* we show the different sizes of bands used for the various sizes of boxes, and in *Fig. 2* the various sizes of collars.

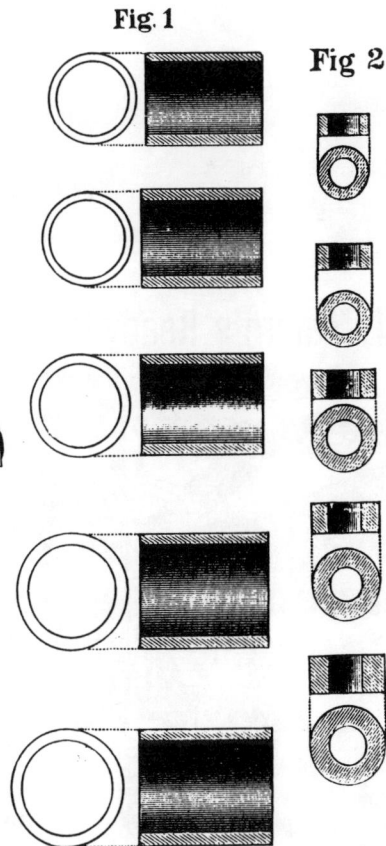

Fig. 1

Fig 2

HUB BOXING APPARATUS.

The Carriage Monthly, June 1884

Hubs

Cement for Axle Boxes.

LANCASTER, PA., July, 1885.

MR. EDITOR:—We desire to ask a question on cementing hub boxes. We have used white lead and japan, also whiting and japan for this purpose, but neither proved satisfactory. Can you recommend something better?

Yours, truly,   H. & W.

Several years ago carriage-builders made a practice of using keg white lead mixed with oil and japan, but as a result, when the axles needed oiling they ran warm, dissolved the oil from the lead, driving the oil into the spokes and hubs, in some coming under our notice the oil penetrating entirely through. Whiting and japan is a very good cement, of sufficient consistency if the wheels are properly boxed so as to hold the box in position without becoming loose. The cement should not bind the box and hub together so as not to be able to drive the box out again; it is sometimes necessary to remove the box in repairing, and if cemented too hard the box will be damaged at the front. Whiting and body-maker's glue is very good, but it sets too quick, which is objectionable. It is our purpose to make some trial tests with various cements without oil soon, and we will give the results in a future number.

The Carriage Monthly, Aug 1885

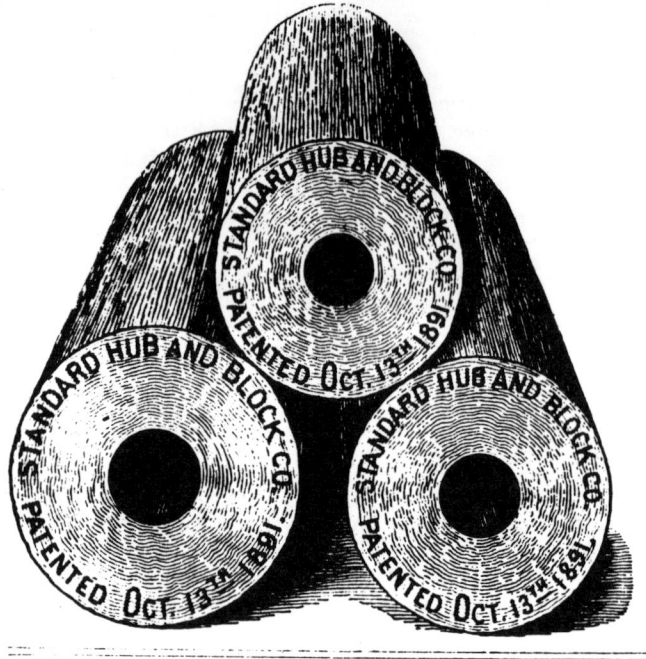

Wm. J. Matern's Ready Hub-Boxing Gauge

Pat. Oct. 20th, 1885.

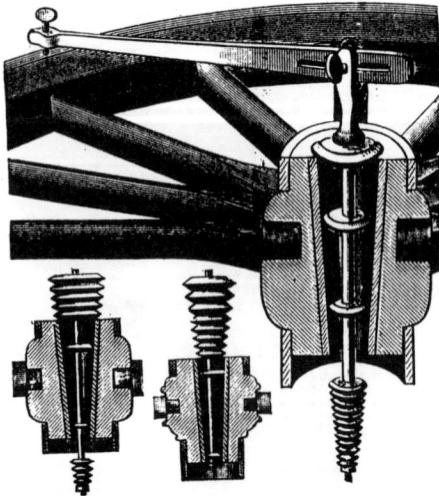

This gauge of graduated rings is perfectly accurate; works quickly and equally well in large, small or worn boxes. It saves time and trouble, cannot get out of order, and gauges a wheel at once. Two sizes are furnished: one for wooden axles from 2 to 4¾ in., and one for iron axles from ¾ to 2 in. The price is $2.75 for either size, or $5.00 for the pair (two sizes).

Descriptive circulars on application. Address

WM. J. MATERN,
306 & 308 W. Front-st.,
Bloomington, Ill.

The Hub, Sept 1886

Boxing Wheels with Hydraulic Power.

Boxing wheels with hydraulic power is by no means new, and has been done during the past forty or fifty years. Why is it that this method has not come in general use? The reason is that fitting the boxes to the hubs was done by hand without a boxing machine, and if the boxes were forced into the hub with hydraulic pressure, the force required could not be gauged, and the consequence was the hubs split. If driven home with the hammer the mechanic knows when the box does not fit the hole, and he will drive the box back, and give it a better fit. Since boxing machines have been invented and used, this difficulty is partly avoided, but we believe that a box driven by blows, combined with pressure from a standard machine, is better than if driven with hydraulic pressure only. This can be seen with a contrivance for driving home the rims on the shoulders of spokes. The blows and pressure combined does the work.

The Carriage Monthly, Oct 1892

Cement for Axle Caps and Boxes.

WARE BROS.

GENTLEMEN:—Will you please inform me through THE CAR-RIAGE MONTHLY what kind of a mixture is used for cementing or fastening boxes in wheels and oblige

A SUBSCRIBER.

ANSWER.

Adhesive cements almost without number have been prepared, and are being prepared constantly. They are used for a great variety of purposes, and a cement which might be admirable for one purpose might not be valuable for another. There are un-doubtedly merits in most, and there are certainly defects in all kinds of which we have any knowledge.

They are however classified as follows:

I. According to their composition. Under this arrangement animal glues would be placed in one class; vegetable gums in another; resins in another, and so on through the list.

II. According to the purposes for which they are intended, such as application to stone, iron, wood, leather, etc.

III. According to the time required for them to dry or harden. Some become hard immediately upon application, others requiring a longer period.

GENERAL RULES FOR USING CEMENTS.

Quite as much depends upon the manner in which they are used as upon the cements themselves. The best will prove utterly useless if improperly handled and applied. Not infrequently fail-ures are reported in the use of cements which, in the hands of other parties, have given complete satisfaction. Carriage builders are aware that animal glue, as used in the shop, will hold two pieces of wood together so firmly that when great force is ap-plied the wood will part elsewhere sooner than at the glued joint. Use the same glue on wood and iron, and the joint will be weakened fully 50 per cent. Let it be used on iron alone, apply-ing it to the natural black surface, and the joint will give way at the least tension, while if it is applied to a surface which has been freshly roughened with a file, the joint will be about one-tenth as strong as in the case of wood.

Looked at in another way, cements may be divided into four classes. One hardens by evaporation; another by cooling; an-other by oxidation, and another by chemical changes. Probably most carriage builders use common glue for securing the axle to the bed or for fastening the cap. The glue must be boiling hot, and the axles and caps must also be heated, if the best re-sults are expected. The glue in this case is thicker than that used by body makers. Such glue is easily prepared and easily handled. It is inexpensive and generally quite satisfactory.

CEMENT FOR AXLE BOXES AND OTHER PURPOSES.

Cement used on axle boxes should be free from oil, and of a kind which sets hard. It should not be of the kind which will be softened by the oil in case of a break or flaw in the axle. For a good binder most carriage builders use dry lead mixed with japan and a small quantity of oil. It is used thick. This makes a good cement, which sets in a few days, and has usually proven satisfactory. Some, however, use keg lead with a small quantity of litharge or a dryer. But the keg lead contains oil, which is likely to penetrate the hub and the spokes and prove injurious to the wheel. There are still other cements used for setting axle boxes and for cementing caps, which are inexpensive and have produced good results. Caseine or cheese has long been used for the purpose of cementing wood to wood and wood to iron. It is sometimes called cottage cheese or skim-milk cheese, and is used in connection with quicklime, borax or silicate of soda. It makes a capital cement. For pitch pine, where the com-mon glue will not hold well, this cement is well adapted, but it will not hold as well on hard woods, or on wood and iron. For axle boxes, however, it is just the thing. It is impervious to water or oil, and has all the properties of a good axle cement. It is important in preparing it to free the caseine from all oily matter. This means that all the cream should be carefully skimmed from the milk. If cheese is used, select that free from oily matter, wash it thoroughly and remove the water by squeez-ing it through a salt bag. Then mix it with quicklime. The quantity of lime can be determined by testing the glue and add-ing more as occasion calls. Mixing it with borax or silicate of soda makes even a stronger cement.

A cement impervious to water or oil, but more expensive, is made by using twenty parts of resin, four parts of beeswax, four parts of red ochre and one part of plaster of paris. The resin and the beeswax are melted together, the other substances being mixed in a finely powdered state. This cement should be applied while hot.

Probably the strongest cement known is what is called marine glue. This glue is made of a combination of shellac and caout-chouc, in proportions which vary according to the purpose for which it is to be used. It can be made either very hard or very soft, the degree of softness being regulated by the propor-tion of naphtha used in dissolving the caoutchouc and shellac.

The Carriage Monthly, Sep 1900

THE COACH-MAKERS' MAGAZINE.—PLATE NO. 8.

The Hayes Hub Boring and Mortising Machine.

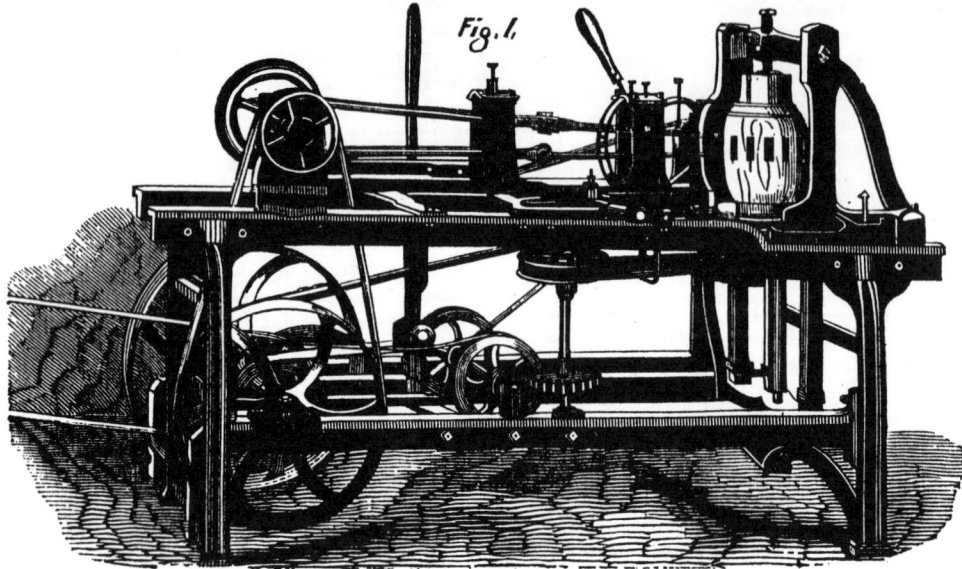

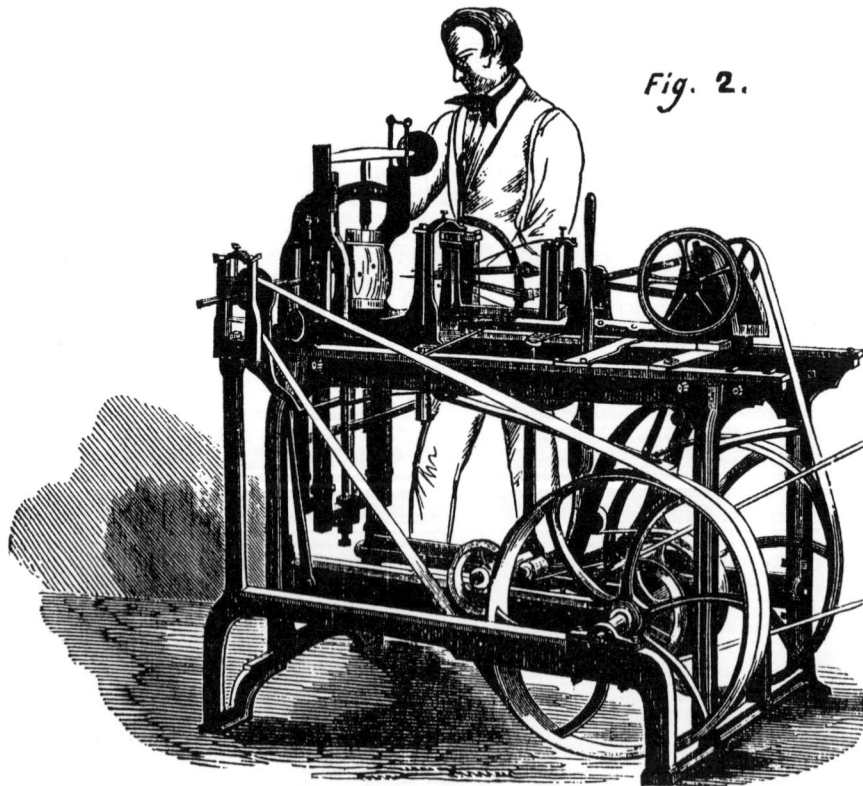

Fig. 1.

Fig. 2.

[For Explanations, see Pages 17 and 18.]

The Coach-makers' Magazine, Feb 1858 — (Description next page)

Plate No. 8.

THE HAYES HUB BORING AND MORTISING MACHINE.

It is our good fortune in this number of the Magazine, to present our readers with an illustration of the " Hayes Hub Boring Machine." While at Quincy, Ill., on the 4th of last month, we done ourself the pleasure of calling upon the extensive carriage-makers, Messrs. Hayes, Woodruff & Co., and " scraping an acquaintance" with them. These gentlemen have now the most elegant and complete carriage factory west of the mountains. The style and finish of their vehicles manifests a degree of originality and neatness in the execution of the work that none but the most scientific and practical men can claim.

But the one thing that interested us most, was the inspection of that ingenious piece of mechanism—devised by Mr. Henry Hayes—the " Hub Boring and Mortising Machine." We have never, as yet, had the pleasure of seeing a machine of any kind which embraces such an unlimited degree of novelty as this invention of Mr. Hayes. The amount and quality of work that is performed by one of these machines is almost incredible, when we are informed that one buggy hub after another is bored and mortised in the short space of *one and a half minutes*. But as we saw the machine at work, we know that this is no exaggeration, but, on the contrary, strictly true. There is probably no machine in the world that can perform the same amount of labor in the same length of time as this. There is nothing strange in this assertion, when we come to consider the manner in which it is operated.

Mr. Hayes has manufactured two sizes of this machine. The largest is intended expressly for heavy hubs, such as are used for wagons carts, &c. The smaller one is adapted for all sizes of buggy and carriage hubs.

Fig. 1 represents the large machine in the act of mortising.

Fig. 2 is a view of the small one as it appears when in the act of boring. The reader will observe that this part of the work is accomplished by the use of *two* chisels at the same time, and here lies the mystery of the great speed with which this machine performs its work. In place of only one chisel, there are two at work in the same mortise, and they move with such velocity that the hole is completed and the chips all drawn out in about the same time that a hole can be made with the auger bitt. The "register" is particularly worthy of note, as the hub does not require to have a mark or line when it goes into the machine; not even the lines usually put on the hub to indicate the position of the face of the spokes, and the chisels are so adjusted that the mortises are as near alike in every respect, as bullets from the same mould. We saw both light and heavy hubs mortised, the latter being intended for large two horse wagons, and it performed its work equally well in either case.

We find the inventor, Mr. Hayes, " an old father in our fraternity," one who has conducted the manufacture of carriages for a great number of years, and now in the " evening of life" his natural genius has rendered him eminently useful as an inventor.

What disposition will be made of this improvement we are not as yet advised, but we think that a factory will be established in Quincy, Ill., where the machines can be purchased, with the right to use, &c. Due notice of this will be given through the Magazine. Mr. H. has two of the small sized machines completed and ready for work, and which he proposes to send to any party that are in want of one, and give them the liberty to set up the same and use it until they are satisfied that it will perform as represented, before they obligate themselves to pay for it.

Mr. Hayes is also the inventor of a very ingenious device for a cheap construction of carriage tops, which is also illustrated in this number. The object of this invention is to obviate the use of all the joints usually applied. We think this is well adapted for a cheap class of work. It saves about $10 in the construction of each top.

For further particulars address HENRY HAYES, Quincy, Ill.

The Coach-makers' Magazine, Feb 1858

Hubs

Hub Boring Machine.

It is a well-known fact to all practical carriage manufacturers, the difficulty they have had in boring their hubs for setting boxes, from the fact that there are so many different tapers and sizes of axle boxes to be set. It is also well known that there are but very few machines yet brought into use that will bore either a straight or taper hole of any angle, from straight to 1½ inch taper to the foot, so that the boxes can be put in the hub and make a perfect fit, avoiding all necessity of gouging or wedging. The tendency has been growing stronger and stronger every year for small hubs and light wheels; consequently the necessity of saving all the wood in the hub possible. Many first-class wheels have been ruined by setting the boxes with improper machines.

The accompanying cut represents the

3d. The hub is bored in exact conformity to both size and shape of box, thereby obviating the necessity of wedging the box so that the wheel will be perfectly true, making the wheel of much greater strength and durability than can be obtained in any other way.

4th. The rapidity attainable in the working of the machine is such that the wheel can be adjusted ready for boring in a minute's time, and any ordinarily smart mechanic can with ease bore from eight to ten sets in ten hours.

5th. The cutters (which sustain the greater part of the wear in the machine) are so simple that they can be made by any ordinary blacksmith at a very trifling cost.

The machine is so simple in its construction, and of such a substantial character, that it will last almost a lifetime if used

Kritch Patent Hub Boring Machine, or Improved Box Setter, patented Feb. 23d, 1864, and improved Oct., 1872. It is claimed by the patentee and the manufacturers to be the *best practical hub borer* now in use, and their testimony is backed by such parties as Brewster & Co., Dusenbury & Vanduser, and R. M. Stivers, of New York City; J. Lowman & Son, Cleveland, O.; Harvey & Wallace, Buffalo; Long & Silsby, Albany, N. Y., and other first-class builders.

The points claimed over other machines are :

1st. It bores either a straight or taper hole of any size and angle desired, cutting the recesses in the end of the hub for collar and nut.

2d. The precision with which the work is done, makes the wheel perfectly true and entirely obviates the danger of breaking boxes in setting.

with care. Messrs. Brewster & Co., of Broome st., N. Y., has had one of the machines ever since 1865, which has been in constant use ever since, and when the last improvement was made (Oct. 1872) they ordered another. It weighs a trifle over 200 pounds, with two mandrels, one for boring small hubs the other for large ones. If required, a power pully can be attached, so that it can be run with power or hand either. The machine occupies but very little room ; it can be placed on a movable bench or stand so that it can be easily set out of the way when not in use.

Price $80.00 boxed and delivered on cars or boat in Cleveland, Ohio, with or without power pully, with full set of tools and directions for using the same, by applying to the Kritch & Crane Mfg. Co., Cleveland, Ohio.

Coach-Makers' Int'l. Journal, Mar 1873

NO. 1 PATENT AUTOMATIC HUB TURNING MACHINE.

WITH FRICTION CLUTCH, ROUGHING AND CUPPING ATTACHMENTS.

THE No. 1. Patent Automatic Hub Turning Machine illustrated herewith is the largest and most powerful machine of its class, designed especially for making carriage and wagon hubs of different sizes and shapes up to 20 in. diameter, 18 in. long, at the largest, having a capacity for finishing 600 heavy hard wood hubs in ten hours, or roughing out 2,500 blocks. This machine receives the block in its rough state, performs the roughing, turning, cupping, finishing the ends, cutting beads and shoulders for bands, making hubs any shape or size complete at one operation, more uniform and perfect and at an immense saving over hand turning. The body of the machine is composed of iron, a massive casting in one piece, of neat design and sufficient weight to stand firm without fastening to the floor, performing the heaviest turning without jar or injury to the working parts. Floor space occupied, 6 ft. 6 in. $\times$ 3 ft. 6 in. The table is built in two parts. The lower half is gibbed and fitted to the frame in V-shaped ways, with adjustment horizontally in line with the mandrel by hand-wheel and screw to center the knives with the hub block. The upper table, with the roughing and finishing knives attached at either end, is mounted upon and gibbed to the lower table, and it slides from right to left at right angles with the mandrel by turning the large hand-wheel to bring either the roughing or finishing knives up to the hub block to be operated upon. The roughing knife, with straight face 18 in. long, is held in the stand at the back end of the sliding carriage, with its cutting edge extending downward, and when in operation removes the surplus material from the hub block in the form of a veneer or ribbon ⅛ in. thick, full length of hub at one cut, requiring no adjustment for length or diameter of block. A gauge governs the depth of cut or feed. The finishing knives are located at the opposite end of the carriage from the rougher, with their cutting edges extending upward, consisting of a body knife with cutting edge shaped to correspond with the style of hub to be turned ; a flat knife at either end upon the same stand, for cutting the front and back bands, with adjustment for cutting bands of different widths and diameters. The cutting-off knives for finishing the ends of hub are placed upon separate stands underneath and in advance of the body and band knives. The cupping attachment is gibbed to the tail stock and provided with gauge to regulate the depth of cup. The shape of the knife governs the style of cup. A special attachment can be furnished for cupping the back end of hub when so ordered. A powerful friction clutch fitted upon a 3 in. diameter steel spindle, driven by an 8 in. belt, communicates power to revolve the hub. The frictions are engaged or disengaged by foot treadle conveniently located to the operator. A single movement of the operator's foot upon the treadle instantly starts or stops the machine without shifting the belt or changing position. The operator has complete control over the machine from the working side. As the material to be operated upon revolves, the roughing knife is first presented to its action by turning the large hand-wheel to the left, to reduce the block to proper diameter. By a reverse movement the roughing department retreats and the finishing knives are brought into service, shaping the hub to desired form and length. The diameter of turning is regulated with graduating screws attached to the carriage, and when once adjusted hubs of one diameter are turned to exact sizes without the use of calliper or rule. The hub is roughed, turned, cupped and polished complete at one starting and stopping. The counter shaft is $2\frac{1}{8}$ in. diameter, 60 in. long, journals turned to $2\frac{7}{16}$ in., two No. 3 hangers 28 in. drop, one shipper complete, one driver 40 in. $\times$ 8 in., tight and loose pulleys 20 in. $\times$ 8 in., speed 400 rotations per minute. Friction pulley on machine 20 in. $\times$ 8 in., speed 800 rotations per minute ; weight of machine complete, 4,000 pounds. Defiance Machine Works, Defiance, Ohio, U. S. A.

NO. 1 PATENT AUTOMATIC HUB TURNING MACHINE.

The Hub, Jan 1892

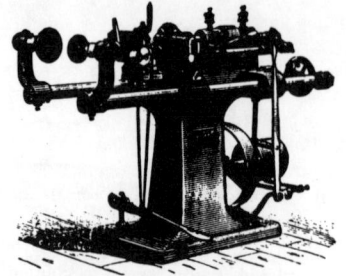

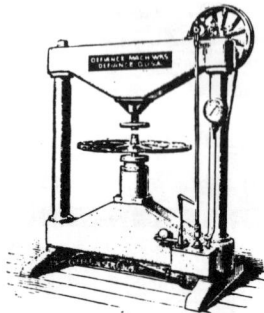

THE HYDROSTATIC GRADUATED STROKE HUB MORTISING AND BORING MACHINE.

THE Hydrostatic Graduated Stroke Hub Mortising and Boring Machine, manufactured by the Bentel & Margedant Co., Hamilton, Ohio, U. S. A., illustrated herewith, embraces many very important improvements not found in the usual "mortisers" offered in the open market.

The active motion of this class machines is reciprocative, in entering and withdrawing a chisel from hard material, producing a square-cornered hole. In order to prevent the breaking of the cutting chisel and the reciprocating mechanism of this class of machines, it is necessary to enter the cutting chisel by degrees, each successive stroke a little deeper, until the full depth is obtained.

Manufacturers of machines obtain this by compound crank,

GRADUATED STROKE HUB MORTISING AND BORING MACHINE.

lever and treadle motion, shifting the full crank motion gradually towards the center line of action, by means of a foot treadle which is at the start about twenty inches above the floor, so that the operator must raise his foot to that very inconvenient height, bringing the foot treadle down by degrees at each stroke, and enduring all this often dangerous stamping and knocking until the full depth of hole is established.

In this the operator does not lift his foot from the floor at all. The point of the foot-treadle is only ⅝ in. above the floor line at the highest, and it requires only a mere touch of the treadle, when the stroke graduates or increases automatically without jar or knock till the full depth of hole is produced.

Operators and owners of this class of machines will hail this improvement with joy; it has been long sought for, and it will do away with the crippled and sore legs of the operators and breaking of machines. This beneficial result is obtained by a specific arrangement of an oscillating pressure cylinder and ram filled with air, water or oil (the latter preferable) further by a special friction pulley, driven from the crank mandrel, and a treadle nearly even with the floor. By touching the latter the liquid in the cylinder will resist the reciprocating blows formerly sustained by the leg of the operator, and escape to the other end of the cylinder until the full stroke of the crank is taken up by the chisel bar. This arrangement is partly visible in the illustration back of the machine, and also the foot-treadle, which is shown at its very highest altitude.

This machine is very strongly constructed, and all the parts are of modern design, providing all possible advantages in mechanical devices for a first-class machine, besides those enumerated above.

Each machine is furnished with eleven mortising chisels and eleven boring augers, viz., ⅜, ₁₆⁷, ½, ₁₆⁹, ⅝, 1¹₁₆, ¾, 1⅜, ⅞, 1⅝ and 1 in. The various hub chuck-cups for the various sizes of hubs needed are also furnished. The machine is arranged to take in hubs up to 14 in. diameter and 18 in. in length or less, down to the smallest. The weight of the machine is about 2,500 pounds. For further particulars address the manufacturers.

The Hub, Aug 1895

POWER CAM PRESS FOR WHEEL HUB BANDS, ETC.

THIS illustration shows a power press used for starting and for pressing on of the various bands and shells on the hub of vehicle wheels, and for other purposes for which a powerful and quick operating press is used.

The press belongs to the class of cam presses, in which a heavy concentric rotating shaft is provided with an eccentric cam acting direct, without the intervention of levers or crank rods, on the lower end of a heavy ram, and raising and lowering the same in accordance with the throw given to the cam shaft, which is in this case five inches.

This machine is elegantly designed and very strongly constructed. The base of the machine contains the heavy bearing boxes and the driving crank-shaft with cam, all of which runs in oil, thus preventing cutting or dry running.

The heavy column rods are strongly bolted to a heavy ribbed support in the base, and carry on the other end the heavy resisting yoke with the powerful adjusting screw. The latter is provided on the head end with a series of handles for raising and lowering the screw quickly and very conveniently, and the heavy counter or fastening nut is also arranged with handles placed within easy reach of the operator, who is able to manipulate both without leaving his place.

Special attention is called to the very important fact that there is no single lever or connecting rod; no use is made of lever cranks and no pins or bolts in the whole mechanism, but it operates in a direct line positively, thus retaining accurately parallel and perfectly straight surfaces of the ram-head and resisting screw, which cannot be obtained by lever presses on account of irregular wear of such devices.

The height of the whole press is 7 feet. The floor space required is 4 feet 6 inches by 2 feet 6 inches. The screw can be raised so as to have a space of 17½ inches between the screw and the ram-head, and the distance between the rods of the table is 21 inches, while the table itself is 30 inches in diameter.

The tight and loose pulleys are 20 inches diameter, 6 inches face, and they should make 500 revolutions per minute, giving ten reciprocating motions to the ram in one minute.

Manufactured by the Bentel & Margedant Co., Hamilton, Ohio, U. S. A.

POWER CAM PRESS FOR WHEEL HUB BANDS, ETC.

The Hub, Oct 1895

IV
SPOKES

SPOKES

In 1850, spokes were still being split by hand from rough stock with froe and froe club, it being the popular belief among wheelwrights that spokes were stronger when split than when sawn. After careful selection of the wood, noting its type (white oak or hickory) as well as its grain, the spoke was fastened in the spokeholder and shaped with drawknife and spokeshave. However, the introduction of the patent hub in 1857, which held the spoke more securely, brought with it the ability to make spokes lighter and more slender, so spoke shapes like the Warner and Sarven were added to the traditional staggered spoke and wagon wheel style.

In 1830 Levi Bissell conceived the idea of adapting the Blanchard lathe to the turning of spokes. Aided by E. Hedenburg he completed the first spoke-turning machine and set it up at Canfield, Hedenburg & Co.'s carriage factory in Newark, New Jersey. This machine was imperfect, giving the spokes a spiral finish. With the development of the spoke turning lathe, patented in 1841, uniform spokes could be made much faster.

A new model in 1888 claimed a record output of 2,695 spokes per 10-hour day, with the operator having only to insert the blank and remove the spoke.

The method of driving spokes by hand with a wooden mallet was replaced around 1870 by a spoke driving machine. About the same time, a hand operated spoke tenoning machine, equipped with a hollow auger, cut round tenons more accurately and with greater speed. Production was increased even more when, in 1894, a patented process to season white oak billets within 2-4 weeks was introduced. Once more, however, despite its time-saving benefit, wheelmakers were slow to accept this process.

Finally, as an indication of the length to which wheelmakers would go to make the "perfect" wheel, a machine was developed to compress the spoke tenon before it was driven into the hub, thus allowing space for the glue, which would otherwise be squeezed out in the process.

Spokes

Wheel-Horse and Mallet.

SPOKE SHAVE, EDITOR.

In my former communications, I referred to a few of the responsibilities of the wheel, the necessity of its having life or spring; also, the importance of the wheelers having a knowledge of the different temperaments or qualities of his material. In the present, I shall, to the best of my ability, try and explain, as intelligibly as possible, how to work these different temperaments so as to harmonize, thereby rendering the wheel serviceable.

In the selection of our hubs, it is necessary that we get them as near one quality as is possible; our spokes we will have to take them as they run in the bundles or sets. Hub, 3½ by 6½; spoke, 1 inch; felloe, 1 inch; tread, ⅞ or ¾ as preferred; 14 mortices back and front; size of mortice, 5-16th or ⅜ scant; depth, to suit the spoke. Height of wheel: back, 4 feet 2 inches; front, 3 feet 11 inches.

In dressing down our tenons to suit the mortice of the above quality of hubs, we should fit them to side of mortice, with draw enough, so you can force the tenon down in the mortice without any extra exertion—the draw on the back of spoke, about ⅛ of an inch. Having dressed the tenon to our mortice, we then dress, or shave, our spokes. The spoke for this size wheel should be 1 inch, as above. The body you will find to be about ⅞ of an inch; such being the case, you will dress out your spoke, if you wish to give it spring, to ½ an inch, or, in other words, ½ of the spoke, this spring should be as near the hub as possible. The face of your spoke should be the same as the body, ¾; the back, at the shoulder, ⅝. Shave your front leaf—bringing it together about 2½ inches from the shoulder. The back leaf, you will observe is not as wide as the front, but you do not give it the same taper; you leave the leaf about ½ an inch, 3½ inches from the shoulder. This gives you a very pretty shaped spoke. It throws the heft of the spoke on the back, thus giving you a spoke of light appearance from the front. After you have your spokes dressed, you will strike off the side edge, at the shoulder, with your file; it must be lightly struck off. The reason for so doing will appear when we go to preparing for the driving.

The next thing to be done, is to deduct the size of the box from the size of the hub. This will give us the length of the tenon for the hub. We then saw them off, and strike off the edges with the file. By so doing, your tenon on the spoke will not force the draw on the side of the mortise when driving before it. If we have a hard hub, as above, and a hard spoke, we will mark off the mortise the size of the tenon at the point, and give our mortise the same draw as our spoke. If our spoke be of a coarse grain, and porous, we will be careful to give it draw enough to compress the pores, so that the spoke will not have a chance to work in the hub; but we must be careful not to give it too much draw, so as to mash—thereby rendering it useless. If we have a soft spoke, we will give our mortise more draw than we do our spoke, with the view of binding it on the bottom of the mortise, and not mashing it at the shoulder. You will continue to discriminate as to the temper of your spokes, upon the same principle, until you have them all fitted separately to your mortise. You will let your spokes down into your hub about one-sixteenth of an inch. This must be let down square, and of the shape or size of the shoulder.

Here you will discover is the necessity for striking off the side of the spoke at the shoulder; that is to say, you will strike off the depth you let into the hub. This is done to secure your shoulder from checking off. You perceive that in doing this, you get the spoke in the hub fair on the face. You are not compelled to putty up this settling of the spoke, and it will also prevent that eye-sore, after the wheel has run awhile, from presenting the appearance of the spoke being loose in the hub. This putty, or paint, will crack around the shoulder of your spoke.

When you are ready to drive, you will deduc the depth of your rim from the height of your wheel, and set your set-screw in its proper place, which is, if your rim is 1 inch, the same below the height. You will also set your set-screw ⅛ of an inch back of the face of your front mortise; this will bring the face of the rim on a line with your front spoke, which will give your wheel about ¼ of an inch disk when it is properly hooped.

When this is done, you will band your hub with slip bands, which will be large enough to receive a strip of leather under the band, so as to prevent the same from bruising your hub. In driving your spokes, you will have your glue as strong as you can work it, and be careful to have the space you have let into the hub for the shoulder of the spoke, well furnished with glue. This will greatly assist in holding the shoulder permanently in its place. You will also be careful to cut around your spoke that is in your hub, with a very thin chisel to prevent the shoulder from drawing the face of the hub down before it. Before you drive your spokes, you steam your hub well so as to soften it. By so doing, you will save much labor in driving; it will also ease the hub, and prevent its checking. You will also find that after this steaming has *dried* out, that the spoke is cemented as it were to your mortise. Having got our spokes in the hub, we will, in our next issue, proceed with the wheel to completion.

LABOR-SAVING MACHINES.

Machines to aid in the completion of almost every kind of manufactured article have become so common, that the question is no longer asked, "shall we employ machinery," but, "where can we purchase the most perfect kind." Probably the rudest machine of any sort ever built was of some value, in saving time and labor, but as in our day perfection in adaptation, speed, and beauty of work are required, it is of the first importance that those who employ machines for any purpose should inquire carefully into the relative merits of those in competition.

The carriage builder is well aware of the value of labor-saving machines, knowing that to do everything by hand, as of old, would add greatly to the first cost of his finished work, and place him at a disadvantage. Not only is it requisite to have machines, but now, while their use is so common, it is necessary to search for the best.

In the accompanying cuts, we have a representation of Dole's Patent Spoke-tenoning Machine, with Deming's improvement, manufactured by Silver & Deming, Salem, Ohio, and pronounced by competent workmen to be the best machine in use for cutting round tenons.

Fig. 1 shows the wheeel in position for cut-

Fig. 1.

MAAS-MANZ-CHI.

Fig. 3

ting the tenons. *Fig.* 3, the Hollow Auger, which is fitted on a mandrel that works through a bearing in the casting A, an arm of which

projects out under the auger, and is provided with a rest and dog *B*, that centres the spoke while the auger is starting on, thus centring all the spokes the same, so that the tenons will cut exactly *true*, and in less time than it would take to sharpen the spokes for the common hollow auger. There is also a feed lever, to use in finishing the shoulder, that the crank may be forced up to the end of the bearing box, thereby making a perfectly square shoulder; the head is raised or lowered in the frame by means of a shaft and pinion to suit different length hubs, being held to its place by tightening the nut on the shaft. This machine may be changed to a Boring Machine, by removing the hollow auger from the mandrel and substituting in its place a chuck for holding auger bits. A block for holding the work to be bored is then put on, producing a complete machine for boring felloes. (Cut No. 2 omitted for want of space.)

Two sizes of this machine are furnished; the small size cuts from 7-16th to 1-inch, and 3 inches long. The large sized machine cuts from ¾ to 1½ inches and 4 inches long.

For full particulars and price list, address Silver & Deming, Salem, Columbiana Co., O.

In the production of a serviceable wheel, much depends on the exactness in which the tenons and mortises are made to fit each other, this being true also of any article made of wood which requires to be joined.

A good machine, adapted to the work in hand, aids the workman in this particular, enabling him to produce perfect fitting parts, with but small outlay of time. From day to day it stands ready to repeat its operations, and finally becomes such a necessity, that when for a short time it may be out of repair, everything appears to have come to a stand still. And were it possible for some magician to wave his wand over the world and banish labor-saving machinery, "its loss would paralyze business of all kinds."

Coach-makers' Int'l. Journal, Dec 1870

Spokes

Wheel Making.

The subject of wheel making has been well sifted through the columns of the JOURNAL, and the different processes of making have seemed to have been almost exhausted, yet there is still a chance of adding a mite to the knowledge that has already been diffused, and which may be of some benefit to the craft.

In the making of wheels, seasoned timber is the first essential point for good work. The principle on which to make a good wheel, so that the spokes will not work in the hub, has been argued upon by many good wheelers, and they have given their theories, and it all lies in the manner of fitting the spokes and mortises in the hub. The generality of wheel makers in dressing the tenon of the spoke, give considerable taper or drive to the tenon; that is, $\frac{1}{8}$ of an inch on tenon, $1\frac{1}{8}$ inch long, and the mortise so that it is a little more at the bottom. Now, I contend that it is not correct. In dressing the tenon, have it approach as near square as possible, giving not over 1-16 inch taper on a tenon $1\frac{1}{8}$ in. long—the mortise same; that is, the mortise should be the same size on the top as the end of spoke, and 1-16 inch smaller at the bottom, as 1-16 of an inch is enough pressure for a good hickory spoke and an elm hub; and having the mortise the same taper as the spoke, the spoke will have the same binding force the whole length of tenon.

Where wheels are made by hand, have a piece of band iron cut the same taper as the tenon of spoke for fitting the back bevel of mortise correct. The manner of proceeding to dress the mortise is familiar to most every wheel maker, but I will give my manner of working. We have hubs which are mortised, and will make the mortise (that is trim) to fit the spokes. After having the guide fastened and set, trim the face of mortise, using a straight edge to find when correct; then take and mark the size of tenon of spoke, and cut to the mark on the end, and with the small iron guide cut the back of mortise until the iron fits exact. In dressing the sides of mortise, scant the lines—have it so that you can hardly force

the point of tenon in, so that when you come to drive your spoke will fit perfectly tight sideways. I suppose some will say that a spoke fitting so tight sideways will split the hub. So it will if you only have a band on the ends of the hub. After the mortises are all finished, have bands made that will fit tight on the hub, and, when ready to drive, have a piece of leather under band, and drive them up close to the mortise, back and front, and there will not be much danger of checking. In fitting my spokes in this manner, I have had them to stick when within 1-16 of an inch of being down to the shoulder, and could not drive them any farther or pull them out, and have always had to mortise them out, where if they had more taper, and not fitted well sideways, I could have drawn them easily.

The theory of having the tenons on spokes nearly square, or with but very little taper, is easy of explanation. Take two pair of shafts; one you have a square tenon on the bar, and the mortise in the shaft 1-16 on an inch narrower than tenon. You will find that in driving together, it drives equally until it is up to the shoulder, and is firm. The other you taper the tenon and mortise, giving the same drive, and you will find that it drives easy at first, and when coming close to shoulder it binds harder, and it is as firm as the first pair. But now take them apart. With the first stroke of the mallet, the shaft with the tapered tenon loosens, and you can take it off very easy. Not so with the other; it holds until it is driven all off; and it is so in the wheel, for the wheel is dished, and the spoke has a strain which is drawing the tenon out of the hub, the face edge of the shoulder acting as a fulcrum. Some wheel makers say if a spoke is not drove in a certain way that the wheel will dish very easy; that is, the tenon in the hub will cause a wheel to dish when being tired. That is all humbug, for if the wheel was made straight when rimmed ready for the tire to be put on, and was to dish by the spokes giving in the mortise, the wheel was good for nothing, for a spoke will not give in the mortise without becoming loose, and if there is much taper to the spoke it will soon come out of the hub.

In conclusion, to make a good wheel, according to my experience, is to give but a little taper, fit exact to mortise, having same taper to mortise as spoke, and the spoke to bind hard sideways as well as edgewise. I have tried this on all size wheels, and have never known it to fail to make a good wheel.

G. H. T.

Coach-makers' Int'l. Journal, Sept 1872

White and Red Hickory.

BRICKERVILLE, PA., April, 1873.

MR. EDITOR:—I have received a great deal of valuable information through the medium of your journal, and would be pleased to have your opinion as to the relative value of white and red hickory, of XX quality, when used for spokes. Which kind would you prefer? My experience leads me to think there is very little difference in the choice between them, but wish you would give us your opinion in the next number of the MONTHLY. Yours,

G. M. J. Y. C.

ANSWER.—The white hickory is preferred by the majority of persons, as it is considered to be much tougher than the heart portion of the wood. The red, however, has a quality which is not to be lightly estimated, viz.: that of stiffness. Toughness, or firmness of cohesion of the fibres of the wood, seems to be more desirable, because of the violent wrenchings to which the wheels are at times subject when put into hard service. The red wood spoke may keep its shape better under ordinary strain put upon it, but as the quality of stiffness necessarily carries with it a lesser degree of pliability, it follows that a red wood spoke will break when a sudden and severe strain or shock is received. It is not, however, necessary to carry the distinction between the two qualities to that extent, as to wholly exclude the red wood, as there is no great liability to fracture of the spokes by the ordinary travel on the road; and when extraordinary force is brought against a wheel, as in the case of a collision, or the mad antics of a run-away horse, the tough spoke may be bent out of shape and the stiff spoke broken, so that both will require to be replaced by new ones.

The Carriage Monthly, May 1873

CASE OF RIM-BOUND WHEELS.

Mr. Editor: A foolish question, perhaps, but I would like an answer on this subject: "How to rim a set of wheels properly?" One man will make them stand all right, while with another, using the same quality of felloes, they will become rim-bound, and split at the end of the spoke in a short time. In fact, after some men have re-rimmed a set of wheels, they are almost worthless.

J. G. P., Woonsocket, R. I.

Answer.—The fact of wheels becoming rim-bound, is due in most cases to the condition of the hub, or to the way in which the spokes are driven. If a wheel is made properly, and the smith sets his tire with equal draft, wheels will never become rim-bound—providing you use hubs of equal hardness. But if one hub out of a set is a little softer than the others, the spokes in that hub will be liable to settle into the hub a little more than in the other three wheels. The rimming of a set of wheels is not so often the cause as the driving of the spokes.

The Hub, Feb 1878

TENONING AND MORTISING WHEELS.

There seems to be a mistaken idea on the part of some wheel-makers, that they must drive their work together with great force in order to have it hold firmly. We have heard of those who claimed that they gave, and no doubt supposed that they gave, $\frac{1}{16}$ in. of what they called side pinch to each spoke in a hub; but if we examine such work carefully, we find it is an impossibility to do this without destroying the fiber of the wood, and thereby weakening its strength.

Let us suppose a hub to be 6 in. in diameter, and 18 in. in circumference on the outside; and, at the inside, or where the spokes meet, 6 in. in circumference. This being the case, and with sixteen spokes driven into the hub, there would be 1 in. driven into 6 in., and the fiber of the wood would be crushed with a force equal to 3,000 lbs. per square inch.

The above illustration is an exaggeration, but it serves to illustrate the point I wish to make, which is this. No tenon should be driven into a mortise with too great force, nor should a pillar or spoke be struck after it is down to the shoulder, for the reason that such excessive force is liable to bring a pull upon the tenon equal to half its tensile strength.

A simple experiment will illustrate the effect of crushing the fibers of wood. Fasten a bar of wood, we will say one inch square, into a vise, and hang a weight upon the end of the bar. Then screw the vise tighter, and you will find that the bar will break close to the vise. F. J. Flowers.

The Hub, Oct 1886

Spokes

Spoke Driving Machine.

J. A. FAY & CO.'S PATENT.

It is well known that in the manufacture of wagon wheels, the spokes are driven into the hubs at the expense of much hard labor, more especially in large heavy wheels, such as are used on transportation and express wagons and other heavy vehicles.

Many attempts have been made to devise some means of accomplishing this by machinery, but heretofore without success. The cause of failure in most cases has been on account of the blow given to the spoke being too heavy, and non-elastic. Those acquainted with this work know that spokes can be driven best with a comparatively light wooden mallet, with a light springy handle, which gives a quick, smart blow; one that does not cripple or jar the spoke.

The machine here represented gives exactly this kind of a blow, only much quicker and more powerful, and the stroke can be instantly varied at the pleasure of the operator, and, in fact, is as much at his command as though he held the mallet handle in his own hand.

The blow is given by a swinging mallet which makes a complete circle at each blow, and from one to one hundred and twenty-five blows per minute, may be given as required.

The machine is very simple in construction, not liable to get out of order, is easily and quickly changed to work large or small wheels, and will do the work of from eight to ten men on heavy wheels better than it can be done by hand. It requires but one man to operate it, and any man of ordinary abilities can in a few hours learn to operate it successfully.

It is manufactured only by J. A. Fay & Co., Cincinnati, Ohio, manufacturers of all kinds of machinery for making carriage, wagon wheels, and all kinds of patent wood-working machinery of the most improved patterns.

For further information, address as above.

The Carriage Monthly, July 1873

99

Spokes

SPOKE LATHE, MANUFACTURED BY EGAN COMPANY, CINCINNATI, OHIO.

A TRIUMPH OF AMERICAN INGENUITY.

THAT there is the greatest appreciative interest in original and improved mechanisms, to save time and labor, there is no longer any doubt whatever. Progressive builders of machinery introducing them do not ever have occasion to say "dull times," or "slack business." The always stirring Egan people are uneasy and restless unless bringing forward the something new to benefit the woodworker, and it is with pleasure we present to our readers an illustration and a brief description of a remarkable advance in this era of progress. Spoke makers have been content in the past with what was thought the best creation for the production of spokes through the employment of a lathe known as the Blanchard Screw feed type, with a capacity ranging close to 800 per day, and this, too, with the aid of a skilled operator ; to-day an improved lathe of the same type turns out 2,200 to 2,400 per day.

The cut above represents a spoke lathe which combines the principal features of what is known as the "Blanchard Screw-feed Type," with greater improvements. As it is now made it is as nearly automatic as it is possible to make it, having a capacity and efficiency larger than ever before attained. It is indeed the decided novelty of the period, the latest, the greatest and best of its class in American ingenuity, its output being the largest ever reached, and is nearly three times as great as the best made by others. It has a record of 2,695 spokes per day of ten hours, and 13,735 spokes per week of fifty-eight hours. The cutter head is entirely new and is a combination of hook and gouge knives The gearing is cut from the solid. the center one being of steel, with double width of face to enable the operator to change the shape of spoke.

The centers are made of best cast steel, fitted to large steel mandrels ; the back center bearing is so constructed that various lengths of spokes can be turned from one pattern, a desirable feature for any factory.

The feed mechanism is very simple and reliable, and is constructed so as to change the speed of feed at the throat of spoke with but a slight change, can be made to feed at one rate of speed the full length of the spoke. The feed can be stopped in any desired position along the spoke for setting the rests or truing the knives. The great improvement whereby the lathe is given such extraordinary production is an ingenious arrangement for automatically lifting the frame carrying the spoke into the cut, thus making the lathe perfectly automatic in all its movements. The operator has nothing to do but put in the rough stock and remove the finished spoke. He does not leave his position, merely lifting a lever, which sets the vibrating frame into the cut ; the carriage, with cutter head attached, travels along the bed. completing the spoke, the vibrating frame throws forward and the carriage and head return to the starting point ready to cut another spoke. On all other machines of this description, this frame must be lifted into the cut every time a spoke is turned, and when it is considered that the frame weighs over thirty-five pounds, and estimating over 2,400 spokes per day, it is not difficult to estimate how much labor is saved the operator by this new improvement. The success of this lathe has been phenomenal, and has almost entirely supplanted other makes.

For further information relative to the above, address the originators and builders, the Egan Company, Nos. 196 to 216 West Front-st., Cincinnati, O., U. S. A.

The Hub, Aug 1891

Spokes

IMPROVED SPOKE TENONER FOR ROUND TENONS.

THIS new improved round tenoner, especially designed for the wheel trade, cutting off and finishing the end of the spokes, ready to receive the felloe, has many points of advantage, and will enable the operator to do a greater amount and a better quality of work than can be done on any other machine of this kind made. It has all of the necessary adjustments for large or small wheels, heavy or light, and all the adjustments are within reach of the operator.

The column is one entire casting, cored out, and with ample floor space, so as to stand firmly on the floor when the mandrel is running at a high rate of speed.

The pulley and journals form a sleeve, and are one entire piece. The mandrel works through the center of same, and is brought forward by a lever handy to the operator. This mandrel carries a chuck on one end, and in connection with a concave saw, cuts off the end of the spoke to the desired length. The mandrel is then brought forward until a stop is reached, which governs the length of the tenon.

The wheel is held between tapered cups to suit the diameter of hub used. These cups revolve upon spindles, one being set in a certain position, the other held by a friction binder, so that a change of wheels is accomplished instantly. The parallel shafts which carry the cup arms are adjusted by a hand wheel and screw, and there is an index placed on the shaft to gauge the different diameters of wheels.

The spoke is held in position for cutting the tenon by means of two V jaws, working simultaneously and operated by treadle. These jaws always bring the spoke to the center of the cutter-head, and enables the operator to have both hands at liberty to facilitate the work.

There is no question but that this machine will be found the most simple and reliable of tenoners, and of great capacity, as it is capable of cutting off and tenoning one hundred and thirty sets of wheels per day of ten hours.

For further information, address the builders and introducers, the Egan Company, Nos. 196 to 216 West Front-st., Cincinnati, O.

IMPROVED SPOKE TENONER FOR ROUND TENONS.

The Hub, Feb 1892

Spokes

Improved Spoke Facing Machine.

MANUFACTURED BY THE BENTEL & MARGEDANT CO., HAMILTON, O.

This new spoke facing machine is a fine tool of modern construction, perfect in its arrangements, and very superior and satisfactory in operation. The frame is well designed and strongly arranged with heavy wide base, so that the operating cutting disc can be run at a high speed without tremble or jar. The strong substantial table on which the spoke is laid is arranged for sliding easily and accurately on the guide bars, and is operated by the foot treadle shown in front. The table returns quickly by spring power. Spokes faced on this machine do not roll or quiver when the cutting is done, while the facing is perfect, with straight surfaces and sharply developed edges without splintered corners.

shaft furnished with the machine when so ordered, is provided with patent differential tight and loose pulleys, the loose pulley being smaller in diameter than the tight. The tight pulley is 10 inches diameter by 5¼ inches face, and should run 1050 revolutions per minute. The weight of the machine without countershaft is about 500 pounds, and with countershaft about 650 pounds.

The apparatus for resting the spoke on the table is arranged to grip the spoke firmly in its proper position and hold it there, at the same time offering superior advantages for quick changes. The angle and position of the knives is such that a shearing cut running with the grain of the wood is secured. Operators who prefer centering gripping jaws with a hand lever to hold the spoke during the operation of cutting, may adopt an ingeniously arranged device of that class now used by many wheel makers in preference to anything else.

The arrangement of this machine is so perfect the disk is run at a speed of 2100 revolutions per minute. It carries a driving pulley 10 inches diameter by 5¼ inches face. The machine is furnished either with or without countershaft as may be ordered. If it is desired to run the machine without a countershaft, this can be done by arranging a belt tightener near the line shaft from which it is driven. The counter-

The Carriage Monthly, Mar 1893

Spokes

HOSLER'S PATENT SPOKE DRIVING MACHINE.

WITH GRADUATED DRIVING POWER, ADJUSTABLE GAUGE, ETC.

(See Illustration Accompanying.)

THIS lately improved machine is designed for driving all kinds of spokes into the hubs, and making any kind of wheel required in wheel, wagon, or carriage shops. A former pattern has been in constant use in some of the very largest wagon and wheel manufactories in this country for the past ten years, giving unqualified satisfaction. It is built in the most thorough manner upon a heavy iron column, with a view to the hard work it has to perform. It requires but one man to operate it, is easily managed, and any mechanic of ordinary capacity can learn to operate it successfully in a few hours.

The frame which holds the hubs will receive any size up to 12 inches in diameter (at the large end), and is adjustable to hubs 16 inches in length. The blows are given by swinging mallet or hammer, similar to blows given by hand, only with much greater force and rapidity, which are obtained by the pressure of the foot upon a treadle. The blows can be changed instantly from heavy to light, and from quick to slow, or any one blow can be struck very quickly and heavily, and the next one lightly and slowly, and as easily governed as if the operator had the mallet in his own hands.

The graduation of the blows is so quickly accomplished that the stroke can be changed after the mallet is started by changing the pressure of the foot on the treadle. There is an adjustable guide for trueing the spoke being driven into the exact position desired, and is raised out of the way when not in use.

This machine is adapted to wheels from two to six feet in diameter, and is very quickly changed from one size to another. It will do the work of from eight to ten men on medium or heavy wheels, and do it in a more perfect manner. For the purposes intended it is unequaled.

The driving pulley is 20 x 5 inches, and should make 125 revolutions.

For further information address the makers, J. A. Fay & Egan Company, 196 to 216 Front-st., cor. John, Cincinnati, Ohio.

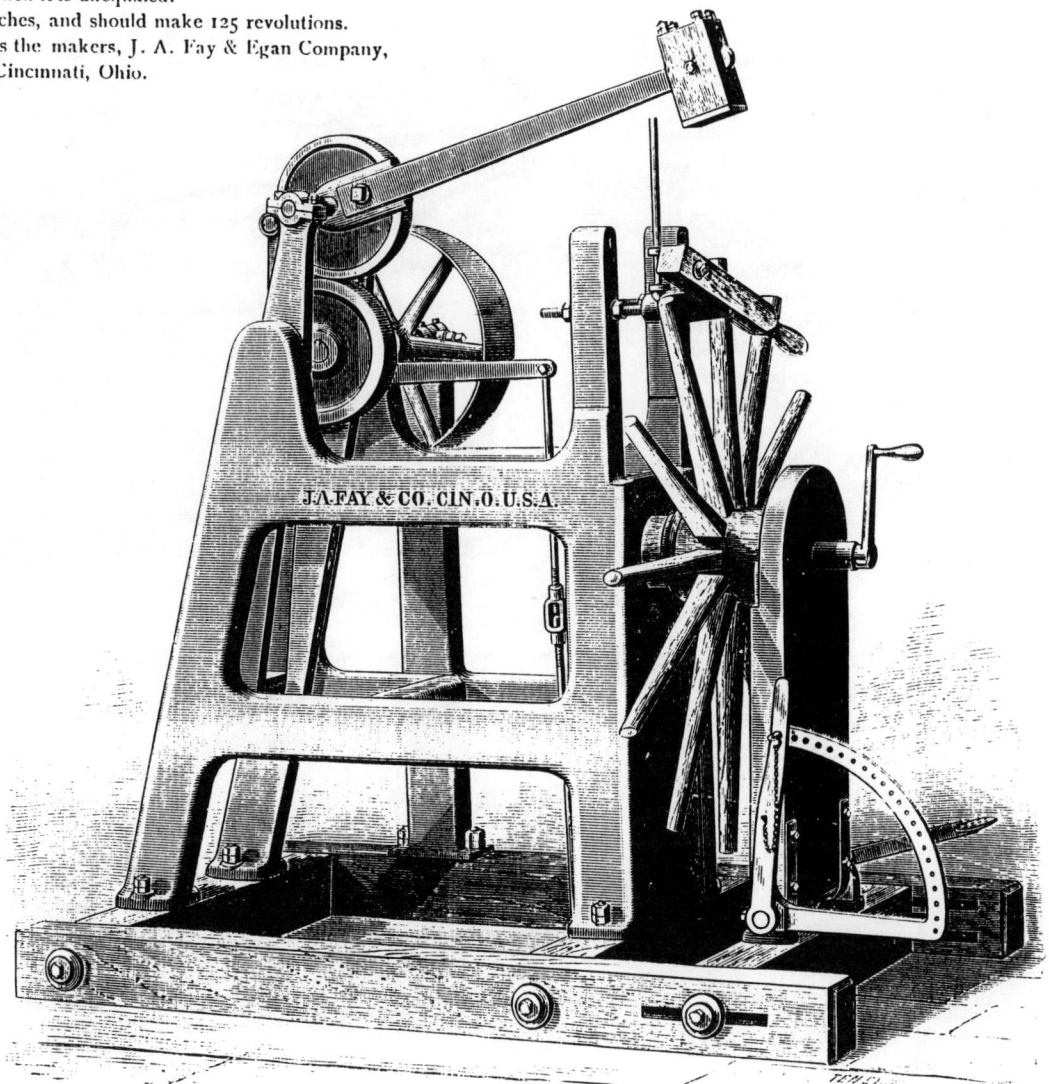

HOSLER'S PATENT SPOKE DRIVING MACHINE.

Spokes

Automatic Combined Spoke and Handle Lathe.

MANUFACTURED BY THE DEFIANCE (O.) MACHINE WORKS.

This automatic machine is used for turning and squaring spokes for wagon and carriage wheels, having necessary adjustments to turn common, Sarven-patent or sharp-edged shapes, making either light hickory spokes or heavy spokes for wagon, truck and artillery wheels, up to 5 inches diameter, 42 inches long at the largest, covering every requirement for rapidly and accurately producing spokes of every variety and size. It has a capacity of 2,500 to 3,000 pieces per day.

The body of the machine, supporting the working parts, is a massive casting in one piece, having cored center and broad base, and is very stiff and reliable, occupying over all 6 feet by 3 feet 6 inches floor space. The cylinder is composed of a sufficient number of cutter heads placed side by side upon a 2¼-inch steel spindle to fill the length of turning. Each head is provided with three cutters, with 3-inch face, which lap over each other, forming a continuous cutting edge over the entire length of cylinder, to turn the full length at one cut. The heads are secured to the spindle by friction binder.

The table is constructed in two parts, and it is gibbed, and slides upon the frame in angle ways moved to and from the cutters by either hand or foot lever; the upper portion supporting the centers is pivoted to the lower half near the tail center by a steel pivot, in one of the several holes through the table, upon which it vibrates for oval turning. At the opposite end on the head center spindle a cast iron cam is placed of whatever shape desired to turn, the cam rides against an upright shoe extending up from the lower table, and is held snug against the shoe by a coiled spring. When the table is moved toward the cylinder to where the turning shall begin, an automatic feed slowly rotates the object to be turned, and the cam revolving against the shoe oscillates the upper table in a path corresponding with the shape of cam. When the pivot is placed directly opposite the tail center, the machine will turn the material round at the tail center end with a gradual change in shape toward the opposite end, at which point the turning will agree with the shape of cam. Long oval or irregular turning, when both ends are required to agree in shape, is turned with the vibrating table locked to the lower half, with the cam revolving against a shoe fastened to the frame, thus vibrating both tables alike at each end. The diameter of turning is regulated with graduating screws, having adjustments sufficient to turn work from ½ inch to 5 inches diameter.

The tail center can be quickly adjusted to the desired distance from the spur center for short or long turning, or at right angles for straight or taper turning. The swinging cutter head advances and retreats from the work automatically, its position is governed by the movement of the table, it is brought down to its work at the same time the turning commences, and when the table is moved backward to remove the turned material from the centers, it is lifted out of the way by spring balance; its action upon the turning is governed by a cam upon the live center spindle, and it will follow the path of either a square cam for squaring the head of spokes, or oval, oblong, hexagon or octagon shapes suited to finishing the eye end of handles, having the necessary adjustments to turn tapering in either direction, as well as the different diameters.

The operation of this machine is very simple; no expensive labor is required; the rough blank, either sawed or rived, is placed between the centers and when presented to the action of the cutters revolves slowly and is turned its full length at one time, very smooth and to exact shape, requiring little if any finishing after leaving the machine. The material is placed into and removed from the machine without stopping. The cams and cutter heads are numbered, rendering changes from one class of work to another simple and easy to effect.

The Hub, July 1893

Spokes

Rotary Feed Double Spoke Throater.

MANUFACTURED BY THE EMPIRE MACHINE WORKS, MOUNT
MORRIS, N. Y.

The empire rotary feed double spoke throater has been on the
market for several years, and is in operation in all parts of the United
States and Canada, on all kinds and sizes of stock. It has given good
satisfaction wherever used.

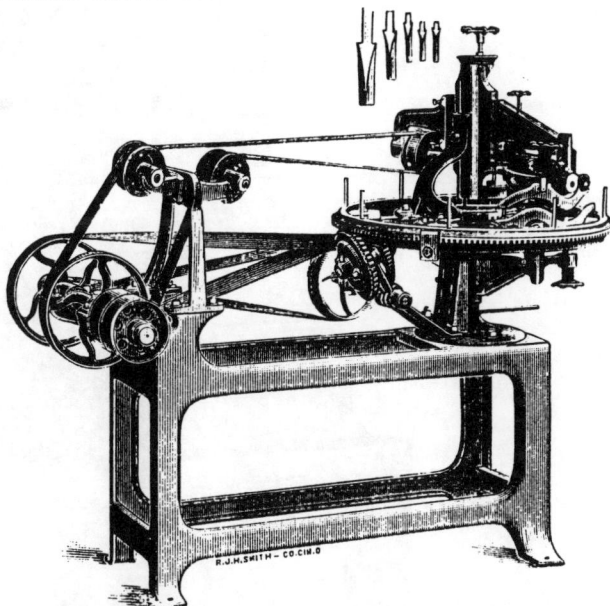

The feed is rotary, the felloe end of spoke traveling much faster
than the tenon end, by which a greater variety of shapes can be produced than in any other way. The spokes pass between two cutter-
heads, the outer end being guided partially around each head by two
spring-pressure guides, which are adjustable in every direction. The
tenon end moves in a horizontal plane, and is held by a double
pressure, each acting independently, holding tenon ends of vari-
able thickness, without cramping the machine.

The cutter-heads are turned on every side true to center:
the knife is made on a small circle, and, when ground to proper
shape, will cut down on the grain and not sliver the spoke.
The machine will throat both sides complete, as fast as they
can be laid on the feed, about fifteen thousand per day.

The Carriage Monthly, Aug 1893

Spokes

IMPROVED SPOKE TENON COMPRESSING MACHINE.

(See Illustration Accompanying.)

AFTER the tenon and miter of a wheel spoke has been planed, and immediately before driving the same into the hub, it is compressed and molded with a series of line ridges on both the face sides. The object of this procedure is at once understood. The spoke has been planed slightly full, so that the driving of same into the hub would be somewhat difficult, and the glue on the tenon of the spoke would be shaved off the face on account of the fulness of the tenon.

The tool illustrated compresses and grooves the two face sides of the tenon, so that it enters easily into the hub mortise, while the grooved surface retains the glue, so that it is brought into the deepest part of the mortise. After seating the spoke, the compressed spoke tenon will expand quickly to its former size, filling the mortise fully and combining the faces of the tenon with the sides of the mortise by means of the retained glue, thus insuring the best and closest fit obtainable.

The machine performs this work quickly and in a most satisfactory manner. The general arrangement is that of a power punching machine. The compressor proper consists of a heavy "power-punch shaped" casting, which rests on a strong frame. The gear wheels and pinions and mode of driving are well shown in the illustration. A heavy steel shaft with eccentric or crank device operates the upper sliding head and dies, which are adjustable and removable for various sizes of tenons. The upper head carries on the right side an adjustable inclined sliding head which operates at its up and down motion on the side die of the form, moving the same back out of the way so that the spoke can be entered, and placing the dies at the down stroke in proper position to retain the width of the spoke tenon while pressure is applied on the two faces of the tenon to compress same.

All these parts are quickly adjustable and exchangable for the various sizes of spoke tenons, and the operation of the machine is quick and positive in action, producing accurate and uniform dimensions. All surfaces which are subjected to wear are of extra length and adjusting compensation for wear provided, and the machine is highly finished and well built throughout.

Manufactured by The Bentel & Margedant Co., Hamilton, Ohio, U. S. A.

IMPROVED SPOKE TENON COMPRESSING MACHINE.

The Hub, Dec 1894

V
FELLOES/RIMS

The Long Rim Plan

Felloe/Rim Machinery

FELLOES/RIMS

The subject of the felloe, or rim part of the wheel, was by no means exempt from the continual discourse and exchange of opinions published on almost all aspects of wheelmaking. The "Long Rim Plan" debate, beginning on p.117, is a good example. The argument was whether to leave a slight gap (usually a sawcut width) where the rims are joined, in order to allow for "drawing up" when the tire is installed. This prevented the wheel from becoming "rim bound." Sectional felloes were handled this way in early shops. With the "Long Rim Plan," invented in 1870 by Samuel Toomey, the bent rims are joined tight, with no gap.

The hand-method of wheelmaking practiced in the early wheelwright shops involved cutting the felloe sections from rough planks with a felloe saw, a type of frame saw with a narrow center blade that cut on a curve. A felloe pattern was used to mark the planks for each size and type of wheel with each section covering two spokes. After cutting, the inside surface of the felloes were smoothed with a compass plane and the edges rounded or chamfered with a spokeshave. When the wheel spider (hub with spokes installed) was completed, the distance between spokes was determined by transferring measurements directly from the spider, as described in the article on p.112, or by the trial and error method of walking the wheel circumference with dividers. As advances in standardization were made, tables were developed that gave these spoke distances for various size wheels.

However, it was the development of the bent rim in the mid-19th century that (along with the patent hub) probably changed the actual mechanics of wheelmaking more than any other technical advancement. According to the letter on p.56, the first bent rim was made by James Hansen of Saugerties, New York, in 1835; the first patent for bending wheel rims was the Reynolds patent in the same year. However, it wasn't until after 1850, with the introduction of Thomas Blanchard's wood bending machine, that the two section bent rim became practical, resulting in less breakage of oak and hickory. The technology of steam-bending allowed rims to be made in two, instead of the traditional six or seven, pieces. With only two joints to be fitted, the result was a stronger wheel.

The article on p.128 notes the interesting fact that by 1895 a rim bending machine was in use which could bend a piece of hardwood with a thickness up to 4", a width up to 9", and a diameter anywhere from 18" to 72". It was claimed that the machine could bend 1800 medium-sized felloes in ten hours.

Wheel-Horse and Mallet.

SPOKE SHAVE, EDITOR.

Having got my spokes in the hub, I shall, in the present article, try and explain, as best I can, how to rim the same correctly. In cutting the finger tenon, we must be careful to get them on a straight line with the face of the wheel; to do this we lay the face of the *hub* on a bench, which is perfectly straight and true; we will then measure the extreme points of the two front spokes from the bench, that is to say, the two spokes immediately opposite each other.

The size of the tenon depends upon the tread and body of the spoke. If the tread is $\frac{7}{8}$, and the body of spoke $\frac{3}{4}$, you will make your tenon $\frac{1}{2}$ an inch; if your tread is $\frac{3}{4}$, you will cut your tenon about 7-16 of an inch. This will give a good supporting shoulder, which is so essential to the wheel, as you will discover when we get to wedging up the rim. You will also observe that this size tenon will give more strength in your rim.

Having our tenon cut, in the preparation and boring on the rim, we take our *felloe*: if it should be winding, which is invariably the case, to some extent, we true them up as best we can; we then mark off for boring; face them off perfectly true; and gauge off, so as to bring our tenon, when dressed off, in the centre of tread; if we have no machine for boring, we gauge off on both sides, and bore accordingly. After boring, we drive on the *felloe*, but before driving it on with a chisel, take off the inner edge of the hole, so as to let our shoulder in, or wedge without throwing the strain on shoulder of spoke, immediately around the tenon. This will prevent your shoulder from chafing; we then drive our rim on, cut out what joint we want, but here you must be governed by the quality of your timber, as some felloes will wedge up more than others. When we have got our joint out, we then take a straight edge and see how our spokes are, and if there are any crooked, we mark them on the rounding side, that we may know how much crook there is in each spoke; we must know what our marks mean; when this is done, we take off our felloes, round them up; when rounded up, we go back and examine our marks; we then shave off the tenon on the rounding, or the side our mark is on.

If our mark indicates the spoke to spring a great deal, say $\frac{1}{8}$ or 1-16, we must exercise our judgment how much we must take off; we must here bear in mind our quality of spoke, if we wish to be correct in our judgment; in shaving off, we must not take it off all the way up to the shoulder, but commence according to the amount to be taken off; when we have completed this, we then drive our felloe on again; next take the wedging chisel and drive it in on the hollow or opposite side of our mark, leaving the heft of tenon on the side of mark; we then insert our wedge; now force the rim up by hammering from the hollow, or opposite our mark on spoke; this will bring the spoke straight, provided we do not take too much off of it. If we should take off too much, we can remedy this by hammering on the rim where the mark is.

In wedging up, we must settle the shoulder up square on the rim; this prevents the spoke from chawing off at the shoulder; it gives permanency to the wheel. Having our wheel wedged up, we then dress off the same. The tread is the next point of interest; this we must be careful about; we take a plane, and dress the tread off by a square, that is, when the square is laid level on the tread, it strikes the face of rim on the opposite side; this is done for the reason that, when the tire comes on, it will draw the wheel the same all around, leaving the wheel as it was when we was preparing to cut our finger tenons.

Having got to the tire, we shall, in our next, have a brotherly confab with V. Ulcan.

Coach-makers' Int'l. Journal, Mar 1869

OILING WHEELS.

After the tire has been set, before the wheels are painted, if they could be placed in a vessel of oil, and the entire wheel completely submerged in in oil, for several hours, it would be of great benefit to them, by preventing every portion of the wheel from shrinking; and would render them very durable. As the wheels are exposed to mud and water more than any other part of the vehicle, they will absorb much water; and even when they are well painted water will enter the felloes at the joints. The result is, the wood swells, and when the wheels become dry they shrink; so that sooner or later, the tire will become loose. But if every part is well saturated with boiled linseed oil, it will effectually exclude the water, and prevent entirely the shrinking of every part, even when exposed to mud and water.

But, as the process would be rather impracticable as a general thing, I have practiced oiling only the felloes, by placing the wheel erect, with the rim in a trough, two or three inches wide, and six or seven inches deep, filled with oil. After one felloe has absorbed as much oil as it will during thirty or forty minutes, or longer, let the wheel be turned around until every felloe has been saturated; this process fills the tenons of the spokes with oil, which would otherwise be filled with water. If the other parts of a wheel received a thorough oiling, as they should, wheels will last an age, and literally *wear out*.

I have often treated wheels in this manner when the tire was loose but little, and the felloes have swollen to such a degree that the tire became as tight as if it had just been re-set. I have often seen scores of new wheels made of poor timber, and poorly made, which if well saturated with oil, would wear more than twice as long as if they were merely painted. Sometimes when the spokes work a little in the hub, this process will render them so tight that they will not give under a heavy load.

It would be as well to paint the trough on the inside, before using it, to prevent its absorbing the oil. A sheet iron trough would be as good as a wooden one, and would cost but little.

The Coach-makers' Magazine, July 1855

SHOULD TENONS BE FLUSH WITH THE RIM.

Mr. Editor: I remember that two or three years ago, The Hub, when in its small form, had a good many letters upon the subject, "Is it best to have the tenon flush with the outside of the wheel?" and quite a variety of opinions were expressed. Will you please tell me what was the result of the discussion, that is, what seemed to be the majority vote?

D. George Emerson.

Editor's Answer.—The discussion was closed by the following note by us:

"We would request that this conclude the letters on this subject. We have had the written opinion of eight practical wheel-makers, besides several talks with others, and the common opinion seems to be this: *Make the tenon of the spoke so very nearly flush with the outside surface of the rim, that you know it will become perfectly so when drawn together by the tire.*"

This position has received no contradiction in the mean time, but was, at the time of publishing it, approved of by a number of new correspondents.

The Hub, Sept 1872

BOILING FELLOES IN OIL.

Mr. Editor: One of your correspondents once expressed the wish to hear from others on the above subject, and having had some experience in the business, and not seeing any other replies, I will briefly state my views. Take a hard, close-grained piece of second-growth hickory; the pores are small and filled with air, and, however dry, with a little water also. Put this in hot oil. The heat converts the water into steam, and expands the air so that it is forced out of the timber, but I think very little oil goes in until the stick is taken out; then, as the remaining air cools and condenses in the wood, the air on the outside presses in what oil remains on the stick. The agitation or boiling that takes place while the wood is in the oil, is produced by the gas escaping, and not, as some imagine, by the oil penetrating the wood. I know that very little oil goes in, as I have boiled wheel after wheel in the same dish, and could see but little difference in the quantity of oil remaining. I think I could put on about as much with a paint-brush. To put the thing to a practical test, I once took a set of wheels, boiled the felloes of two wheels in oil, and the remaining two left unboiled. I then painted them, and put them on a business-wagon. When the tires needed setting, I tried to find the two boiled wheels, but failed to discern a particle of difference. The oil also loosens the spokes in felloes. I agree with L. Henry, that just as good wheels can be made *without boiling* as with. We sometimes make use of the above process, however, to straighten felloes at the joints when they are bent in too much.

Theresa, Jefferson Co., N. Y.   M. S. Stotler.

The Hub, Nov 1875

Fastening of Felloe Joints.

Undoubtedly the weakest point of a well constructed wheel, is in that portion of the rim where the two ends meet, when made as is the common mode. First, the wood-worker inserts a wood or iron dowel, the wheel then goes to the smith shop, where it is further weakened by the insertion of two bolts, for the purpose of securing the plate which is ostensibly put on for the purpose of strengthening and more securely fastening the joints, hoping the natural tendency of the ends working to one side or the other, by this means will be obviated, the achievement of which is secured, but at the expense of having cut away so much of the wood, first, parallel with the grain, and then across, severing the parallel grain through the whole depth of the felloe, thus forming a very weak portion of the rim. The natural consequence of which is, the tendency of the ends to split, and eventually to bulg out, notwithstanding the malleable clips, which are turned down from the plate, for the purpose of counteracting this inclination to split. The felloe is then occasionally further weakened (though seemingly for the time being strengthened,) by the insertion of a rivet, passing through from either side of the rim, which shows after a few weeks usage, by the giving way entirely of the felloe, necessitating its being replaced by a new rim.

This source of trouble has been experimented on to quite an extent. Perhaps among the first was the placing on the ends, covering the joint, a snugly fitted ferrule, which answered admirably the purpose of securing the ends from the natural tendency for splitting, but necessitated the extra outlay of time and labor, and unless great care was observed, the smith was in the dark as to the required amount of draft to be given the tire. This ferrule was used considerably a few years since, but gradually fell into disuse, being superseded by the cheaper and easier mode of using a plate secured by bolts.

One other plan was the doing away with the dowel in common use, by using one that should pass through the joint of the felloe, from the tread to the inside of rim, similar to the bolt B, Fig. 1, not having however a head. It was supposed that a dowel placed in this manner would answer every purpose of the parallel dowel, in such wheels as were properly made. The smith was then to further secure it by means of the common plate and bolt.

But without a doubt, the best, as well as the cheapest mode for fastening these joints, is the one now being used by C. S. Caffrey, of Camden, N. J., and we understand was originated by him, but not secured by patent. The manner will readily be understood upon consulting the accompanying cuts. A, Fig. 1, shows sectional view of rim. D, D, the plate. B, the bolt

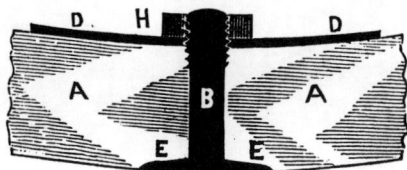

Fig. 1.

E, E, the head. Fig. 2, shows the manner in which the bolt should be constructed. B, the flat portion, about ¾ inch wide, made quite thin as represented at E, E, Fig. 1, and of sufficient length between lips A, A, Fig. 2, to suit the required size of felloe or width of tread.

In using this commendable fastening, the wheel maker having finished his wheel, giving it the required opening, and having prepared a mixture of glue and sawdust, or chalk, a sufficient quantity is inserted in the joint, to insure its holding the ends in position, when it shall have hardened; time having been given for the accomplishment of this, he bores the hole through the joint the required size, for the insertion of the bolt, as shown at B, Fig. 1, and cuts away a portion of the rim on the tread, each side of the joint, to insure the outside of the bolt head sinking in, so as to be even with the tread. The bolt is then inserted, the plate D, D, Fig. 1, which is of malleable iron, similar pattern to those in common use, with exception of having but one hole, is placed in position and confined by the nut H. After the tire bar has been set, the nut is tightened if required, the sides of the plate pounded or pressed (and also the lips A, A, of the bolt, Fig. 2,) firmly against the wood.

By the use of this fastening, the dowel, as well as one bolt is done away with. The whole

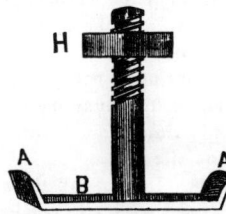

Fig. 2.

strength of the wood is retained, and as the tire can be screwed to the rim further from the joint, there is produced a wheel of greater strength, at so important a point, and to a great extent if not wholly doing away with the liability of splitting.

The Carriage Monthly, Oct 1877

HOW TO RIM WHEELS PROPERLY.

WOONSOCKET, R. I., Nov., 1881.

TO THE EDITOR—DEAR SIR: Some time ago we asked the important question, How to rim a set of wheels properly ? It seems an easy matter to spoil a set by improper rimming. Suggestions will be thankfully received.

Yours truly,   J. G. & J. B. PROCTOR.

ANSWER.—To properly rim wheels it is necessary first to have tools adapted to the work, the most important of which is one to cut the tenons upon the spokes. To make a durable wheel these tenons must be cut true with the face of the wheel, and have their shoulders square with the tenon. This is the first and most important part in rimming.

Next in importance is *dry* material, well bent to the *proper height*, which should be about one inch higher than the wheel upon which they are to be used. The rims should be of a perfect semi-circle, and not twisted sidewise. By casting the eye over the edge of the rim, any deviation in this direction can be seen. Each piece should be carefully faced up "out of wind" with a fore-plane, and the inner surfaces squared with this face.

Having brought each rim true, the wheel should be firmly secured to a solid horse or bench, and each half rim opened out evenly to pass over the ends of the tenons. Each spoke should be marked on the rim, that its exact position may be maintained when the rim is bored and driven on, a point midway between the spokes being selected to make the joint in the half sections.

Boring the holes in the rims is another vital point in rimming. In no case should more than one-half the width of the rim used be taken out in boring ; and less than this amount is advisable when the spoke tenons will admit. These holes should be the exact angle of the spoke upon which they are to rest ; any deflection from this line will cause the spoke to spring under the draft of the tire. The holes should also be bored nearer the face of the rim than the back, the amount of which should be governed by the tread given the wheel, enough being allowed to bring the tenon in the center of the tread when finished.

After boring, the rim can be rounded. In doing this the wheelmaker follows his fancy, the important point being—not to let fancy interfere with the *strength* of the rim. After rounding and shaping, small holes are bored in the ends of the rims for *wooden dowels*, which can be cut by the smith in setting the tires.

The rims can now be driven on. Do not tap them on with a light hammer, but settle them down firmly with a heavy hammer, and see that every spoke has the rim *solidly* down upon its shoulders all around. If tenons and holes have been properly fitted, the rim will go to its place as if it grew there. No wedges should be needed to keep it there, though there are occasionally cases where wedges are necessary.

Keep the joints up tight, and dress off the rims with a plane and file, to suit the style and the price.

To recapitulate the main points of the above, I would say, Be sure you have accurately cut tenons upon spokes, rims of *dry* material, properly bent, and holes in rims absolutely true with spoke tenons ; rims hard down upon the shoulders of the spokes, without straining them from the position given in driving. The observance of these conditions, together with an average amount of mechanical skill, will produce durable results in rimming wheels.   H. M. DUBOIS.

The Hub, Dec 1881

Oiling Wheels.

SYRACUSE, N. Y., June 8th, 1883.

WARE BROS.

Gentlemen :—The question has come up with us whether or not it is a good plan to oil carriage-wheels *before* setting the tire. In contending *against* the practice, a good wheel-maker argues that oil has a tendency to soften the timber, allowing it to split easier and also causing the spokes to become "upset" at the shoulder. Give us your opinion, and oblige,   Yours, truly,   INQUIRER.

The claim that a coat of linseed oil softens timber is new. That softening makes it split more readily, is refreshing. That oiling the bearings for the rim so as to keep the water out should cause spokes to upset, astonishes us.

The effect of oiling wheels with hot linseed oil when they are taken from the finisher is to keep the grain of the wood from raising, and to protect them from absorbing moisture. When dried, it hardens the surface of the wood so that it takes a better polish, filling as it does to some extent the cells of the timber. The hot oil penetrates the shoulder-bearing of the spoke under the rim, and keeps the water from softening it, which causes the wood to bruise up. It also goes into and around the round tenon, and preserves it from the effects of water.

The only objections to oiling are that it stains the wood ; that those who make patent fillers object, because it interferes with their business. Some painters prefer other filling. The wheeler who prepares the wheel for the smith objects to being compelled to wedge the round tenons to keep the rim down to the shoulder, because the oiled tenon does not hold it well.

The Carriage Monthly, July 1883

METHODS OF DOWELING AND TANGING WHEELS.

(See Illustrations Accompanying.)

THERE are no branches of vehicular construction to which inventive talent has directed itself to with more persistence than in that of wheel construction. The solid iron hub, the metal and wood hub combined, the sectional rim, the bent rim, the India rubber tire, the pneumatic tire, and many other devices, all tending to and claiming perfection at their point of application, have for years been knocking at the ears and senses of the trade. Yet still the inventor comes forward with still further

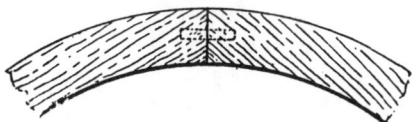

FIG. 1.

claims of perfection. Such persistency shows vitality, but it frequently happens that the inventor has hit upon an idea which is old in its application but new to him, and the patent laws give us ample proof of the fact of new men patenting old and worn out ideas and plans. In the illustrations on the doweling of felloes are some of those ideas which have been patented, but which are almost as old in their applications as wheels are themselves.

In Fig. 1 is shown the common method of doweling across, square to the joint. Now this is the straight way of doweling when the felloes are got out of a cross grained plank, so that the dowel rests upon severed layers of the grain of the wood; a felloe fixed in this way with a dowel will never split across, because it is supported by the sections of the grain in a contrary direction to the weight acting upon it. In Fig. 2 is shown a felloe cut to the grain of the curve of the wheel and the dowel bored square to the joint. The grain in this case is at right angles to the dowel and acts like a wedge to the weight acting upon it, causing the felloe to split in the line of the dowel's fixing. In Fig. 3 is shown a *vertical dowel*. This acts

FIG. 2.

neutrally, and simply holds the *felloes* in their place sideways, which is the function of a dowel, and cannot act as a wedge nor work any other effect than that of a holder. The dowel is *bored* down the center of the joint of the felloes, the joint being held firmly with a clamp and wedged up to its place for *boring;* then the dowel hole is bored and the dowel driven in in this way all round the wheel when fellowed. The plan has good points, say in jointing felloes before tiring. The joint can be cut right through and a fresh dowel inserted without

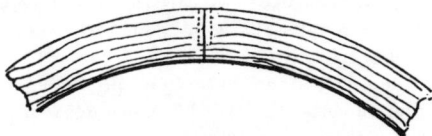

FIG. 3.

driving the felloes back with the chance of breaking them at the spoke. This idea formed the subject of a patent, but it had been done by wheelwrights before the patentee was born. In Fig. 4 is a method of tanging wheels, which commends itself for maintaining strength over the shoulder of the spoke, preserving a greater strength of timber at this weakened part of the felloe. This has also been patented, but old wheelwrights and those who are not so very old have repaired wheels in this way, years before the idea was patented. Such patents have no val-

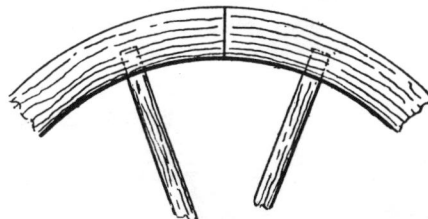

[FIG. 4.

idity at law, and are tolerated by the trade because of the expense of fighting against such encroachments on the rights o the trade. The idea is a good one, and when well and carefully done, so that the shoulders and end of the spoke take a bearing together and the fit of the tang of the spoke in the felloe is perfect, a good and lasting job is obtained, but there is no necessity to be afraid of a patentee, as the ideas are the common property of the trade, and of the world.

The Hub, Aug 1895

New Features in Wheel Making.

LAMINATED RIMS.

BY T. H. MITCHELL, U. S. PATENT OFFICE.

Wheel making is one of the oldest of the useful arts, but is none the less interesting by reason of its honorable age. The great diversity of uses to which wheels are put, and the constant changes necessary to adapt them to new requirements, together with their close relation to the using public, has made them the subject of much thought and many patents.

Of late years a new field has been opened in the construction of wheels, made necessary largely by the introduction of the pneumatic tire on light wheels, such as velocipedes and sulkies, where the object is maximum strength with minimum weight, coupled with durability and resiliency.

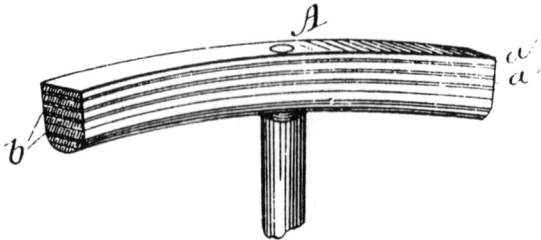

No. 235,176.

This style of wheel rim does not seem to be one of the many so-called "fads" to which the cycling world is so much given, but has merits of its own, and it seems has come to stay.

Laminated or multiply wheel rims seem to be the favorite form just at present, as they are alleged to possess certain merits on account of their peculiar construction. This form of wheel rim is not, however, new, they having been used in wheels with the old form of thrust spokes as early as 1869, of which year patent No. 87,823 has this claim : " Forming the rim of a strip or strips of wood

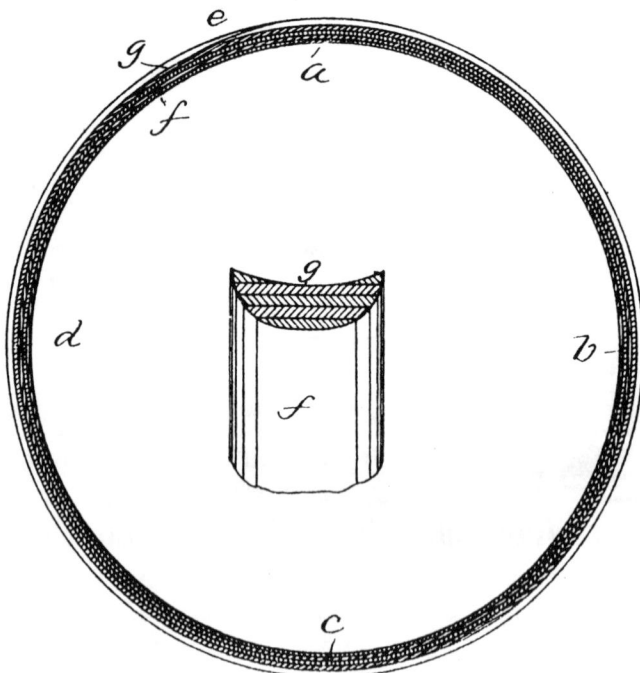

No. 496,971.

wound around a form of required shape until the rim shall have attained the desired size or thickness, the successive layers or folds being united by glue or cement as they are layed on each other." And again in 1880, patent No. 235,176 shows a wheel rim or felloe of the laminated type, and claims :

"As a new article of manufacture a wheel felloe or rim made up of a series of thin layers or wood saturated with strips of elastic substance interposed between them and the whole glued together."

Such rims as the above were not thought to be adapted for use in suspension wheels until recently. Suspension wheels differ from thrust wheels in that in the former the load is supported from the rim by the tension of the spokes, while in the latter the load is supported by the hub and from below by a direct thrust of the spokes on the felloe.

Metallic rims are objectionable on account of their rigidity, and liability to bend and batter. This defect is overcome to a great extent in the wooden rim, which is said to be equally as strong with less weight and possess a very desirable resilient quality. The popularity of this form of rim is attested by the vast number that has gone into use in such a short time. Almost the entire output of velocipedes for this year have been equipped with this form of wheel rim.

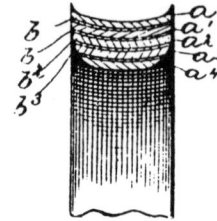

No. 509,259.

There were many defects and difficulties to overcome in the practical use of the wooden rim on wheels with tension spokes, principally on account of their tendency to split along the line of the spoke fastenings or nipples. It was to cure this defect that the laminated, or built-up rim, was again suggested in this connection. In patent No. 496,971, of 1893, illustrated below, is shown and described a wheel rim similar to the one shown above, but intended for use in a suspension wheel, and in which the grain of the different laminæ are arranged at different angles.

It is described as a rim for wheels comprising in its construction a series of sections or plies of wood of varying course or direction of grain cemented together, the ends of each section breaking joints with the ends of adjacent sections, and the inner surface being of convex form and the outer surface of concave form.

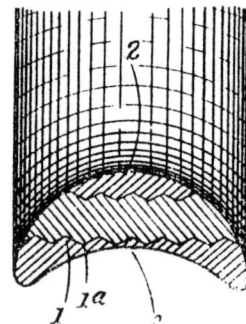

No. 528,666.

Another manner of constructing the laminated wheel rim is shown in patent No. 509,259. In this rim the layers or coils are shaped separately, the outer rim being coiled within or upon a suitably shaped mold or mandril, on which it is pressed to the desired shape in cross section ; the second layer or ply being pressed upon or within the first one after the two have been coated with a proper cement or glue. This rim is preferably built from the outside ply inward, as the strips or ribbon can be better held in place in this way. Alternate layers of wood and textile material, as canvas, are used to prevent warping and splitting.

In patent No. 528,666, of 1894, the laminæ constituting the rim are grooved and corrugated to prevent lateral motion and better hold the cement.

continued next page

Felloes/Rims

It has also been proposed to make the rims of these wheels of paper, in strip form, wound upon itself to form a plurality of layers or plies, and cemented together and compressed to form a tire groove. In such a rim the first layer is usually made broader, and is folded around over the other layers, forming an envelope. This rim is described and the machine for making it is shown in the patent No. 522,047, of 1894. It has also been proposed to construct these rims of vulcanized fibre, by molding and compressing it in an endless form, and treating it with some waterproof material. Again, the ends of a single strip of wood are joined up by splicing, forming a hoop, which is afterwards turned to the proper and desired form.

placed under the spoke-head or nipple, having spurs at the corners which sink into the wood, and, as the spoke is tightened, tend to draw the sides of the rim together.

Patent of March 19, 1895, No. 536,089, shows a crown-shaped washer, with toothed or serrated marginal flanges, and a countersink on top to receive the head of the spoke nipple; this washer is set in a counterbore in the rim, and is drawn to place by tightening the spokes, and, is claimed, has a tendency to prevent the splitting of the rim by the spokes.

A number of other devices have been proposed and some patented with this same end in view.

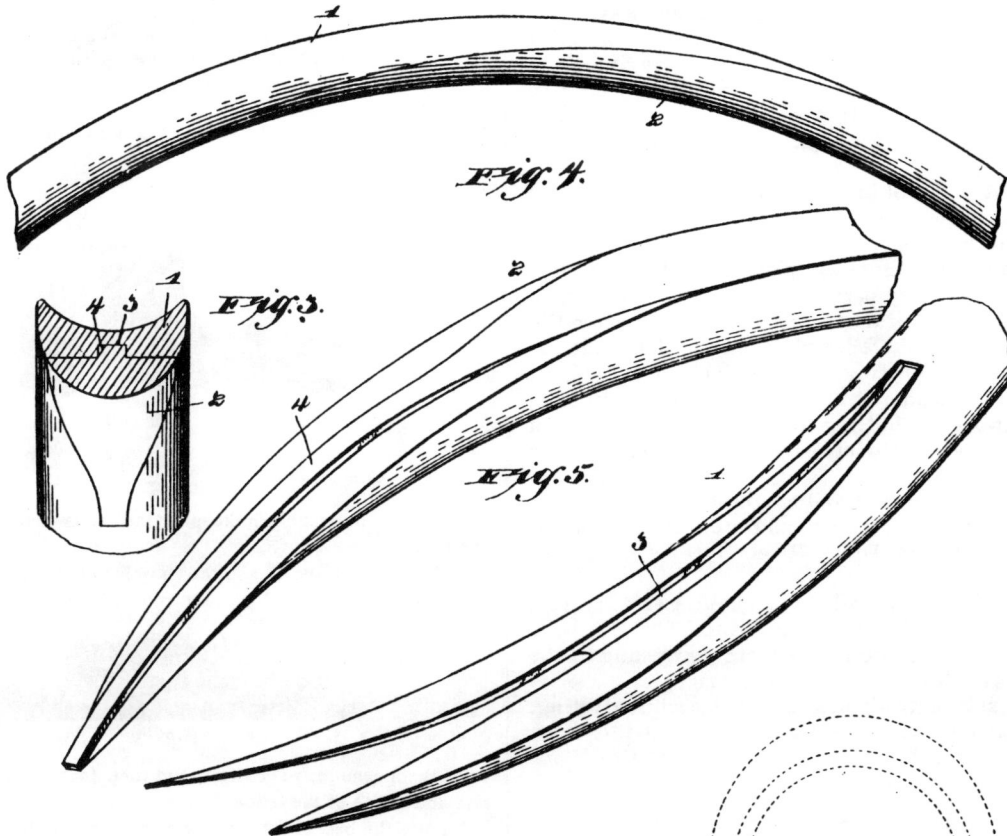

No. 521,187.

Where these rims in suspension wheels are made of a single piece, or hoop, joined at the ends, it has been found very difficult to preserve the true circular shape, and prevent sagging or bulging or lateral displacement at such joint, and it has been the object of several recent patents to remedy this evil. One form is shown in patent, No. 528,741, of 1894, in which a number of interlocking tenons, convex in cross section, fit a like number of sockets in the opposite end of the rim, which sockets are concave in cross section. This tends to prevent either lateral or up and down movement of the ends of the wheel rim.

Patent No. 521,187 shows a long scarf-joint, with corrugated meeting faces to better hold the glue. Cloth is sometimes cemented over the joint to strengthen and protect it. Another, and rather peculiar form of joint has recently been patented, No. 541,119, which can be better understood from the accompanying cut than from a mere description. The contiguous overlapping beveled portion at the ends of the rim have a corresponding groove and feather, or tongue and groove, which also tapers so as to fit each other from point to base, forming an interlocking key on the meeting faces, which tends to hold the ends in proper relation.

It has been found that in use these wood rims have a tendency to split or crack along the line of the spoke nipples or attachments, and to prevent this a number of devices have been resorted to and several patented.

Patent No. 509,260, of 1893, shows several modifications of a device for the purpose, the leading feature of which is a metal washer, to be

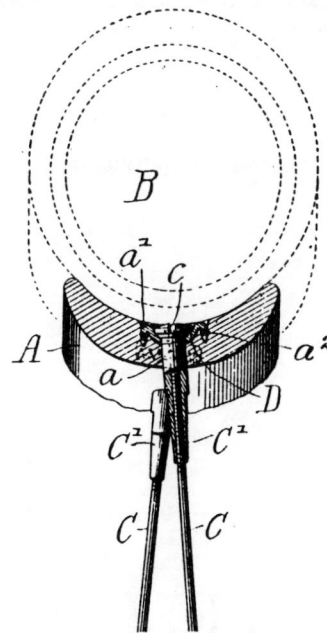

No. 536,089.

That the end is not yet, and that perfection has not been reached, notwithstanding the antiquity of the art of wheel building, is evidenced by the large number of applications constantly filed in the Patent Office for improvements in this line.

There are now over one hundred patents in the sub-class of felloes and rims alone, besides 650 in wheels proper and 600 more in tires, which relate largely to the same art.

The Carriage Monthly, Dec 1895

STEVENS' COMBINED FELLOE PLATE AND JOINT SUPPORTER.

HERE is something new in the line of wheel making, something that is intended to strengthen one of the weakest points, a support for the rim joint that is strong, attractive and cheap. The application is in using the strength of the short side of the rim, which will not break down, in combination with the plate, the principle of which is that of the steel beam; i. e., the metal in such a position that the strain comes edgeways. By the use of this plate a perfect support is obtained for the joint in the rim of the wheel. It obviates the knocking in of the joint, which is particularly noticeable on wheels run on city pavements, and in addition to this admits of the plate bolts coming near the spoke, thus to a great extent lessening the jar on both bolts and plate. In resetting the tires, when it is necessary to cut a little out of the joint, the plate can be turned back, giving the saw a clear chance.

By actual experiment it has been found that the long side of the rim will break down before the plate and before the short side of the rim gives in in any part.

It will be observed that, although the bolts come nearer the end of the rim than in the old felloe plate, on the short side the bolt holds the rivet from splitting, and the rivet holds the bolt. The rim on a wheel run a short time will sometimes split sideways, due to the two ends butting together, but by the use of this plate the splitting is obviated to a great extent, as the bolts come nearer the joint.

The plates are made of the best and toughest steel obtainable, which is rolled expressly for them, and are struck with flat surfaces on which the nuts bear. Rivets furnished without extra charge.

Directions.—The joint to be made from 1¼ in. to 2½ in. from the spoke, depending on the size of the rim. The tire bolts should be put through first, forcing the plate down snug to the rim, then, after what little shaping is necessary, it is a good plan to clamp the side of the plate to the rim while riveting. Care should be taken to have the rivet fit nicely, as this is an important point. For sale by A. E. Stevens & Co., Portland, Me.

The Hub, Mar 1896

A NOVEL FELLOE-JOINT-PLATE.

(See three Illustrations accompanying.)

THE following description of a novel form of felloe-joint-plate appears in a letter from one of our subscribers doing business in Rhode Island, who requests that his name should be suppressed. He says:
"Please accept the inclosed for what it is worth. During the few

Fig. 1.

years I have been in the carriage business on my own hook, I have tried many devices for holding up the joints of the felloes of my wheels; but, until I hit on the present plan, the results were only partially successful— or, I might say, total failures. By the aid of the present device, how-

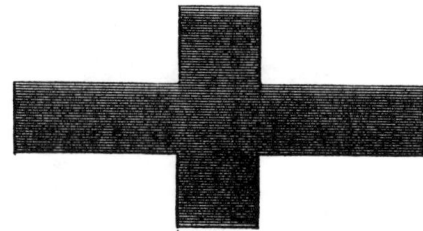
Fig. 2.

ever, I have no trouble in keeping the joints as full as any other part of the felloe.

"Sketch No. 1 shows you the combined felloe-plate and band. Sketch No. 2 shows you how I cut them out. Sketch No. 3 shows you how I

Fig. 3.

bend them around, weld them, and then form them upon a tool just the size and shape of the felloe.

"I use the best sheet-iron I can get, and my blacksmith knocks out a set in short order."

The Hub, Sept 1896

Felloes/Rims

THE "LONG RIM" QUESTION.

At the earnest request of friend PROGRESS, I must set aside my previous determination to delay noticing the question of the *Long Rim*. In doing so I shall simply declare the proposition, in my opinion, ridiculous. I do not wish to be understood to deal lightly with my friend's proposition, for the reason that, from his question, I do not understand that he is himself an advocate of such a theory. I am disposed to accredit PROGRESS with more scientific and practical knowledge than is represented in said proposition. I cannot for one moment think that any member of the trade, who has paid any attention to the construction in connection with the required labor of the wheel, would so far do violence to his mechanical genius as to go back on the principle as handed down to us by the earliest tradesmen in this particular. What I mean to say is this. It has in all times, as far as I am informed, been the custom to leave a certain amount of *joint* in the rim; and now for the wheeler, under the title of PROGRESS, to assume to the contrary, is to my mind preposterous. I have said that it has been in all time, so far as I am informed, the theory of our forefathers to allow for the draw of the tire. Now the question comes up for our consideration. We have been taught this fact, but the want of a circulating medium has prevented the diffusion of the reasons why.

Having the advantages for discussion presented by the JOURNAL, it is now our duty to examine carefully into this all-important question, and that in a spirit or *temper* which will reflect credit on ourselves, as well as throw light on the proposition as made by friend PROGRESS. My experience, abstracted by close observation, has confirmed my opinion that it is all important the wheel should be constructed with a certain proportion of joint. My reasons for so thinking are based upon the following causes and effects. In the wheel we find that it is necessary that it should be harmoniously put together, or, in other words, each and every part should bear its own proportion of labor. Should this point be conceded, the question of the long rim will be buried.

What are the effects or the result of the rim being left one-quarter of inch longer than the net circumference of the wheel? Is it not that the rim must stand off at the shoulder of the spoke? If so, we must, in order to bring the rim down on the shoulder, give the tire the necessary draw to press the rim together. We find we have produced the cause that has brought about this effect, *viz:* we have joined the rim; in doing so we find that the weakest points in the rim have been called upon to bear this extraordinary pressure. Now, where do we find the weakest point? I claim that the points where the rim is bored for the tenons are the weakest points. If so, we have got to buckle or bilge the rim immediately where the holes for the reception of the tenons on the spokes are inserted. If this be a fact, we must admit that the rim has been damaged to a great extent; consequently the whole wheel has been affected. The above we conceive to be the effect upon the rim directly. Should this be admitted, then it becomes our duty to inquire what are the effects upon the other portions of the wheel? To do justice to the rest of the wheel it will require a more thorough investigation, for upon this point rests the reputation of the wheeler; for the reason that the condition in which we find the rim when hooped is not to remain alone in the rim, but sends its effects directly to the tenon on spoke, thence to the hub or tenons on the spokes as inserted in the hub; for it must be borne in mind that the labors of the wheel are not supposed to be in the position as being suspended above the base, in which simple revolution was required. In this condition it would matter but little to the wheel, whether the rim was a quarter of an inch long or short; the fact that the wheel has the weighty responsibility of the entire carriage with all its varied weights of freight resting upon it.

Did I say that the wheel was taxed to do all the work? If I am to be so understood, I wish to correct. Is or is not one-half or two-thirds of the wheel idle as to real service? Does not the burthen of all the weight rest upon the one-third immediately under the axle. If so, what is the effect upon these upper two-thirds?

Having located the part that has to do the work, it next becomes us to consider the effects upon the other portions of the wheel. With the theory of the long rim, we must not lose sight of the fact that it is supposed that the wheel is performing revolutions with the weight upon the axle—the result is the burthen resting on this portion is all the time pressing on the centre of the wheel. Now, under the long rim theory, *and in all other cases*, the rim being too long, it does not support the tenon on spoke and the tenon in hub; or in other words the surplus on rim has all been forced to the upper portion of the wheel, causing or tending to force the rim from the shoulder of the spoke, either on rim or from the hub. If the spokes are firmly set in the hub, it may resist this raising or lifting from the weight below, and the long rim above; if your spokes are proof, or equal to the work, the tenon as in the rim must give way to meet the demand of this lifting or raising of the rim. Could it be possible to get an equal amount of weight on the upper part of the wheel to that which is carried to the lower, then the two portions running parallel with the base would suffer or consume their own portion and that part above, which we have supposed to have an equal amount with lower or burthen part of the wheel. This being in our opinion impossible. We must defer further remarks until our next.

———

Coach-makers' Int'l. Journal, Nov 1870

117

For the Coach-Makers' International Journal.
CANAL DOVER, O., October 10th, 1870.

"LONG RIM" PLAN.

MR. EDITOR:—Having been for some time a reader and a close observer of your worthy journal, particularly in the wheel department, I find by the different questions asked, and arguments produced on rimming wheels and setting tire, that it is becoming my duty as the patentee of the Toomey improvement in the mode of constructing bent rim-wheels, to give your readers part of my experience in rimming wheels and setting tire, and a few explanations in defence of the "long rim."

I commenced my trade as a smith in 1847, and in setting tire, the first thing I was taught, was that if the rim was close on the joint the wheel would rim bind, which was perfectly correct on the *plank* rim. But time brought the bent rim in use, and we believing that all wheels would rim-bind, of course set the tire as we had done on the plank or sawed rim; the result was the wheel would dish and the tire be loose. This we found was involving a great principle, which should be investigated. By the aid of proper glasses, and close examination of the wood after it was bent and well seasoned, the matter was plain that a sawed rim was in its natural condition and formed a solid arch, and as soon as the tire drawed all the joints close, the wheel would stop dishing, then it would become rim-bound. But in the bent rim we find no such arch; the grain of the wood being broken, the rim will adjust itself to the tire without the least injury to the rim. What I understand by rim-binding is simply this. So long as the wheel can be dished without cutting out of the joint, so long it is not rim-bound. All ironers of my age know that sawed or plank rims must have open joints, or they will not dish. This proves that the sawed rim is in its natural condition, and if the tire be set and drawn, so as to close the joints tightly, it will form a solid arch; while in the bent rim we have a broken grain and no arch, consequently can set no tight tire on the old system. When we used the sawed rims the question was asked by the jour, "How much shall I saw out of the joint to obtain the proper dish?" He knowing that if the rim was left too long the wheel would rim-bind. But this form of question has changed materially since the bent rims were introduced. Now he asks, "How much draft can I give so as not to dish too much?" We have, then, lost control of the wheel, and can set the tire only so tight as to avoid too much dish. If we set the tire and get the proper dish, we are compelled to let them go, whether under this old rim-binding system they set tight or not. But not so with the TOOMEY mode of constructing wheels. *It* advises to set the tire tight in all cases, and dish much or little.

A word to my fellow ironers. Is it not the most critical job you have—that of setting hard steel tire on light wheels, and take all kinks out, and still have them lay on ringing tight without dishing too much? I venture to say nineteen out of twenty will answer in the affirmative.

A few words in reply to H. D. F. In Vol. V, No. 11, in his reply to Progress, he says' "If a heavy Boston coach-wheel should be tired on this plan, it would become rim-bound in two weeks by running on pavements, and a light wheel would bilge or wrinkle each side of the spokes, and tend to squirm from under the tire. This assertion is similar to the Infidel's faith, "it will do to live by, but will not do to die by." An assertion or an opinion will do as well as a fact, when people will take it. But we are not willing to take any opinion or assertion after experimenting for five years.

This squirming out from under the tire is just what we are overcoming. For the last five years, by this process, we have dispensed with tire bolts except at the joints, using six wooden screws in the front wheel and eight in the hind wheel. We do not claim that a heavy rim should be as long in proportion as a light rim, because heavy wheels require more draft on the spoke to produce the same amount of dish.

In reply to the question asked by Progress in Vol. V, No. 10, I would simply say if his rim is of common quality and very open grain, and ¼ inch too long, drawn down with the tire so as to produce from ¼ to ⅜ dish, the result will be good. But in case the rim is of first quality or second growth, the rim should not be longer than ⅛ inch, so as to require from ¼ to 5-16th draft. As to the question asked by Progress in No. 11, viz: "What would we do with old wheels that were rim-bound," we reply, if they were plank or sawed rims, I would take out of the joint enough to let them down, and have the joint open and then set the tire, and if bent rims I should give them sufficient draft to bring the rim down on the shoulders, and produce slight dish just enough to show that the tire was binding on the spokes; in case the old wheel was not rim-bound, I would open the joint and fill up so as to produce a tight tire.

We want a tight tire in all cases. I am aware this new theory of upsetting bent rims appears erroneous, because we were taught differently. So were the men of old, when the great philosopher Galileo asserted that the earth or globe was in motion, making a revolution every twenty-four hours. For this he was imprisoned and compelled to recant, but notwithstanding the globe still revolves. So with the bent rim, it will upset, let men's views be what they may. Hoping in the future to give explanations on different points claimed, I will stand ready to answer questions concerning my invention at any time fair and honestly.

SAMUEL TOOMEY.

Coach-makers' Int'l. Journal, Nov 1870

HAMMER SLEDGE & ANVIL

V. ULCAN EDITOR

HOOPING WHEELS RESUMED.

I suppose I shall have to answer my friend Spoke Shave, or he and others may think I have been silenced. He says, the more he investigates my arguments, the more he becomes convinced that to practice the same would be detrimental to the durability of the wheel. I would ask my friend what he supposes the word investigate means? Does it mean hearsay? From his communication we would suppose so, for he said, on page 152, that he (meaning H. D. F.) is reported to have said that very light buggy wheels should be rimbound 1-32 part of an inch. Now, bear in mind, he heard it reported, did not read it himself, but sits down and records a denial, then calls it an investigation.

I will state the qualifications I made on page 152, after saying what he heard I said, which is true. A straight wheel cannot be dished, but for a wheel made straight across the face and perfectly tight, give the same draft as in the straight wheel. I said nothing about the strength of a straight wheel or of its durability. Spoke Shave ought to remember the time when the straight wheel was the prevailing style, and if one or two of the spokes happened to be sprung a little, the wheel was condemned, and in order to have the tire hug the felloe properly without springing the spokes, we gave the rim 1-32 part of an inch longer then would allow to rest on the shoulder of the spokes. This style of wheels reached its climax some eight or nine years ago, probably before my friend Spoke Shave went to learn his trade. He says he objects to closed down joints by any means, other than the heated tire. I would ask Spoke Shave how he would determine the precise draft to give the tire with loose joints. It would be impossible to tire wheels left in that condition and have them dish uniformly. I should take Spoke Shave to be like a great many workmen I have seen, depending upon the tire to do away with all the imperfections caused by bad management in putting up the wheels. Now, this will never do, Spoke Shave; if you wish ever to be an accomplished workman, make your joints all tight, so the smith will know how to find your wheel. I did not know, before you told me, that you had been schooled at this branch of trade, namely, working the rough material; you also say you know how to run machinery; well, I suppose some saw out spokes by machinery, and you have doubtless become quite handy at it by this time. The worst feature I see about it is,

that such labor does not demand very high wages. You say, too, that you have had the advantage of repairing some wheels constructed by some body else, or in some other shop in the United States, and have always examined the work well before you let it go out of your hands. Now, after having all these advantages, and having availed yourself of the lessons therein taught, you did not say, but might as well have said, that you were about ready to set up business for yourself.

The correspondent that signs himself PROGRESS, said on page 171, No. 11, that V. Ulcan has given us a rule by which we should govern our draft. His rule might work with him, but will never work in this country. He then gives the draft they use on different sized wheels, keeping in reserve any points that would shed any light on the subject whatever. We will take one example he gives. He says, take from $\frac{7}{8}$ to one inch $\frac{1}{2}$ tire, give $\frac{3}{8}$ draft, now where is his starting point? Does he mean on the long rim plan, or does he mean to have a tight felloe, or does he open the felloe $\frac{1}{8}$ or $\frac{1}{4}$ and then give $\frac{3}{8}$ draft? I consider this very important in giving information. Neither did he say whether he gave $\frac{3}{8}$ draft, in addition to the heat that remains in the tire after welding, or whether $\frac{3}{8}$ after cooling off the tire. He says, $\frac{3}{8}$ draft will dish their wheels from $\frac{1}{4}$ to $\frac{3}{8}$. Now a wheel of the dimensions my friend speaks of on his long rimmed plan, will dish $\frac{1}{4}$ inch or more with $\frac{3}{8}$ actual draft, and if it should be the fact that he was letting his wheels run on the long rim plan, I think it would be best to have some of our industrious mechanics go out there and set up a repairing shop; I think they would do well. He wants to know what we should do with a set of old wheels that had become rim-bound. I will say take off the tire, make the wheel tight in every joint and on every shoulder, then trim the ends of the spokes a trifle below the tread, and if it is a light buggy wheel, rim the traveler to the mark; give no draft save what heat there is in the tire after welding.

Again, my friend PROGRESS appears on page 186, in No. 12, in a great query why his questions were not more generally answered in the August number? Probably the hints that Abraham gave him will be as much as he will want to dispose of, whether Spoke Shave comes to time or not.

He refers to the spirit in which George R. Groot takes hold of it, when he said he had made a discovery, experimenting on the long rim. He found in knocking off the tire from the wheel that the extra length in the rim had diminished from a full quarter to a scant eighth, showing that the wood had permanently upset 3-16 of an inch. Now, as Spoke Shave said, to that I record a denial; all the upsetting of the rim spoken of is at the spoke.

All timber is more easily indented sidewise

continued next page

than endwise. Admitting this fact, then the tenons of the spokes must give a trifle, and it would not be but a trifle, to only upset $\frac{1}{8}$ of an inch in the felloe of 18 spokes, or in other words 18 weak points. Now he says again, in measuring for resetting, he gave the solid draft of the tire only $\frac{1}{4}$ inch, which, when reset, brought the dish the same again, viz: $\frac{1}{2}$ inch. We learn, then, by this, that the first time he put the tire on it dished the wheel $\frac{1}{2}$ inch, and after knocking off and resetting and giving $\frac{1}{4}$ more draft, the wheel had the same dish as it had before, that is, $\frac{1}{2}$ inch dish; then adding the draft of the first setting to the second will bring his draft on the long rim plan $\frac{3}{8}$ of an inch. Now this appears to me absurd in the extreme, and is most certainly contrary to all sound philosophical reasoning.

You see, by my friend Groot's philosophy, that a $\frac{3}{4}$ draft will dish a wheel just as much as a $\frac{1}{4}$ draft, and he will pardon me for the assertion, that it was a dangerous experiment to tire 20 sets of wheels on the long rim plan, with only the experience acquired by experimenting on wheels in the shop.

In No. 11, I see an editorial headed, "get out of the ruts." I agree. This is an age of improvement, but there are times and places when to get out of the old ruts proves destruction. There are men that degenerate so far from the true principles of philosophy, good sense and reason, that by getting out of the old ruts they tumble over an abyss, which often proves their utter ruin, while they sink to rise no more. The good Book says, prove all things and hold fast to that which is good, and many shall try to climb up some easier way than the old principles laid down, which have stood the test of ages, and will continue to stand in all time to come. Just so in this case, there are thousands studying and trying to find some easier or better way to construct wheels than the old principles our fathers adopted, and this is what I have been proving for more than 20 years, to construct wheels and set the tire in a manner to prevent their becoming rim-bound, and twisting from under the tire, bilging or splitting at the spokes.

We have tried the long rim plan, also the short rim plan, together with the closed down joints, and with 25 years of hard labor and unrivaled experience, I stand firm by the old principles. It is an old saying that experience is the mother of science, and has cost many of us a dearly bought lesson. By experience we find that the closed down joints are far preferable, and have never had any trouble with them. Still we vary the principle according to the make of the wheel, as I have stated in former communications. Progress was not the first one that tried the long rim plan; but time and space forbid me to dwell.

I have recently tried one or two experiments myself. I took a wheel for a Boston coach, $1\frac{1}{4}$ inch felloe, and forged a piece of iron just the shape of the end of said felloe $\frac{1}{4}$ of an inch thick, and placed it in a tight rim, then gave the tire $\frac{3}{8}$ actual draft, and the result was that three spokes between the shoulder and the rim showed an open space of scant $\frac{1}{8}$ of an inch, still the wheel dished $\frac{1}{8}$ of an inch more than the original dish, which was three-eighths, being the dish of the wheel when the tire was on to one-half inch dish. Now how will my friend account for this. I am very well satisfied in my own mind that he would not answer the question, and under this conviction I will try and answer it myself.

Suppose then we take a wheel with closed down joints, lay it on a stone prepared for setting tire, then while the tire is hot, drop it on, then pour on the water, and as soon as the tire begins to hug the felloe, the rim of the wheel begins to hug the shoulders of the spokes, and what dish we get in the wheel in this way is actual dish and will not run out by use. Now, then, before we proceed to put the tire on the long rim, my friend must admit first that the end of the grains of timber are stronger than they are sidewise; regarding this, then, a fact, we will drop a heated tire over the long rim, and as soon as the tire begins to hug the rim instead of it immediately hugging the shoulders of the spokes, the rim commences a process of upsetting to meet the demands of the tire.

The question then arises, where does the felloe upset? I will answer. In the holes made in the felloe for the tenons of the spokes, the grains of the rim coming endwise against the tenon of the spokes being so much stiffer and stronger, they yield to the superior strength of the felloe, and thus the holes become oblong, which prevents the felloe resting firmly on the shoulder of the spokes; for two reasons more, first, because of the powerful resistance the felloe makes to the tire; second, because of the indenting or mashing the tenons of the spokes. Now, then, if the felloe indents the tenon of the spokes, it must necessarily hold them so tight that the spokes will draw over before the rim will set down firm enough on the shoulders to prevent it. Again if the holes are made or become oblong, caused by a long felloe and an extravagant draft in the tire, the felloes must wrinkle or bulge each side of the spokes, and also check each side of the spokes, and more than half of the dish got into the wheel on the long rim plan is a false dish, and will run out of the wheel by steady use in dry weather in less than three weeks time, in some form or other. If very light wheels, it may be in the felloe, squirming from under the tire, or wrinkling or splitting at the spokes. If heavy wheels, they will be more apt to lift from the shoulder of the spokes, also broaching up the shoulders and wearing less in the rim, and the wheels straightened up to their original dish. Again I tired one wheel on the long rim plan with one inch felloe, which dished it $\frac{1}{4}$ of an inch, over one-half of it I considered a false dish. We want no better proof than in the case of the heavy coach, dishing one-eighth, when on three of the spokes the felloe would not come down into $\frac{1}{8}$. I hear, by the way, that some one has got a patent on the long rim plan.

Coach-makers' Int'l. Journal, Nov 1870

Felloes/Rims

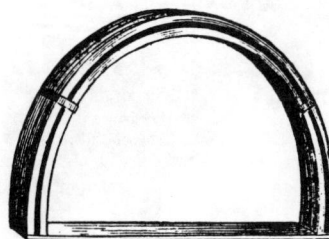

For the Coach-Makers' International Journal.
"THE LONG RIM PLAN."
CANAL DOVER, O., Dec. 14, 1870.

MR. EDITOR:—A few explanations in reference to rims upsetting at the spoke tenon, as some of our oponents would have it, and seem to fear it would be an injury to the wheel. On this point of contracting the rim endwise on the spoke tenon we claim a decided improvement. If we could tenon the bent rim as tight as we did the plank or sawed rim, then upsetting the rim would not be of so much importance, but experience has taught all of us that the bent rim must be tenoned comparatively loose, or the rim will check or crack on each side of the tenon. (This checking of the rim only demonstrates the weakness of the wood after being steamed and bent.) In view of this cracking of rims the oval tenon was introduced, which we think is a decided improvement. Wheels made with oval tenons can be tired with less draft than those with round tenons, simply because the oval tenon fills the mortise tight endwise, so that the rim is more firm on the tenon before the tire is put on. But take the wheel constructed with either tenon, and set the tire on the old plan, and you will leave the tenon in the rim just in the condition the wheeler put it, and this we claim is loose, and unless we upset the rim on to the tenon it will stay loose. Let the objections be what they may, we all agree that tight joints are better than loose ones, and if we can get the tenon in the rim tight, as well as all other joints, we should do it. We prevent rims from flattening between the spokes by upsetting the rim on to the tenon; by so doing we get an arch, the rim standing firm against the spoke tenon. But when the tire is set on the old plan we cannot expect that the rim will stand a hard blow, because it is loose on the tenon, and the wood being weak on each side of the tenon, it will give way and flatten between the spokes; this is the reason why the tire gets loose so soon when set on the old plan. Hard roads will settle the rim against the spoke tenon, and the tire must give. Knowing this to be the result, why not lay our peculiar prejudices to one side, and upset the rim before it goes out, and hold the tire firm? Why not take the long rim and, while the tire is drawing, use the hammer and settle the rim firmly down against all shrinkage? This should be preferable, rather than set the tire on the old plan and let the roads do what we should do with the hammer, and then have loose tires. SAMUEL TOOMEY.

Coach-makers' Int'l. Journal, Jan 1871

The Long Rim Plan

There are many methods of tiring buggy wheels in the United States. Hardly a smith but has his own theory, while builders may have half-a-dozen. Perhaps—nay, without doubt—the "long rim plan," as invented, we believe, some years since by Mr. Toomey, of Canal Dover, Ohio, is the best known method. It has been practiced by a leading house of this city for years on all classes of wheels, and always with the most favorable results. The old and customary method had been to draft or remove so much from one end of one of the half-rims as to leave a space presumed sufficient for allowing the spokes to settle in the hub; and the rim to settle on the outer tenons of the spoke; which had the effect of increasing the dish of the wheel more than was desired. By the "long rim plan" a reverse action is had; the rim of the wheel is so constructed that when one joint is closed and the rim set firmly on the outer tenon of the spokes, the other joint will overlap from $\frac{1}{16}$ to $\frac{3}{32}$ of an inch, according to size or caliber of rim. The rim is then placed back to its proper position, the wheel measured, and the tire given the proper draft, as per size or caliber of tire, which we would say for tire $\frac{1}{4}$ inch thick should be $\frac{1}{16}$ or $\frac{3}{32}$ of an inch—measured when cold. Heat the tire to proper heat and apply to the wheel, using as little water for cooling as possible, and avoid wetting the rim.

The question arises: What becomes of the extra length of rim? does it not allow of the wheel becoming "rim-bound?" The wheel does not become rim-bound. The extra length of rim is lost by contraction, or "upsetting," and if the rim be not thoroughly seasoned, you will find on removing the tire—two days after setting—that your extra length of rim has disappeared, and instead of overlapping as formerly, that an open space exists. Also, when wheels are found to dish too much, by reason of too much openness of rim, if the tires are removed and the outer tenons cut down sufficiently to allow of the rim overlapping as above, or if a piece of leather of the proper thickness be inserted between the joints, and the tire given the proper draw, the dish will become greatly reduced.

In offering this we are not theorizing in the least, but we are giving such results as have been found by a continued practice of many years, during which period not a single reverse has resulted from such practice.

We feel assured that if our Australian friends will follow the foregoing instructions they will have no further trouble in tiring. M.

The Hub, Oct 1878

NEW WHEEL-TREAD SANDING AND EQUALIZING MACHINE.

THE illustration shown on this page represents a new single wheel-tread sanding and equalizing machine, very simple in construction, manufactured by the Bentel & Margedant Co., Hamilton, O. This machine has lately been introduced in order to perfect and economize work in the manufacture of wagon and buggy wheels, and has already met with great favor.

The troublesome and uncertain process of finishing the "tread" or tire side of the rims of wheels is accomplished automatically and perfectly on this machine, and the placing of the wheels in position does not require the employment of skilled labor.

The heavy iron disk, 30 inches in diameter, ground accurately to a true surface, is covered with sandpaper (no cushion intervening) in such a manner that changes of the paper can quickly and easily be made between working hours. Running at a high speed, this disk is carefully mounted on a strong, heavy frame, well spread on the floor to resist the momentum of the disk.

In connection with this stand is a strongly braced double swinging frame, with elbow joints to adjust to different sizes of wheels.

To the end of the swinging frame a vertical mandrel is attached, with a center piece and drivers, shown through the spokes of the wheel in the illustration.

Motion is imparted to this vertical shaft through chain gearing, and a worm gearing not shown. The latter is carefully encased, and, runing as it does in lubricating oil, will, with ordinary care, never be apt to require renewing.

A wrought-iron vertical resting frame holds the wheel exactly horizontal in the act of sanding, and is adjustable for various diameters by being lifted partly from its socket and moved in or out.

The whole apparatus for supporting the wheel is adjustable in a manner admitting the use of the entire surface of the disk before a change of paper is necessary.

The illustration shows the single machine, but a double machine is made on the same general plan, with a wider frame and two disks on the same mandrel, with a driving pulley between them.

The double machine is also supplied with two independent feeding and supporting arrangements, which can be operated together or separately. The attention of but one operator will be required. The wheel is dropped on the center pin, where it will rotate at a slow speed in contact with the disk. The changing of the wheels is accomplished without stopping the motion of any part of the machine.

The weight of the single machine is about 900 pounds; and of the double machine, about 1,300 pounds.

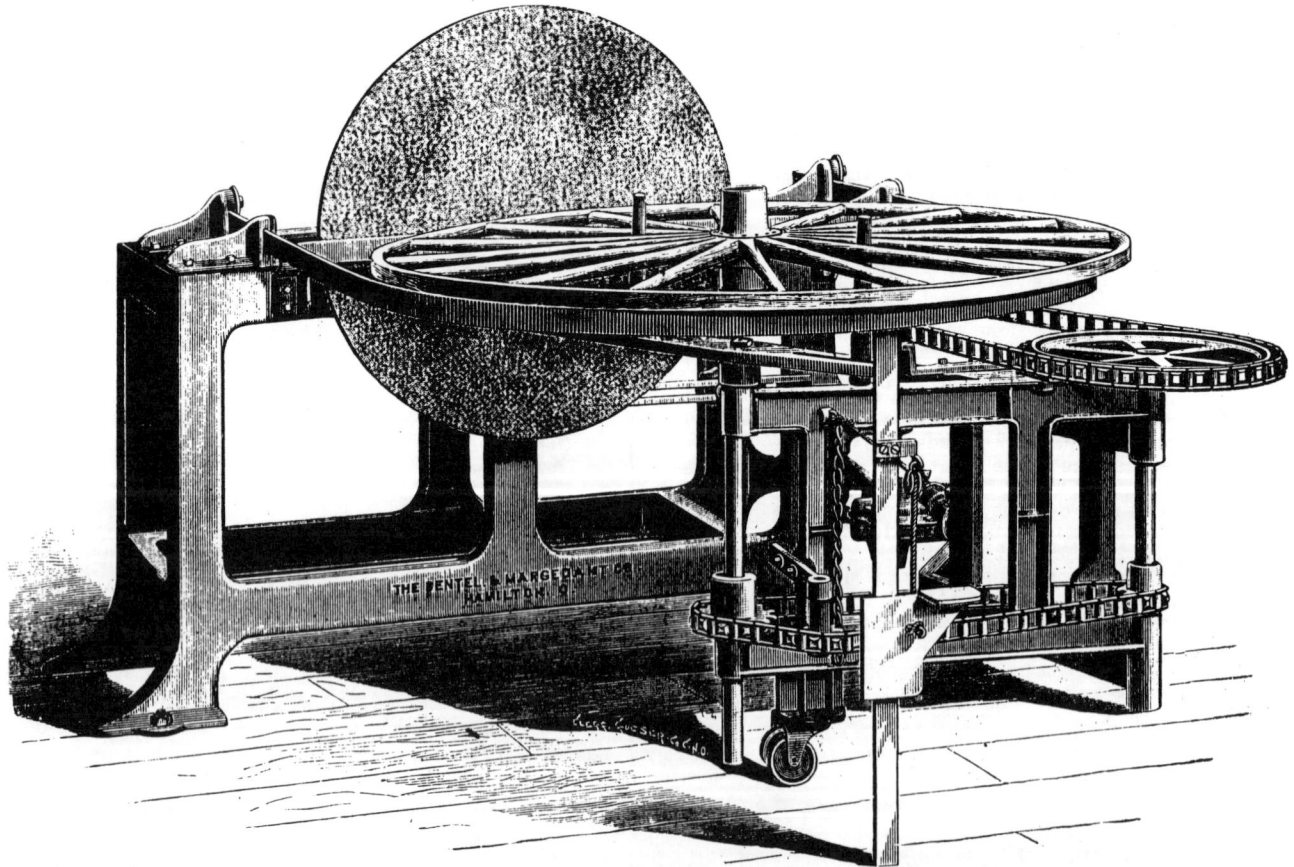

NEW WHEEL-TREAD SANDING AND EQUALIZING MACHINE.—(See description on this page.)

The Hub, Jan 1887

A BRACE OF NEW AND INDISPENSABLE WHEEL MACHINES.

MANUFACTURED BY THE BENTEL & MARGEDANT COMPANY, HAMILTON, OHIO.

WHEEL-TREAD SANDING AND EQUALIZING MACHINE.

AMONG the various tools we have lately introduced to perfect and economize work in the manufacture of wagon and buggy wheels, none have met with greater favor than that which forms the subject of our illustration.

The otherwise troublesome and uncertain process of finishing the "tread" or tire side of the rims of wheels we now accomplish automatically and perfectly, and the placing of the wheels in position does not require the employment of skilled labor.

The wheel is finished to a perfectly true circumference, with an accurate straight surface, sharply developed edges or corners, and very smooth throughout, regardless of joints in the rim and projecting spoke ends.

The machine is very simple in construction, the illustration showing clearly the method of operation.

A heavy iron disk, thirty inches in diameter, ground accurately to a true surface, is covered with sand-paper (no cushion intervening) in such a manner that changes of the paper can quickly and easily be made between working hours.

Running at a high speed this disk is carefully mounted on a strong, heavy frame, well spread on the floor to resist the momentum of the disk.

WHEEL-TREAD SAND EQUALIZING MACHINE.

In connection with this stand is a strongly braced double swinging frame, with elbow joints to adjust to different sizes of wheels.

To the end of the swinging frame a vertical mandrel is attached, with a center piece and drivers shown through the spokes of the wheel in the illustration.

Motion is imparted to this vertical shaft through chain gearing, and a worm gearing, not shown. The latter is carefully encased, and, running as it does, in lubricating oil, will, with ordinary care, never be apt to require renewing.

A wrought-iron vertical resting frame holds the wheel exactly horizontal in the act of sanding, and is adjustable for various diameters, by being lifted partly from its socket and moved in or out.

The whole apparatus for supporting the wheel is adjustable in a manner admitting the use of the entire surface of the disk before a change of paper is necessary.

The illustration shows the single machine, but we manufacture a double machine also, on the same general plan, using a wider frame and two disks on the same mandrel, with driving pulley between them.

The double machine is also supplied with two independent feeding and supporting arrangements, which can be operated together or separately.

The attention of but one operator will be required. The wheel is dropped on the center pin, where it will rotate at a slow speed in contact with the disk.

The changing of the wheels is accomplished without stopping the motion of any part of the machine.

The weight of the single machine is about 900 pounds ; of the double machine about 1,300 pounds.

WHEEL POLISHING MACHINE.

The introduction of our Wheel Tread Sanding Machine sheet 377, now largely used by the principal wheel manufacturers, marked a revolution in the finishing of wheels and caused a demand for a machine to sand and polish the two sides of the rim as quickly and perfectly as that machine operated on the tread.

After a careful study of the requirements of such a tool we are pleased to present the machine shown by our illustration herewith, which will sand and polish the two sides of the rim at the same time, both uniformly and rapidly.

It is so arranged as to take in and admit of quick changes for various sizes of wheels, which makes the machine equally valuable for constantly changing job work in filling running orders, or for larger orders on wheels of the same size.

The arrangement of the machine is very simple, the operations performed automatically, and the construction strong and durable, as plainly shown by the illustration.

All wearing parts are well protected by dust proof devices, and may be run many years without requiring any renewal.

The machine consists of a heavy stand, well spread at the base, and supported on six legs. The top of this stand carries the two heavy adjustable housings in which the mandrels of the two sanding disks rest. These housings are adjustable at various angles to correspond to the bevel of the rims or fellies, an index scale showing the bevel per foot. They are also adjustable to and from each other by means of a compound lever and handle for admitting and accommodating various thicknesses of rims.

Each mandrel is arranged with a spiral spring acting toward the sides of the wheel, thus securing uniformity of angle for bevels, and pressure of disks and against the sides of the wheels. If desired, the left hand disk may be set stationery as to its longitudinal movements, so that the left hand disk may form a standard from which to sand. This fixing or adjusting is accomplished by a single turn of the set screw at the end of the mandrel.

The sand disks are arranged to admit of quick removal of the sand paper, being provided with a brass clamping rim, with a fine screw thread cut on the inner side, corresponding to one on the disk. The sand paper is placed and clamped in position firmly, no glue or other attachment being necessary.

The lever moving the disks to and from each other is placed on the right hand side of the machine, and is not shown by the illustration.

The carriage holding the wheel is a solid double casting, fastened by two

WHEEL POLISHING MACHINE.

pivots to the main frame, and is thus afforded an accurate and very easy motion in changing wheels or adjusting for different sizes. The operator places his foot on the treadle shown, and raises the adjusting handle, thereby at once withdrawing the wheel from contact with the sanding disks. In again entering the wheel between the disks, the hand lever is raised, the wheel returns automatically to the operating position, and the operation of sanding or polishing is at once resumed.

The top of the wheel carriage is provided with the necessary means for centering, fastening and rotating the wheels. It consists on the left side of a self-centering spiral chuck, which centers the hub by gripping it on the outside, regardless what the diameter may be. The automatic motion of the wheel and the resistance of the sand disks tightens the clamp quickly, and the opening of the clamp requires only a quick return of the wheel ; thus it is not necessary to make a stop.

The other end of the wheel carriage is supplied with a cone center quickly operated by a lever. The rotating motion is given to the wheel by means of a series of flexible worm gearings, working under close cover, in oil. Once started, the whole machine is kept in constant motion without stop for change of wheels.

The Hub, Sept 1891

FELLOE CUT-OFF, BORING, AND DOWELLING MACHINE

ILLUSTRATED on this page has the three devices for cutting off of felloes, boring the spoke tenon and the felloe dowel holes, all mounted on one long frame, which also carries the countershaft with all necessary driving pulleys, thus making the machine tool complete and self contained. The cut-off saw is placed on the right end of the frame which it overhangs in the rear. The sliding table is provided with the necessary stops and fences, which are adjustable for various diameters and thicknesses of felloes. The purposes of these devices and their application for the work are clearly shown by the illustration.

The spoke tenon boring device is arranged to give a firm rest by placing the felloe on the long table, which is adjustable for height by means of sliding bearings and hand screws. The mandrel rests in a sleeve journal mandrel, which also carries the pulley, and it will be understood by the above that the mandrel proper does not rotate in the journal boxes or bearings, but is incased by the sleeve mandrel in which it has a reciprocating motion only, thus preventing irregular wearing. The handle shown on the top of the machine is the one which is operated to bring the mandrel forward, while it is returned to its first position by the weight.

The clamping device for adjusting and holding the felloe in proper position is very powerful and positive in its action. It consists of a combination of slides and compound levers and of wedged shaped supports, all of which are adjustable for various sizes. It is operated by the foot and relieved by the heavy weight and pivoted lever shown on the back of the machine. The illustration shows on the left side of the table the adjustable spacer for spacing off the tenon holes, thus making the laying off of the holes unnecessary.

The dowel boring device is arranged on the opposite and left end of the machine. The boring mandrel is not reciprocated in the operation, and it is therefore firmly imbedded in its housing. The sliding table and the righting and clamping device are quickly moved towards the housing, aided in this by the closing action of the jaws which centre and hold the felloe while the boring is done. The relieving of the clamping treadle and the withdrawing of the felloe replaces the table in its first position. The clamping device is very powerful and positive in centreing and holding the end of the felloes and is adjustable for various sizes.

The extreme length of this machine from out to out is about 8 ft. The distance from the centre of mandrel in the middle of machine for boring the tenon holes to the centre of mandrel is 3 ft. 3 in. From the centre of tenon boring mandrel to centre of dowel boring mandrel 3 ft. 6 in.

The weight of this machine complete is 900 pounds. The countershaft with its tight and loose pulley should run 575 revolutions.

It is manufactured by the Bentel & Margedant Co., Hamilton, Ohio.

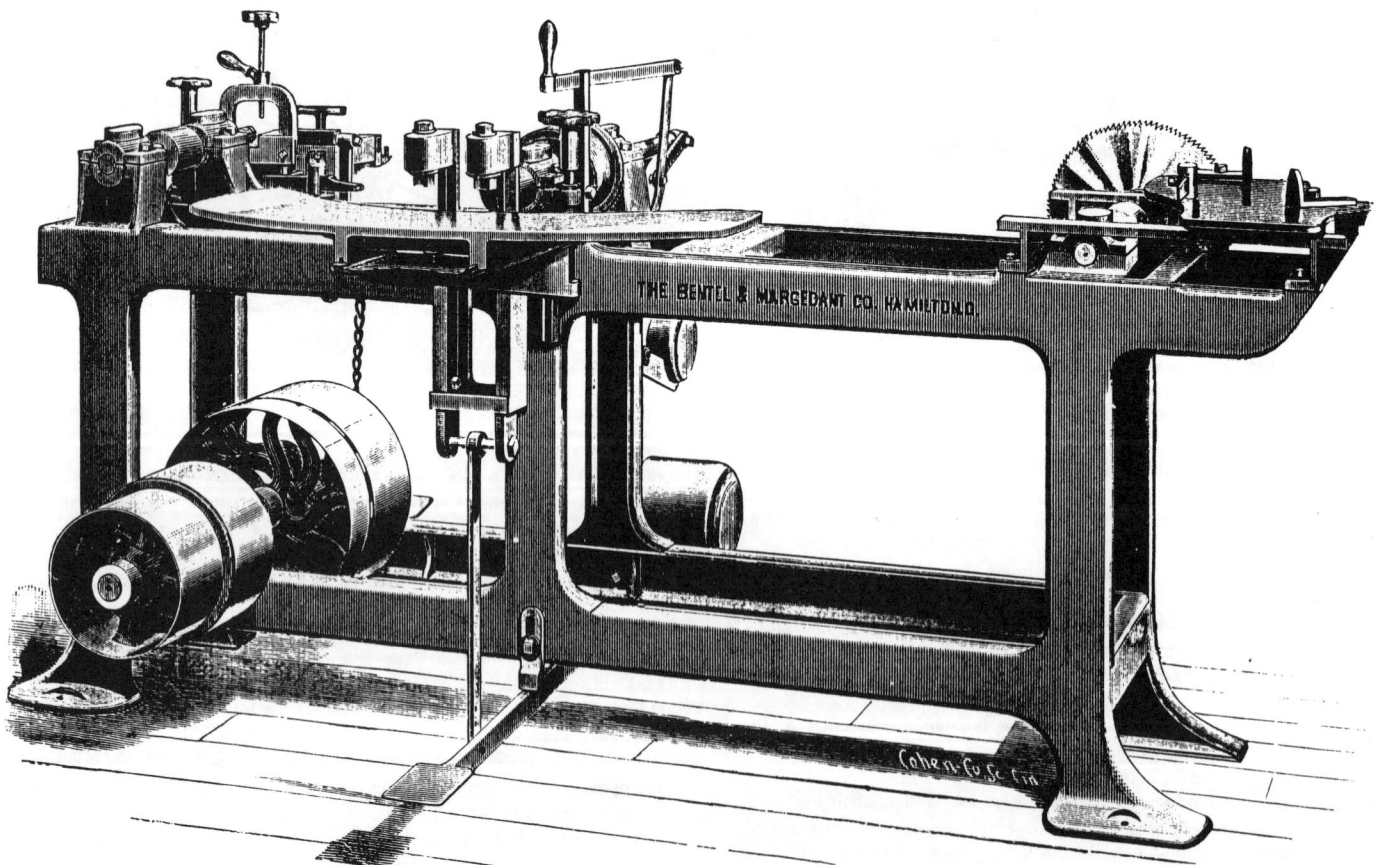

NO. 1 CUT-OFF BORING AND DOWELING MACHINE.

The Carriage Monthly, Apr 1892

IMPROVED FELLY ROUNDING MACHINE.

(See Illustrations Accompanying.)

THE improved machine we herewith illustrate is one that we have design-ed and perfected only after a careful study of the requirements of our best wheel manufacturers and consultation with them, and naturally we feel jus-tified in claiming for it that it is, in the construction and arrangement of its parts, its strength, quickness of adjustment and convenience, the best and most reliable machine of the kind yet placed on the market.

The frame of the machine is a heavy column, cast with the journals all in one piece, with wide base well spread on the floor to prevent jar and vibra-tion.

The countershaft is mounted in such a way that the driving belt is in the center of the column, with the mandrel pulley between the two journal boxes, while the tight and loose pulleys are outside of the frame, so the belt connection can be made from above or below, or from either side.

The front journal box is placed close to the cutter-head, forming a firm support for the fine tool steel mandrel. Two horizontal turned bars are placed one on each side of the top of the frame, forming the support for the half cir-cle side guide, on which it can be adjusted for wide or narrow fellies. Both bars are firmly fastened by set screws, and they will not slide back and for-ward, as in former crude constructions.

Each circular side guide rests on the right side on fine screw sleeves with fine notched heads, and the supporting bar passes through these screw sleeves. To adjust the circular side guides for greater or less distance apart, it is only necessary to turn the screw sleeves right or left. This can be done while the machine is in motion, the circular side guides remaining hinged to the left side bars.

By loosening a milled screw set on the right bar, the circular side guides can be swung back on the left bar, retaining the given adjustment and re-quiring no resetting.

The circular side guide on the back is wider than the front one, but both fit close to the circle of the cutter-heads.

The front circular guide has a double face plate, the inner one or that near-est the felly being self adjusting to any irregularity in the thickness of the felly. The boss projections shown on the front are arranged with springs through which the dowel bolts of the adjusting guide pass, thus constantly pressing the felly against the back guide, and adjusting itself to variations in thickness.

The center guide or rest between the two cutter-heads, on which the felly rests, can be raised or lowered at will during the operation of rounding. This steel guide or rest plate is attached by a lever hinge at its lower end to a sliding bracket, permitting the center guide to be raised and lowered, or swung back out of the way, retaining the adjustment given.

It will thus be seen that the removal of the guides or cutter-heads, or their adjustment, can be accomplished quickly and accurately.

To remove either cutter-head for sharpening, etc., the front guide is swung back to the left, the front head removed, the steel guide rest swung toward the front end, the back guide swung to the left same as the front one, all be-ing very quickly done and without disturbing any of the adjustments in the least.

Moreover, the entire arrangement forms a very firm and substantial sup-port while operating.

The cutter-heads used are principally of the well known "Dennison" pat-tern, but greatly improved, especially in the seat and adjustment of the knives.

Provision is made in rounding the fellies so that the flat seat or shoulder for the spoke is produced in the most accurate and uniform manner.

We furnish with each machine one pair of heads of any size ordered, the two halves of the head constituting a pair.

If there is considerable flatness on the tread, one pair of heads can be used to round various widths of fellies.

The weight of the machine is 800 pounds.

The countershaft is on the machine, and carries tight and loose pulleys of our patent deferential pattern, the loose pulley being smaller in diameter than the tight. The latter is 10 inches in diameter by $4\frac{1}{2}$ inches face, and should make 950 revolutions a minute.

The Bentel & Margedant Co., Hamilton, Ohio.

IMPROVED FELLY ROUNDING MACHINE.

The Hub, Apr 1893

NEW SECTIONAL AND BENT FELLOE BORER.

(See Illustration Accompanying.)

THE above engraving shows a new machine designed and constructed for rapid and accurate boring, and yet simp'e to operate, making a very desirable and profitable machine for wheel-making, also suitable for general work.

The column is cast in one piece, being cored at the center. The front of same has planed ways to receive the adjustable table. The table may be adjusted the proper height by means of a crank handle, and held in angle

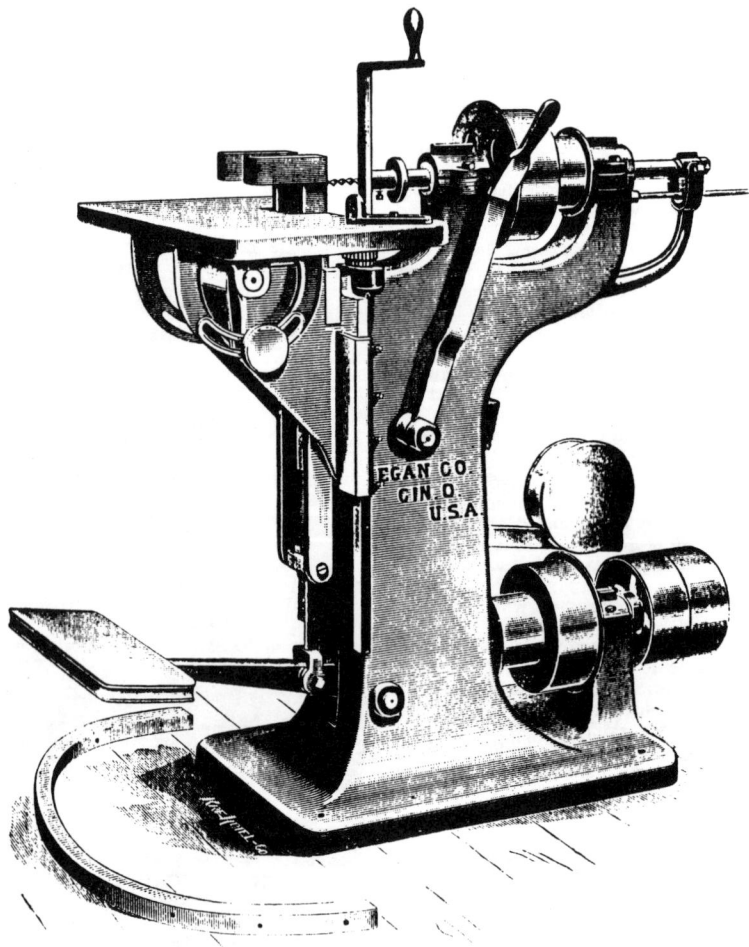

position by a bolt and hand-wheel. The table can be adjusted instantly to suit bevel or straight planed felloes.

The clamping device is very simple and reliable, and powerful enough to atraighten warped stock, holding the felloe down straight to the table while being bored.

The mandrel is of steel, passes through a sleeve pulley, and consequent'y does not come in contract with the journals; and the bit is brought forward by a lever convenient to the operator, while the felloe is clamped firmly to the table by a foot lever.

For further information address the makers and introducers, the Egan Company, No. 196 to 216 West Front-st., Cincinnati, O.

The Hub, Dec 1893

BENDING MACHINE

PATENT AUTOMATIC RIM AND FELLOE BENDING MACHINE.

The Hub, July 1895

continued next page

continued from previous page

BENDING MACHINE.

THE machine shown by the accompanying engraving is used for bending felloes, wagon hounds, carriage reaches, sled runners, chair stock, etc. It will bend successfully the lightest rim used for carriage wheels, up to the heaviest work required on farm wagon, truck and artillery wheels, bending hard wood up to 4 in. thick x 9 in. wide, having the necessary adjustments to bend from 18 in. to 72 in. circles. The frame, of iron, is heavy and substantial throughout, standing 9 ft. in height, occupying 13 ft. 6 in. x 5 ft. floor space over all. The principle involved is the bending by levers from the center outward. The cable chain, which operates the bending arms or levers, is fastened to their outer ends, passing over the sheaves at the top downward to a drum on which the chain is wound. The drum is driven by a worm wheel and screw. The outer end of the screw shaft is fitted with a double friction clutch, driven by two 5 in. belts in opposite directions for the up and down movements. A convenient foot lever, as shown at the base of the machine, is used for controlling the position of the frictions. A slight movement of the operator's foot upon the lever instantly raises or lowers the bending arms. The form, over which the material is bent, is made of iron, with face and edges turned true, and when bending remains on the machine. A wooden cap is used on top of the form, which is of the same length as the diameter of the form, and it is always taken off with the bent wood, requiring one cap for each batch of timber bent, and it must be left in until the batch of timber is cold and thoroughly set, so as not to spring when the shackle is taken off. The bending arms or levers are of cast iron with cored centers, making them very stiff, their inner ends supported on fulcrum pins projecting from the lower ends of two links pivoted to the front of the main frame. The entire upper surface of the arms is covered with a steel strap 174 in. long, 9 in. wide, No. 10, and when lowered forms a level table upon which the bending strap is laid to receive the straight timber. Each end of this strap is fitted with a heel casting, which rests against eccentric head blocks fitted to each of the bender arms.

The material to be bent should be equalized to exact lengths and placed between the heel castings on the strap. The hand wheel beneath sets the wood up to the form at the point of commencing to bend, and as the arms are lifted and approach the completion of the bend an automatic releasing attachment operates the eccentric on the head blocks, which relieves the end strain upon the timber and allows it sufficient freedom to prevent fracturing or buckling. It is the first successful bending machine offered, and will effect a saving of 33⅓ per cent. over machines previously used. The automatic releasing attachment is positive proof against buckling or breakage of the stock, and the iron frame or form overcomes springing, which enables true work to be accomplished. It will bend, on an average, 600 wagon hounds or 1,800 medium sized felloes in ten hours, and other classes of work in proportion. At each bending a sufficient number of pieces to fill the strap 9 in. wide can be placed into the machine. Weight of machine complete, 5,000 lbs. Manufactured by the Defiance Machine Works, Defiance, Ohio, U. S. A.

The Hub, July 1895

PATENT DOUBLE DRUM FELLOE POLISHING MACHINE.

THIS machine has been designed for polishing, at one operation, both sides of vehicle wheel felloes before they are rounded and placed onto the wheel. It is arranged to polish the material lengthwise with the grain of the timber, so that a perfectly smooth surface can be secured. By this system a saving

working parts are mounted upon a heavy iron frame cast in one piece, of sufficient strength to keep all the parts in perfect alignment, and avoid jar or vibration when in motion. The sandpaper drums oscillate, and they are covered with heavy, endless rubber bands, over which the sandpaper is stretched to form a cushion, and their spindles of heavy steel are attached to substantial frames, which can be quickly adjusted for felloes of different widths, or tilted to any angle to polish felloes with par-

PATENT DOUBLE DRUM FELLOE POLISHING MACHINE.

of about 5 cents per set can be effected in manufacturing carriage wheels, and a larger gain is claimed for heavy wheels, over the cost of finishing felloes after they are fitted to the wheel. Another feature of importance claimed is, that when the sides are finished before the rounding is performed, the sand belt used for polishing the round portion will not show a line or mark where the round and flat surfaces meet, which it is impossible to prevent when the side finishing is performed last. The

allel or beveling sides. All the wearing surfaces are enclosed to prevent the admission of dust or dirt. The vertical feed rolls are powerfully geared, and have two changes of speed, and they are adjustable for work of different sizes. The counter is furnished as follows: Shaft, $1\frac{5}{16}$ in. x 74 in. long; two No. 2 floor stands; two driving pulleys, 24 in. x 5 in.; tight and loose pulleys, 10 in. x 6 in.; speed, 425 rotations per minute. The Defiance Machine Works, Defiance, O., U. S. A.

The Hub, Apr 1896

IMPROVED RIM PLANER.

WE present herewith a machine for manufacturers of rims and wheels which is claimed to be a radical improvement for planing the two sides of rims at one operation. In the arrangement of the parts, chiefly as to the adjustability of the feed rollers for various diameters of rims, and as to the planing tables, the difficulties heretofore encountered in the dressing of rims are removed. As shown by the illustration, the machine is a very strong and heavy one, and well adapted to the largest work.

The main part is a solid cast column well spread at the base, giving a good rest on the floor. In addition to this support, it is provided at the back with projecting wings, also resting on

depth of cut on either upper or lower side, and for support of the rim. The tables in raising and lowering obliquely, keep close to the cutting edge of the knives, without danger of contact.

The width of the table tops is eight inches, and the combined length fifty-six inches. The upper cutterhead rests in two strong bearing boxes, with the driving pulley between them, as with the lower head.

The feed rolls are four in number, two on each side of the cutterhead and quite close to it, and all four inches in diameter. They are so arranged as to grip the rim in a line running toward its center, and feeding it in the true line of its circle, thus avoiding all undue friction and strain.

The angle of the feed rollers can be changed instantly to suit different diameters of rims. The feed shafts pass through to the

IMPROVED RIM PLANER.

the floor and affording perfect steadiness in running. Two V-slides on the front of the column, one on each side, support the main table with the lower cutterhead and housing, all of which can be raised and lowered to suit various thicknesses of rims, by means of the large hand-wheel shown in the front, connecting with the screw. This, and a similar heavy adjusting screw in the rear of the machine support the main bracket and cutterhead housing, and are operated together by a worm-gearing connection with the hand-wheel. One turn of the hand-wheel raises or lowers the cutterhead and tables $\frac{1}{64}$ of an inch. The two tables, one on each side of the cutterhead, rest on inclined ways and are adjustable for depth of cut, and for support of the rim back of the cutterhead by means of the two small hand-wheels shown at either side of the bracket. This affords complete and accurate adjustment for changes in thickness of rims,

back of the machine and rest in strong vertical housings which slide together or apart on the two wings, and are operated by the hand-wheels shown on the extreme ends of the machine. Those who have experienced the difficulty of feeding rims to and from the cutterhead will at once recognize the value of this novel arrangement of the feed rollers, as it enables the operator to adjust the angle of the rollers to conform to the circle of the rims, always pointing toward the center of the circle, and thus feeding with ease and without interruption.

The cutterheads are of our well-known patent triangular shearknife pattern, the knives being set at an angle for a shearing cut, thus producing smooth work and preventing any tearing. The cutterheads are also made tapering to correspond to any bevel of rim, by the Bentel & Margedant Co., Hamilton, O., U. S. A.

The Hub, June 1896

VI
TIRES

TIRES

The traditional method of tiring wooden wheels involves five main steps:

First, the wooden wheel is measured by rolling the traveler around the circumference and transferring it to the iron strap.

Secondly, the iron strap is bent to fit the wheel. In the early rural shops, an old wheel or millstone was used to bend the iron tire into a circle. The post tire bender described in the article on p.136, "Cheap Tire Bender," is a rather primitive, but effective, tool for this process. It was later replaced by the roller type bender, which was turned by hand.

In the third step, welding a tire, the ends are scarfed or upset to avoid a weak butt joint, then drilled and pinned, in preparation for forge welding.

The fourth step of the process is tiresetting or hooping. It is interesting to note that in the early shops, in preparation for tiresetting, spoke ends of the wooden wheel were deeply gouged to prevent the possibility of their contacting the tire; later methods, however, involved setting spoke tenon ends as nearly flush as possible. (See "Should tenons be flush with the rim" p.110).

In the early shops, the tire was heated to expand it, in the blacksmith's forge or a fire pit, both methods producing uneven temperatures. While early tire furnaces were fueled by coal, the introduction of the more efficient gas furnace in 1864 allowed expansion of the tire without burning or charring the rim. A discussion of the merits of the vertical versus the horizontal furnace is included in the article "Heating Tires" on p.143.

In the cold-setting process, the tire is welded large enough (in standard sizes) to drop easily over the wooden wheel and inserted (cold) into the hydraulic pressure tiresetter. This eliminates the need for accurate measuring as well as the burning, charring and wetting which accompanied the old method. The dish can also be controlled more accurately with cold-setting. In 1870, J.B. West's American tiresetter was introduced, and in 1891 a hydraulic cold-tiresetting machine more evenly set the tire and was more accurate in controlling the dish.

In the final step of tireing, the tire is fastened to the wooden rim with tire bolts between each spoke. The heads are countersunk in the tire, so as to be flush with the outer surface. The article "Drilling Tires" on p.161 discusses a bit that drills and countersinks in one operation. In 1890, a patent was issued for fluted tire bolts to prevent turning when tightening or loosening the nut.

On a historical note, the article on "Straking" by Ron Vineyard of Colonial Williamsburg describes this early method of tiring wheels with short metal straps which were nailed to the rim. The advantage of this method, used in England and America until the early 19th century, was that it allowed for an easier tire repair than the continuous tire, a benefit on the trail.

the question, and to remove the whole required much tact and skill, to say nothing of the labor, on the part of the smith.

The next we have is the small iron rollers, the advent of which was considered among wagon and carriage-makers as a vast achievement. To-day how different! The small rolls of 1½ inches have given away to those of 4 and 5 inches in diameter, and where two good strong men were required to bend an ordinary sized tire, an ordinary boy does the same work with the most perfect ease. But with all this improvement, nothing has ever been done, until recently, to guide the tire through the rolls, making rather a task to prevent the tire from moving to either one side or the other, and preserving it from the small winding places which must necessarily occur from such action.

An Improvement for Bending Tires.

It is with infinite pleasure that we look back to our earlier days and witness the many improvements which have since been made in the simple matter of bending tires. Our earliest memory carries us back to the day when the tire was measured by the wheel, cut in two at the center, heated in a huge fire upon the roadside and bent around an upright log; and fortunate indeed was the "son of Vulcan" who was the possessor of an old mill-stone, which was used for the double purpose of bending the tires around its periphery and for setting the tires upon the wheels. The two pieces were duly marked, and in due season the two welds were made. By the time the two last sections were bent, the first were cool enough to handle; and, as before mentioned, were welded and laid upon the wheel, and the places for the holes marked off; the tire was again placed in the fire, and the nail holes punched. By commencing at daylight on the longest days in summer, and working real hard until dark, with neighboring help, a set of tires could be put on in one day. In fact, Vulcan always notified his patrons, far and near, that on such a day he would be engaged in putting on new tires, which meant that nothing else would be done that day. It also meant, if on an old vehicle, that the owner furnished the material, three bars of iron, and paid five dollars cash, gave the smith the old tires and the ends of the new ones, and in a majority of cases one gallon of "old rye." This method of bending tires was still in vogue in remote districts in the Southern and Western States just prior to the war of the rebellion.

The next improvement was made by securing a sawed felloe to a post erected for the purpose, or to one of the joists of the smith shop. At one end of the felloe was affixed a strong iron staple, against which the iron rested; over this felloe the tire was bent, bending just the length of the felloe at each pressing down. Lighter tires were bent by securing one end to the wheel with a pair of tongs, and rolling the wheel upon the tire.

How defective either of these methods are or were the modern smith will readily perceive. To have a tire without more or less wind was out of

It is of this long-needed improvement we now speak, which is the invention of a Mr. Metzger, of New York, and is described by the following illustrations:

Fig. 1 represents the tire rolls, with the attached tire guide. B, the block upon which the tire rolls are secured: A, A, are the vertical plates on sides of the rolls, cast solid on the bed-plate K; C, C, C, C, C, are the boxes in which the journals of the rolls rest; M, M, are the upper section of the central boxes, secured by the screws L, L, L, L, into which is fitted a sliding box, which is lowered or raised by the screws

N, N, so as to accommodate any particular sized tire; D, D, are the bolts which secure the tire guide to the bed-plate K; E, is the block to which we secure the tire guides F, F, which Are fitted with the slot H, which permits of their being moved or changed to suit any width of tire. all of which, with the exception of the rolls, are minutely described in the subjoined drawings:

A, *Fig.* 2, represents the plate which is secured to the tire roll bed-plate; B, B, are the holes by which the plate is secured to the rolls, the end of the bed-plate being at the dotted line D; C, C, are the holes by which the guides are secured to the plate A, by means of two T-bolts, as per *Fig.* 5. A, shank of bolt; B, thumbnut; C, top of head; D, D, turn'down on each end of head.

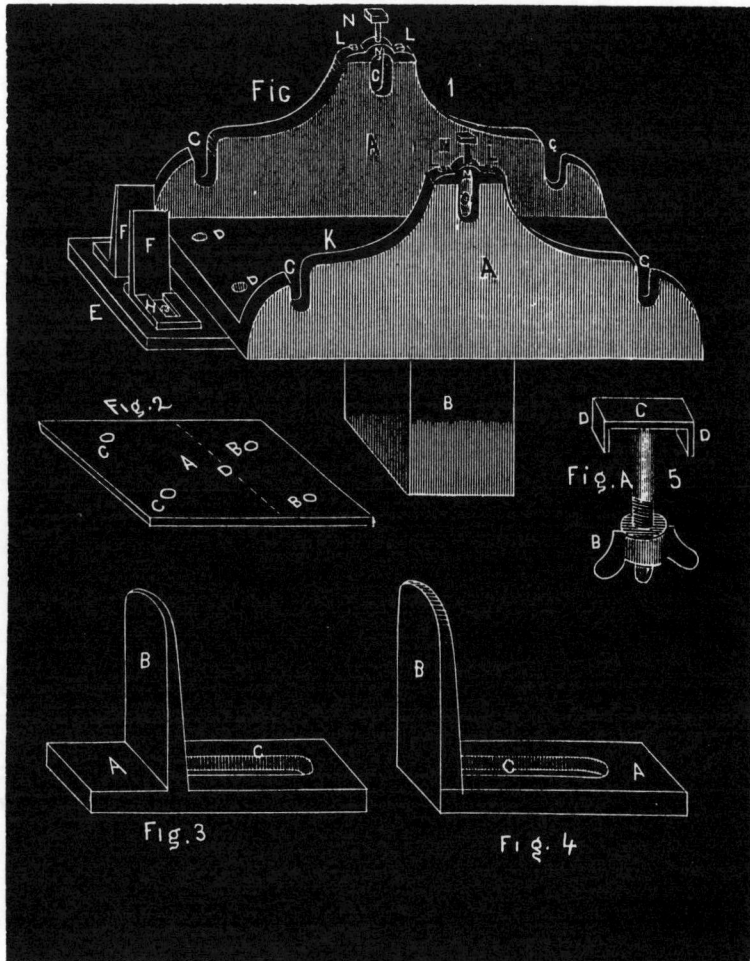

Fig. 3 represents one guide, the lower one. A, the bottom plate; B, the guide; C, the slot in the horizontal bed A.

Fig. 4 represents the upper guide; A, bottom plate; B, upright or guide; C, slot. Through the slots C, in *Figs.* 3 and 4, the T-bolts, as per *Fig.* 5, pass and secure on under side to allow of adjusting to suit any shaped width of tire: also, to permit of moving either way to prevent wearing of rolls all at one point.

Our friends of the smith shop will readily perceive what an advantage one of these guides applied to each end of the tire bender will give them (it requires two); and we are confident they will gladly accept the news that it is not patented, and is therefore free to the use of all carriage builders.

A Tire-Bender for One Dollar.

MR. EDITOR:—There are many shops where the high-priced tire-bender cannot be afforded that would hail with delight a cheap machine for this purpose that would do good work. Having made such a machine, and after using it six months and finding it did as good work as any bender I ever saw, I send you a drawing and description of it for the benefit of those situated as I was.

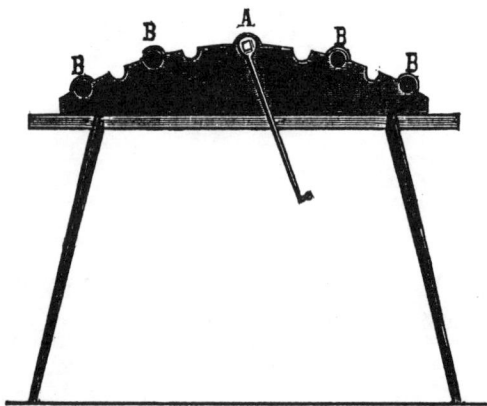

The portion resting or secured to the trestle represents two planks, about 1½ inches thick, or two 1¼-inch planks, made exactly alike, with places cut at intervals for the reception of the rollers *B, B, B, B,* which should be lined with leather to enable the rollers to turn easily ; these rollers are about 4 inches in diameter, made of good hard wood, with bearings at each end, about two inches in diameter to fit into the boards. These boards should be about 4 inches apart. *A* is a roller about 5 inches in diameter, with one end long enough to take the handle. This should be held down to the boards by suitable straps of iron. With this simple and cheap device, the cost of which need not exceed one dollar, any tire that is to be found in a blacksmith shop can be bent perfectly true, and by simply moving the rollers into different notches, any size wheel can be fitted. These rollers will last a year in constant use, and can be replaced at trifling expense.

              Yours, truly,   H. A. F.

The Carriage Monthly, Feb 1882

Cheap Tire Bender.

B. W., of Milwaukee, says he has become tired of journey work, has saved a few dollars, with which he has bought all the tools required but a tire bender. When he gets one, he wants a good one, but can't afford one just now. With a friend who is a wood-worker,

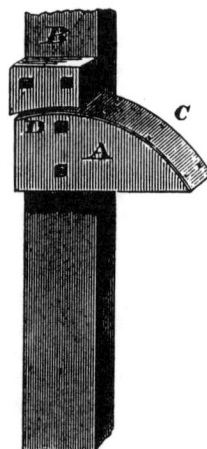

he is going to a country village to try business as boss. Can we tell him how to make something that will serve as a makeshift ?

The following, as per our simple sketch, will serve you. Describe a circle of about three feet and make a block of three-inch stuff, as per *A.* Conform to ⅙ or ⅛ of the circle; fasten it to one of the upright timbers of your shop ; then fasten a strap over the whole, as per *C,* passing out at *D;* then bend down on the tire until it conforms to the block *A,* and continue the process until your tire is bent. With a little practice you can bend quite accurately, and in a few minutes, a full set of tires.

The Carriage Monthly, Feb 1885

Contrivance for Squaring Tires.

MERRIMAC, MASS., Dec., 1886.

MR. EDITOR:—The accompanying illustrations represent tools for

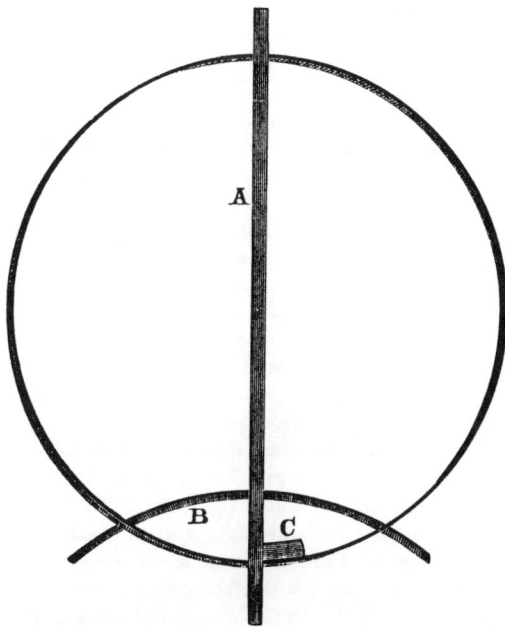

Fig. 1.

the squaring of tires. *A, Fig. 1*, is the long end of the square laying edgewise on the tire; *B* is the circle tongue welded on *A* edgewise; *C* is

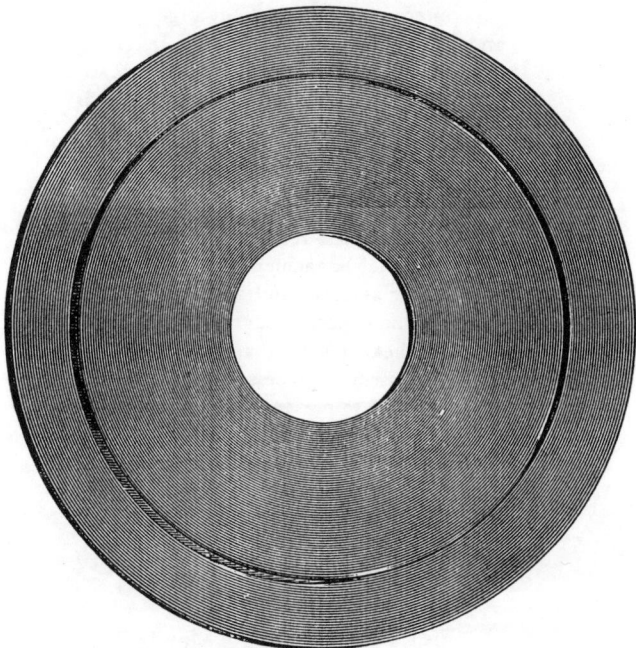

Fig. 2.

the tongue that takes out the twist of the tire; the whole head of the square can be got out of a piece of steel 1 by 1½ inches, and welded on.

The next thing is to roll the tire true, so that it will not twist, as tires twisted when rolled will throw the hub out of center. Take any common rolls and put guides on both sides in the center, as shown at *A, A, Fig. 2*, chalking the ends of the bottom ones so as to see when they become level with each other; roll one at a time, as the guides are to be set for a single tire.

Fig. 3.

The next thing for the tire setter to do is to take the twist out of the tire where it is welded with the square; then lay it on *Fig. 3*, and it can be seen at once whether it is true or not, after which take the square a second time to take out all the twist, as it will become some twisted in straightening on the plate; if there is any twist at all, it will be sure to twist the wheel. With my square you can easily ascertain if the rim is true; neither must one side of the rim be higher than the other to make a good job. I always cool my tire before trucking, so as to get the exact draft and draw the hardness out. After it is trucked and the weld filed, *Fig. 3* can again be used to set the tire on if desired.

Yours, truly,   D. E. B.

The Carriage Monthly, Jan 1887

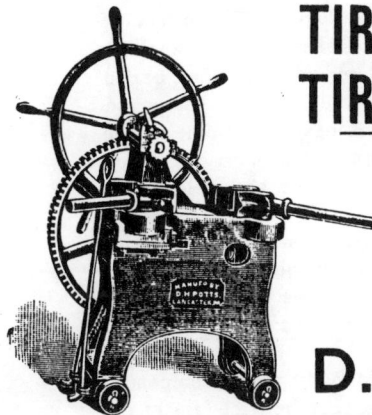

Tires

COMPOSITION FOR WELDING AND RESTORING BURNT STEEL TIRES.

We present our readers with a useful recipe for making a composition to be used in welding and restoring tire steel. Take 2 pounds of borax, ½ pound of salammoniac, ¼ pound of potash; reduce these in a mass to powder, by pounding, and make them into a paste by adding one gill of water and one gill of alcohol. Put this into an iron vessel, steaming the contents slowly, and stirring them all the while, until it forms a cake, which is then to be used in the same way as borax. A longer application of heat to the cake produces a capital article for restoring burnt steel. Try them.

The Coach-makers' Magazine, May 1863

Cherry-Heat Welding Compound.

In our January issue we made brief mention of a new flux for welding steel, or iron to steel, the invention Mr. H. Schierloh, of Jersey City, N. J. As then stated we had not made thorough tests of the compound, and could not, fairly, give anything like real practical results. Since writing the former notice, Mr. Schierloh, at our request, favored us with a five-pound package for making experiments with. Our experiments were quite numerous, and as severe as could well be performed by one of the old fogies of the old school, having a kindred feeling for borax, and rather biased against all new fangled compositions calculated to relieve borax as a silicate or heat neutralizer. We stood for three or four hours watching the different feats performed with the justly celebrated flux. The thinnest plates of iron, steel, and iron and steel, were welded with the greatest ease; broken punches, old files and saw blades, were united as readily as a painter could join together two pieces of putty. Thin pieces of old tire steel, and thin pieces of iron, were laid up in piles, heated to a dark cherry red heat, the compound applied, and again heated, this time to the compound welding heat, cherry red, when it was drawn from the fire and welded with the greatest ease. This particular piece we had taken to a grindstone, and ground off completely smooth on all its six surfaces, after which process we failed to detect where fusion occurred; in fact, it appeared like a solid mass.

Mr. Schierloh in sending us the compound, in his letter says: "Let the steel fuse in the fire, then try the compound in welding the same; also, try what effect it would have if made into a cutting tool." We caused a piece of steel to be literally melted in two pieces in the fire, applied the compound, and made a perfect weld; cut it in two pieces at the center of the weld, made a chisel point, tempered and ground the same, and tried its effects upon iron, and found no difference between the improvised chisel, and one made for the purpose. We next caused some iron tools to be faced with steel, using the compound as a flux with the greatest success. The last test was upon three different sizes of tire steel, ¾ by 1-16 inch, 1 by ¼ inch, and 1½ by ½ inch, and in neither case could we detect anything which pointed toward imperfection. One thing, however, did not escape our notice, which was, that unlike tires welded with borax, there was no wasting away of the metal at either side of the weld, which we believe is mainly due to the low heat at which steel fuses with the new compound. The compound is used in the same manner as borax.

We make the following extracts from a letter received from Mr. Schierloh while writing the present article: "I have been trying my compound in a new sphere. I visited one of our iron foundries in Jersey City, and requested permission to experiment with the compound upon molten metal, depositing a sum sufficient to cover any losses the factor might be subjected to in case of a failure. I selected one hundred pound ladles, caused the metal to be drawn or flowed into the ladle, and for each one hundred pounds of molten metal added five ounces of the compound, which at once united with the metal, and caused all the deleterious compounds, such as sulphur, graphite, phosphorus, and slag to rise to the surface, the same as dross on molten lead. The metal was then poured. The experiments were made on a gas retort and smaller articles, and resulted in giving a clear, smooth and quite malleable casting. Unlike other flux employed, it is not necessary to apply the same to the cupola or blast furnace, thereby destroying or injuring the linings of the same. I shall make other experiments in season, and inform you of their results."

Mr. Schierloh has achieved a great success, and if he is as successful as we anticipate, we may soon expect to have a much superior and cheaper steel than that produced by the Bessemer process. Mr. Schierloh would do well to present the same for sale at carriage furnishing houses, and the trade would be wise in using the new compound in welding steel tires.

The Carriage Monthly, Mar 1875

Tires

WELDING STEEL TIRE.

One of our subscribers complains of having much trouble welding steel tire, and writes in the following strain: " With some tires I get along as easily as though nothing was in the way. Then again I experience all sorts of bad results, when it seems as though I cannot possibly weld the stuff. I would be ever so well pleased to overcome the trouble. Perhaps you can tell me how." From the tone of our correspondent, we are led to believe that he is dealing in divers grades of steel. He may at one time have had Bessemer process steel from one ore, and from a different ore at another time. On some other occasion he perhaps may have had low grade crucible steel, and at another time a higher grade of crucible steel. All steel tire makers do not make steel of the same grade. One maker by the Bessemer process may allow the metal to remain longer in the heating process, and get a tougher grade of steel than would another maker. The manufacturer of crucible steel may employ pure magnetic ore, and get a high grade steel as the result, while another maker as a matter of profit gives a lower grade.

There are just as many cheats among steel and iron makers, comparatively, as there are in other trades. It, therefore, becomes necessary for the iron or steel worker under all circumstances to use great precaution when working a grade of steel to which he is unaccustomed. A smith who has worked crucible steel all his life, would be lost on his first attempt with a softer steel by the Bessemer process, and *vice versa*. A man who has always worked a low grade of iron is able to manipulate it at will; and can apparently perform wonders with the same, but would be lost on his first attempt with a finer or tougher grade. While on the other hand, a man who has always been accustomed to the better grades of iron would give up the ghost if called on to work the commoner iron productions.

The lower the grade of steel, the greater amount of time necessary to produce a union heat. While a high grade fine steel welds at about the same degree of heat that is required to melt cast-iron.

We will say to our correspondent, that it is always wise to try the ends of the tire which are chipped off before trying the tire. Learn about the amount of heat required for a union, then try it with the various fluxes, first with borax. If that fails, add a few steel or iron filings with the borax. If this fails to produce the union looked for, add about one-fourth in weight of dissolved ammonia with the whole, and apply the same as borax. If this fails, perhaps your forge fire is out of sorts, or there is much sulphur or other deleterious substances in the coal, which destroys the chemical action of your welding compounds. If such is the case, mix a little chloride of sodium, (common salt) with grate coal, or better still throw it in the fire. If that fails, get black oxide of manganese, dissolve it in water, and moisten the coal with the same, which will cause immediate combustion of all the deleterious matter, and send it off in gases up the smoke stack. A little charcoal mixed with your bituminous coal will also be of use.

A good clear coke fire is always necessary to weld steel thoroughly. You can not be too careful in getting your heats. Any flux has a better effect on steel, when applied at a heat just below the welding heat, than when at a low heat. You can not be too particular in preparing the points. While you may scarf iron by corrugating it, should may not serve steel in the same manner, just because steel does not yield so readily to the manipulation of the hammer as does iron. The closer and nearer your steel at parts to be welded is together, the better are your chances of making a perfect union, because the union begins to take place in the fire, or else you could never weld it. Steel requires coaxing. Your first blows must be quick and light, so as to not throw away the heat you have gained. After you have begun the union, then add harder blows until you know you are safe, and the steel is welded.

Steel chills and heats more rapidly than iron. If your anvil, hammer and other tools are warmer at that point used in working the steel, so much greater will be your success in welding. While you may edge up an iron or a thick steel tire, such an operation with the lighter steel is not to be commended. The better method is to have a sharp chisel at hand, and remove the superfluous material on the sides, and dress up the same with a file.

If you fail to weld your tire on the first heat, clean off all the former flux, and add new, because the first flux has performed its work, and anything that remain behind is detrimental. We sincerely hope that this item or brief lecture will fill your bill.

The Hub, Oct 1883

Welding Steel Tire.

ANNVILLE, PA., March, 1884.

Mr. Editor :—I have noticed but little in the Monthly about welding steel tires, and I would be pleased to have some experienced blacksmith give his method for doing this. I have tried many different ways of welding pieces of steel tire for sleigh shoes, and encountered considerable trouble in some cases.   Yours, truly,   WILKINS.

Much depends in welding steel tires upon the kind of heat the blacksmith has. It should be a heavy borax heat, with a little fine white sand mixed in with the borax. The fire should be cleaned out once to every fourth tire welded, so as to avoid green coal; the blacksmith should, by all means, have his fire as short as possible; a short heat on steel tire always welds better than a long heat.

To weld pieces on sleigh shoes, punch a small hole in each end and rivet them together; have a slow fire and heat the pieces to a cherry red, then use the borax mixed with the white sand; take a white heat a shade or two darker than a welding heat on iron; by this process I have never failed to weld the hardest kind of cast-steel. There is no better flux than borax and white sand mixed. It makes an elegant weld for broken spring plates, and they never break where welded. The borax and sand should be used very lightly; a small quantity will answer the purpose better than a shovelful; I have seen one pound of borax and a half pint of sand weld 14 sets of steel tire, 1½ inches wide and ½ inch thick, and weld up the pieces that came off the same tire, without the least trouble. In taking the heat there should be no borax used on the outside of the tire, as it draws more or less dirt from the bottom of the fire; it should always be put on the inside of the tire, through which it will penetrate.   W. T.

The Carriage Monthly, Apr 1884

HINTS ABOUT WELDING IRON AND STEEL.

[Received from an expert, in reply to questions propounded by a *Hub* correspondent.]

Editor of the Hub—Dear Sir : In response to your inquiry of recent date, I am pleased to give you the following answer to your questions regarding iron and steel, so far as my knowledge extends.

All irons are composed largely of oxides, which carry with them in a primitive state many other substances that come under the head of minerals. Iron pyrites, or sulphuret of iron, if rich in iron, carry from 30 to 35 per cent. of iron, the balance being largely sulphur. Ordinary good hematite ores carry from 30 to 40 per cent. of iron, the balance being of various compounds. The best magnetites carry from 35 to 60 per cent. of iron.

To remove the impurities, chemical action must be resorted to, which we obtain by artificial heat. To produce this heat two agents are necessary, carbon and oxygen, which produce a combustion giving off a third agent, carbonic oxide gas, and a fourth, known as carbonic oxide.

Bituminous and anthracite coals are largely carbonaceous, but carry with them much other matter which is volatile. Charcoal is the nearest approach to carbon until we reach the precious stones, and charcoal is therefore the best agent in the manufacture of iron, because of its great powers of decomposing other substances. By coking bituminous coals we relieve them of much of their impurities, and they produce a better agent for the manufacture of iron.

Carbonic oxide gas is the only known substance that will melt iron. See certain costly experiments made by means of electricity, which have yet to be analyzed. Carbonic oxide gas exists only where it is formed; when it reaches a higher point in the cupola or melting furnace it becomes carbonic oxide, and, while it has heating or burning powers, it will not melt iron.

Some iron-workers still hold the opinion that iron can be heated to the melting point in a strong flame. Such, however, is not the case. The iron may be consumed, but not melted. A hydro-carbon flame or heat will do the same. A certain amount of oxygen is necessary to consume a certain amount of carbon, which, in turn, gives off a certain amount of carbonic oxide gas, sufficient to melt a certain amount of iron, or to eliminate the iron from the ore, send off the volatile matter, and deposit the slag.

continued next page

Tires

Both iron and steel give off much oxide while in a state of fusion. The gray, thin flakes or scales which fall from iron or steel while in a state of heat, are oxides. It is always necessary, when welding two separate pieces of iron or steel, to remove the loose particles of oxide which are formed while conveying the heat from the furnace or forge to the anvil, which is done by tapping the piece to be welded on the anvil. If the pieces, as thus relieved from the loose oxide, are instantaneously placed together and impacted, the union ought to be perfect, that is, if the method described below is observed and adhered to.

By way of digression I would here remark that many iron-workers, in making welds, first weld the points, thus allowing loose oxide to form on the center, which will prevent all possibility of welding at that place. When the pieces to be welded are lapped, let the impact or blows be given at the center and then toward the points, which has the effect of producing a union at the part impacted, and working out at the ends any loose oxide or small particles of slag remaining between the laps.

I would also here mention a noteworthy fact which may prove of service to workers of iron. If the anvil is worked up to 100 degrees or over before beginning to weld, much better results may be obtained in welding. Iron is a good conductor of heat, and an anvil will condense all the heat in its region, and consequently remain a number of degrees colder than the atmosphere surrounding it. Consequently, when the parts to be welded come in contact with the cold anvil, the heat is partly absorbed and the union of the metal retarded. Where the part to be welded can be frequently changed on the anvil, it is of great service to do so, as this allows the inner or pent heat to come to the surface. The blow of the hammer is but momentary, and has not the chilling effect of the anvil. For proof of this, I would refer to the outside points of welded tires.

There are many fluxes which are of value in the welding process. The borax of commerce is the base of the majority of these; and borax, either alone or mixed with small particles of iron or steel filings, is about as good a flux as is known. A solution of permanganese and water sprinkled on the coal has a good effect in welding steel; and a blast passing over permanganese and sulphuric acid in compound also has a beneficial effect. In fact, welds can be made with steel by such a blast without the use of a flux.

In cases where it seems impossible to make a perfect union, or, to use a common phrase, where the workman "can't make the weld stick," the cause is usually due to impurities in the coal. The parts to be welded have probably been placed in the fire before the coal has become sufficiently coked, by burning out the sulphurous matter and earthy or slaty substances.

In many cases, the cause of failure to create fusion is due to the insufficiency of coal above the fire pot, which prevents the generation of a sufficient amount of carbonic oxide gas to produce a proper melting point. The amount of coal above and below ought to be nearly equal, and also the same at the sides.

The amount of carbon in steel can only be told by analysis, and the variations cause many iron-workers to be led astray. I should consider 2 per cent. very high, while the average might be placed at 1¼ per cent., or even less. The greater the percentage of carbon, the more care required in making welds, although, to the expert, it perhaps matters but little, as he knows his subject and handles it accordingly.

Yours truly,   C. H.

The Hub, Jan 1887

Welding Iron Tires.—I see, in the March number, W. S. Groves wants to know how to weld iron tires. I will tell him how I weld them. First upset the ends well, as shown at D, and punch hole for pin, and then put some welding compound between the laps and pin together. Now put in fire, heat, and hammer lap together snugly. Then clean fire out nicely and put your tire in and put on borax. Don't hurry your fire at first. Use lots of sand on bottom and edges of tire, and don't be afraid of heating too hot. I have had 25 years' experience with them and have little trouble with welding them this way. If they are new tires, cut the right length, upset and scarf before bending, as shown at A.

E. A. Buzzell.

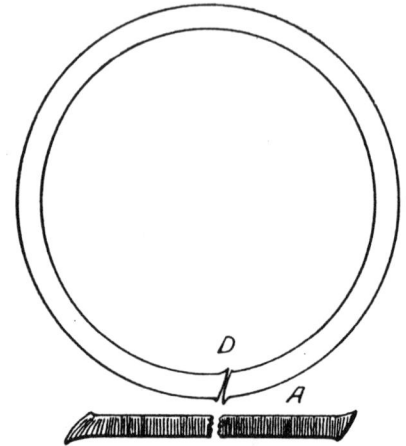

WELDING IRON TIRES

How to Weld Wide Tires.—I will give my way of welding wide rings and bands, which will do as good on tires, I think. In the first place have a clean fire, and a wall of wet green coal on bottom side of your tire. Now place tire in the fire, and cover the lap with a fire brick, letting it come over each edge about one inch. Cover this with good coke, heat slow, scarf and place in fire again the same way, and heat slowly and weld. I use borax on all of my rings which are from 3½ to 5½ inches wide and ½ to ⅝-inch thick, Bessemer steel, and O. H. I have welded about 150 in the last year, and haven't burnt one, or heard of one breaking at the weld, and they are all machine finished to ⅜-inch thick.

W. D. N.

Welding Wide Tires.—I noticed in the last number that there was a brother smith that wanted to know of the best way to weld wide wagon tires. There are different ways; one is to scarf the ends, be sure there is no dirt between laps, and, with a good welding compound, a good clean fire, heat slowly till nearly welding heat, then throw a little sand or borax on the tire. If you have got the proper heat, you are sure of a good weld. Another way is to fit your tire same as before, and use a piece of Laffitte welding plate. The plate should be somewhat larger than the tire, of course, and it must be put between laps. Use the heavy plate and you will have an easy and good job of it.

J. M. Gaza.

The American Blacksmith, May 1907

SMITH-SHOP.

PROPOSED IMPROVEMENTS IN HEATING TIRES.

BY J. L. H. MOSIER, OF NEW-YORK.

*Gas and the Compound Blow-pipe—Heating of Tires in Water—
In Oil—By means of Steam—The Molten-Lead Bath—Will Molten
Lead deteriorate Steel?—Hardening Files by Lead Bath—Read's Gas
Tire-heater—Its Advantages and Defects—Proposed Improvements—
Heater with Rotary Table and Stationary Burners—Heater with
Stationary Table and Rotary Burners.*

[On page 120, Volume XV., we published an article by Mr. J.
L. H. Mosier, on "Fusing Iron by Electricity," which was based

FIG. I.   FIG. II.

on facts that had been gathered by the writer from a visit to
the manufactory of Messrs. Tiffany & Co., jewelers, of this city.
In this article he described the use of gas in connection with
the compound blow-pipe; argued the impossibility of fusing
the denser metals thereby, but expressed the opinion that a
heat of at least 700° or 800° Fahrenheit might speedily be produced on metal in small sizes. Mr. Mosier has since given
the subject much attention with relation to heating steel tires,
and below he gives a *résumé* of his experiments, and the infe-
rences which he has deduced therefrom.—Editor.]

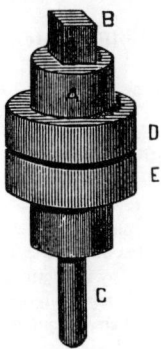 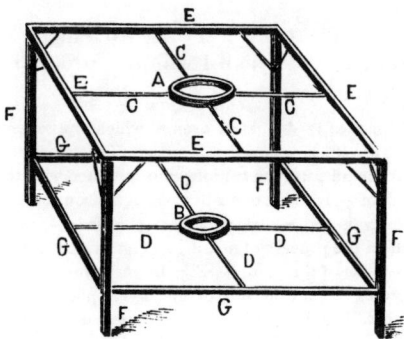

FIG. III.   FIG. IV.

The subject of tire-heating is one to which I have given much
attention during several years past.

My endeavor has been to produce by some means—first, a
heat to cause expansion of the tire sufficient to encircle the
wheel, without burning or charring the rim, and, second, a heat
which would cool without the application of water as a cooler,
in order to prevent swellage and shrinkage of rims.

My first theory worthy of notice was the heating of steel
tires in an immense caldron of boiling water, or oil, or lead.
The idea of employing lead for the purpose, I have not yet
wholly discarded, but I await the time to make experiments as
to chemical results. The cost of appliances for heating by
water was too great to make, at my own expense, any attempts
beyond heating hub-bands, which gave fair results, but not
sufficiently striking to impress me strongly in its favor. Three
hundred degrees Fahrenheit, I estimated was as much as I could
produce under the most favorable circumstances. The matter
of heating with steam next presented itself, and its use impressed
me rather more favorably than that of hot water. With steam,
after a careful study, I made my estimates upon a basis of about
four hundred degrees Fahrenheit, with a fluctuation of **fifty**
degrees either way. But the process appeared to be rather **slow**;
and this fact, together with my lack of appliances for making experiments, caused me to turn my attention in a different direction,
which resulted in the molten-lead bath. With this I am confident that I could readily produce, at a moderate cost, the necessary heat, six hundred degrees Fahrenheit or more; and that
evenness or homogeneousness of heat which is so requisite. The
only serious obstacle which presented itself to me in connection
with the theory was this: "Will the molten lead have a deleterious

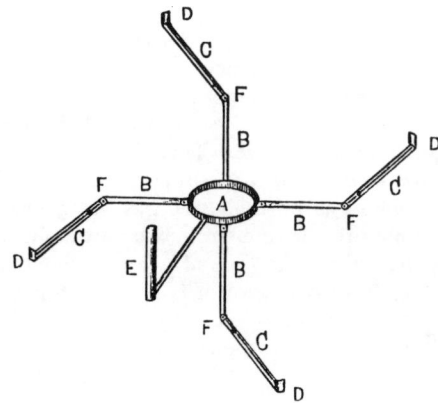

FIG. VI.

effect on the steel?" But I have learned that some of the file-
cutters harden their files by first heating them in the lead bath,
and this has led me to believe that the thing is really practicable.

By the use of a moderate-sized blast-furnace, and a flat bath-
tub, having the requisite diameter and about six or eight inches
depth, the lead could be heated to any desired temperature,
from the melting-point of lead to as high as twelve hundred
degrees Fahrenheit. Into this bath the tire could be placed, and
left for any period without danger of fusion. This plan, it
strikes me forcibly, would cheapen very much the cost of heating heavy iron tires, and might prove of much importance to
builders of locomotives or railway-cars, for heating locomo-
tive-tires. As before mentioned, I shall await chemical results
before taking any further steps in the matter, and if I obtain

continued next page

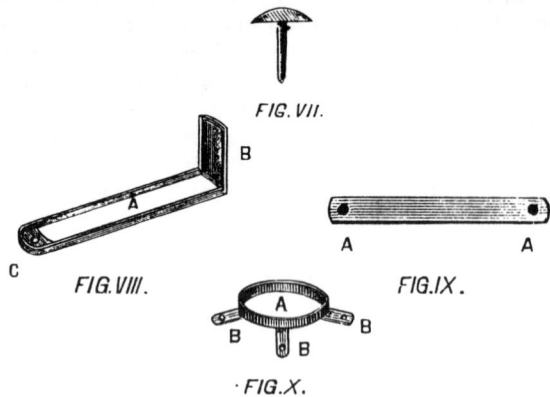

FIG. VII.

FIG.VIII.

FIG.IX.

·FIG.X.

satisfactory results, I shall then consult further with Mr. Gray, file-cutter, of Newark, N. J., and proprietor of the patent for hardening files by molten lead.

I can not think that all my theorizing on the subject of heating tires has been without result, as the following practical illustrations will bear evidence:

For the past ten years, the carriage trade has been acquainted with Mr. Read's gas tire-heater, illustrated in cut 13, which consists of a piece of gas-pipe bent in a circular form and furnished with fifteen or more brackets, each of which is supplied with three jets or burners, and so constructed as to be easily adjusted to suit any size of tire. This circular pipe rests upon an iron platform, which is, in turn, supported by a wooden trestle or table. The whole is portable, the gas being connected, by means of a flexible tube, with a stop-cock on the main gas-pipe.

Mr. Read's invention has proved a great boon to carriage-makers in heating light tires; but it fails to meet all the requirements. First, the amount of gas consumed is too great, and the time required to get the proper heat too long, to prove very profitable. Again, while it is an improvement over the old method of heating in the furnace or smith's fire, it fails to produce that

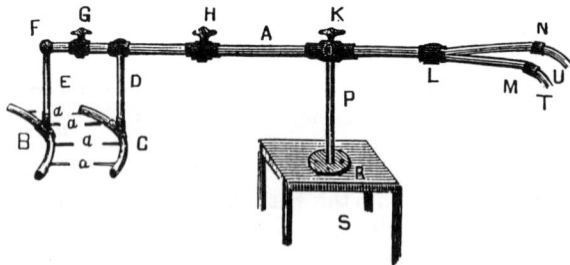

FIG.XI.

evenness of heat which is so necessary in heating tires, from the fact that the jets are too far distant from each other, leaving the spaces between the jets open to the atmosphere, which must have the effect of leaving the tire colder at such points than where it is directly under the flame. And, thirdly, in the case of heavier tires—say, above seven-eighths by one-eighth inch—it is hardly of any use. Now, in connection with the above, I have

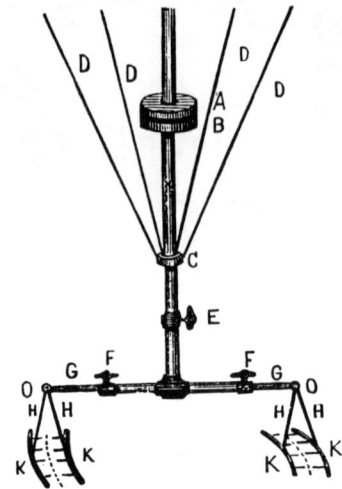

FIG. XII.

deduced the following: that, with a rotary heater, and the use of the compound blow-pipe, in conjunction with gas, a homogeneous heat can be produced with a few burners in much less time, and at a greatly reduced cost, than by Mr. Read's invention. It must not be thought that I am speaking disparagingly of Mr. Read's invention, for I am willing to give him the credit of having produced a very valuable invention; and to

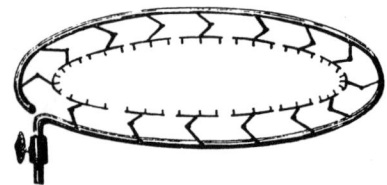

FIG.XIII.

it I am indebted for the improvement which I have illustrated herewith, and which I will now proceed to describe:

In Fig. I., A represents a circular table; B is a hub, furnished with a hole eight inches in diameter, around which the adjuster, Fig. VI., is placed; C, C, C, C, are the holes for the insertion of the elongated head-bolt, Fig. VII., which holds the adjuster in position while operated with the crank E, Fig. VI. This, or the plate A, is made of cast or wrought iron, one half inch or more thick, planed to an even upper surface, so as to prevent any bending or warping of the tire while in the process of heating. Fig. II. is the bottom hub, and is bolted to the under-side of the plate A, Fig. I. A is the hub, and C the flange by which it is bolted to the bottom-side of the rotary plate. B is the square hole for fitting on the shaft, Fig. III., at its squared upper end, B. The hub A is of sufficient depth to hold it securely to the shaft. Fig.III. represents the rotating shaft; A body of shaft; B upper end for insertion in bottom hub of rotating plate; C bottom end for fitting in B, Fig IV.; D is a loose pulley, and E a fixed pulley for rotating the shaft and plate. Fig. IV. is the table for the support of the whole machine, C, C, C, C, the supporting arms of upper shaft-bearing or socket A; D, D, D,D, being the supporting arms of bottom shaft-bearing or socket B, and E, E, E, E, being the upper bars of frame; G, G, G, G, are the bottom bars of frame; and F, F, F, F the legs or uprights of the frame. It will be further noticed that there are corner stays at each leg for further support. The table is secured to the floor, the shaft being placed vertically and the plate placed in position on the shaft.

continued next page

Fig. VI is the adjuster, which is simple of construction, and can, by operating the crank E, be made to fit any tire, and hold the same in position while the plate is rotating. A is the circular portion which fits over the hub, B, Fig. I.; B, B, B, B, are the plain inner bars of the adjuster, as shown by Fig. IX.; A, A, rivet-holes. C, C, C, C, are the slotted outer arms, as shown by Fig. VIII.; A, slot for insertion of bolt, Fig. VII.; C, rivet-hole; B, the right angle which presses against the tire; E is the crank by which the adjuster is operated when the tires are placed on the plate; D, D, D, D, are the right angles, as shown by B, Fig. VIII.; F, F, F, F, are the joints where B, B, B, B, are joined to C, C, C, C. Fig. X. is a sectional view of the center of the adjuster. A is that portion which fits over the hub, B, Fig. I.; B, B, B, are the projections to which the arms B, B, B, B, Fig VI., are riveted.

Fig. XI. represents the heating portion of the apparatus. A is the compound tube; B is the inner bracket, convexed and furnished with four jets to fit inner part of tire; C is the outer bracket, concaved, and furnished with four jets to fit outer side of tire. A, A, A, A, A, A, A, A, are the jets; E is the pendant of the bracket B, and is hinged at F; in the event of its not being required, the supply being shut off at the cock G; D is the pendant for the bracket C; H is the general stop-cock; O is the adjusting and supporting socket of supply-pipe A; P is its standard; R a flanged base secured to the resting-table, S; K is a set screw for holding supply pipe A in position, when adjusted to the tires; L is the outer or supply end of compound supply-pipe, A; where the gas-pipe M and air-pipe N are inserted, U and T being the flexible tubes for conveying gas and forced air from the blower or blast-pipe.

After having illustrated thus far, the rest may easily be told. Our tires being welded, we place them, two at a time, one above the other, on the rotating plate, first elevating the bracket B, and turn the adjusting clamps to fit the tire and hold them in position. We next adjust the compound supply-pipe, turn on the cock H, light our gas, and set the plate in motion, at least two hundred revolutions per minute; this motion is almost equivalent to keeping the tire in a continuous flame, and must produce almost constant and regular heat, or a heat by which tires one and one half inches by three eighths of an inch (two at a time) may be heated to six hundred degrees Fahrenheit in about four or five minutes. In the invention of this machine, a light sheet-iron cover suggested itself to me, for the retention of heat; but if this were made of the lightest material it would be cumbrous and unwieldy, and could not retain more than five per cent of the heat, which would not pay for the labor of its frequent removal.

The same lever which shifts the belt from the stationary to the loose pulley can be made to operate on a brake applied to the bottom surface of the rotating plate, and thereby instantly stay the motion.

Fig. XII. represents a different and much cheaper method, which dispenses with the lower machinery as above described. In this invention, the lights revolve, while the tires remain in a fixed position.

X is the compound supply-pipe (or it may be used without the compound pipe), which is secured to the ceiling, or to an upper frame, and furnished with a strong journal coupling; A is loose pulley; B fixed pulley; C is a lower collar, to which are fitted the stays D, D, D, D, which prevent vibration when in motion; E is the stop-cock of supply-pipe; G, G, are the sliding adjusting portions of the main brackets; H, H, H, H, are the pendants of the brackets K, K, K, K, into which are fixed the jets. The pendants H, H, H, H, are jointed at O, O, to allow of their being removed or turned up when placing the tire on the plate. F, F, are set screws, for holding slides G G in position when adjusted. This apparatus will also produce a quick and general heat, but not so quick as the first one described, as a portion of the heat will be lost by the flare of the jets in rotating. The brackets K, K, K, K, must be convexed and concaved, to fit shape of tires.

I have made models of horizontal tire-heaters, for using wood or coal, but have condemned them because they prove too costly in construction and occupy too much space; but in the above, I have presented what is moderate in cost, and what I consider thoroughly practical for heating from the lightest even to the heaviest tires for carriages, wagons, or locomotives.

Heating Tires.

A correspondent asks of us the best method of heating tires: "Whether to heat them in the smith's forge, in a vertical furnace, or in a horizontal furnace?" Our answer is, the heating of tires on the forge is a rather costly method, and does not allow of the generation of an even temperature. By this method one portion of our tire will have a temperature of 300 degrees, and the other portion fluctuate all the way between that point and 800 degrees. By placing a tire on a wheel thus heated we can readily imagine the results: First, the trouble occasioned in getting the tire upon the wheel. Second, the next to impossibility of setting up to its proper position the colder spots, and, again, the action of the wheel at certain points, caused by the sudden contraction of the extra heated points of the tire. Tires heated in this manner will appear to be perfectly tight at different points, while at other sections they will appear loose enough to require resetting.

Heating tires in a vertical furnace is a decided improvement over heating them on the forge, and yet this method has not much to commend it unless it be in the matter of heating heavy tires for ordinary heavy work. The chief objection being the extreme heat at the lower portion of the tire, or that portion immediately over the fire, or where the heat is the most intense, showing between the two extreme points a difference, sometimes of 700 or 800 degrees, which is much worse than when heated on the forge. This action may be overcome by what is termed "turning the tires," by means of the "turning dog," or changing their position in the fire, in order to produce

continued next page

a more even temperature. This also has its evils. The application of the turning dog to the warm tire with the necessary amount of force required to change the position of the tire, and the haste with which it is done by the smith, in order to avoid the scorching heat of the fire, frequently fills the tire with those short "pinks," which are only removed with the greatest difficulty.

Then again, the great amount of water applied, necessary to cool the tire, in order to prevent its charring the wood, will have anything but a good effect upon the wheel.

The horizontal plan in our estimation is the only reliable one, and yet this is not without its demerits. In heating tires in a horizontal furnace or horizontally in the open air, we generally bolster the tires up with a few bricks or other indestructive material, and in each instance have them so far apart that when the tire reaches an advanced degree of heat there is an inevitable sagging down at those intervals between suspension, which it is impossible to remove if our setting flag be furnished with the most level surface. The effects produced by this action are more visible to the wheel boxer than any other person connected with the trade, and cause him very much trouble and loss of time in his fruitless attempt at boxing the wheel, or in inserting the box so that the wheel will run with that amount of trueness necessary to allow the vehicle being propelled with the requisite amount of ease.

The question of heating tires is not a new one, but, nevertheless, we claim the privilege at this instance of saying, that we are at present the first to have stated or published the many faults connected with improper tire heating, and shall offer in the present chapter some good suggestions relative to the matter.

In communities where gas is used for illuminating purposes, the tire heater, as invented by Mr. Read, of Boston, and for many years in use in several city factories, may be found to be a valuable acquisition; but when it arrives at a matter of heating heavy tires, it is found wanting in the balance; its heating capacity not extending beyond a 1 by 3-16 inch tire, and then the time and gas consumed are considerable.

Mr. Mosier's (foreman of smith shop, Brewster & Co., of Broome street, New York) inventions, two rotary heaters, with either of which he proposes to use the compound blow pipe, seem to us the most favorable thing yet offered in the way of a serviceable tire heater, and, we doubt not, can be applied with success in large factories where gas can be had and power applied, and, we believe with him, the same would not only prove a valuable acquisition to extensive carriage-builders, but also builders of locomotives and railroad companies.

We are also inclined to believe that the heating of tires by means of magnetism or electricity would prove as good and as cheap as any other method, provided, that the action of the current would not have any deleterious effect upon the metal of which the tire is constructed.

A favorable theme of ours has ever been the construction of a horizontal heater, whereby the tires might be heated without coming in direct contact with the flame, by means of inclosed stationary or rotating furnaces, where the tire might be heated to about 600 or 700 degrees evenly. This we believe to be one of the best plans which might or could be adopted; and we may, on some future occasion, present an illustration of the same to our readers for their appreciation or endorsement.

The Carriage Monthly, Sep 1874

Tires

HEATING TIRES WITH ILLUMINATING GAS.

WE have from time to time favored our patrons with various designs of tire-heaters and furnaces, which for all purposes of that character, where illuminating gas may not be had, are highly sufficient; but where illuminating gas may be had at ordinary rates, there is not, in all probability, so good a method of heating tires as by Reid's improved gas tire-heater, of which we give a description in detail. The heater has a reservoir consisting of a piece of wrought iron pipe bent in a circle, to which are fitted a number of swinging or jointed brackets, each of which is furnished with two or more burners; the whole resting on a circular iron plate, which is in turn supported by a wooden or iron trestle. The burners may be adjusted to any size of tire. A compound blast may be used where there is a power blast; or atmospheric or oxygen burners may be applied with good effect. It is economical in many ways. It heats quicker than a furnace, or smith's forge, or a horizontal fire. The tires may be heated while the smith is at work. The heat is uniform and does not destroy the wheel. Tires which are to be removed from the wheels may also be heated sufficiently to remove without destroying the wheel. On the whole it is a great success, and quite a stride in the matter of tire heating.

The Hub, Sep 1878

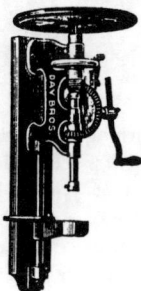

FURNACES FOR HEATING TIRES.

CLEVELAND, O.

HUB PUBLISHING CO.—GENTLEMEN: We build a great deal of heavy work, and we want a furnace for heating tires. Could you let us know if they have any improved kinds down East, and how they are built?

Yours respectfully, JACOB HOFFMAN.

ANSWER.—Fig. 1 illustrates a good, durable and moderately cheap tire-heating furnace. A is the outside wall of the furnace; B, the top; C C, main wall of building; D, section of chimney; E, door-way; H

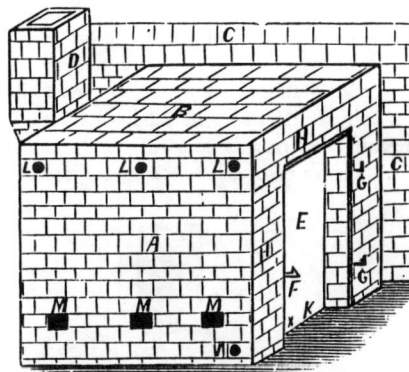

FIG. 1.

H, front end wall; G G, hinge hooks; F, door-catch; X, iron frame for door; K, iron anchor at bottom of door-way; L L L, anchors at top; M M M, cast-iron rests for the tires, four inches square, and about eight inches above the bottom of furnace, to allow of draft; and N, end of anchor K.

Build the furnace six feet long, six feet high, and two feet eight inches wide. Make the walls eight inches thick, lined with fire brick and fire clay. Let the chimney have an 8 x 12 in. flue. Make the door of No. 8 sheet-iron, and rivet to a frame of $1\frac{1}{4}$ x $\frac{1}{4}$ flat iron. This kind of furnace may be built separately, or at the side of the building, and if properly built it will last for years.

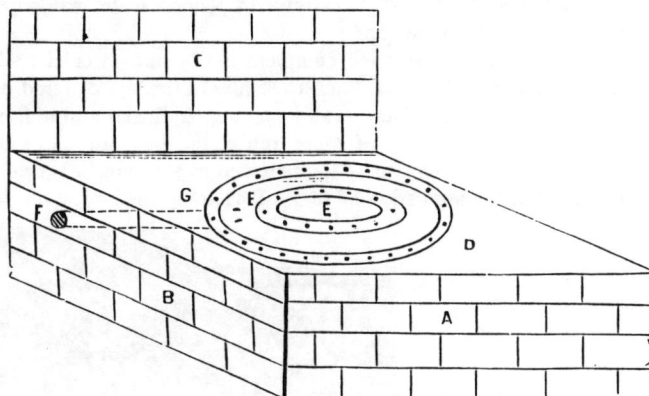

FIG. 2.

Fig 2 represents a novel pattern of tire-heating forge. In this A represents the face of an ordinarily built forge; B, one end; C, wall of building; D, top of forge; and E E, two pipes bent in a circle and perforated on the upper side, at intervals of one or two inches, with one-eighth inch round holes. The dotted lines, G, represent the pipe for conveying wind from the bellows, to be inserted at F. The smaller pipe circle is for small tires, and the larger one for large tires.

The air pipe should be fitted with valves, so as to allow of wind entering either tube, as may be desired.

In some cities gas is now used very successfully for heating heavy as well as light tires, and has the advantage of being economical in space, clean, and giving a uniform heat. Blow-pipes must be used in connection with the gas, to supply air and increase the heat. It is expected when the new Domestic Heating Company becomes established in New-York, supplying a gas much cheaper than the present illuminating gas, that its use for this and similar purposes will increase largely.

The Hub, May 1882

Our Tire Heater.

In our January number for this year, we gave, in reply to one of our Western subscribers, a plan of a tire-heating furnace, without going into any great detail, thinking at the time that it was clear enough to be readily understood in all its features. Since then we have had numerous inquiries as to size, details of construction and cost of same. When we inserted, wrote and sketched the affair, we

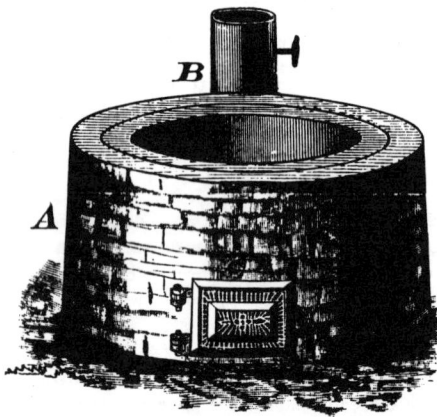

Fig. 1.

did it off-hand, and now, in reply to numerous inquiries, republish the original sketch, with additional drawings to ½-inch scale, with a detailed statement of construction.

Fig. 1 represents the furnace complete, *A* the outer circular wall, which is 16 inches at the base and 12 inches at the top, battered or tapered on the outside, the inner wall being perpendicular and of fire-brick. *B* is the smoke-stack of sheet iron at the back, to lead into chimney, or, if out-of-doors, to be high enough to give sufficient draft. *C* is the door for the admission of fuel. Presuming the largest tire to

be 4 feet 6 inches outside, the heating-pot ought to be 5 feet inside and ½ inch thick; then, with a hot-air space of 8 inches all around, the width in all would make the brick-work 6 feet 4 inches inside. Allowing 6 inches for ash-space at the bottom, the grate bars to be 3 inches deep, 12 inches fire-space, and pot 24 inches in clear, we have a total of 45 inches from the base to the top of the wall. If the furnace is to

be built out-of-doors, a stone foundation-wall going below the frost-line ought to be built to rest the furnace on. The ash-pit can be submerged or sunk below the level of the floor or earth, provided there is clear space enough to admit of a proper volume of cold air for a proper draft to facilitate combustion. This will bring the furnace down lower, and facilitate the lifting out of the tires.

Fig. 2 represents the tire or heating-pot—*D* the inner part; *E* the flange, of sufficient width to extend to the outer circle of the brick-work of the furnace; *F* the body of the cylinder.

Fig. 3 is a sectional view of the whole less the cover, door and smoke-stack, showing the flanges which are omitted on *Fig. 2*. *G* the flange bedded on the wall with fire-clay or plaster of Paris; *H* the interior; *K* the bottom of the tire-pot; *L* represents the wall; *M* the interior of the brick-work; *X, X* the grate-bars; *O, O* are fire-bricks projecting from the wall all around and touching the pot, less at proper places, as hereinafter explained, to form the flues; *P* and *R* for the circulation of the heat and the transmission of smoke to the smoke-stack; *S* the ash-pit.

The fire extends up to and around the bottom of the pot until it reaches *T*, where it enters an open space at the back and then circulates around the pot to *U*, where it again passes through another opening at the back and crosses around and through the flue *R*, and from there enters the smoke-stack at the back.

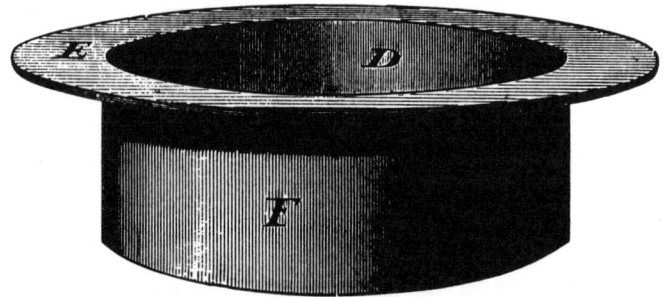

Fig. 2.

Build up the wall to the grate-bars, which set from ½ inch to 1 inch apart, and so set them that they can readily be replaced or removed. Next build up to the bottom line of the pot, and place two or three cast-iron bars across, masoned in, to support the bottom of the pot. Then set the pot and proceed with the brick-work, forming the flues—8 inches—at the proper intervals.

The door of the furnace ought to be large enough to facilitate the

Fig. 3.

burning of wood and the admission of a medium-sized person to make repairs. The door or cover of the pot had better be in two pieces, and of 7/16-inch boiler-plate. Where wood or shavings are used, the grate-bars may be dispensed with. A better heat is raised by keeping them in. For coal they are indispensable. The doors or covers for the pot are to retain the heat.

continued next page

We are unable to estimate the probable cost. The pot would be made of what is known as a loam casting, and would require no pattern. The cave would be built of brick. A bricklayer can give cost of brick-work very readily, as can also the foundryman the cost of the iron. When once complete, it would last an age; and, if built within the factory, would materially lessen the hazard of insurance over the fire-traps now in use.

The brick wall would prevent a circulation of heat outside, thereby making it more pleasant for the workman employed in putting in the tires. If the pot is firmly braced on the bottom, it would hold 6 sets of 2-inch tires. There would be a uniformity of heat over the whole surface of the tires. One man or a boy could attend to the heating until the tires were hot enough to put on.

Owing to the pot being still warm after a first batch were removed from it and reset, a second batch could be put in and heated much quicker than the first batch. All smoke and dangerous flame is dispensed with. There is no possibility of turning the tires, no matter how hot the pot may become.

With a steam-tight wooden cover, the same could be readily utilized for the motion of generating steam for bending purposes. There will be many things to commend it which can only be learned after having put it in practice.

The ashes ought always to be removed before starting a new fire, so as to facilitate the draft or combustion.

The Carriage Monthly, June 1885

A Practical Coal-Burning Tire Furnace.

M. ALBRIGHT.

I notice several of the readers want to build a tire furnace. The accompanying engraving shows one I have used for two years with success. The tuyere is of 3½-inch gas pipe. Fifteen ½-inch holes for a distance of eighteen inches being bored for blast. The end of pipe must be closed with a removable plug for cleaning. The rollers are made of ¾-inch gas pipe, cut so as to space the washers or collars far enough apart to allow the widest tires to run between them. These pieces of pipe are fastened by means of a ¾-inch shaft running through them. I use a 16-inch blower and this fire heats tires as fast as one man can put them on. The furnace should not be over 12 or 14 inches wide inside and should have a 10 or 12-inch smoke pipe directly in the center of the top. The rollers should be about ten or twelve inches above the tuyere and two feet apart. Any kind of coal can be used.

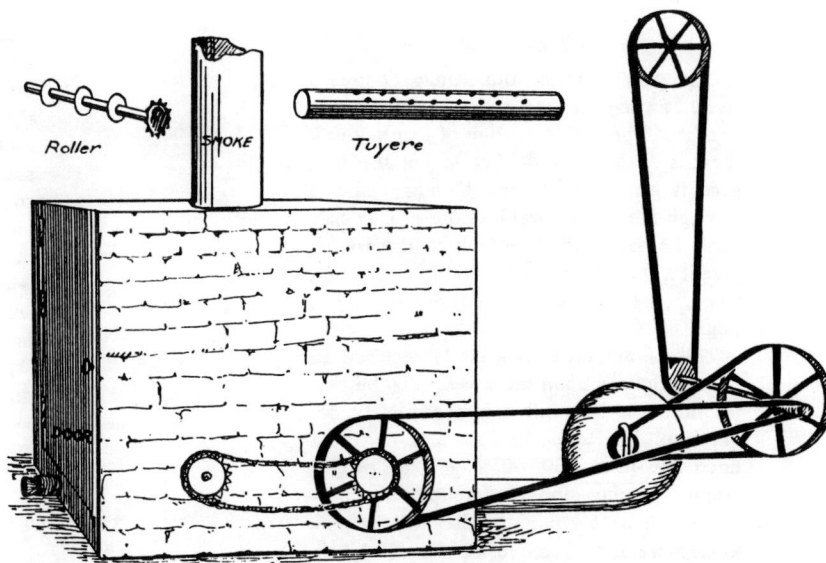

AN EASILY BUILT TIRE-HEATING FURNACE

The American Blacksmith, Apr 1911

Tires

HOOPING WHEELS.

After having the tire rolled out according to the size of the rim of a heavy coach wheel, we joint our tire, ¼-inch thick, so as to have it ¼-inch shorter than the circumference of the rim. This will, when cooled off, draw the shoulder and joint tight, and leave the dish unaltered. Before welding a tire, add three times the thickness of the same to the joint; upset, then rivet together and weld; measure again and see that your tire has ¼-inch less width than the rim. This rule for welding applies in the same way for a ½-inch tire intended for a buggy wheel; consequently, have it ⅜ths, or three times the thickness of tire longer than the joint, which is, in this instance, 1-16th of an inch plumb shorter than the rim. If the wheel is not dished, joint the tire 3·16ths of an inch shorter, which will draw the rim over ⅜ths of an inch, or the required dish. Should the tire draw less, take it off and draw out until, when sighted across your tire and a mark on your hub in conformity with a ⅜-inch dish, it ranges with said mark. Heavy wheels, with six or more short felloes, require a tire 1 inch shorter than the circumference of the rim to draw the joints and shoulder of spokes close.

Coach-makers' Int'l. Journal,
June 1869

Fitting Tires.

The proper fitting and setting of tires is more of a science than many are inclined to accept. The major portion of smiths think there is nothing to do but to cut the bar, give it an ordinary straightening, pass it through the rollers, weld to the proper size, heat the same and apply it to the wheel, give it a few sharp raps with the hammer, true up the face side, cool it off, and all is done.

This is but little of what is required in putting the tire upon the wheel. To be better understood we will commence at the beginning. We take the bars and lay them upon the floor, roll off to get the proper length, leaving enough for upsetting purposes; we next straighten them on the parallels; we next remove all the wind; we then upset the ends, believing it better to do so before welding than after. The next process is to pass the same through the rollers or benders. In performing this last action, it is next to impossible to have the ends assume a perpectly round shape, no matter how we may regulate our rollers or benders. After bending, it again becomes necessary to ascertain if there is any wind or twist in the tire, which should be removed before welding. We next weld the tire, and after welding, while the tire is yet warm, straighten the same on its edges. We next proceed to have the tire the same shape at the weld as it is at any other portion. To perform this feat, we have a series of patterns made of band iron or tire steel, of the same shape or size—sectional—of the rim of each wheel, which we apply to the weld and adjoining, to assure us that the same is harmonious shape with the other sections of the tire. These patterns we generally make about one foot or eighteen inches long, and have one for each sized wheel, duly marked and numbered.

After having fitted the tire to its proper shape, or a shape corresponding with the shape of the wheel, we heat to an even temperature throughout, avoiding having any place heated to redness when applying the same to the wheel. This action prevents the use of water in such volumes as to drench the wheel to such an extent as to set it to swelling. We next give the wheel a fair trueing on the face side, after which we apply a concave wooden mallet, covered with leather, to the inner portion of the rim, which allows of our setting up the tire, in its off places, without marring the rim. If there are spots that fail to yield to the hammer, we heat a piece of iron, apply the same to the tire in the obstinate spots, which soon generates heat enough to allow of their yielding. Tires fitted and adjusted in this manner retain their position, preserve the wheel, and give the wheel that appearance which pleases all carriage builders. We have performed our duty to you in explaining how. It but remains for you to do your duty to yourself.

The Carriage Monthly, Nov 1874

Hints to Smiths.

SETTING AND MEASURING TIRES CORRECTLY.

Your first-class smith never allows his wheels to leave him until the dish is regular; that is, to have both front wheels dish alike and both hind wheels alike, with the extra amount of dish in back wheels over dish in front wheels, as per difference in hight of wheels. " How can it be possible that this wheel dishes more than the other? I am positive that I gave both the same amount of draft." We will tell you how it is: Wheels are singular things to deal with, the longer we are acquainted with them, the less we understand their habits, unless we look at and deal with them scientifically.

Before setting the tires or putting them upon the wheels, we ought ever to examine the wheels thoroughly, and ascertain if the wheels dish alike, also if the tread be at right angles with the face of wheel, which may be ascertained by placing the straight edge across the face of the wheel, and then apply a try square to the straight edge and tread of the wheel, as shown by *Fig.* I. *A*, straight edge resting against the face of the rim *D D; B*, one angle of the square resting against straight edge; *C*, the other angle of the square resting on the tread of the wheel. If the wheel be properly constructed there will be no space discernible between the square and tread of the wheel. But if there is any discrepancy existing, the same should be rectified before putting the tires upon the wheel.

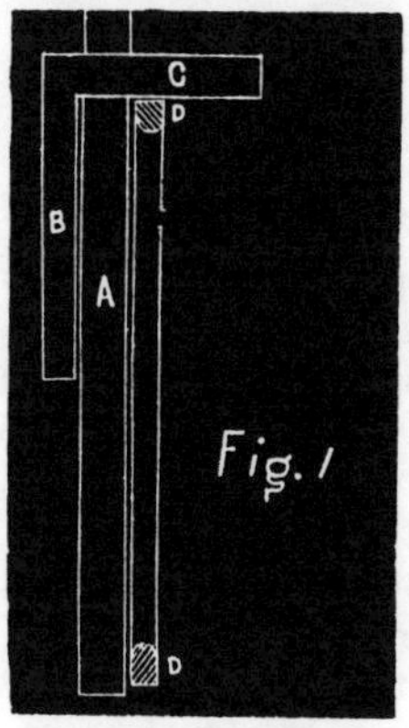

Fig. 1

Again, if drafting wheels by the old method of so much open, be sure that the same space exists in each wheel, and in proving this measure with a sharp pointed pair of compasses or dividers. Another thing to be remembered or to be considered is the temperature of the smithing room. To bring the wheels from the wood room to the smith shop, and subject them to an at-

mosphere ten or twenty degrees colder, is to cause much difference in the dish. That is, if the wheels be brought to the smithing

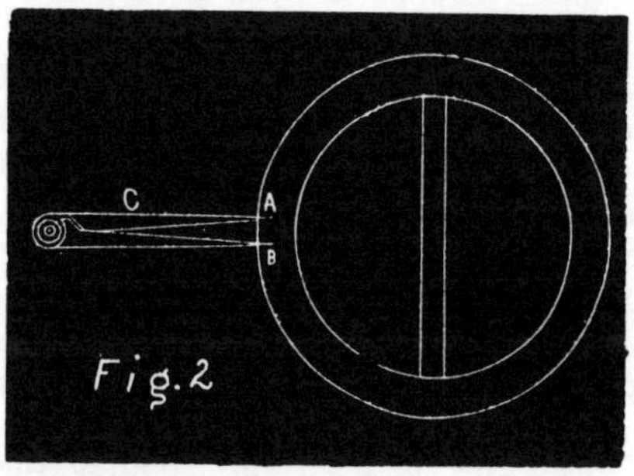

Fig. 2

room at 10 A. M., and by 10.30 we have the first tire on, and twenty minutes later, the second one finished, and so on to the last, we must expect to find a difference in dish. (This portion of the chapter applies to the cold seasons, and damp days.) But if we were to allow the wheels to remain in the smithing room for a matter of two hours, or, to coin an expression, until they have become acclimated, before putting the tires on, we will meet with much better success; from the fact of the wheels having had a chance to assume a position conformable to the surrounding atmosphere.

The next thing to be considered is, do we measure our tires correctly? It is a very rare occasion that we find two wheels which measure exactly alike by the traveler. Hence, how impossible it is to guess whether the draft given to the tires is exactly the same. The only method by which we can possibly be sure that both are alike is before measuring the tire to have the same perfectly cold throughout, and then in measuring the tire with the traveler to take the distance or amount of draft with the sharp pointed compass, as in *Fig.* 2. *A*, size of wheel; *B*, size of tire; *C*, compass showing exact draft of tire. By adhering closely to the principles here set forth you will meet with better success in your tires setting.

LOUIS HOWARD.

The Carriage Monthly, Mar 1875

A DIFFICULTY IN HOOPING PATENT WHEELS.

TO THE EDITOR: I would like to say a few words about the dish of patent wheels. Being a carriage-smith myself, I have experienced a great amount of trouble in hooping light wheels. It is very seldom that you find a patent wheel but what has too much dish before the tire is put on. The majority of patented wheels have from $\frac{1}{4}$ to $\frac{3}{8}$ inch dish before the tire is put on; and what I want to explain is this: How can a smith tire a light wheel properly, when the wheel is dished so much in the first place? It is impossible to hoop a dished wheel properly. For instance, you are hooping a very light set of wheels, and your wheels are all dished in the first place. We will give the tire $\frac{1}{8}$-inch draw, and the wheel is dished already $\frac{1}{4}$ inch. You will find, after the tire is put on that wheel, that you will have $\frac{3}{4}$-inch dish, which makes the buggy look very bad. Let such wheels be run one year, and at the end of that time the tires should be reset. And now the main difficulty comes, for the wheels have already so much dish that it will be impossible for the smith to reset the tires without spoiling the wheels. If patent-wheel manufacturers will take a little more pains with their wheels, and have them made perfectly straight, without any dish whatever, there will then be less trouble for smiths in hooping them. I have seen smiths trying to screw the dish out of a wheel, thinking that would take it out; but you might as well try to make water run up hill as to take the dish out of a wheel by screwing it down on a platform. Different carriage-makers have different opinions. Some want $\frac{1}{4}$-inch dish, while others claim that a wheel will not stand unless the wheel is dished $\frac{3}{4}$ inch. If a wheel is properly made, I think it should not have more than $\frac{3}{8}$-inch dish after the tire is set, and then, when the tire has to be set the second time, you will find that the smith will have but little trouble in setting the tire without springing the spokes. SCOTT SMALLWOOD.

QUINCY ILL., October 6.

The Hub, Nov 1876

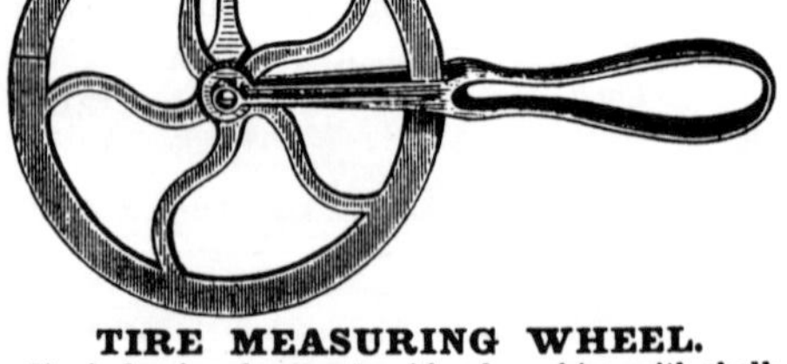

TIRE MEASURING WHEEL.
The index hand saves trouble of marking with chalk, and the whole, made of metal, is light, accurately fitted, handy and strong. Price, $1.50. Postage, 15c.

HOOPING OR TIRING WHEELS.

TO THE EDITOR: I would like to say something more about setting tires. The durability of a wheel depends very largely—indeed, principally—on the setting of the tire, and for the benefit of the trade, I will explain what I consider the proper way to do this. The first thing to be done is to prepare the tire for welding. Some smiths are in the habit of upsetting their tires and riveting them to keep them from slipping, but I consider that all this is time thrown away. First, cut the tire off a little shorter than the circumference of the wheel, say about the thickness of the tire that is to be welded; next, prepare for welding, by making an ordinary scarf, about $\frac{3}{4}$ inch long, and fit the laps closely together, to prevent dirt from getting between them; now slip the top lap to one side, and take the center-punch and make a small impression in the bottom part, about $\frac{1}{2}$ inch from the point of the lap, and after this is done, put the top lap back in its place, and insert another impression in the top lap, directly over the one in the bottom lap. In this simple way you form a wart, which answers the place of a rivet and keeps the tire from slipping. Now place your tire in a very small fire, get it cherry red, take it out and apply your borax, place it back in the fire again, and blow very steadily until you have a light welding heat; bring it on to the anvil and tap it a few licks on the point of the scarf, very lightly and slow, until the smith is satisfied that the laps begin to knit together, and then strike quicker and harder.

Steel of any kind is easy to weld, providing the smith will coax it a little at the start. The main point for a smith to observe is to be particular in running his wheel and tire, so that he may be sure he has got the necessary draw in his tire.

He will have to be governed by his wheel. It is very difficult to put a tire on a dished wheel and make the tire stick to the rim without slipping, unless you want to dish your wheel about $\frac{3}{4}$ inch. I think it is a wrong idea for wheel manufacturers to make their wheels with so much dish. I think if a tire is properly put on a wheel, and put on tight enough, there need not be room for an argument about the dish springing out of a wheel. Why is it that wheel manufacturers are making their staggered spoke wheels with about $\frac{1}{4}$ to $\frac{3}{8}$ inch dish? This undoubtedly is wrong, for the simple reason that the inside spokes will have a tendency to draw out. By giving the tire the ordinary draw, you will have too much dish for a staggered spoke wheel. A staggered spoke wheel should not have over $\frac{1}{4}$ inch dish after the tire is set. I will explain how to make a tire fit close to the rim. The rim should have a coat of lead before the tire is put on, which protects the felloe from getting rotten; next, warm the tire evenly all around with just enough heat to pull the tire on; now the smith wants to have a pair of tongs large enough to clamp the tire and rim together, and one jaw of the tongs should be wrapped with rags, to prevent the rim from getting bruised; let the helper take the tongs and clamp the tire and felloe together wherever he can see daylight between the tire and rim, and let the smith strike light with a small hammer close to the tongs, and by going around the wheel in this way, he will find that he can fit the tire up close without any trouble, and then set the wheel one side to cool off gradually. The cause of smiths having so much trouble about fitting up their tires is this: they let their helpers get too much heat in the tire, and when they go to put it on the wheel, it has expanded so much they have to chill it off with water before they can commence to fit the tire up to the rim. Now the difficulty comes, for the heat is irregular in the tire when thus cooled off, and it is impossible to get it to fit up. SCOTT SMALLWOOD.

QUINCY, ILL., March 7th, 1877.

The Hub, Apr 1877

DISCUSSION ABOUT SETTING TIRES.

TO THE EDITOR. SOUTH BEND, Nov. 17th, 1877.

DEAR SIR: I should like to ask you a few questions in regard to setting tires. I have had an argument with a fellow blacksmith, and he says that a ⅛ or ¼ inch dish is enough in a Sarven wheel, and that a Sarven wheel will not go back with a ⅛ dish. He also says that if the wheel has more than ⅛ inch dish, when it comes from the factory, that he puts leather between the joints. Now I claim, that a tire will not be tight with ⅛ inch dish, and that a wheel will not stand with loose tires on. And I claim that if you put leather between the joints, that it will be rim bound, and that the Sarven wheel should have from ¼ to ¾ inch dish when it came from the factory. I also claim that a wheel is straighter with ½ inch dish than it is with less. Please answer what you think on this subject in your next issue.

S. S.

WHAT TUBAL CAIN THINKS.—Controversies have occurred ever since—well, at any rate, ever since I can recollect, and it is quite right that they should. Augument brings out many fine points that otherwise might never have been presented. It is amusing to listen to a number of young blacksmiths explaining and extolling their individual modus operandi of doing this, that, or the other thing. This is well, but they forget that "A man convinced against his will, is of the same opinion still." Of course we all know best, and because *I* know best, for certain, I will give my views on the subject. If experience is of any value in regard to setting tires, then I may claim to know a few facts at least relative thereto; for my experience extends over a period of about thirty years, in as many different shops, and over half as many different states. I find that the same wheels are differently influenced in different states and localities. In localities where the atmosphere is moist, wheels will not stand as much dish (except patent wheels) as where it it dry. In Chicago, wood-hub wheels take more dish in running, while in some places the dish will come out. No Sarven wheel will stand properly with ⅛ or even ¼ of an inch dish, unless, perhaps, a very heavy wheel, and even then it should have ½ at least. I have never seen a patent wheel of any kind (iron flange) that stood with less than ⅝ of an inch dish, without going back straight, at first, and then dished backward. I have reference to light wheels, up to 1 or 1¼ inch tires. The Olds, however, claim that theirs should not have more than ½ inch, and warrant them to stand at that; as more dish bends the spokes. People who claim that ⅛ is sufficient dish for the Sarven have evidently had but little experience in the wear and tear of them. Some prefer a straight wheel, thinking it is better; but of course it can not stand or wear as well as one with more dish. Three-eighths of an inch will not look bad, and will, for a wood hub, make the wheel give much better satisfaction. But a Sarven *must* have from ½ to ¾, or it will not stand. This I know from experience and practical observations, as I have Sarven wheels retired every day, of which the dish has come out and the wheels "gone back," as we say. My experience has taught me that this straight Sarven wheel theory is about like that of setting the wheels of a carriage so that they will line with a straight-edge, while both axles are the same length, having ⅛ or ¼ gather, plumb spoke, 5 inches swing, etc.—not worth a cent for practical purposes. Let us theorize less and practice more, and we will have better results.

TUBAL CAIN.

The Hub, Dec 1877

Breaking of Tires.

In reply to a very pertinent inquiry as to why heavy tires break more frequently than lighter ones, we would say that it is owing to the fact that they carry a greater proportionate weight or strain. Heavy wheels are generally of smaller diameter than lighter ones, and the spoke being driven with less "dodge" or "stagger" are more nearly perpendicular, and will not spring or yield to concussion, and the amount of pressure necessary to draw the spokes into disk is much greater in proportion to the weight of tire than in lighter wheels.

The Carriage Monthly, Apr 1882

Setting Tires on Light Wheels.

RED FALLS, N. Y., March, 1885.

MR EDITOR:—Do you allow the craft to ask any questions through the MONTHLY pertaining to all kinds of work? If so, I would like to ask the opinion of some of the craft in regard to tire-setting on light wheels. We have a man in our shop who has all of his light tire set on a frame, and has his wheels screwed down; and when the tire is on, they shower the wheel with cold water until the tire is cold; and I claim it is not right; for if you use water to cool the tire, the wheel takes a certain amount of water, and that swells it, and gives it more dish than you intended to give. Besides, if the tire is hot, it will harden it in spots, and cause it to break, and when the water and steam leave the wheel, it will cause the tire to loosen. Now, I claim that after you have measured your wheel and you know what steel you are using, you can tell about what draft you want, and then fit your tire around and heat it hot enough to go on without burning the felloe, and, as it cools, set it up if need be, and not put any water on at all I have set a great many that way, and I can make tires stay tight about one-half longer, and not check felloes at the spokes. Let me hear your opinion in the April MONTHLY. Yours, truly, G. M. O.

The safest and surest way of tiring light wheels is to first get the wheel and tire of proper size; the tire should be bent around perfectly, with just enough heat to get it on; if made too hot, you cannot set it up to the rim, but if properly done, no trouble will be experienced. It is not necessary to use water for cooling off the tire, nor to screw light wheels in a press, as this has a tendency to loosen the spokes in the hub. If wheels are placed in a press to keep them from dishing, the tire will be found to be too tight, and should therefore be taken off and made to the correct size. Some smiths try to make the tire larger while it is on the wheel by putting it on the anvil and striking it on the edge; but this should not be done, as it kinks the tire and makes the wheel crooked. Steel tire ought to have a regular heat all around, not have it red-hot in one place and black in another, as this is what makes hard spots. If water has to be used, the wheel should stand straight up and down, and not more than the thickness of the tire allowed to remain in the water; a good way to do this is to get a wine cask, saw it in half, take one half and put an iron hoop around 2 or 3 inches from the top, and cut a piece about an inch deep out of each side of the butt, thus preventing the water getting on the wheel. We do not think it best to allow water getting on light wheels, and this is easily prevented by being careful and not getting your tire too hot.

G. M.

The Carriage Monthly, Apr 1885

Platform for Tire-Setting.

In a late number appeared an advertisement for an iron tire-setting flag. Since then we have received a number of communications asking where the same can be had. One from "Z. B.," of Sherbrooke, Can., is as follows : " I saw your advertisement in relation to an iron flag for setting tires. If you know where the same is to be had, please inform myself and others through your valuable journal. I have never seen anything but a plain board or plank platform or an old millstone used for the purpose; which answers

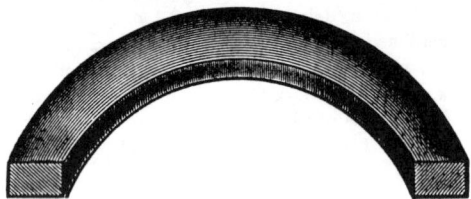

Fig. 1.

quite well as long as the face side of the wheel is down. The dish ot the wheel then protects the spokes from being bruised; but when the position of the wheel is reversed and the hub inserted in the hole, it is impossible to preserve the spokes from bruises, the nails always drawing out and tearing the rims. Can you devise and inform me of some good plan on which to build one which will insure safety to the wheel? I feel certain that such a platform would be highly appreciated by the craft, more especially by those living or doing business at remote distances from large iron-works prepared to make such castings."

We are not in possession of the address of any concern that makes such articles and keeps them on hand for sale. All we can learn from those who have them is : " We had it made to order. We made the

patterns and had them cast." About the cost: " Well, it may appear fabulous, but ours cost us, in a rough state, nearly sixty dollars. We then sent it to a machinist and had him take off one cut, for which we paid twenty dollars. It has saved us more than that during the twenty-five years we have used it."

When we compare the cost of patterns, the molding and the iron, together with the machine-labor, we are not surprised at the cost.

Since receiving the communication above-quoted, we have set our inventive powers to work, and have found them fertile enough to give the trade and our patrons the following :

First lay out a circle 1 or 2 feet longer in diameter than your highest wheels. Then make a pattern, and saw out pieces enough as per *Fig. 1* to make your circle. Have them 2 inches thick and 3 inches wide. After this do the same for the inner circle or hole for the hub, and have it about 18 inches across and ¾ inch thick. Midway between these two place an intermediate circle, thick enough, when all are properly beveled on the inner periphery, to give you a straight line from the outer circle to the inner one.

Fig. 2.

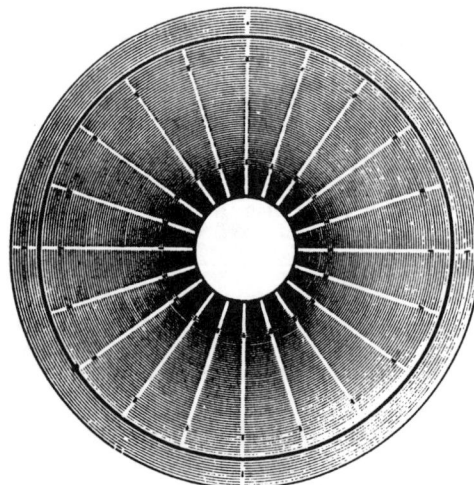

Fig. 3.

When your circles are complete, spike them fast to the floor, or, better still, beams or sills prepared for that purpose. Before securing or fastening the same to the floor, saw out one or two waterways in each piece, to allow for the drainage of the water used in cooling, and for a free circulation of air to prevent decay.

The next process is to prepare the top pieces from 2½-inch yellow pine, or clean spruce, cedar, or locust, as per *Fig. 2*, tapering the pieces to ½ inch or ¾ inch. Then taper the same pieces on each edge to suit the circle. *Fig. 3* illustrates the top view of the platform when finished ready to set the tires. Use seasoned timber all the way through. Dowel the circular pieces and leave the joints about ¼ inch open. Paint all your slats on the under side, also the edges, with some good, reliable metallic paint and raw oil, and when dry secure them to your circular pieces with hickory dowels or treenails, leaving the joints about ¼ inch open, to allow for swelling and the free passage of water used in cooling. As to the iron-tire platforms, of which we have previously spoken, they are considered superior to all others for general work, and, if cast true, are not necessary to be turned in the machine-shop. We will give further information in regard to these platforms in a future number. B. B.

The Carriage Monthly, Dec 1885

Tires

HOW TO TIRE WHEELS WHEN OUT OF CENTER.

(See Illustrations accompanying.)

EDITOR OF THE HUB—DEAR SIR: Hoping that I may in some way enlighten the tire-setter, the apprentice and perhaps the older ones in the business, I will endeavor to give my attention to what has long seemed a mystery to tire-setters.

I will give a short remedy by which hubs which are out of center after tiring can be improved to some extent.

I have often heard it said that heating the tire unevenly would throw the

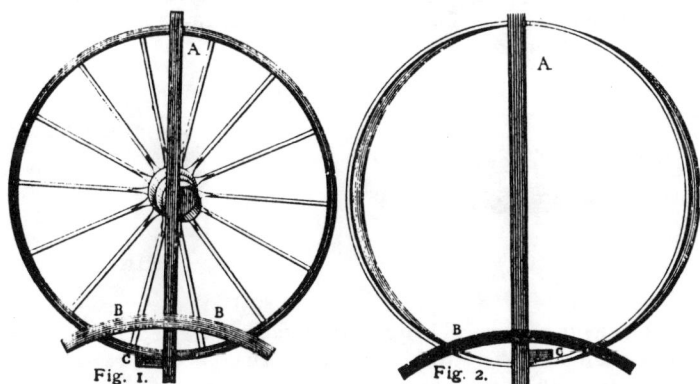

Fig. 1.   Fig. 2.

hub out of center; this, I think, is a mistake, as I have tried by heating one side of the tire and leaving the other half cold, and I could not throw it out. I have found out that when a rim is higher on one side than on the other it will throw the hub out of center nine times out of ten.

How can this evil be remedied? you may ask. It may be set right by the tire-truing square, as seen in the illustration, Fig. 1, placing the square next to the hub on the face of the wheel, the inside of the square C for the rim of the wheel, and the outside for the tire, which takes out the twist C of the welded tire B B, lying on the face of the wheel.

Wheels must be perfect to set a tire right, and you must have the right tools to do it with or you can not do it properly, it being impossible to make a perfect wheel of an imperfectly gotten-up wheel, but you can improve one if it is not unreasonably imperfect.

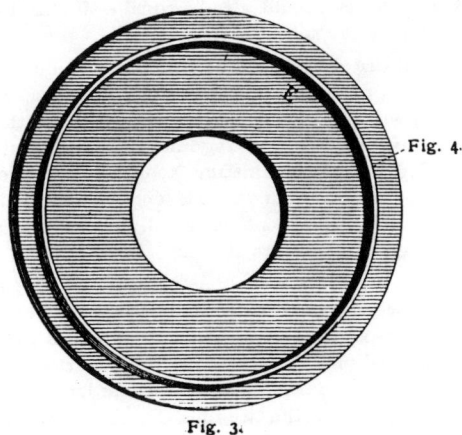

Fig. 4.

Fig. 3.

The first thing is to see that the rim is in readiness for the tire. Take the square and try the rim all around. Take the square, as seen in Fig. 1, and proceed with a spoke-shave to level the rim, making everything true all around. Then lay the wheel on Fig. 3 to ascertain if the rim is not twisted. If it is it can be seen at once while on the plate. If you find it twisted it will be hard to get a true wheel out of it. It can be improved by shaving on the front of twist. Shave according to how much it is twisted, using your own judgment. To find out how much to shave off, set the tire, then place the wheel in the straightening wheel-horse, set the gauge as seen in Fig. 5, touch the rim in the nearest place, chalk

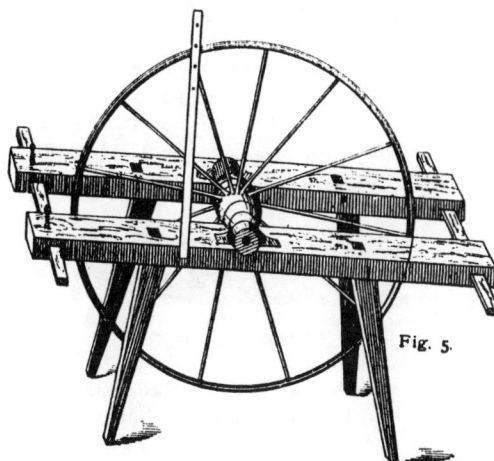

Fig. 5.

the rim of the wheel, take off the tire. If you want to throw the tire back take off in front, but should you want to give one-half of rim more dish with the same draft of tire take off a little of the back of the rim.

There are other troubles for the tire-setters to look after. Some rims are softer than others; if one-half of a rim is soft and the other hard it will twist the wheel. Every wheel should be tried in the wheel-horse to see if it is all right before it is given to the jobbers to box. You may say, perhaps, all ought to be looked after; I say not quite all, yet see that the tenons on rims bear evenly, if not, this will twist the wheel. I believe as long as there is a tire to set, just so long there will be something to learn about wheels.

Fig. 2. The next thing is to have a true tire to make a true wheel. First, rule it true and not twist it, as a twisted tire is hard to get true again and will throw out the hub somewhat, if it does not entirely spoil the wheel.

For a neat job take any common rolls and place movable collars on the bottom rolls for guides to be set for a single tire to be rolled; chalk the bottom ones so that you can see more readily when they become level with each other. The next thing is to weld it and leave it true. This is the easiest thing of all. After welding take the square as seen in Fig. 2, with straightening tongue you can see any twist is in the tire. Then place the same tire on Fig. 3 to get the side twist all out, then again try the square to see if it has been changed by any truing on plate. Then try the wheel in the wheel-horse to see if true. If not, take off tire, before the spokes get time to set and so spoil the wheel, and shave off the edge to take out the twist, as directed, and never lay the blame on uneven heating as many do. Wheels are thrown out by twists in the rims and other causes too numerous for the good of carriage-makers.   Yours truly,   D. E. B.

The Hub, Aug 1888

153

Tires and Tire Tools.

BY UNCLE STEVE.

About all the preceding chapters of this course have been devoted to the metal which you work, or are to work, and the tools and appliances for working the same. After three hours' study and thinking I cannot think of any thing which (in the way of tools or appliances) has been left out. If, as we move along further in the discourse, I think of any tool of import which has been forgotten or overlooked, I shall place it where it belongs with full explanations.

Having made all the tools necessary and also having told how the metal is produced and how to work it, it is now time to take up the parts of the work on which you are to engage, and explain plainly and thoroughly how to proceed with the same.

Fig. 1.

About the first thing to be done when we commence a job, is to place the tires on the wheels, and bands on the hubs, in order that the boxes may be placed in the wheels, or hubs of the wheel, and thus enable us to set our axles to the proper pitch before filing them and adjusting them to the gearing, or, as it is usually termed, "the carriage-part."

To begin, place the tire on the floor, mark one joint of the wheel and at one end of the tire place the marked joint, and roll it along the tire until marked joint again comes in contact with the tire, at which point you mark the tire. You also give the wheel and tire a corresponding number, as 1. 1—2. 2—3. 3—4. 4. Having marked your tire at the point marked, you have only marked the exact length of tire. You must now increase the length by extending the mark as much in advance as will be necessary to make up for losses by oxidizing in upsetting, welding and finishing up the weld, which will increase or necessitate increase of length as the metal increases in caliber. Having concluded how much extra metal you require, you cut off the tire at such point and proceed to prepare it for the wheel.

First of all proceed to straighten it. It is not wise to straighten by hammering on the edge unless the tire has been bent after cooling off at the mill. After the bar leaves the roll and strikes the earth it is subjected to various currents of air that will cause the bar to cool quicker at one place than at another. If the bar cools quicker on one edge than it does at the other, it necessarily shrinks on that edge which causes the other edge to draw or elongate just enough to throw the bar out of a straight line at that point, if the tire be straightened on the edge. When rolled or bent in the tire rolls or bender, welded and heated, it will again resume its original shape, that is, the short side bends will be reproduced. To straighten the tire, draw the concave sides until straight at that point.

Having removed all the crooks from the edges, you next proceed to take out the winds; unless you do, all such places as are winding will make themselves apparent when on the wheel, and all the hammering you may do in a year will not help you over your omission. After the winds are removed, you will upset and scoff each end and scoff so that the tire when lapped shall be a close joint. It is better to punch a small hole in each end of the bar. When all ready put them through the rolls, and if any winding appears remove the tire and adjust the rolls so as to dissipate such winding. If you were to allow the tire to go through winding and then make the points meet by twisting or putting over on the tire, you place a twist in the tire which you cannot well remove, and which will cause the dish of the wheel to be awry.

Presuming that the tire has gone through the rolls all right, insert a rivet in the hole, heat the points and fit together closely while hot with the hammer, apply the flux and heat and weld, being careful to to weld the outer point first, before it becomes chilled by coming in contact with the anvil. It is a wise precaution to heat a heavy piece of iron, and to lay the same on the face of the anvil sufficiently long to remove the chill from the surface.

In welding thin tires it is well to have an iron block on the forge on which to weld the tire before it loses its welding capacity. Do not attempt to edge up light tires; remove the surplus metal or flash with a chisel, and round or finish with a file. [The riveting of the joints prevents slipping. From the fact that a traveler may be or can be bought for about one fourth of what it costs to make one, I do not describe how to make one.]

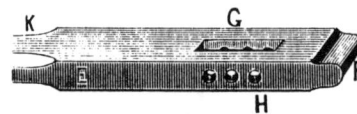
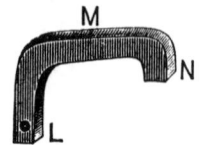

Fig. 2. **Fig. 3.**

After the tire is welded hang it up or stand it to one side to cool, and measure the wheel with the traveler. Then, if the tire is quite cold, measure the tire; to measure a hot tire is but guess work. Presuming the tires have just the necessary amount of draft to close up all the joints thoroughly and to bind securely on the rim, we proceed to heat them, being careful that the heat throughout is uniform and not a red heat, and when hot enough to place them on the wheels, placing the weld near but not over the spoke tenon; and, if necessary when the rims are stubborn, to use a tire drag to draw the tire to the rim all the way around. When the tire is on be sparing with water, cooling only where the tire burns the wood.

If the heat is not uniform there will be discrepancies in the dish. If the tire is true throughout and heated uniformly, there will be a uniformity of dish throughout.

A concave iron plate is the only table upon which tires can be properly set. One from board soon loses its level surface. Stone ones mar and cut the rims.

The dish of the front wheels should or ought to be equal. The dish of the back wheels should also be equal—all this in order to avoid complications when setting axles.

Fig. 1 is a plain tire drag made of iron; *A*, section of handle or lever; *B*, end of bar or fulcrum. *C* and *D*, the cant or drag hook. *Fig. 2*, a lever made of wood; *E* and *K*, section of handle; *F*, the fulcrum; *G*, mortise for the insertion of the drag hook, *Fig. 3*; *H*, shows three holes for adjusting pin when adjusting hook. *Fig. 3*, cant or drag hook; *L*, the part inserting in mortise *G*; *M*, the neck; *N*, the hooked end of the drag fitting on tire. For light work a pair of tongs, with hook on upper jaw, answers all purposes.

Avoid stepping on the spokes. As soon as the tire is well on, with a light hammer rap the tire smartly at the spoke ends, to set the rim up on the spoke shoulder. Then remove the tire from the plate, and with a wood mallet, held on the inner side of the rim, with the light hammer set the tire to the rim where it may be off, using or taking care to not indent the tire into the rim, so as to cause an abnormal pressure at any one place or places.

When the tires are set the wheels too are set until the tires are cold and the wheels dry. Then, with straight-edge laid across the face of the wheel, measure the wheel at different points near the hub to see if the dish is equal. If not, learn the cause, and remedy the evil. The exact amount of draft to give the tire in each instance is hard to determine. An old rule, and perhaps, quite as good as any, is, that the amount of draft shall equal the thickness of the tire. In a majority of cases this will be found equal to the occasion. But there are cases, when the wheel is firmly made, that a less amount will answer, or where the wheel is loosely made a greater amount is necessary.

continued next page

Tires and Tire Tools.

BY UNCLE STEVE.

Space did not permit, in the September number, to deal thoroughly with the matter of tires. In the present chapter I will try and complete this part of the series.

There are some few points to be considered before we begin to manipulate the tires. In every instance the tread of the rim or felloe ought to be at right angles with the face of the wheel, as per *Fig. 1*, as explained. *A, A*, outline of rim at opposite sides; *B, B*, tread of rim; *C*, the short angle or tongue of a wheel try-square; *D*, the long angle. Place the long angle across the face of the wheel, and bring the short angle down on to the rim. If the rim is perfect, the try-square will fit closely to the rim all the way around; if not perfect, have it made so.

Fig. 1.

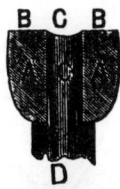

Fig. 2.

Fig. 3.

The spokes and rim ought to fit close enough to not necessitate any wedging of the spoke tenon. See that the rim sets up thoroughly on the spokes all the way around, and that the joints of the rim are close together. It is a serious mistake to leave the joints open. Such action increases the dish, and causes the spokes to buckle either on the sides or on their faces.

Do not allow the ends of the tenons of the spokes to project through the rim, as such action would cause the tires to rest on the spokes only, or with not sufficient pressure on the rim to act as a perfect binder. Do not gouge out the end of the spoke tenon, as shown at *C, Fig. 2. A, A*, rim; *B, B*, tread of rim; *C*, spoke tenon; *D*, section of spoke. The proper manner is as shown at *C, Fig. 3; A, A* rim; *B, B*, tread of same; *D*, the spoke tenon; *E*, section of the spoke. Take a narrow chisel and remove just as much from the end of the tenon as will cause it to be flush with rim after the compress of the tire sets the rim up on the spoke shoulder.

It is advisable to touch up the ends of the tenon with metallic paint, and allow the same to become thoroughly dry before setting the tires.

In the prior chapter I did not dwell long on the matter of tire plates. The average tire plate is flat. Some few are made with a disk of perhaps one inch or more, the object of which is not to benefit the wheel, but to allow the water used in cooling the tires to gravitate to the center and to run off through the hub hole. The tire plate ought to be made so as to allow the tire only to rest on the plate, the swell of the rim to clear the plate. To accomplish not less than 10 inches disk would serve well all grades of wheels with which one has to deal in general work.

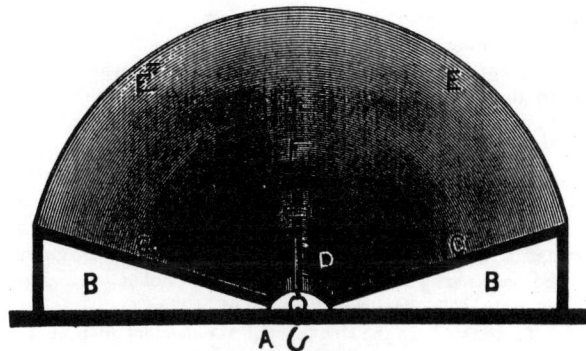

Fig. 4.

Fig. 4 shows a section of the plate; *A*, the floor; *B, B*, the practical rim of the plate; *C, C*, the dished line of the plate; *D*, the hub hole; *E, E*, the outer periphery of the plate; *F*, the concave or dished surface of the plate. A plate of this kind will be an expense at the beginning, but in the end will pay for the cost in saving the rims from becoming bruised.

It not infrequently becomes necessary to piece a tire after it is bent. A piece may be broken or otherwise unfit to remain in the tire.

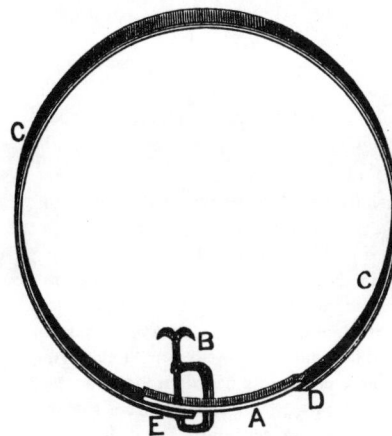

Fig. 5.

To put in a piece, just get a piece longer than you require, scoff the end, and also the tire, lap and rivet as at *D, Fig. 5. C, C*, the tire; *A* the piece; then with a small clasp as per *B*, secure the tire and piece as at *E;* now proceed and weld the tire at *D*, remove the clamps, measure the tire, and cut off to the right length, scoff and weld as usual.

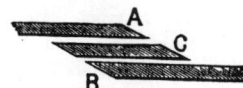

Fig. 6.

Should a mistake occur in running off the tire, and the tire become too short by 1 inch, more or less, proceed as per *Fig. 6. A*, section of tire; *B*, section of tire; *C*, a piece inserted between the two ends of the tire and riveted. A good mellow heat will allow you to weld all at one time. Care should be taken to only insert as much, as will give the required increase in length.

The Carriage Monthly, Oct 1889

SETTING TIRES.

THE durability of a wheel depends largely upon the way in which the tire is set. Too much draw in the tire bends the spokes, and when they get set in a crooked shape there is no way of straightening them out. It seems to be the universal rule to give a buggy tire 1/8 in. draw, no matter how much dish is in the wheel. It seems to prevail in the minds of many smiths that a tire must have draw. Even if the wheel should have 3/8-in. dish, the draw in the tire must not be changed. I find from experience that the average factory wheels will stand better with about 7/16-in. dish. Some shops want less, while others want more. Some smiths will say: "How can I get 1/2-in. dish in my wheels when some have 1/4-in. dish, and others have 3/8-in. dish before the tire is put on?" The way I manage this is to give the tire of the wheel that is straight 1/8 in. draw; when the wheel has 1/4-in. dish I make the tire without any draw, and when the wheel has 3/8-in. dish I make the tire 1/8 in. larger than the wheel. I do not approve of this screwing down of a wheel and hammering it to pieces. In order to get out the dish remove the tire and stretch it; this saves time and trouble. Don't guess at the amount of heat that you have in your tire. Cool it off and get your draw just right. To keep the tire round will save a good deal of trouble. To do this, run the tire through the benders so that it will be smaller than the wheel, and then pull the tire over the wheel (before it is welded) and hammer all around on the tire. The smith will be well pleased with the result. Heating up a tire on the forge is a matter that should be looked after. A light tire requires only a very small amount ol heat to put it on. To do this, keep turning it in the fire all the time until it is warmed evenly all around. Don't cool it off with water, as that makes it difficult to fit it close to the rim. SCOTT SMALLWOOD.

The Hub, Mar 1889

TIREING WHEELS.

THIS is a subject to which too much attention cannot be given. In the first place, a blacksmith must use his own judgment as to the amount of draw the wheel will require to give it the proper dish. Then the question arises: What is the amount of dish a light wheel, with a 7/8 or 3/4 in. tread, should have? This depends largely on the make of the wheel. But take the average wheel, Sarven patent; this should have from 3/8 to 1/2 in., not over half an inch dish; if it is more there will be sprung spokes, and nothing mars the appearance of a vehicle more than two or three sprung spokes. A compressed band, or, more commonly called, wood hub wheel, with the average stagger, should have from 3/16 to 5/16 in. dish from the front spoke; 1/8 in. or even straight for a full stagger will do. The dish in a staggered wheel very seldom runs out as in a patent wheel; this is because it is got from the shoulder of the tenon, whereas in the other it starts from the throat. If a compressed band wheel or shell band wheel is given 1/2 or 5/8 in. dish when new, after it runs awhile the back spokes will pull out of the hubs a little. The hubs are made small, and, of course, do not allow for a heavy tenon; and if the wheels are straight, as they should be, the tenon being the weakest point, naturally draws from the back and ruins the wheel.

Another point is to get the wheels through the paint shop without having to reset the tire. If the wheel is properly made, the tire can be set tight enough without dishing the wheel too much. Do not heat the tire too hot in putting it on, as it has to be cooled off, and the hot water and steam enters the wood and injures the rim. Let the tire cool without water, if possible, as it sticks to the rim better when it is cold. Many wheels are ruined by tireing, and the wheel-maker is charged with having made a poor wheel.

The Hub, July 1892

SETTING AND RESETTING TIRES.

(See Illustration Accompanying.)

EDITOR OF THE HUB.

Dear Sir—In the October number of THE HUB I noticed an article entitled "Setting and Resetting Tires." I would like to add what little I know about setting tires, old and new. In the first place, it is a pretty delicate case to handle, as the longer a practical man in that line works at it, the more he may learn.

I am a man of 23 years' experience in the carriage business, and know something about it. The first thing is to see if the face of the rim is level; if it is not, the tire will twist the wheel more or less. This can be overcome by the little machine planes and levels as shown by Fig. 1. Having the rim of the wheel ready for the tire, the next thing is the trimming of the end of the spokes just right. Some wood workers cut the spoke ends below the rim too much, which causes the wheel to become "rim bound," as the end of the spoke must rest on the tires when set. About the best way to get over this difficulty is to head the end of the spokes down with a riveting hammer, if but a little sticks through the rim. I will warrant the wheel will go longer without being rim bound than by any other treatment, as this sets the shoulder of the spoke on the rim solid, and the workman can tell just what joint requires to be cut out. I always want my wheels cut so that the joints will just pass each other without striking.

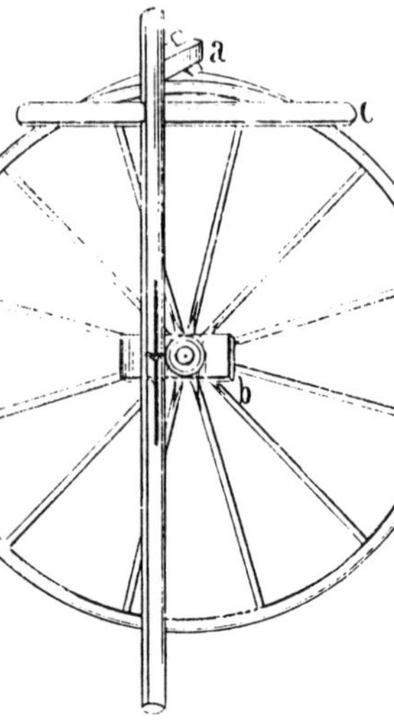

FIG. 1.

Now to set a tire right, and so that it will look well. With an ordinary depth of rim I used 1 x 3/16 round edge tire, the round edge to extend over the side of the rim to the extent of the round on the tire. This size of wheel with spokes driven straight on the face will hold 1/8 in. draft for 5/16 dish of wheel; the amount of draft to give wheels varies, as every smith knows, while wheels from different manufacturers stand different drafts, poor wheels will stand less than good. Any good tire setter can testify to this by putting on a lever; using it to take out the dish is a bad practice except to use it just enough to hold the wheel tightly on the tire stone. To use it to take out the dish where there is too much in, loosens the spokes in the hub, rendering the wheel unfit for long service. A better way for taking out the dish of a wheel, is either to drive off the tire and draw it a little, or you can take out quite a portion of dish by driving the tire nearly off, and then driving it on again. I have seen 1/4 of an inch taken out of a heavy wheel in this way. A 1 1/8 in. rim tread wants, in my estimation, 7/32 in. tire; 1 1/4 in. tread, 1/4 in. tire; 1 3/8 in. tread, 7/16 in. tire; 1 1/2 in. tread, 1/2 in. tire or 3/16 in. according to the quality of the wheel. This means first class wheels; poor wheels, if anything, take lighter tires than good ones.

The accompanying cut shows a face hand planer to true rims before setting tires. It is simple in its construction, as it has an attachment to the arm made as a plane and resting on the rim of wheel. A shows a slide which can be adjusted to fit different heights of wheels, and secured by means of two set screws. C shows a cross arm that rests on the face of the rim to carry it steady. D. E. B.

The Hub, Feb 1893

Tires

FITTING WHEELS FOR THE TIRES.

EDITOR OF THE HUB.

Dear Sir—Will you be so kind as to inform me whether you consider it the best practice to cut off outer ends of spoke tenons, so when the tire pressure is on the ends of tenons will not bear on the tire; or, if this is not the best practice, do you prefer the spoke tenon to bear its proper proportion of weight and pressure against the tire when tire is first put on? Will you please be kind enough to give us your reasons for whichever plan you approve, and your objections against the other plan? Also, whether your rims are left a trifle long before setting tire, and how much?

Yours truly,
MANUFACTURER.

[Answer.]

Probably there is no question in connection with wheel making that has caused so much discussion without reaching a definite conclusion as that propounded by our correspondent. There are wheel makers who insist on having the spoke ends flush with the tread of the felloe, and the ends of the felloes firmly together, so that the draft of the tire cannot set the felloe hard against the shoulders of the spokes. There are others who insist upon cutting the spoke tenons off a little below the tread of the felloe, and shorten the felloes so that before the tire is set the aggregate space between the felloe ends at the two joints is about three-sixteenths of an inch; they argue that unless this is done the wheels will be rim bound and the tire cannot perform its full service.

THE HUB holds to the opinion that the wheel, like every other manufactured article, must be so treated that each and every part will bear its own proportion of the strain. To insure this, we like the use of the machine which grips the felloe and sets it down snugly against the shoulders of the spokes sufficiently solid as to insure a firm bearing, but not so as to break the grain either of the end of the felloes or the shoulders of the spokes; and when so set down trim the ends of the spokes with a plane, so as to preserve the exact form of the tread of the felloe; then set the tire so that it will draw all as snugly together as when the wheel was on the machine.

If this is done the wood is not weakened by the fibre being broken, and the ends of the spokes bear their proportion of the strain, and the tire, in addition to taking the wear from wood, shares with it the strain that is put upon it.

We do not believe in excessive draft to the tire. An amount sufficient to draw all parts of the wood together and set the iron snugly against it, to insure a solid bearing, is all that is necessary. Anything more or less than this will prove injurious to the wheel.

The Hub, Apr 1893

In Favor of Cold Setting.—Much has been said about cold tire setting, and I have just read about Brother Richard Loades, of Kansas, who says it does not seem right to him to cold stove iron or to kink it. I say that there are no reasons for kinking a tire in cold setting it, as a man has it in his own hands to keep the tire straight and not kink it on any machine. I own a Scientific hydraulic cold tire setter and I set over 1200 tires in 1910. I can set four buggy tires in about twenty minutes and do a good job. I have set four in eight minutes, too, for a man held the watch. I have set several set of wagon tires in twenty minutes and in no case need more than thirty minutes to a set of ordinary wagon tires. One brother says that he wants some cold tire setter man to explain how he can make any tire smaller than the wheel and do it with the tire on the wheel. Well, here's the answer: in the first place you understand, Mr. Huff, that a wheel won't go to dish any before the tire is off the rim. My tire setter drives the two heads together about $\frac{1}{8}$ of an inch on an average. Therefore I have a gauge made as in the engraving and I set the thumb screw against the spokes, the straight edge being against the rim or tire and commence to pump, watching the thumb screw at the spoke or hub. When it begins to leave the spoke I count one, two, or four for wagons. I have set a tire in this way and taken it off the wheel and having measured it found it exactly that measure on a 1⅛ by ¼ tire on a spring wagon wheel. Of course, on a wagon wheel it will usually take one extra stroke of the punch handle, say five strokes to make it ¼ of an inch on a thick tire. To the simplicity and rapidity of setting tires cold add the lessening of the blacksmith's expenses and discomforts: no burnt arms, no hot fires to bend over in hot weather, no kindling to cut and carry out and no smoke in the eyes.

I wonder what our readers think I get for writing this letter. Well, I get just what every reader of THE AMERICAN BLACKSMITH gets by reading the article. If he will buy a cold tire setter, he will profit by the reading, but if he does not, it will pass out of his mind after a time and no good will come of it. It does seem to me that any fair-minded man can see the advantage of the cold setting process. I have worked at the trade all of my life, so to speak, and am running a shop of my own for over twenty-six years. During this time I have set tires in every known way; cut and welded them and shrunk them with two pairs of tongs and a helper and have used a hot tire shrinker and now I am using a cold tire setter, which I do earnestly think is the best way. One says stoving a tire cold might cause it to break in cold weather. Out of the 1200 I set last year I have not had one single tire to weld on account of breaking in cold weather. I did break three tires by gripping them in an old weld, but that was my fault. Now I think I have said enough about cold tire setting for once, but if any brother wants more I can tell him. This letter is unsolicited and I am not writing it for any reward, but I do it for the benefit of our brother readers.

My shop is, I think, well equipped: Two brick forges with Champion 400 blowers, Easy trip hammer, Fairbanks-Morse Gas Engine, Champion power drill, Reese No. 203 emery wheel, rip, cut-off and wood saws, disc sharpener, grindstone, cold tire setter, hot shrinker and Balingers hot tire setter, and all the small tools which are regularly needed in an up-to-date shop. I have also a Barcus horse stock which I like very much. My shop is 20 by 78 feet. I own my shop, four lots, a seven-room dwelling and thirty-eight acres of farm. I farm as a side line.

J. W. JEFFRIES, Missouri.

The American Blacksmith, July 1911

Tire Upsetter.

One of the MONTHLY correspondents sends us from Stockton, Pa., drawings of a novel tire upsetter, originated by himself. He sends the following description.

"I send you drawings of my tire upsetter, having made two of them, which have been found to work well. I can upset a tire $1\frac{1}{2}$ inches wide, by $\frac{7}{16}$ inch thick; $\frac{3}{4}$ of an inch at one

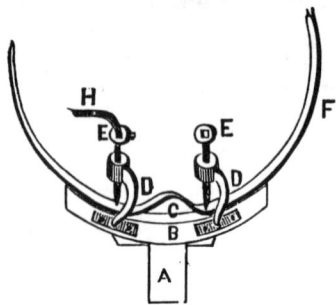

FIG. 1.

heat. I do it all by standing behind the anvil, no running from the anvil to the upsetting machine. The very small outlay this upsetter will cost is two dollars, any smith can make one in two hours. I made this one in an hour and a half. I have never seen anything of this kind in the MONTHLY. I believe many a young smith through the country would be glad to see a drawing of it in your valuable magazine.

The following are explanations of the cuts:

Fig 1. A, shank to go through anvil hole; B, block, $1\frac{1}{4}$ inch square; C, top of block; D, D,

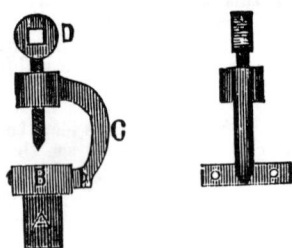

FIG. 2.

neck, with T rivetted on the block; E, E, set screws pointed at lower end; H, lever to turn set screws; F, the tire in position.

Fig. 2. Gives two views of the clamp; A, shank to be placed in the hole of the anvil; B, block; C, neck, 1 inch round, Swedes iron; D, set screws, $\frac{5}{8}$-inch round steel.

FIG. 3.   **FIG. 4**

Fig. 3. Lever.

Fig. 4. Top of block, cut same as file, to prevent tire from slipping.

Distance between neck-head and block, $1\frac{1}{4}$ inches. The T at bottom of neck rivets on to the block, with $\frac{3}{8}$-inch rivets; square hole at D, *Fig 2*, for entrance of lever, $\frac{7}{16}$ inch, Yours, etc.

HENRY BLEECK.

The Carriage Monthly, Nov 1877

A SIMPLE FORM OF TIRE UPSETTER.*

TUNBRIDGE, VT., May, 1882.

EDITOR OF THE HUB—DEAR SIR: Accompanying please find a rough sketch of a tire upsetter of my own make; there is no patent about it. The cost of it is trifling, and any blacksmith can make one. It is made of 2-inch square iron; and where the dotted lines are shown, that part is caught in the vise when in use; or it could be cast for a wider tire, and bolted to the bench, as shown below the dotted lines, and it would

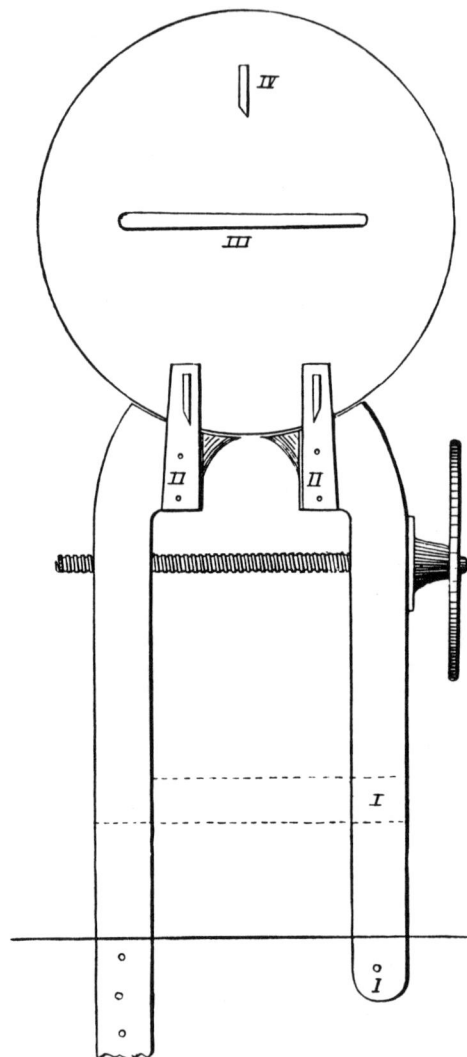

open and shut from the pivot I in either case. The parts 2 2 are made of flat iron and bolted to the main part, and two keys, made as shown in sectional cut 3, are driven through them to hold the tire. The key is made of steel, with a sharp edge, shaped as shown in sectional cut 4.

When a blacksmith gets used to this upsetter he can upset, or, by reversing the keys, draw the tire the least fraction of an inch, and fit his tire the first time.   Yours truly,   T. JENKYN.

The Hub, June 1882

Tires

SCHOU'S PATENT TIRE & AXLE UPSETTING MACHINE.

COMBS & BAWDEN, Freehold, N. J

It is simple in Construction

It is durable.

It requires only one Person to operate it.

It is a valuable Machine for every blacksmith to have.

The most practical Upsetter in use.

Price $ 35

Address for Circular or Machines

EXCELSIOR TIRE AND AXLE
UPSETTING MACHINE

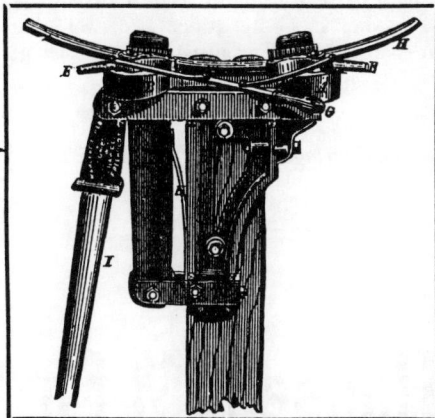

Schou's Patent.

Price, $25.00.

COMBS & BAWDEN,
Circulars furnished on application. **FREEHOLD, N. J.**

The Carriage Monthly, July 1887

A HANDY TIRE SHRINKER.

(See three Illustrations accompanying.)

EDITOR OF THE HUB—DEAR SIR: Would you kindly describe and illustrate in your valuable journal a handy device for shrinking tires. By so doing, you would confer a great favor upon an old reader.
J. B.

ANSWER.—We think "J. B." will find the tire-shrinker herewith illustrated an excellent machine in all respects.

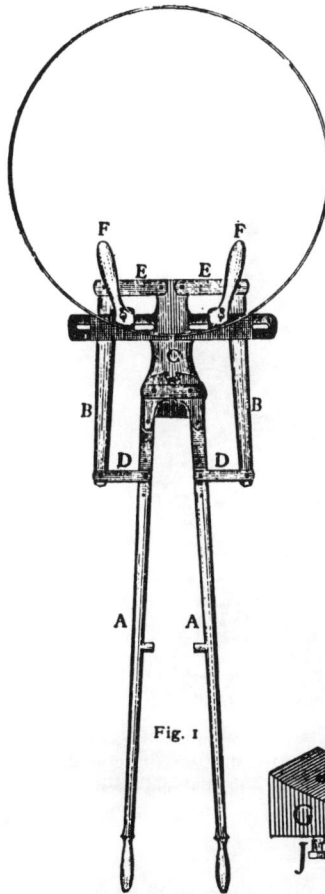

Fig. 1

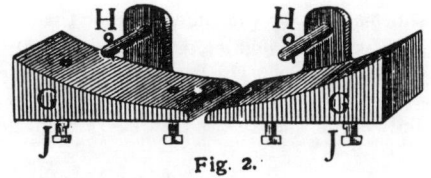

Fig. 2.

Fig. 1 shows the complete machine. A, A, the handle levers which are made of 1½ × ½ in. refined iron; they should be 3 ft. in length. B, B, the first, or power levers; length, 12 in. C (see also C, Fig. 3) is the main bed-piece of the machine which carries the slides by which the tire is upset, it is ½ in. thick and 13 in. long. The slides that do the upsetting are 12 in. The other connections where the handles D, D are

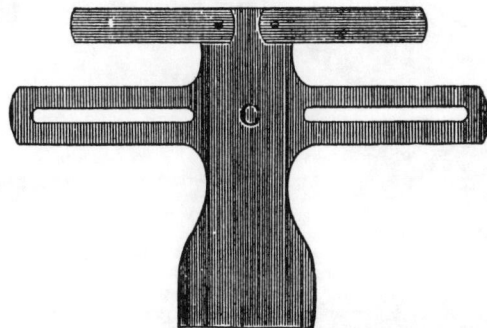

Fig. 3.

attached to each other are 5 in. long of ⅞ × ¼-in. tire steel. E, E are the front connections. F, F, are the handles which keep the tire from slipping whilst being upset.

Fig. 2: G, G are the bottom slides. H, H are the pins which hold the eccentric in place; they are made of steel. J, J, the bolts holding on the purchase levers.
D. E. B.

The Hub, May 1890

159

A New Tire Upsetter.

MANUFACTURED BY BUTTS & ORDWAY, BOSTON, MASS.

The accompanying cut represents a new tire upsetter, which is described as thoroughly constructed of best materials, and is easy of operation and adjustment. The size shown will upset 3 by ⅝ inch, and larger sizes are in preparation. The method of operating this machine is simple. One downward motion of the lever throws the

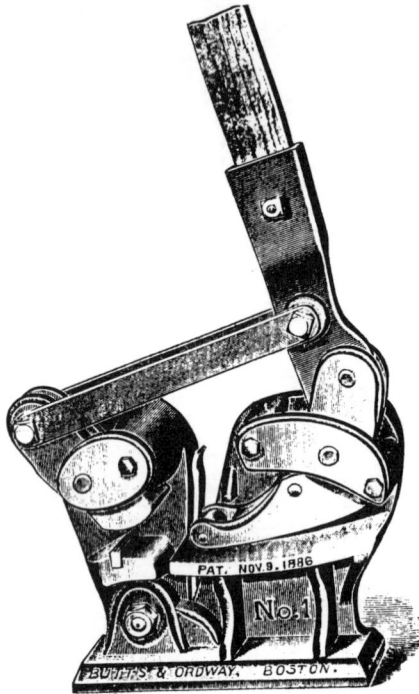

shoe-cam on the tire and upsets the tire ¼ inch. Raising the lever half way and making another downward motion upsets the tire another ¼ inch, which operation may be repeated as often as necessary. The back notch on the lever is intended for use on heavy tires, and will do the work as easily as on light tires.

The Carriage Monthly, May 1892

A New Tire Shrinker.

MADE BY S. D. KIMBARK, CHICAGO, ILL.

A new tire shrinker, called the " Ideal Mole," has been put on the market, which, it is claimed, has all the advantages of the old " Mole " shrinker, to which is added the new and important adjustable or floating shoe for holding down the tire when upsetting, so as to prevent its kinking. The improvement consists of a shoe attached to the bed place, and connected by cogs to one of the dogs in such a manner that the movement of the lever operates the shoe and holds

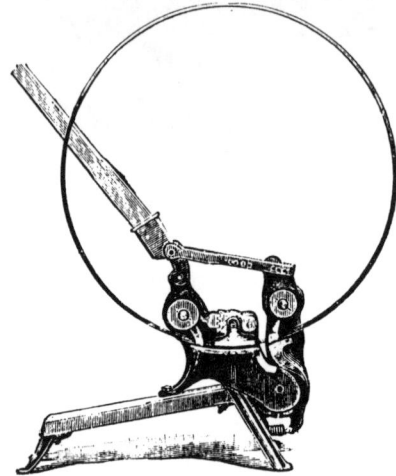

the tire firmly until upset. The shoe also adjusts itself to the different thicknesses of the tire, or can be quickly removed if necessary.

The new shrinker has been strengthened at every point, so that breakage will be almost impossible. Its manufacturer, S. D. Kimbark, Chicago, Ill., claims that it has been adapted particularly for truck tires, and will shrink a tire with a 2-foot-10-inch circle, or larger, while one man can operate it. A tire cannot kink, and it will shrink a light buggy tire as well as that of a heavy wagon.

It is made in one size only, 4-inch bed, weighing 350 pounds. Every machine is tested and warranted against breakage.

The Carriage Monthly, June 1895

Tires

RIVETING vs. BOLTING TIRE ON THE RIMS.

To the Editor: Why is it that some builders, chiefly European, rivet instead of bolting on the tires, and what, if any, is the advantage gained by the riveting process?   T.

Answer.—In former times all tires were nailed on. American builders were the first to discard the nailing process and to adopt the method of securing tires with tire-bolts. But few European carriage-builders have adopted tire-bolts, probably from fear of losing the nuts and thus also the bolts, or having the latter come part way out, and cause other damage. A better finish may certainly be given rims when rivets are employed, and this might sometimes lead us to use rivets, but as relates security, we consider the bolts quite as safe; and as to cost, the bolting process is much cheaper. It is only with the heavier grades of work that rivets are likely to ever become popular with American carriage-builders.   M.

The Hub, July 1877

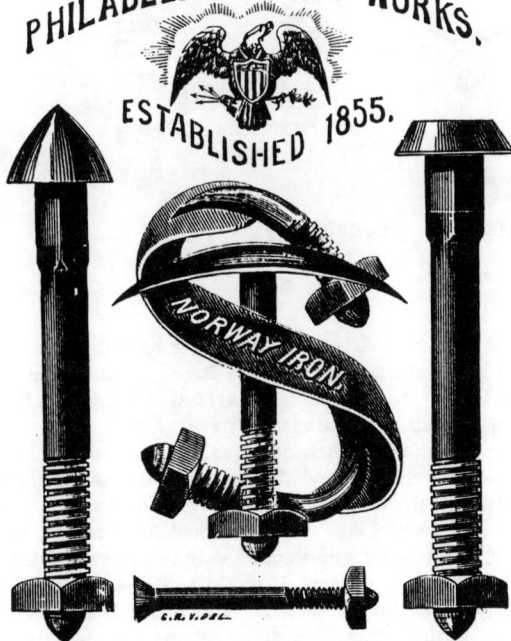

The Hub, Sep 1879

A TIRE-BOLT WRENCH.

We have recently received several inquiries from workers in the smith-shop as to the use of various tools of precision, and we now have before us a question from T. D., concerning tire-bolt wrenches.

The simplest and clearest way to answer such a question is to show an illustration of a tire wrench. The worker in tools will then see at a glance how it is constructed, and whether it is adapted to his wants.

The tool shown in the accompanying engraving is known as the Green River tire-bolt wrench, and is made by the Wiley & Russell Mfg. Co. It is simple, strong and durable, and modeled in a way that makes it convenient to handle. These conditions seem to fulfill the requirements of a wrench of this description, and as the one illustrated is now found in many of the shops we visit, we conclude that it has met with favor and approval, and we hope it may be just what our correspondent is looking for.

The Hub, Jan 1882

Drilling Tires.

The illustration below shows a tool for drilling tires and counter-sinking them at the same operation. This countersinking tool is pierced to carry the drill, which may be adjusted according to the depth of hole wanted. The drill piercing the tire and felloe, and the countersink following, the job is finished at a single operation without the trouble of adjusting tools and work twice. These tools are new, and but recently introduced by the Wiley & Russell Manufacturing Co., of Greenfield, Mass. They furnish sizes and prices on application.

We believe tires of mixed metal, half iron and half steel, are not as yet upon the market. It is claimed they have the hardness of steel without any liability of breaking. The introduction of such tires will be beneficial, providing they will not break in cold weather, as our steel ones often do, will weld readily and are sold at the same price as the regular steel tire.

The Carriage Monthly, Oct 1883

Securing Tires.

"Will you have the kindness to inform me the best method of securing tires to the wheel?" is the very pertinent question sent us by a rural builder.

Believing the question to be one of much import, we set out with a view to getting the desired information by making a personal visit to a number of builders of good reputation. In our visits we failed to find any two who agreed in all particulars. Some were in favor of bolts; some preferred screws alone; others preferred screws and bolts; some secured the tires with rivets and burs; and some builders of heavy work never have used anything but nails. All had had various experiences in the matter. Some secured their tires by a certain method because "so and so" did them so. In our meanderings we called on a builder of humble pretensions, who always has plenty to do, and who has the reputation of giving better satisfaction with his wheels than any of the others. His remarks were so intelligent, and his expressions and illustrations so logical and plain, that we have accepted his statement in preference to all the others, and quote nearly as possible *verbatim*:

"Half a hundred years ago, when I was about the height of a new chopping-block, and spent my 'after-school' hours in the shop of my daddy, tire-bolts were a novelty, in fact, were only used when a customer called with a broken tire and could not wait to have it welded and reset. In such a case we would drill a hole, near the break, on each side of it; countersink the hole, and put in bolts to hold the tire temporarily. The custom at that period was to nail all tires on to the felloes. I say felloes, because bent rims were not the fashion at that time. The tires were drilled over each spoke and nails, varying in size from $\frac{1}{16}$-inch diameter up to $\frac{3}{8}$-inch diameter, as per the size of the felloe and spoke and the caliber of the tire. The object in driving the nail into the spoke was to preserve the felloe against splitting, which would have followed if the nail was improperly driven.

"When the tires required resetting, we cut and broke them at the weld, pressed off the tire, and then removed the nails from the felloes, welded and reset the tires and again drove in tire nails.

"Later on, when I had been at school long enough (according to dad's way of thinking), and went to work in the shop, tire-bolts began to show up, but they were all $\frac{1}{4}$-inch diameter and were used for light wheels only. The buggy wheel at that time seldom ever carried less than 1 by $\frac{1}{4}$ tire. The bent rim became conventional about the same time. The rims of the light wheels were made in about this shape. [Showing a section of rim. See outline L.] This one [showing an old, short felloe] was about the shape of the felloes of the heavy wheels for coaches, chariots and caleches. We hadn't begun to use the bent rim on heavy wheels.

"The bolts (tire and carriage) were then chiefly made down East, and were called Yankee bolts. They were none the best, having been made of a short, waste iron; the threads were poor and the nuts of common plate iron, cut out and punched cold, but poorly fitted and broken or stripped very readily. Then came the Philadelphia bolts, of all lengths and caliber, from $\frac{1}{8}$-inch diameter up to $\frac{1}{2}$-inch diameter, which permitted the carriage or buggy-maker to get his rims a trifle lighter and to use a much lighter tire. Then came the rage for very light work. The buggy which formerly was considered light at 300 lbs., with 1-inch axles and 7-inch hubs, and a tire 1 by $\frac{1}{4}$, was tabooed by the sporting class.

continued next page

"We then began to use ⅞ by 6½ axles and ⅞ by ₁³₆ tires, and managed to get buggies that weighed 250 lbs. Later on a ¾ by 6 axle was considered about right, and ¾ by ⅞ compound tires just the thing; the weight was then brought down to about 225 lbs; and when the half spring came along we dropped down on the weight to about 200 lbs.

"With all this whittling down in sizes, the drivers of fast nags were not satisfied Then came the whittling down of the rims, until they were about this shape [showing a section of rim; see outline *K*], which prevented or precluded using tire-bolts of even the lightest caliber, because of no base on which to rest nut or washer; the only bolts were at the joints and held the plates. Screws were then substituted for bolts, a screw being placed on each side of the spoke, about three inches from each side of the spoke. Of course, the wheels looked rather natty, whittled up in that shape and showing no nut on the rim, but it was beauty at the expense of safety; just where the breaking strain came

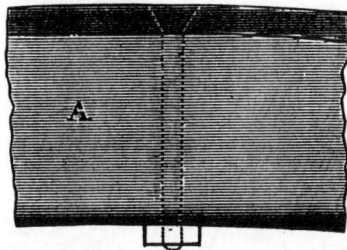

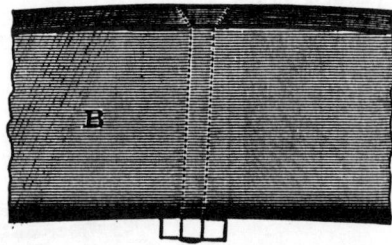

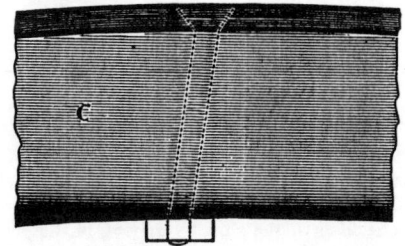

the stuff was the weakest. I made one or two sets that way and stopped short off.

"Besides the weakening of the rim from whittling down so much, the screws were a torture when tire-setting time came. The crease in the screw would be about worn out, or so much so as to bid defiance to the screw-driver to remove them. A common device for removing the screw was to insert the point of a broken spring leaf between the tire and rim and then try to turn out the screw with a hammer and cold chisel. This kind of treatment would break off about every other screw at the first part of the thread from the head.

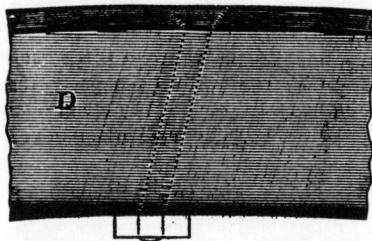

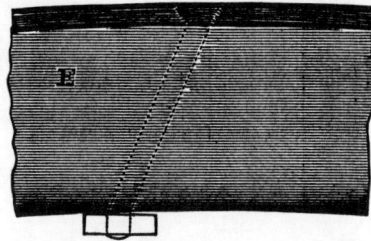

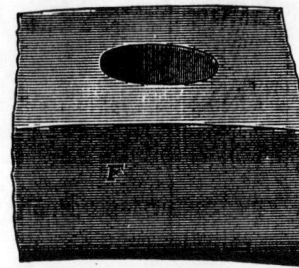

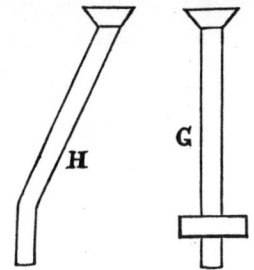

"When all failed the tire would be cut and a lever applied to remove the screws, which oftener broke the screws on the tires at the hole than it removed screws. The dragging out of the screws by force always took more or less of the wood with them, which left the holes so large when the tires were reset as to make it impossible to make a screw hold on without drilling larger holes in the tire and using larger screws, alike detrimental to rim and tire.

"The broken screws which remained in the rims were a great annoyance; they were in the path of the new screw. You had to go to one side and had to use the greatest precaution to prevent splitting the rim or breaking off the new screw. The three dollars which we got for resetting such a set of tires only about half paid for the labor. The cussing and extra labor, stock and broken bits, were never paid for.

"I really could not invent a better method for destroying rims, wasting time and exhausting patience without recompense than fastening tires to wheels with screws; and a careless or willful hand or workman could do more injury to your reputation with two sets of wheels than you could pull up in a whole year. No man ever made better wheels or better buggies than Dunlap. He always put two bolts between spokes and about three inches from the spoke.

"I said I made but two sets of wheels with the pinched or whittled-down rims, and they gave me as much of that kind of work as I wanted. I fell back to the old style, as I showed you (see *L* and *M*). In all my

light wheels, from ⅞-tire down, I put a bolt three inches or four inches from each side of each spoke, and manage, with my common-sense rim, to keep my rims and tires pretty near round. My wheels will run from six years to eight years before they begin to show a break or yield in the center of the rim.

"Now, then, I want to show you a few features about tires and rims, a little of the pure cussedness that is always going on. I keep these models on hand to show my customers when they growl about my prices of resetting and rebolting tires [showing specimens on subjects as per *A, B, C, D, E, F, G, H*]. Here [showing *A*] is where the tire-bolt goes through tire and rim in a direct line; and we have a good rest for the head, also for the nut and washer. There is an equal strain at all points; the tire is held fast and will not slide off.

"Here [showing *B*] is where the mischief first begins; the bolt is just a little out of line. The upsetting or shrinking of the tire has changed the position of the holes; they do not match. Here [showing *C*] you get just a little more. The brace and bit help you out in this case. Here [showing *D*] is a little more crookedness; the brace and bit and burning-iron come to your relief, and, after much cussing, the bolt gets there. With patience and a few hammer-strokes on the head and nut, you possibly get through without breaking the bolt or splitting the rim.

"This [showing *E*] is the last stage, or the quintessence of cussedness and foolishness combined, a very reckless and unprofitable combination to a carriage-builder who tries to get the credit of doing good

work. The brace and bit is of no value here; the burning-iron or nothing; and if the burning-iron does not break off in the hole you are lucky. Next you must bend the bolt and file the point, like this [see *H*], which is not a bit like its original shape as this is [see *G*]. It gets there after awhile; but what a get! When you wish to remove it, if you

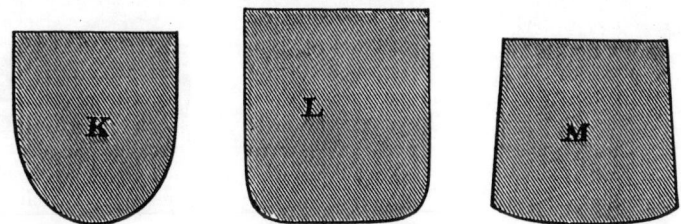

do so without splitting the rim four or five inches you are the luckiest cuss extant.

"Now, then, just look at this [see *F*]. You can see just about the shape of the holes in the rim of the average light wheel after the tires have been reset three or four times, a hole on the outer part of the rim fully ¼ inch long and ⅛ inch wide, tapering down to the inner edge of the rim to a trifle larger than the bolt.

continued next page

163

Tires

Securing Tires.

" Just as soon as tires are taken off in my factory for resetting, and are welded, a survey is made for all holes in the rim which will not correspond with the holes in the tire. In each of these, after first removing the ironrust with a file, I glue in a plug of wood enough softer than the rim to not cause the rim to split. Should a hole be missed, I cause it to be plugged from the inside after the tire is on. By this little precaution I preserve the rim, and also save in the matter of time, which would be lost by following in the ' cussed way.'

" Another great advantage gained in this system of plugging up the holes is that there are no cavities for the water and dirt to lodge in and cause the rim to decay. A small brush and a little lead is always at hand to coat over the plugs as soon as inserted.

" No, the blacksmith does not put in the plugs. The plugs are made by the hundred, the same as dowels, by driving them through a hole in a piece of spring steel ; they are made by an apprentice overtime. When the plugs are driven in, the rim is placed in a vise cushioned with hard leather, which prevents the bruising of the wheel ; the vise prevents the checking of the rims by driving in the plugs.

" I have always followed or used this method, and always plugged the holes when I used to fasten tires with nails. Here is a rim that has worn out two sets of tires and is now having the third set put on, and does not show a sign of a split or break in it. Such is the great care I take and demand with all my wheels and tire-setting."

The lecture was a very interesting one to us, and, we believe, will prove both interesting and profitable to our subscribers, and well worth studying and practicing.

The Carriage Monthly, June 1887

Improved Lever Feed Power Drill.

MANUFACTURED BY GEO. BURNHAM & CO., WORCESTER, MASS.

This drill is specially designed for drilling tires in carriage shops, and is one of the cheapest and most useful machines in use for that purpose. It is large enough for the largest tires, and will do as much work as a heavy drill. The lever feed can be adjusted to any position of the drill. A new wheel holder has been made especially for this drill, that holds the wheel firmly so that there is no danger of broken bits. The table arm to which the holder is attached can be raised or lowered, and allows the drilling of the largest or smallest tires in use. All the holes can be drilled without removing the wheel. The claim is made that because of these features more tires can be drilled with this machine than with any other.

This drill is a very useful boring tool for wood workers and for a general class of work. Being a sensitive drill it can be used for very light work, and is adapted to cotton and woolen mills, paper mills, stove shops, hardware shops and many other lines of trade where a light and cheap drill is needed. It can be used as an overhead or suspended drill for bicycle shops and other works requiring a drill of that nature.

The diameter of the spindle is $1\frac{5}{16}$ inches ; length of spindle, 28 inches ; greatest distance from spindle to table, 23 inches ; vertical movement of table, $17\frac{3}{4}$ inches. It will drill a hole to the center of a 16-inch circle, $5\frac{1}{4}$ inches deep. The drill socket screws to the spindle and takes drills $\frac{11}{8}$-inch round shank. This socket can be removed and a No. 1 or 2 Hartford Universal Chuck put in its place. The counter shaft, with hangers, has a three step cone pulley and a 7-inch tight and loose pulley for $2\frac{1}{2}$-inch belt. The machine weighs 200 pounds.

The Carriage Monthly, June 1892

Coach-makers' Int'l. Journal, Mar 1873

REED'S PATENT
Gas Tire Heating Machine
FOR HEATING CARRIAGE TIRES.
EASILY ADJUSTED.   ALWAYS READY.

REQUIRING BUT FROM ONE TO SIX MINUTES TO HEAT.

Tires Ready as Fast as a Man can Set Them.

Burns only 1½ feet of gas per minute, and much less when used in connection with blower.

In use and recommended by Brewster & Co., N.Y.; Abbott, Downing & Co., Concord, N. H.; Emond & Quinsler, Boston; Henry Hooker & Co., New Haven; Frank Ivers & Co., Cambridge, Mass.; McLear & Kendall, Wilmington, Del., and many others. Address,

SHERBURNE & CO.,
53 Oliver St.,   **BOSTON.**

The Carriage Monthly, Mar 1887

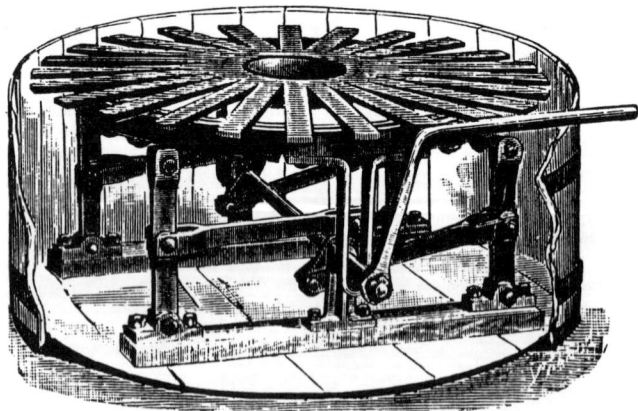

THE BOKOP PATENT TIRE SETTER *AND* COOLER

H. D. BOKOP, Patentee and Manufacturer,
DEFIANCE, OHIO.

This machine has already found its way into One-hundred and Twenty-three shops.   The **advantages** of this Machine are:

1st.   Instead of springing wheels and tires by setting, all kinks and crooks are taken out.

2d.   The tire being cooled instantly, there is no chance of the felloes getting charred.

3d.   The tire can be moved on the wheels when finishing it without tearing out the edges of felloes as the tire does not become imbedded in the felloes.

4th.   The tire can be set with one-half less labor.

5th.   They can be used in any part of the shop as well as outside, as there is no slopping of water and but little smoke.

6th.   After the tub is once filled it requires no handling of water; they are made of nearly all wrought iron and cannot be broken by the heaviest tire.

Sent on trial to any responsible party.

For prices, reference and descriptive circular, address as above,

Yours Truly,

H. D. BOKOP.

The Carriage Monthly, Apr 1887

PROCESS OF TIRING WHEELS-

The Carriage Monthly, Aug 1889

Steam Power Tire Setter and Hub Bander

MANUFACTURED BY J. B. WEST, ROCHESTER, N. Y.

This machine was patented as early as 1870, but was not brought to a full measure of success until recently. The present machine is a departure from all former attempts, and it is said excels all ten fold in speed and perfection of work. It upsets equally all around the circle, having twelve equal sections or jaws, that advance uniformly towards the center, with sufficient power for the work intended. It has been used on wheel-barrow tires at the rate of twelve per minute.

The machines for carriage and wagon tires are supplied with a gauge that is operated by the dishing of the wheels as the tire tightens. It is adjusted to any required amount of dish, which, when reached, trips the valve automatically, reverses the throw of the piston and opens the machine instantly, giving the same "dish" to all wheels, and effectually preventing injury to any. This is an entirely new and important feature, adding greatly to its value.

A new automatic cut-off allows no more steam to the return or up stroke than is necessary to open the machine, thus saving fully half the steam it would otherwise require. This machine is capable of unlimited range in sizes, has no pulleys, belts or gears, and is ready for business when the steam and exhaust pipes are connected. It is strong and durable, unlikely to get out of order, and does not require skilled labor to operate it. A new clamping apparatus holds tires and rim evenly down to the turned surfaces of the machine, so that if the tire should be "winding" or in any way out of true when it goes in, it will be held tightly down while the upsetting is being done, and will be perfectly true when finished, and perfectly even with felloe, saving trueing and evenning afterward with hammer.

It sets tires cold, and saves the running or measuring of wheels and tires, and the heating and cooling and the lost time in transporting from the smith shop, wetting, time required for drying, prevents burning of felloes, and swelling of the wood by reason of hot water and steam, and subsequent shrinkage. Tires can be ordered of sufficiently accurate length at the mills, allowing ½ to ¾ inch to the usual length, and any number of tires can be rolled and welded, and wheels need not go to smith shop, any tire fitting any wheel.

It is claimed to set four light tires a minute.

The ease and rapidity with which tires can be set with this machine by simply laying the wheel (with the tire on) in the machine and pushing the handle down to the catch, the truer wheel, the accuracy and uniformity of dish, none having to be reset, and all in the short space of from two to four seconds to a wheel, keeping them all clean and dry ready to paint, are sufficient advantages to recommend it to all manufacturers of wheel vehicles.

The hub band machine is the same, except the broad top and automatic gauge. It is operated in a similar manner to a steam hammer, the operator having only to depress the valve and raise it again to set a band. It saves measuring every hub at every place where a band is to be put on and fitting a band for that place and driving it on. Bands can be cut by gauge and welded standard sizes, and set as fast as they can be handled. It actually saves measuring, keeping each band for its particular hub and place on the hub and driving on.

The center (or spoke) bands, that loosen when the wheels are dried, can be set on this machine before the spokes are driven and can be reset without taking them off, after the wheel is made and dried, by laying the wheel on the machine, making them as tight as you wish in two seconds to each band.

The Carriage Monthly, Mar 1892

THE M'FARLANE TIRE FURNACE.

(See Illustration accompanying.)

ONE hardly realizes the magnitude of the improvements made daily ; they press on so rapidly, that the novelty of yesterday is the necessity for to-day. This is true to as great an extent with the carriage industry as with others, which come more directly under the observation of the public. These improvements are quite as noticeable in machinery and methods as they are in vehicles and their attachments ; but until a comparatively recent date, the heating of tires was performed in a primitive manner, and even now there are few carriage factories where modern improvements have been introduced. We see the wood fire, or the brick oven, in places where all other work is performed by the aid of modern appliances. Why this is so is a problem. The complaint has been that there was nothing better; if this was the case, that time is past, as there is a furnace that will perform the service more cheaply, more cleanly, and with a rapidity that is astonishing as compared with other methods—we refer to the McFarlane Tire Furnace, for which the inventor was granted letters patent, August 28th, 1891. It is in every respect an innovation over old methods, and deserves, as it will surely receive, the attention of the carriage trade. The illustration on this page explains more fully than we can by words the method of heating—the fuel being oil or gas—while the advertisement on another page explains fully all the claims made for it.

Claims are often put forth that cannot be sustained, but the practical tests made, demonstrate without a shadow of doubt that the inventor has been modest in stating its merits.

Though but a short time on the market, it has found its way into some of the largest factories, and wherever introduced, it has won over the most skeptical. The following testimonials are from houses that would not approve of an article that was not considered by them as the first in its class :

"CHICAGO, June 8th, 1892.

"Office of T. W. Seelow, 461 467 22d-st., Manufacturer of Wagons and Carriages.

"H. McFARLANE—DEAR SIR : Your Mr. Cline was here this morning and set twenty-five tires in forty minutes. The furnace is giving the best of satisfaction. I would not part with it for one thousand dollars, providing I could not get another one. Yours very truly,   T. W. SEELOW."

"RACINE, WIS., April 15th, 1892.

"H. McFarlane, Clinton and Harrison-sts., Chicago.

"DEAR SIR : Replying to yours of the 12th inst., we will state that we have tested your tire furnace in our factory, and find its heating capacity fully up to the standard you claim for it. It is capable of heating from seventy to eighty sets of ordinary tires per day, and heavier and wider tires in proportion.   Yours truly,   FISH BROS. WAGON CO."

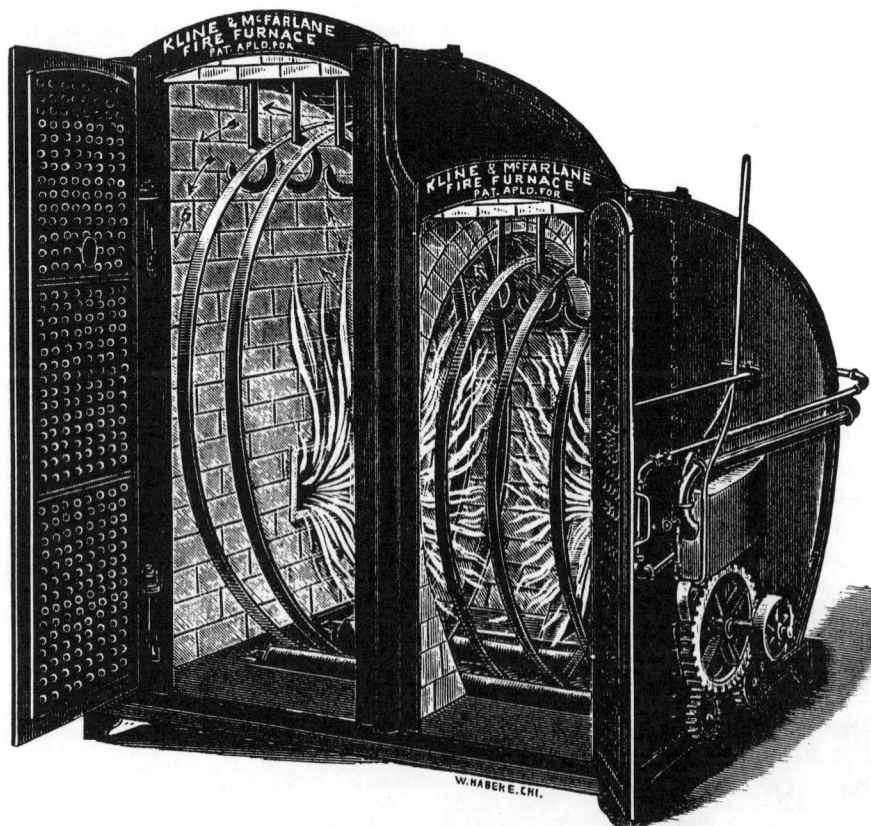

THE MCFARLANE TIRE FURNACE.

The National Combination Drill Press.

MANUFACTURED BY THE NATIONAL HARDWARE CO., CINCINNATI, O.

The accompanying illustration represents a recently patented machine which is destined to do the work of several men by any former method, and with an accuracy and such a saving of time that it at once will be seen to be a most valuable addition to any shop where many wheels are tired.

One-half of the entire circumference of the wheel (or an equivalent to nine holes) can be drilled and countersunk at one operation. Each hole is automatically made in the exact middle of the tire, an advantage assured over the hand press; also, the distance from the spoke, either the usual three-inch space, or equi-distant from the spokes, or at whatever point it may be set, all holes will be exactly at same point.

The machine is manufactured with cast iron frame and of the best tool steel, and upon scientific principles. The cog wheels are cut and

The Carriage Monthly, Nov 1892

A PERFECTED COLD HYDRAULIC TIRE SETTING MACHINE.

IT is the opinion of those acquainted with the new tire setting machines that all carriage and wagon tires will before long be set on these machines. The rapidity and economy is unquestionable, and it is coming to be known also that a good operator will also do better work than is done by hand.

The machine, of which the cut is here shown, is a machine that has passed through its initial stage, and is being used by a number of factories with great satisfaction. Several years' use, and improvements, have combined to render the machine nearly perfect in its construction. All the arrangements and parts are especially ample and massive, and solid in operation. The pistons, which are arranged radially around the machine, are nicely and accurately combined in their motion, so that they all travel exactly together. Their com-

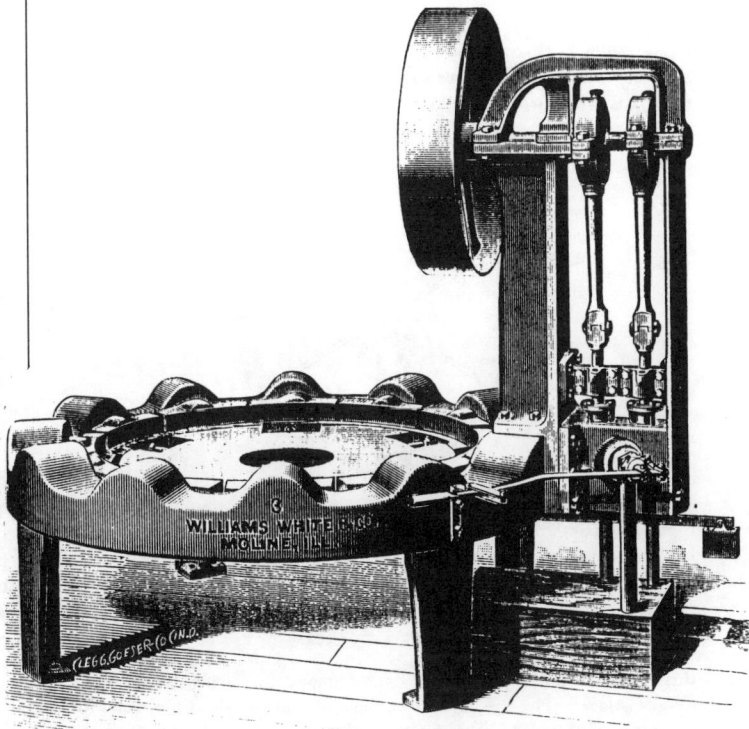

A PERFECTED COLD HYDRAULIC TIRE SETTING MACHINE.

bined motion is that of the expansion and contraction of a true circle. The hand lever which controls the machine, is so nicely balanced, that the slightest movement stops, starts and retreats the machine; thus bringing it under the absolute control of the operator, and permitting the stopping of the machine at exactly the proper point. The pump is of an especially strong and perfect pattern, having double plungers; thus giving a continuous steady motion. It is worthy of note that the pistons of this machine are positively guided in their stroke, so that they will not lift, no matter how wide the tire. The importance of this will be appreciated by those having had experience in this direction.

This machine is made by the firm of Williams, White & Co., of Moline, Ill., who make different sizes for every kind of tires.

This firm has recently also brought out a tire welding hammer, which gives a side as well as vertical blow, and promises to be a great attraction to the carriage and wagon world. The firm solicits correspondence regarding these machines.

The Hub, June 1894

Tires

"STREAK TIRE."

(See two Illustrations accompanying.)

It may be news to many that the "streak tire" is yet used in some parts of this country, and, without doubt, there are many in the carriage trade who do not know what the streak tire is. As a matter of curiosity and for the information of the trade, we present two illustrations of a wheel with a streak tire furnished by a correspondent who

STREAK TIRE IN PLACE.

finds this form yet in use to a limited extent among the farmers of Newton, Lower Falls, Mass. The tire, as will be seen, is made of short pieces bolted to short felloes, with the laps of the tire midway between the joints of the felloes. This is an extremely primitive method of tiring wheels.

The Hub, Sep 1891

The Evolution of the Tire.

BY J. B. WEST.

In these wonderful days of accelerating progression in the arts, it is surprising when we note how soon the old way of doing things is forgotten after the new is adopted, and how few of those now interested in tiring wheels for the millions, know that it was not always done in the way they are doing it, with a tire in one piece, and that that way was new and a great invention within the last century.

In the writer's boyhood there was an old cast off cart wheel lying in father's shed that was tired in the good old-fashioned way, the best that was known at the time it was made, the tire being in pieces, same length as the (sawed) felloes, and nailed on, with the ends of the tires in the middle of the felloes, "breaking joints," the nail holes being punched in oblong form, similar to those in horse shoes, only much larger, the longest way across the tire, three at each end, and others along the center the entire length, the nails being forged to fit, with a cutting edge across the grain of the felloes. It was common for the teamster to be furnished with nails on starting out on a trip, to replace those that might be lost on the road, thus becoming his own "tire setter." But early in the present century an improvement was made, claimed by many, but who the real inventor was is uncertain.

A smart smith thought of a hoop for a tire,
  Welded it up when the boss was gone,
Heated it well in a circular fire,
  Doused it with water and shrunk it on.
Many a smith would not believe it,
  Many a head was shaken "no,"
Many a one would not receive it,
  Nevertheless it was a "go."
It was a great thing,
  'Twas a wonderful day,
When tires were shrunk on
  In this new fangled way.

But the news and the general adoption of the new way spread rapidly over the country, all wheels being tired that way, and all have suffered the burning, the pounding, the wetting and all the other injuries incident to the imperfect "guess work" system that it has proved to be. Tire and wheel were accurately measured, to be sure, but the amount of draft necessary to give the required dish to the wheel has remained a matter of guess work, with its ever varying results; but like the way first above described, it was the best that was known, till now within a few years another long stride in the right direction has been taken, reaching perfection, in the production of a machine for compressing them on without heat by hydrostatic pressure, doing away with the accurate measuring of wheels and tires, the burning, wetting, sledging, etc.

The tires are simply welded large enough (standard sizes), to go easily on the wheels "cold," laid in the machine and clamped down, perfectly true tire and rim even with each other, and set in a few seconds, the machine being provided with sufficient brains to know just when to stop, to give the exact required dish, adjustable to any amount, and when so adjusted will make them all alike; also resetting loose tires without taking them off. Like the first described improvement, many don't believe it, saying " no."

But so many are now in use in the leading carriage, wagon and implement factories of the country, giving entire satisfaction, that there is no longer a doubt but that it will go into general use. Those who use it say that the work is certainly superior to that done by the old method, and at a minimum of cost that they never expected to be able to reach.

The Hub, Dec 1893

STRAKING

An Experiment to Recapture the Technology of Straking a Wheel

by Ron Vineyard

A method of fitting iron strakes (sectional tires) to wooden wheels was described by Dr Gordon S. Cantle in "Fundamentals of Fitting Tires," in THE CARRIAGE JOURNAL, Vol.27, No.1. The present article describes the actual straking operation using the tools and technique of eighteenth century Virginia and it is taken from a report prepared by Ron Vineyard, Master Wheelwright, for the Colonial Williamsburg records.

Documentation of the Process

Research into "shoeing a wheel with strakes" reveals a number of documents on the subject. Perhaps the most authoritative and detailed work is George Sturt's "The Wheelwright's Shop." Sturt draws a clear distinction between "tyring a wheel with hoop-tyres" and "shoeing a wheel with strakes." He writes:

For the felloes were longer for a straked wheel than for a hoop-tyre, there being fewer of them to make up the rim. In a wagon for instance, instead of seven felloes for each hind wheel (as with a hoop-tyre) there were only six if strakes were to be used, while for the front wheels there were but five ... But this was not the only difference. In the newer way of making a wheel, more than half the dish was produced by the shrinkage of the hoop-tyre, ... As strakes did not do this to any great extent, and dish was very important, it had to be secured when the stock (nave) was being mortised. From the start the wheel was made more hollow. It was useless to depend on the strakes. The wheelwright himself must drive his spokes with the required slant right away. Nor was this all. As I have previously explained, the contraction of the hoop-tyre drew all the wood-work of the wheel together with a matchless tightness. Sledge hammers and wedges could

Master blacksmith Peter Ross, assisted by Ken Swartz (right), driving strake nails.

not do half so well. But as the strakes were not expected to give such aid, the wheelwright was then at pains to put his wheel together more strongly than was thought necessary afterwards ... Therefore, to avoid any looseness or "give" in the wood-work, the wheelwright preparing for strakes was not contented with two round spoke-holes in his felloes. He chiselled out mortises in them, and then tenoned his spokes into the mortises, as exact as in cabinet work.... Soon, with strakes, square-tongued spokes gave way to cheaper work; indeed I am of the opinion that the wheelwright's craft began to decline when square-tongues went out of fashion. But they were always too costly, which is one reason why, to reduce their number, the felloes had been fewer.[1]

From the foregoing, it is obvious that a wheel to be straked was designed so from the start. Felloes were fewer to reduce costs and the mortise and tenon joint between spoke and felloe provided the necessary strength

by being "as exact as in cabinet work." Also, the wheelwright was more exacting in building the required dish into the wheel by "driving his spokes with the required slant right away." Further, in all respects the wheel to be straked was put together more strongly and with greater exactness than for wheels to be "tyred." Thus the method for installing metal to the circumference of the wheel was not a decision to be made after assembly, but rather was to be considered not only in design, but assembly, where specific techniques were used.

Preparation

The wheels to be straked for this project were wheels depicted in John Mullen's "A Treatise of Artillery," in his discussion of constructing a Galloper carriage.* Specific dimensions for components of the wheels came from this work.[2] Mullen called for straked wheels with six felloes which is consistent with Sturt's considerations for straked wheels. Joinery was made as accurately as possible at both the nave to spoke and spoke to felloe joints. Square-tongued mortise and tenon joints were used in the spoke to felloe joint as recommended by Sturt.

An extra felloe was made to be used by the blacksmiths as a form for curving the strakes. Each strake was made ¼ in. shorter in length than the felloes themselves. This was necessary to ensure that the heated and expanded strakes would fit between adjacent strakes. The ¼ in. dimension was chosen because that distance between strakes is depicted on the drawing of a wheel to the Lister Post Chaise in the Architectural Archives at Colonial Williamsburg. This ¼ in. gap between strakes appears consistent with photographs of other straked wheels, although no other specific measurements have so far been found.

Nail holes were punched in the strakes using the square and tapered configuration described by Sturt. Two holes were punched at the ends of each strake with three holes equally spaced between, along the length of the strake.

*Galloper—a gun limber for the "Flying Artillery" of the British army in the eighteenth century.

THE CARRIAGE JOURNAL
Vol. 28, No.1, Summer 1990

Tires

Ron Vineyard (center) with Ken Swartz and journeyman wheelwright John Boag, installing the samson.

Sturt described strake nails thus:

A strake nail was about as long as a sardine and a little thicker, but it was square-cornered and forged out to a thin end, wedge-shaped. A big thick head it had, battered over, and for half-an-inch or so under the head it was made tapering to fit into the holes that had been punched, also tapering, in the strake. The idea was that, as the strake wore thin, still the tapering nail might hold tight in the iron."

Partington, in his "Coachmakers' and Wheelwrights' Complete Guide" also provides a description of strake nails for carriage wheels as follows:

The nails used for fastening the iron-tyre felloes (strakes) on wheels are made on purpose for the work; they have square heads, forming a cube of about three-quarters of an inch each way. Heads for strake nails used on some military wagons rise as much as an inch above the surface of the strake. The driving part is about five or six inches in length, made quite flat and wedge-like, and they are larger or smaller in proportion to the size of the felloes and tire to be perforated and fixed by them.

With respect to the length suggested by Partington, it should be noted that his method of straking includes clenching the nails on the inner curvature of the felloes.[3]

A strake nail shank and a strake nail heading tool are shown in Salaman's "Dictionary of Woodworking Tools."[4] This sketch depicts a nail

which agrees with the nail described by both Sturt and Partington.

Lastly, and most important, in the National Park Service Ironwork Collections from the archeological digs at Jamestown Island there is a remarkably well-preserved strake nail which is identical to the above descriptions. This strake nail is the basis for the specific design used by Colonial Williamsburg blacksmiths in making strake nails for this project.

In addition to strakes and strake nails, there are special tools necessary for the straking process. As previously mentioned, a clamp or "cramp type" device called a samson is required for pulling the felloe joints as tightly together as possible. James L. Kirkland in his Diderot Translation from "The Anvil's Ring"[5] informs us that the

With the samson in place, Ron Vineyard uses a wooden mallet to close up the felloe joints.

samson is an English device, whereas the French used a device called the "chaine." The French chaine, quite similar in use but different in design, is depicted in the "Encyclopedie of Diderot and d'Alembert."[6]

Sturt provides his usual interesting, but non-specific, description of the samson. He writes, "The samson was a stiff parallelogram of iron specially contrived to use in shoeing. It was about as long, and half as wide, as a sheet of "The Times." Welded square to one of its ends stretched two long arms fitted with heavy nuts; which nuts, screwed down or up, tightened or loosened a ponderous moveable cross piece—a sort of shackle from arm to arm—at the other end."

The most useful information on the samson and its use is contained in Salaman's work. He informs us that the samson measures "up to four feet nine inches long by about nine inches across." He continues, "But the purpose of the samson is clear: it was to draw adjacent felloes together in the absence of the tyre." Salaman provides two illustrations of different style samsons and one illustration of the samson on a wheel. The first depiction of the device is similar, if not identical, to that of Sturt. The second samson, from a Sussex smith's shop, is essentially two L-shaped arms fitted with adjusting nuts on the long arm of the L.[5]

Salaman also shows a tool he refers to as a Strake Nail Claw. He describes it as, "A heavy iron claw about ten inches long used for extracting strake nails when taking off old strakes." This tool was made by the Blacksmith Shop and proved its worth in actual use.

It was decided that the tongs normally used for handling hoop-tyres would be adequate for straking. These tongs have one portion of the jaws straight and one with a lip. When the straight jaw is placed between the felloe and the tire or strake, the tongs can be easily withdrawn after the tire is in place. The lip on the opposite side provides excellent security for the hot metal since the lip is hooked around it.

Once the strakes were completed by the blacksmith, each was numbered for a specific place on a given wheel by marks on the front or face side. In preparation for boring pilot holes for the strake nails, each felloe was divided in half using large dividers. A mark was struck at the mid-point of

Tires

each felloe. The samson was used to draw the felloe joint as tight as possible. The strake destined for that position on the wheel was then centered between the mid-point lines on the felloes and each strake hole was marked on the felloes. Pilot holes were then bored at the marked positions using a 3/8 in. bit to a depth of 1¾ in. Size of these pilot holes was chosen based on results from experiments with various diameters and depth of holes bored in a spare felloe. Nails were driven into these test holes and checked for holding power, splitting of felloes, etc.

Thus, we had accomplished what Sturt described as follows:

While the nails were a-making, the wheelwright had to take a hand again, fitting newly-punched strakes and carefully marking their destined places on the wheel, so that the right one should be taken, and by the right end too, when at last it had to be nailed on to the wheel. Most important, it was to get the exact place then, because every felloe had to be bored beforehand for those big strake-nails to enter. There was no other way of driving them into the hard timber; they would have been too big and blunt. Below every one of the holes in the iron, therefore, a corresponding place for a nail in the felloe was wanted. Accordingly, as I say, the wheelwright tried on and chalk-marked every strake, and bored its nail hole in the rim of wood, against the critical time, soon to come now, for shoeing the wheel.

A shoeing-hole was dug with a post on either side so that the wheel could be hung on a "stout iron bar thrust through the central opening of the wheel."[1] The height of the posts was such that the bottom portion of the wheel was submerged in the water of the shoeing-hole.

Now all was in readiness for shoeing our Galloper wheels with strakes.

First Wheel

The first wheel was straked on June 13, 1989 in an area prepared behind the Anderson Blacksmith Shop. Since this project was a joint effort between the Wheelwright and Blacksmith Shops, personnel from both shops participated in the actual straking.

Strakes for wheel number one were laid out on bricks in proper order for installation. They were "nested" together approximately 3–4 inches apart. The fire was laid using a generous amount of wood shavings and dry kindling. Steve Mankowski, a blacksmith, started the fire using flint and steel. This open wood fire technique was used in place of the "strake-chimney" described by Sturt since no such chimney was available. The open fire technique had proven quite satisfactory in heating hooped iron tires.

While the strakes were "hotting," the wheel was hung over the shoeing-hole on the posts previously described, with the joint to be bridged by strake number one at the top. Strake-nails were partially but firmly tapped into pilot holes outside the area to be covered by strake number one. Over these nails the samson was placed and tightened to close the felloe joint. With a wooden mallet, the felloes were driven down tight on the square mortise and tenon joints with the spokes. As the samson was further tightened, the mallet was used to align the felloe joint. A nail was driven into the rim of

Ron Vineyard uses the tongs to hold strake number one in place while John Boag hammers in the nails.

The smoke begins to rise as Peter Ross and Ron Vineyard drive the nails home.

John Boag splashes water on the hot strake as Peter Ross drives in the last nail in strake number one.

THE CARRIAGE JOURNAL

173

Tires

the felloe at the mid-point line (previously marked) to aid in placing the hot strake exactly in the proper position for nailing. Strake nails were placed in two buckets filled with water and positioned on each side of the shoeing-hole.

During these preparations the fire was tended for rapid heating of the strakes. Also, the strakes were tested for temperature by touching with a long wooden stick. When the stick burst into flames upon contact with the strake, we knew the desired temperature had been reached. Using the special tire tongs, strake number one was removed from the fire and quickly carried to the wheel where the wheelwrights and blacksmiths waited with hammers and handfuls of nails from the buckets. From this point no description is better than Sturt's:

> At double-quick time the smith hooked out a red hot strake, seizing it with his tongs by the middle, hurried it across to the wheel-joint to the place marked by the one tapped in nail, and held it down while the wheelwright banged in a nail. Forthwith the other man, at the other end, striking down the iron all along to fit the rim, got his nail; flames and choking smoke leapt up; the men, burning their fingers and wrists, dipped their hands hastily into the pail of water, smote in their nails —whack, whack, bang, bang— with deft sledge work, the smith caught up the spare sledge to help, someone exclaimed fiercely 'He's burnin'!' as the flames shot up higher; the coughing wheelwright puffed the smoke aside to see that all was right. Then (lest 'he', the wheel, should burn too much) pulled the now fastened strake round into the shoeing-hole, where the water began to pobble and boil with the sudden heat, the clouds of white steam mixed with the blue wood smoke. The men stood back panting for a minute's rest, and watched the heaving turmoil in the water.

> But only for a minute. Before the water was quiet, each was stooping for another handful of nails from the bucket, spitting on his hands for gripping the sledge better, watching where the smith was already hooking out a second strake. Then the same thing again

—crackle of burning wood, hiss of water, rising of steam and smoke, all to the loud clatter of sledge blows. This for only a minute, followed once more by the pobbling noise of boiling water, as strake number two was pulled round into the shoeing- hole.

Using the process described by Sturt, strake number two was put on the opposite side of the wheel from strake number one. This process allowed the first strake to continue cooling while the next was being nailed on. The strake nails at each end of the strake were not fully driven in. This gave secure points for attaching the samson to close the adjoining felloe joint. However, due to the large heads on the strake nails, this practice of leaving the nails partially out proved unnecessary. The samson could be secured very nicely with the nails fully driven. With smaller strake nails, as used on lighter wheels, leaving the nails partially driven would be necessary.

Not knowing how much compression could be achieved with the combination of heated strake and samson, the amount of gap left between the felloes proved to be too little, resulting in a felloe-bound wheel. That is, the felloe joints were tightened very well, but there were small gaps between the shoulders of the spoke and the felloe itself. This was corrected by removing the strake nails for one strake with the claw, removing the strake and "kerfing" with a tenon saw to give an additional $1/8$ in. gap. Upon re-installing the strake, all joints closed very nicely. The felloe gaps on wheel number two were adjusted accordingly and closed very well on the first try.

Wheel number two was shod in the Wheelwrights' Yard at the Palace Stable on June 20, 1989. This time another layout for heating the strakes was used which proved much easier and efficient. This layout used two semi-circles of strakes about a foot and a half apart as opposed to the "nesting" layout for wheel number one. Heating strakes in the semi-circle pattern reduced heating time by almost one half.

Shoeing of wheel number two went very quickly and smoothly. The experience of wheel number one had given us a high level of confidence in the procedure. As stated previously, all the joints on wheel number two closed tightly with the increased felloe gaps.

Summary

Straking is a very old method of attaching metal to the outer circumference of a wheel. It was commonly used in the eighteenth century, not only on carts, waggons and military vehicles, but also on coaches, chariots, phaetons and chairs. The building of a Galloper carriage provided a rare opportunity to experiment with straking because these wheels will be used almost exclusively on the greens and not on our asphalt streets where the large-headed strake nails could do great damage. The project also provided a unique interpretive opportunity. Most visitors were totally unaware of the straking process and its use in eighteenth century life. This is understandable as straking was rarely, if ever, used past the middle of the nineteenth century since advanced technology allowed for very large hooped iron tires to be produced with relative ease, and also the brake, using a shoe against the outer surface of the tire, was becoming common by this time. Such a brake could not be used with strakes and their large protruding nail heads. These two factors saw a rapid decline in the use of straking. thus our visitors had the opportunity to learn something of this type of wheel and its components.

Our experiment gave great insight and practical information on this process. This has expanded our knowledge of our trade. In addition, the "hands-on" work necessary provided a greater appreciation of the writings of those who actually performed the work. Written descriptions by Sturt had clearer meaning after one had experienced the smoke, steam and "pobbling" water which are so much a part of straking. Further, we are confident of our ability to produce additional straked wheels, in an accurate and authentic manner, when the need arises.

References:

[1]Sturt, George—The Wheelwright's Shop, Cambridge, England, 1963.
[2]Muller, John—A Treatise of Artillery, London, 1780.
[3]Partington, C.F.—Coach-Makers' and Wheelwrights' Complete Guide, London, 1825.
[4]Salaman, Raphael A.—Dictionary of Woodworking Tools, London, 1977.
[5]Kirkland, James—"Diderot Translation," THE ANVIL'S RING,Fall 1983.
[6]Diderot, Denis—Encyclopedie, Ou Dictionnaire raisonnes Des Sciences, Des Arts, et Des Metiers, Paris, 1764.

VII
RUBBER TIRES

175

RUBBER TIRES

The rubber tire was a vast improvement over the conventional iron or steel tire, resulting in a more comfortable carriage ride. The original idea for the pneumatic tire was patented in England in 1845 and was reinvented in 1888 by an Irishman, John Dunlop, for his son's tricycle. (See Ken Wheeling's article on the history of the rubber tire, p.182.)

Once developed, the styles of rubber tires, as well as tire channels, proliferated. The year 1895 saw the introduction of the Rodgers Rubber Cushioned Vehicle Tire. In the same year, the Standard Tire Co. of New York made a wheel with a rubber cushion between the wooden rim and the steel tire channel. In another procedure the tire channel was installed using the cold tiresetting method and secured with rivets. Then a solid rubber coil was laid in the channel and the wheel was put in the tire end setter shown on p.181. When the wheel was removed from the machine, the cemented rubber joint was closed.

The article on "Comparative Tests With Steel and Pneumatic Tires," p.178, describes a test done to show the resistance of steel versus pneumatic tires carrying various loads on different types of road surfaces.

————————————

A subject not covered in the articles is the popular belief that rubber tires were black for horse-drawn vehicles. The first rubber tires made used zinc oxide as a reinforcer for the rubber, which colored the tires white or gray. The rubber tire industry notes that it wasn't until 1904 that the first patent was taken out, by S.C. Mote of England, to reinforce rubber with carbon black. It was not practical for rubber tires to be reinforced in this way until 1912, and the first commercially produced black tires were made by the Diamond Rubber Company, which was bought by B.F. Goodrich in the same year. It still took several years of a mass advertising campaign by the rubber tire industry to convince customers that black tires were just as good as white, or better. (Information taken from "Evolution of Today's Pneumatic Tire," *Rubber & Plastics News*, Vol. 18, No. 2, August 22, 1988, pp.26-28.)

RECENT ENGLISH NOVELTIES IN WHEEL TIRES.

AMONG recent English novelties in reference to wheel tires is a grooved felloe with a ribbed tire made to fit into the groove, as shown in the annexed sketch No. 1, which sufficiently indicates the idea without description. Great advantages are claimed for this device, on grounds of strength and absence of bolts or rivets through the felloe ; but I do

FIG. 1.

not anticipate that it will ever make much progress in England. A similar principle, I believe, is involved in the Silvester patent tire, which I see advertised in your columns, the only difference being in the manner of its application, and the position of the rib which holds the felloe fast.

Another notion, even more vigorously pushed, is the Moss Patent wheel, which was shown at our Sportsman's Exhibition last February. This was lavishly praised by one of your contemporaries, and your regular English correspondent " Spes " also noticed it a few months since, but it has not been explained in your columns.

This Moss wheel is made about as usual, but with a thin, light tire, and outside and beyond this is another tire or hoop having sides made as shown in Fig. 2, forming a continuous groove in which the wheel proper runs. This outer tire is made in two parts, as shown in Fig. 3, which are fixed together by screws all around the rim. This outer tire is about 3 inches larger in diameter than the inner tire, and the sides or flanges are wide enough to allow of a little play between them and the felloes, which latter they overlap.

FIG. 2.   FIG. 3.   FIG. 4.

Between these two tires are placed a series of springs, shaped as shown in Fig. 4, which are fastened in the center by a bolt or screw opposite the end of each spoke, through the inner tire and into the spoke itself. The points press against the outer tire. At each end is a block of specially hardened rubber to prevent vibration.

The claim of the inventor of the Moss wheel is, that the resistance of the road is virtually destroyed, and that these springs are really sufficient alone, without other springs, to secure easy riding. I cannot admit that they are as effective as this, although they doubtless reduce the vibration ; but, on the other hand, they look heavy and clumsy, which I think will prevent their extensive adoption.

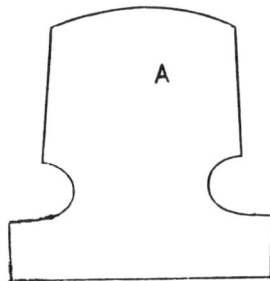

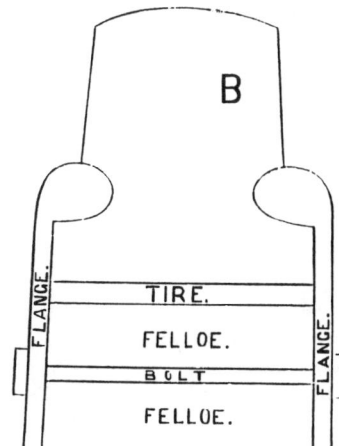

FIG. 5.   FIG. 6.

Another notion which promises greater success, and has been more favorably received, is a new patent India-rubber tire. A company has been formed to introduce this, and I believe it has a future in England. There were already two competing systems of India-rubber tires, but this new one seems likely ere long to eclipse them both.

One of the previous systems consists in the employment of very hard rubber for the part next the iron tire, which holds firmly the heads of the bolts or rivets by which it is fastened through the tire and felloe, by a burr or nut on the inner side of felloe. A softer rubber is used for the outer portion of the rubber tire, which, by fusion, becomes one all through.

Another system, known as Mulliner's patent, consists of a rubber tire shaped as shown in Fig. 5, which is secured by iron flanges outside the felloe, bolted through the felloes, and fastened by nuts at the back of the wheel as shown in Fig. 6.

The objections used against these prior inventions were as the first, its costliness, and the fact that wheels so tired could not be repaired in case of damage, but must be entirely replaced ; and as to the other, the great increase in weight it involved, and its clumsiness of appearance.

The new plan, which the Noiseless Tire Co., of Manchester, is now introducing consists in the use of an iron tire of the shape shown in Fig. 7, in lieu of, and not in addition to, the ordinary tire. This tire is rolled in two sections, which are then welded together down the middle of the tire, as complete, and then contracted to the wheel. Into the groove thus formed a specially prepared rubber tire is pressed, as shown in Fig. 8, in which the dotted lines show the original size of the rubber tire, and the other lines the section of the tire as a whole, when finished.

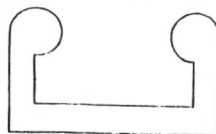

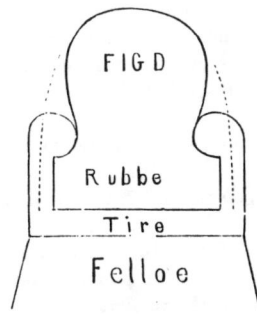

FIG. 7.   FIG. 8.

This plan is free from all the objections above named. The tire can be easily repaired or replaced ; the increased weight is very slight, and it is less unsightly than either of the others, and considerably cheaper in first cost ; and it necessitates no additional bolts or rivets through the felloe, either laterally or transversely. With all these advantages in its favor, I feel quite secure in predicting for it the first place among India-rubber tires, and consider it one of the most important introductions in carriage building of the year 1883.   SPIRO.
LONDON, ENG.

The Hub, Feb 1884

Rubber Tires

COMPARATIVE TESTS WITH STEEL AND PNEUMATIC TIRES.

I will first consider the matter of draught or road friction, and give you the result of some tests which I have recently conducted, and which are now made public for the first time. Two box buggies were employed, one having the usual steel tired wheels, 44 and 48 inches in diameter, and weighing 254 pounds; the other having pneumatic tired wheels, 32 and 34 inches in diameter, and the vehicle weighed 232 pounds. The cross diameter of the tires was 2 inches. An amount of weight equal to the difference was placed in the lighter vehicle, and care was taken to see that the front wheels of the two vehicles bore exactly the same weight.

The surface upon which this first test was made was a new, hard pine floor which was as smooth as such a floor could be, and the wheels were drawn lengthwise of the boards. The amount of power required to move these vehicles under the following conditions, was carefully noted by means of a registering spring balance which was attached alternately to the king bolts by means of a long cord.

In each case several tests were made, and when all did not exactly agree, owing to slightly varying conditions at different points on the tire, we took the average pull.

The same tests were made with the vehicles empty, and afterward when they were loaded with 300 pounds each. It was found that the power required to start the pneumatic tires from a standstill was 4 pounds, and the power required to haul them at a slow walk was 3½ to 4 pounds.

The power required to start steel tires was found to average but 3 pounds, and when started, the power required to draw them was but 1½ to 2 pounds, showing an average difference of about 50 per cent. in favor of the steel tires.

Next, an obstruction $\frac{5}{16}$ of an inch high was placed in front of and against the wheels of each vehicle. To haul them over this obstruction from a standstill required, in the case of the steel tires, 25 pounds; with the rubber tires, but 11 pounds. Then they were drawn at a slow walk over the $\frac{5}{16}$ obstruction, and it was found that the power required to draw the rubber tires was 5 pounds, and the steel tires, 8 pounds. An obstruction ⅞ of an inch high was placed against the wheels, and the power required to haul over it from a standstill was as follows: Rubber tires, 24 pounds; steel tires, 44 pounds. At a walking speed the power required to go over the ⅞-inch obstruction was 16 pounds for the steel tire, and 12 pounds for the rubber tire.

The two carriages were next loaded with 300 pounds each. It was then found that the power required to start the rubber tires on the smooth floor was 8 pounds, and to haul at a slow walk required practically the same force. To start the steel tires, loaded, required 12 pounds, and to haul at a walk, 4 pounds.

The $\frac{5}{16}$-inch obstruction was then placed in front of the wheels, and the power required to haul over it was 13½ pounds for the rubber, and 40 pounds for the steel tires.

Over the ⅞-inch obstruction the power required to haul the two loaded carriages was 36 pounds for the rubber, and 69 pounds for the steel.

The two vehicles were then taken out of doors, and placed on a fairly good gravel road.

The power required to haul the rubber-tired vehicle, loaded, 300 pounds, *averaged* 20 pounds, and the extreme power required at any point was 26 pounds. With the steel-tired vehicle, over the same road, the average was 41 pounds, and the extreme 79, or three times the resistance of the rubber.

To haul these two carriages *empty* over a moderately sandy road, the extreme power required for the rubber was 26 pounds, the same as when loaded on gravel with an average of 16. The steel-tire vehicle required an extreme of 40 pounds, and an average of 22. With a load of 150 pounds, the steel tire required an extreme of 57, and an average of 40; and the rubber an extreme of 38, and an average of 16.

After these tests had been made over this particular piece of road, the rubber-tired vehicle was again tried, empty, and it was found that the hauling of it six times over the road had so improved it that instead of the extreme pull of 26 pounds, and an average of 16, the extreme was but 16, with an average of 8.

If we had a perfectly true surface for a road, and perfectly true metal wheels, nothing better could be desired; but the words "perfectly true" mean, in the case of the roads, an impossibility. Even the surface of the best race track is made up of small obstructions.

If the metal-shod wheel meets a gravel stone ¼ of an inch in diameter, and that stone is resting on a hard foundation, the wheel, with its entire load, must be lifted bodily ¼ of an inch high to pass over it, and this takes horse-power; but when the rubber tire meets the stone, the vehicle is not raised perceptibly, if at all, but the stone is imbedded in the rubber, while most of the weight is borne by that part of the rubber which is still resting on the ground; and the power required to go over it is only that needed to dent the rubber is one spot, or if it is a pneumatic tire, to slightly compress the body of air which it contains.

THE IDEAL SIZE FOR PNEUMATIC WHEELS.

The question has been repeatedly asked, are not large wheels better than small ones, and if so, how do you account for the present revolution in trotting gigs?

My answer is, that large wheels are certainly better than small ones in theory, and within certain limits they are better in practice.

A pneumatic tire is better than a steel tire, for reasons which I will shortly explain, but it is comparatively heavy, and has required a flanged metal tire under it which will weigh fully as much per running foot as would a standard steel tire for the same vehicle. And besides, the smaller a pneumatic tire is, the more practical it is to make and maintain.

The pneumatic tire is an advantage, the small wheel is a disadvantage, or, at least, it has been found so when used with a hard tire. From the advantages of the one, we subtract the disadvantages of the other, and find that we have a balance in favor of that combination which warrants and has accomplished its universal adoption, viz., 28-inch wheels and 1¾-inch tires.

It should be remembered that the steel-tired front wheels, under which the obstructions were placed, were 12 inches larger in diameter than were the rubber tired ones; and it is a well established fact in mechanics, that a large wheel will go over an obstruction with a less expenditure of force than will a small one.

When the first pneumatic sulky, with its 28-inch wheels, began to lower the trotting records, thousands of horsemen and mechanics at once began to reason that if there was such advantage in the pneumatic tire as to make the little wheels win, what *couldn't* be done if the same tire were placed on a large wheel.

For obvious reasons we shall never know how many experiments were tried, but enough of them have come to light to prove that the 28-inch wheels, which are universally used to day, were not accepted blindly.

WHY THE PNEUMATIC SULKY IS SPEEDY.

There are four reasons why a pneumatic trotting sulky is faster than the old style; and every one of these reasons applies with more or less force to an elastic tire of any kind for any vehicle:

First. It draws more easily.

Second. Owing to the small diameter of the wheel, the sulky is much stiffer sidewise, and the soft tire stays on the ground and does not slew in rounding curves.

Third. When a horse is being crowded to his utmost limit, as all horses are supposed to be in a race, his nerves are under an intense strain, and he is kept near the point where he will no longer hold to the unnatural motion of trotting, but will adopt the run which comes perfectly natural to all four-footed animals. The noise and jar of any hard tired vehicle, when going at speed, has much to do with the action of any horse.

Fourth. The more securely and comfortably any driver is seated, the better attention he can give to the trotter, and a nervous driver tends to make a nervous horse.

RELATIVE ADVANTAGES OF CUSHION AND PNEUMATIC TIRES.

Elastic tires are at present of two kinds. One depends upon the yielding nature of a mass of rubber, and the other uses compressed air, rubber being employed only to render the confining fabric air tight. For racing purposes where extreme speed is required, that tire will be fastest which has the least between the compressed air and the ground,

continued next page

Rubber Tires

as compressed air is much more elastic than rubber; but as the ideal tire would not be practical, owing to the danger of bursting and puncture, it is necessary to compromise on a quality and thickness which may combine as much as possible of the air spring with the important element of durability. Pneumatic tires are made and used on vehicles, which give no trouble, and I know of two four-wheeled carriages which have run on pneumatic tires every day since last Spring without the necessity of pumping the tires once during the time. I have known solid rubber tires to run over a year without trouble. If speed is wanted, the pneumatic tire only is to be considered.

For ordinary road driving, however, the solid rubber tire has advantages which may, for certain vehicles, offset its lack of resiliency. Neither tire, as made to day, is as durable and safe as we might wish. Both require more care and attention than the old metal tire, and it is not probable that any elastic tire can ever be made which will compare with the other for all round results on all sorts of roads, and in the hands of ignorant men.

It is undoubtedly true that an inferior vehicle with elastic tires will outlast the best one made which has metal tires, and by the same reasoning any vehicle will last much longer with than without the rubber. And many intelligent men claim that enough will be saved in repairs and gained in the extended life of the carriage to warrant the expense of new rubbers each year. If that should turn out to be true, the question is settled, for rubber tires which will last more than a year are now obtainable.

The item of comfort to the passengers is most remarkable, and this is obviously the strongest point in favor of the rubber. In fact, the difference is so great, that to many people the questions of cost and durability do not cut any figure. A person with nerves who has ridden much on elastic tires is never quite as well satisfied with hard tires afterward.

Another point which is greatly in favor of the soft tire, and which is important, if anything is, it not only does not inflict injury upon the road, but is a positive benefit to it, for the reason that while the metal tire in the direction of its length, is round and touches the ground hardest at a single point, the soft tire lies perfectly flat upon the road for several inches and does *not* tend to press any part of the loose surface out of place but *does* act to press it more firmly together.

I have noticed where a large number of bicycles have passed any common point on a muddy road, they naturally look for the best place. I have seen a path a foot or two in width, that looked perfectly dry, caused by the tires, which had passed over it and smoothed it, while the mud showed water on each side.

A LARGER AND WIDER FIELD.

The signs of the times, as they look to me, are, that there will at no distant day be a considerable demand for pleasure vehicles fitted with elastic tires. The thicker and more elastic the tires, the smaller may be the wheels, down to, say, 30 inches. Many intelligent carriage users are beginning to save up their money to buy a set of rubber shod wheels, and I believe that the carriage builder who keeps his fingers on the public pulse, and is in a position to know what is up to date on these matters, will find, when the inevitable demand comes, that he is several lengths ahead of the builder who sneers at the new tire, and tries to talk the progressive customer into taking something else. But there is by far a larger and wider field which seems hardly thought of at present.

I feel safe in predicting that the time will come when the intelligent truckman, the teamster, and the expressman may reap direct financial benefit from the use of elastic tires. The poorer the man is, and the harder he has to work to make a living, the more reason there will be for his adoption of the highest quality of elastic tires.

It is entirely consistent with the facts to say that the intelligent application of the latest wisdom in tires and bearings will enable one horse to haul a larger load than can be hauled by two horses on the best wagon as at present made.

This is surely enough gain to revolutionize wagon building, even though there were no other advantages. But we must put on the other side of the ledger the items of first cost, risk of damage, additional attention, and cost of renewals. What the balance will be no man at this date can tell you. But the problem is by no means as difficult as are many which have been successfully solved.

The Carriage Monthly, Nov 1894

The Rodgers Rubber Cushioned Vehicle Tire

———

THE cut shown herewith of the Rodgers Rubber Cushion Tire (patent applied for) shows a cross section and side perspective view of the vehicle rubber tire used by the Tricycle Mfg. Company, of Springfield, O. The tire is in the steel rim and attached to the wood felloe.

The rubber is in two parts, the base part fitting in cross section, the steel rim, and the round (or nearly round) wearing part that fits into the concavity of the base, this round part only coming into contact with the road. The base part is first secured in the steel rim, and in the larger sizes is held firmly by the wires, as shown in the cut. The round part is then compressed upon a steel core, which core holds it securely in its place. In addition these two parts are cemented together by a rubber solution, making them practically one. The round, wearing portion being of less diameter than the base, forms shoulders at the top of the rim, making it impossible for the rubber to be crowded over on the edges of the rim to be cut. And it is because of this arrangement of the parts that the company is enabled to claim that they use a much more elastic and resilient quality of rubber than is used on any other vehicle tires made.

When the round part becomes worn, the surface can be turned down into the base and the unworn surface brought into contact with the road, thus securing practically a new set of tires. The base part being inside and protected by the steel rim, practically wears forever, but when it is necessary to renew the wearing portion this can be done at trifling expense, as compared with the cost of replacing the whole tire.

The introduction of rubber tires is one of the leading innovations of the day, and it behooves every carriage manufacturer to study the matter in all its bearings, so that no loss will accrue from the use of impractical or defective kinds. The Tricycle Company cordially invite correspondence, and are prepared to give detailed information regarding their tire. Their advertisement, which appears on another page, should be read carefully, and if further information is desired accept the company's invitation and write.

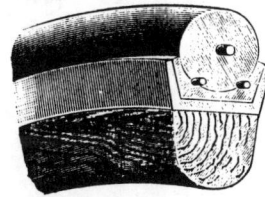

The Hub, May 1895

PERFECT RUBBER TIRED

———

LOCKWOOD & BROWN, of Amesbury, Mass., call of carriage builders to their rubber tired wheels, and ask a trial, feeling that having once tested them they will place them on their vehicles. The claims for their wheels are: The absolute simplicity of construction; there is nothing to get out of order; they have the greatest strength; they are the handsomest wheels; there are no tire-bolts to get loose, and the rims are not split or weakened by them; if broken, they can be repaired without being returned to us; the rubber and steel are welded together so that no mud or grit can work between the rubber and steel; they will wear longer than any others. The up to-date carriage builder don't allow a good thing to get away, and he will have these wheels anyway. But there are others who are slow, and sure to lose because they are slow. It is to these we appeal directly. Write to the manufacturers for information and then try them.

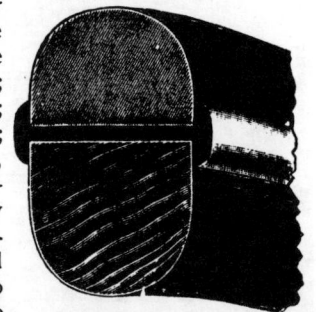

The Hub, Aug 1895

Rubber Tires

RUBBER TIRES.

If we ask the carriage manufacturer what subject most interests the trade, the answer will be rubber tires and ball bearings. Both are with us and both have come to stay, but the rub is the increased cost, an obstacle that must be overcome if the use is to become universal. How it will be with ball bearings we cannot say, but it now looks as though the tire problem is likely to be solved by the Standard Tire Co., of New York. This company furnishes a tire that will cost little if any more than an ordinary steel tire. It is so constructed that it gives as easy motion as the solid rubber and is practically noiseless. It is more durable even than the ordinary steel tire, and in addition to the many good qualities mentioned it adds greatly to the durability of the wheel. The accompanying cut shows a section of a felloe with the rubber and steel tire complete, the different materials being shown by the engraving. The rubber cushion between the metal tire and the wood is compressed, not stretched, and when pressure is applied, as when the tire is rolling on the road, the yield of the rubber is lengthwise and thus forms a continuous cushion. Wheels fitted with these tires have been in use several years, and the result has been to more than carry out the claims of the inventor. As the tires are held in place without the use of bolts or screws, and are put in place by the use of machinery, they can be furnished at a price that differs very little from that paid for the ordinary steel tire.

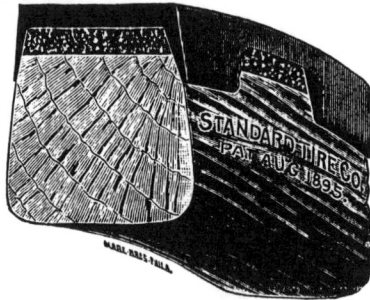

The Hub, May 1896

TIRED? ARE YOU
TIRED OF PAYING TRUST PRICES?

THERE'S REALLY NO NECESSITY FOR PAYING THE EXORBITANT FIGURE THAT GOES TO COVER ROYALTIES. JUST SIMPLY INVESTIGATE WHEN WE TELL YOU THAT WE CAN SAVE YOU MONEY.

THE SIMPLE FACT OF THE MATTER IS WE PAY NO ROYALTIES, CONSEQUENTLY WE FURNISH BETTER RUBBER FOR LESS MONEY.

We have no connection with the CONSOLIDATED RUBBER TIRE CO., and do not handle their tire, but a BETTER TIRE. We guarantee every length, provided it bears OUR FULL NAME and TRADE MARK, as shown in cut.

KELLY SPRINGFIELD RUBBER TIRE Co.
ROBT. KERCHEVAL, President **DAVENPORT, IOWA**

The Carriage Monthly, July 1901

Rubber Tires

A RIM ROLLER FOR SHAPING PNEUMATIC TIRES.

GEO. W. HEARTLEY, of Toledo, Ohio, has recently brought out an improved rim roller for shaping rims of pneumatic tires for the rubber; at the same time roll it up ready for welding; and the same machine can be used to weld tires on it by means of roller dies, that are turned out to fit the stock, and the welding heat is instantly applied, securing a good job easily and quickly; this machine can be used to roll up buggy tires and weld them, and to true the rim under the tire after the tire has been set without injury to the felloe. He also manufactures a self-oiling punch chuck for holding punches that are made of stub steel wire saving three-fourths the cost over the old way; it is indorsed by several of the leading carriage builders, and has proven a great saver of expense in that line; it can be fitted to any punch. See advertisement.

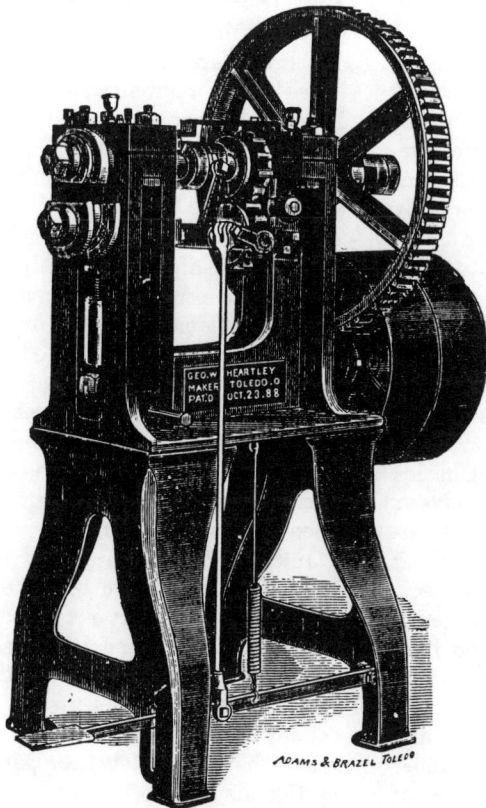

ROLLER POWER WELDING MACHINE,

For Welding Tires and all Special Work. The Weld Swedged same as Stock Instantly from Forge, Furnace or Electric Heating and Drawing out Iron to Taper, such as Dash Feet, Shaft Iron Ends and Seat Irons.

The Hub, Nov 1894

Tire End Setter.

In order to facilitate the bringing of the ends of vehicle tires together after they have been fastened in the channel and riveted. Morgan & Wright are manufacturing a little tool for this purpose, which they are now furnishing the trade.

As shown in the accompanying cut, this tool consists of a small grooved wheel attached to a lever, this lever being four feet in length, thus giving plenty of leverage. (Only part of lever is shown in cut.) This lever is fastened to an iron frame with four sets of holes to make it adjustable to any size wheel.

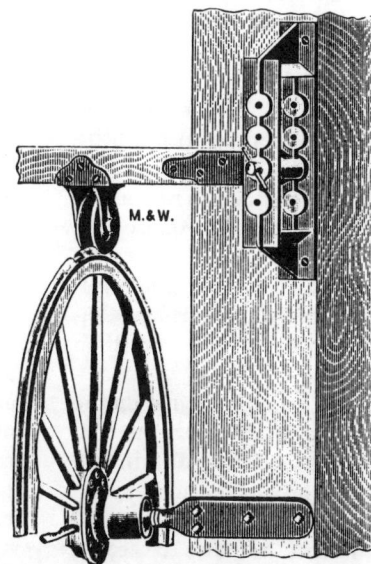

Tire End Setter.

The method of bringing tire ends together is very simple. Place the wheel on the axle fastened to a beam, as shown in cut. Adjust the lever to the diameter of wheel. The grooved wheel is then placed on the tire opposite the opening, and while an assistant bears down on the lever, turn the wheel half a turn each way, which will bring the ends of the tire together, when the rubber solution will soon form a perfect union and a continuous tire.

The ends of a tire can be brought together in thirty seconds by use of this little tool.

The Columbus (Ohio) Buggy Co. is having a business beyond anything in its previous history. They are now employing over 600 men and the pay roll at the close of a recent week aggregated more than $13,000. Not only is the demand for their goods rapidly increasing in this country, but it is likely to do so in other countries, for they are making a splendid showing at the Paris Exposition. Mr. Firestone intends sailing for Paris in the latter part of July and will probably spend a month in that city.

The Timken Roller Bearing Axle Co., of St. Louis, Mo., have gotten out what may be called a descriptive catalogue and price list which it would be well for manufacturers to have on hand. It contains a deal of information of various kinds, and is liberally sprinkled with cuts which make apparent the strong point of their axles.

The Carriage Monthly, June 1900

181

Rubber Tires and the Carriage Industry
Part I—Solid Rubber Tires
Ken Wheeling

They called it caoutchouc! It was a substance suspended in the form of very fine particles or globules in an aqueous liquid in the bark of certain tropical plants. The Spaniards first wrote about it in 1525 when they saw the Aztecs playing games with a little ball with the great ability to bounce. Oviedo y Valdes, in 1536, wrote about the many articles that the Amazon Indians made from it in his history, *Historia General y Natural,* and the Europeans began a lively and curious fascination with the substance from then on.

The French seemed to pick up the gauntlet first, about 200 years later. In 1725, they sent La Condamine to South America on a scientific expedition which returned with a small amount of "rubber" to Paris. His countryman, Fresneau, was the first to describe the process by which this milky substance was extracted from the tree, a process much like that of collecting maple sap. However, this was not the sap of the plant or tree from which it came, but rather, it was exuded from the bark of the tree.

By 1768, chemists were trying to find a way to use this substance and two French chemists, Herissant and Macquer, wrote to the Paris Academy in that year that the rubber was soluble in oil of turpentine and in pure ether. There wasn't much commercial use for rubber at that point, although some experiments were carried out and several inventors developed garments made from it, Nadier being one of these.

In 1770, Joseph Priestley, a distinguished English scientist, suggested that it might have some use as an eraser, for he discovered that it easily and quickly removed the markings made by graphite pencils. However, pencils were rare or nonexistent, but his suggestion gave the substance the

name by which it is known today, rubber.

It was left to an Englishman, Charles Mackintosh, from Manchester, England, to develop the process which was to revolutionize the fledgling rubber industry. Mackintosh found that he could dissolve crude rubber in coal-tar naphtha and that he could then apply the resulting mixture to cloth and make it waterproof. He took two pieces of cloth and coated them with this mixture and joined them together under pressure. This compound fabric immediately became successful for the manufacture of waterproof garments which were christened "mackintoshes." It was welcomed especially by those who rode on the top of English coaches day after day.

Nine years later, in 1832, a small business was established in Roxbury, Massachusetts, to make such cloth. The Roxbury India Rubber Company had as one of its employees a man whose name was to become synonymous with the tire industry, Charles Goodyear. It was Charles Goodyear who was to discover the process of vulcanization, the hardening of the rubber compound without losing its resiliency.

Actually, in 1832, two other gentlemen, Hayward in England and Luedersdorf in Germany, had stumbled onto the process which they didn't really recognize. These men had discovered that they could dissolve rubber in sulphur to prevent adhesion of rubber articles to each other. The process was discovered by Charles Goodyear three years later, in 1835. He did not, however, protect his discovery by securing a patent.

Vulcanization was a process by which the properties of rubber were stabilized when sulphur was combined with it under heat and pressure. Prior to the discovery of this process, rubber became hard and stiff in cold weather

and became soft and sticky when it was hot. Thomas Hancock, an Englishman, discovered that he could achieve the same effect as Goodyear had by immersing crude rubber in melted sulphur and then heating it to 302 degrees Fahrenheit. Hancock did secure his discovery by patent, No. 9952, granted in 1843, and subsequent patents, No. 7344 (1837) and No. 11, 135 (1846), granted by the British Patent Office.

It appears that the idea of tiring a wheel with rubber arose in many heads at the same time and it is somewhat of a question as to whose head should carry the laurels for inventing rubber tires.

In 1835, a Frenchman by the name of Dietz patented a vehicle tire which consisted of a rubber cushion applied to the conventional iron hoop, over which a second iron hoop was fixed, so that the rubber would not have to withstand the wear and tear of rocky roads. Mackintosh also was producing a solid rubber tire for horse-drawn vehicles as early as 1846 and the credit for this goes to Thomas Hancock, who was Mackintosh's partner. In a book written by Hancock in 1856, he states, "these tires are about one and one-half inches wide, and one and one-fourth inches thick. Wheels shod with them make no noise and they greatly relieve concussion of pavements and rough roads; they have lately been patronized by Her Majesty."

In the United States, it appears that the first patent to deal with solid rubber tires was one granted to C.K. Bradford in 1868. Patented that same year, in December, was a curious technique which called for bolting a flat strip of rubber to an iron tire with bolts that passed through the iron tire and the felloe. These bolts had springs on them to take up the movement created when the rubber was compressed by the weight of the vehicle as it moved along the roadway.

In England, Robert W. Thompson (1822-1873), about whom we shall see more in the subsequent chapters of this article, was experimenting with solid tires since his pneumatics did not seem to find much favor among carriage owners or builders. Interestingly enough, these Thompson tires were used on road steamers, vehicles which intrigued both Thompson and Hancock. The Edinburgh *Scotsman,* in 1868, hailed such tires enthusiastically, saying, "the application of vulcanized India rubber to the tires of road steamers forms the greatest step ever made in the use of steam on the common roads." *Scientific American* (September 24, 1870) describes the Thompson tire. They were mounted on a rim that had flanges and were covered over with a series of steel plates connected by means of a chain at each side of the wheel.

In America, there was a rather limited attempt at tiring carriage wheels with rubber. In 1856, the Boston Belting Company was making a few solid rubber tires which were designed by George Souther and George Miller. Various singular efforts were made, both in England and America, but nothing really came of it until 1885. W.H. Carmont had secured patents in 1881 and 1883 for a solid rubber tire which was held in a rolled steel channel in a state of compression, between converging flanges. The Shrewsbury and Talbot Noiseless Tyre and Cab Company, Limited, decided to try the Carmont tires on their new Hansom Cabs on the streets of London. It has been estimated that their experiment was so successful that within a very few years, some 10,000 sets of rubber tired wheels were in use in London.

Such a tire came to America through the efforts of Channing M. Britton. While he had been a student at the University of Berlin, one of his fellow students had been the Earl of Shrewsbury and Talbot. Britton became head of Brewster and Company, in New York City, and Brewster acquired the American rights to the Carmont patent. However, the tires, which were made for Brewster by several different firms, were restricted to Broughams, and several vehicles heavier than that particular carriage. Healey and Co. of New York and C.P. Kimball and Co. of Chicago also were

using rubber tired wheels on some of their production carriages.

The streets of New York and other city streets brought to light an inherent problem with the new rubber-tired wheels. The tire did not stay in place. This characteristic was the cause of much research and experimentation and most of the subsequent patents have to do with the manner in which the tire was attached to the steel channel rather than the particular make-up of the tire itself. There were, of course, some modifications of the tire designs to accommodate whatever the type of fastener each tire patent specified.

In February of 1869, L.W. Cheever had acquired a U.S. Patent for a rubber tire which was D-shaped and seems to have been mounted in a rim that was made with a groove in a thick iron tire. The rubber tire fitted into this dovetailed groove and, since it was flat based, it is presumed that the tire was cemented onto the rim itself. It proved quite satisfactory for light loads.

The formation of the tire itself was brought about by forcing the rubber through a die.

Rubber is cast to a degree, that is in the form of a plastic dough. It is pressed into a mold and baked under pressure. But how is the solid tire shaped to fit the mold? The answer is, by means of a machine known in England as a spewing machine, and in America as a tubing machine. It is simply a great screw revolving inside of a steel tube that looks not unlike a small cannon. At the mouth of this cannon is an opening shaped for the tire, while through it run spindles that make the holes for the wires. The rubber dough is fed into the breech of this cannon, and "spewed" out of the muzzle, in a form that just fits the mold in which it is baked.

Henry C. Pearson, *Rubber Tires and All About Them.* New York: The India Rubber Publishing Company, c.1906.

The molds and presses in which these rubber tires were vulcanized took many different forms.

As processes became refined, carriage tire stock was run in almost continuous lengths and, when finished, wound upon reels from which the carriage maker could cut the proper

lengths required for the wheels he was tiring. Only the lighter weight tires were so made as the mass of rubber involved with larger tires precluded such simple manufacture. Towards the end of the horse-drawn vehicle era, c. 1922, manufacturers were even adding chemicals to the rubber compounds to make colored tires, zinc oxide for white tires, carbon black for black tires, and antimony sulphide for red tires.

There were two major areas of concern in the use of rubber tires on carriages. The first was attaching the tire to the wheel itself and the second was "creep"—that ability of the rubber to move around the rim of the wheel. It appears that most of the patents granted from about 1890 onwards concern themselves with these two "problems." Efforts centered on the shape of the steel rim or band that encircled the felloe of the wheel, or on the use of wires, pins, or staples to attach the rubber tire to the rim.

In 1888, the Dubois Mfgr. Co. of Philadelphia, Pennsylvania, patented a D-shaped tire that had a wire running through it, through a hole near the base of the tire. This idea was picked up by the Springfield Rubber Tire Company of Ohio in 1892. Four years later, Arthur W. Grant patented a D-shaped tire which had two wires running through it and this became the renowned Kelly-Springfield tire. This method and variants thereof became so popular and so widespread that it led to infringements of patents and involved the Springfield Tire company in a morass of litigation that required the expertise of engineers and lawyers to unravel.

This Kelly-Springfield tire was soon followed by so-called improvements. The "Victor" tire also used two wires running longitudinally through it but the wires in this tire were encased in leather to prevent them from cutting into the tire. The rubber that had been used in earlier tires was of such a consistency that it was easily destroyed by hard use over gravel roads and cobblestone pavements or internally from the pressure exerted against it by the various devices used to attach tires to the rims. In the early 1890s, Howard M. DuBois (the Dubois Mfgr. Co. of Philadelphia) succeeded in solving this problem and a better grade of rubber began to replace the earlier formulae.

In England, using the Carmont patent, the tire with longitudinal wires

Rubber Tires

was modified by Burgess only in that the wires were now wrapped with small spiral wires, in order to protect the rubber.

In 1894, rubber tires first appear at the carriage-maker's convention and at least two concerns were making rubber tired wheels in that year. The previous year, 1893, saw the celebration of the Columbian Exposition in the United States and there were a number of carriages displayed there which had rubber tires on them; for the most part, they were heavier carriages by quality makers.

A significant change in the method of attaching the rubber tire to the rim was patented in America by Harvey S. Firestone and in England by Byrider and Swinehart. This was the "sidewire" tire which was held in place by two longitudinal endless wires sprung over the edges of the channel as a retaining "spring." There were embedded wires or bars in the tire which were clinched down by this retaining wire and held in place. There were two important advances in tire mounting that were brought about by the introduction of the sidewire. The first was that they could be fitted more easily and the cutting, brazing and setting of the internally embedded longitudinal wires were avoided. The second, and more important, advancement was that this new method of attachment prevented the movement or creeping of the rubber tire in the metal channel.

There were several varieties of these sidewire ideas, another by Swinehart himself in which the sidewire was used along with the embedded longitudinal wire. The "Republic" used plates or bands which were underneath the tire and under the retaining wires at the side.

The Motz tire did away with the sidewire and used the clincher system, using embedding crosswires in the tire which were slightly larger than the rim of the wheel. By bending the rubber tire in upon itself, several of these could be engaged and by moving around the rim of the wheel, the entire tire could be gradually seated in the channel.

Rees Almond Alson, writing in *The Blacksmith and Wheelwright,* (July 1904) wrote that essentially, by 1900, there were two major types of tires which seemed to be standard in the carriage-making industry: the sidewire and the internal two-wire. This appears to be a correct analysis as the

subsequent patents in the rubber tire industry really appear to be attempts at improvements on one or another of these two types of tires.

Patent No. 646,274 (H.S. Firestone), dated March 27, 1900, changed the wires to a metallic strip or band. W. Christy's Patent No. 817,975, in 1906, did away with the wires and substituted a series of canvas layers as stiffeners. G.B. Drydent, Patent No. 788,824 (May 2, 1905) tried to prevent creep by embedding "transverse binding wires" which resembled the staples used at that time for joining machine belting together. J.W. Carter, Patent No. 733,540 granted on July 14, 1903, did not use the sidewires but bent in the rim of the channel to fit a groove in the tire. The tire was then wired in with staples which passed through both the rim and the tire.

The litigation which surrounded these various inventions eventually protected the patent granted to Arthur W. Grant (Pat. No. 554,675, February 18, 1896) for a rubber tired wheel and the patent he obtained for the device to put them on (Pat. No. 555,480). These rights were assigned to the Rubber Tire Co. in Akron, Ohio, which became the Consolidated Rubber Tire Company. Grant also held a British patent (No. 24386), a patent of which Edwin Stuart Kelly was the registered owner. France granted him a patent, No. 252,731, on April 10, 1896.

A United States patent protects the "sidewire" tire, No. 686,556 granted November 12, 1901 to James A. Swinehart and W.A. Byrider. Three companies in the United States were licensed to make tires under this patent, one of which was the B.F. Goodrich Company.

Of course, the intriguing challenge to come up with a new solution to the inherent problems of rubber tire usage and to achieve recognition for some new advancement of this fledgling "science" remained. C.W. Wheeler, who had patented a "sidewire" tire in 1900 (No. 652,989) produced a new and novel method for attaching the tire to the channel. To avoid the constant problem presented by having to cut the rubber and splice it together to make a tire, done by compressing the rubber and brazing the wires together and allowing the rubber to spring back upon itself and close the gap, Wheeler, in consort with Turner, produced an endless tire. This was a tire which was vulcanized to fit

exactly upon the wheel and had a fixed perimeter. No retaining wires were used, but projections were embedded within the rubber tire which would be engaged by the rim of the channel. To mount this tire on the wheel, one side of the channel was removable. After the tire was mounted, the rim was set back against the channel and bolted on.

One further advancement of this came with the "Diamond" wire mesh base endless tire. The bottom of this tire was hardened and stiffened by woven wire embedded in it. To put this tire on, one had to drive it over the rim with a mallet.

Henry W. Meyer, in *Memories of the Buggy Days,* states that 45,000 sets of rubber tires were in use in the United States in 1897 and that over one million dollars would be spent in that endeavor in 1897. By 1902, according to the *Carriage Monthly,* the carriage industry was buying five million dollars worth of rubber, most of which went for the making of rubber tires.

The rubber tire provided a smoother, more comfortable ride. Aside from that, it prolonged the life of the wheel, a situation which was commented upon by Henry P. Jones, *The Hub,* (October 1908): the rubber tire had cut "wheel trade twenty-five percent, because (it) increases the live of the wheel (and) reduced demand about twenty-five percent." Mr. Jones was only re-echoing a remark noted in the Twelfth Census of the United States: "The introduction into such general and increasing use of rubber tires for light vehicles has had the effect of prolonging the stability and life of both wheels and wagons, and of largely reducing the cost of repairs." (Part IV, v. X, *Special Report on Selected Industries,* p. 306). William B. Adams, in his book, *English Pleasure Carriages,* wrote, "It is clear that elasticity is a necessary ingredient to ensure durability in wheels. . . ." That was in 1837. He, of course, was making a comment upon one of Mr. Hancock's latest "absurdities"—iron wheels.

The story of solid rubber tires does not end with the inventions of the Wheeler-Turner tire and the Diamond tire. They are still made today and their use and application to wheels will be the subject of Part III of this article. Part II will discuss the pneumatic tire, whose development was somewhat simultaneous with that of the solid rubber tire, as we shall see.

Rubber Tires and the Carriage Industry
Part II-Pneumatic Tires
Ken Wheeling

The streets of Belfast, Ireland, in 1888, were rectangular stone paving blocks and a goodly number of streetcar tracks wove amongst them. Gee! This was a horrible state of affairs, especially for a young cyclist whose father was a Scotch veterinary surgeon there. It was tough going for a tricycle.

Dr. John Boyd Dunlop, hearing the complaints of his young son, was moved to alleviate the situation. He bound a rubber tube onto the wheels of the tricycle by wrapping both tube and wheel with tape. Dr. Dunlop was fascinated with his success. so much so that he developed the first pneumatic tire, or what was called in those days a cushion tire. Later on, the term "cushion tire" would be applied to solid rubber tires. Dunlop's tube was a true pneumatic tire, i.e. the air was pumped into the tire under pressure, either into a single cavity or a series of cavities. There had been previous attempts to use air but, prior to Dunlop's experiments, the air had been merely trapped or captured.

The Dunlop experiments were so successful that Dr. Dunlop secured patents for his invention, in Great Britain, Pat. No. 10,607 (1888) and Pat. No. 4,116 (1889). The tires which Dunlop patented were described as follows:

Dunlop's tubes were enclosed in a jacket of Irish linen which was held to a rim almost flat by linen flaps pasted with rubber solution to the inner surface of the rim. The linen jacket was then protected by a tread of vulcanized rubber pasted on its circumference and lapping down almost to the rim or even up on it as did the linen flaps.

The Automobile, June 22, 1916

This description presents the main issue which caused so many early objections to the pneumatic tire. It could not be removed easily for repair. Quite surprisingly, many of the first tires of the pneumatic type were extremely difficult to remove. Copiously derided with names like "pudding tires", "mummy tires", and "rags", it wasn't long before a remedy appeared.

Before that remedy appeared, a forgotten patent was resurrected which proved to be a thorn in the side of many a would-be tire inventor. This was the British Patent No. 10,990, granted in 1845 to Robert William Thomson, of Middlesex, England. This patent was so comprehensive that it encompassed much of what the later inventors wanted to do; it was the conceptual basis for the pneumatic tire.

The tire and felloe are made much broader than usual and project considerably at both sides beyond the supporting spokes. The elastic belt is made as follows:—A number of folds of canvas saturated and covered on both sides with india rubber or gutta percha in a state of solution, are laid one upon the other, and each fold connected to the one immediately below it by a solution of india rubber or gutta percha, or other suitable cement. The belt thus formed is then sulphurized by immersion in melted sulphur or exposure to fumes of burning sulphur which renders it more pliable and prevents its getting stiff on exposure to cold; or the belt may be made of a single thickness of india rubber or gutta percha, in a sheet state and sulphurized, as aforesaid, and then enclosed in a canvas cover. A strong outer casing in which to hold the elastic belt, is then built up (so to speak) around the tire by rivetting together a series of circular segments of leather and bolting them to the tire. The segments at two of their edges are made to overlap each other, as shown, and then secured in their place by passing bolts through the tire and felly and making them fast by nuts. The elastic belt is then laid upon the portion of the segments thus made fast to the tire and secured in its place by bringing the two remaining and as yet unjoined edges of the segments together over the casing, and connecting them together by rivets. A pipe through which to inflate the elastic belt with air is passed at one place through the tire of the wheel and fitted with an air-tight screw cap.

The description goes on to suggest that horsehair or sponge or some other such elastic material could conceivably

be substituted for air, or used with air. A number of separate tubes was also suggested. Thomson's "aerial wheels" anticipated many advances in the pneumatic tire industry and, as a matter of fact, was the protected patent on the Continent, much to the chagrin of the owners of Dunlop patents.

Of course, one of the major questions that arises for historians of the carriage industry and its allied trades is whether or not such tires were ever really used on carriages. Mechanics Magazines, August 22, 1846, mentioned that Thomson tires had been affixed to a Brougham. Actually, what the magazine was really reporting was the results of some experiments Thomson made, proving that the air-inflated tire could actually reduce the energy required to move a vehicle over road surfaces which were not perfectly smooth. "He demonstrated that air-inflated tires can reduce tractive effort by 60 percent when rolling over a smooth surface and as much as 300 percent on rough roads, when compared to the iron tires of his day." (100 Years of the Pneumatic Tire, August 1988, p. 15). The statistics are:

Weight of carriage with its load - 15 cwt On paved streets the common wheels require a force of . . . 48 lbs. The patent wheels . . . 28 lbs. On clean, smooth, hard, Macadamized road the common wheels require a force of . . . 40 lbs. The patent wheels . . . 25 lbs. On broken granite newly laid down the common wheels require a force of . . . 130 lbs. The patent wheels . . . 40 lbs.

(Ibid, p. 15. Extracted from Mechanics Magazine, 1849, Vol. 50)

Carriage makers, dealers and owners did not really see the advantages of using such a tire, nor did they care since they were not the ones pulling the carriages. Pneumatic tires had to wait until the bicycle came into being. However that was to be achieved, it was to Robert W. Thomson, (1822–1873) that such tires owed acknowledgement.

The British Science Museum has in its possession some of the last surviving examples of the Thomson tire,

Vol.28, No.3, Winter 1990

one a leather-covered one which is riveted together and a second one of leather which is laced together. The Brougham tires, 5 inches wide, which "traveled 1200 miles without the slightest symptoms of deterioration or decay," seem to have disappeared.

While Dr. Dunlop was working in Ireland, Amos W. Thomas, of Philadelphia, PA, was issued a series of patents for inflated tires, the first one in 1888. Thomas invented a single tube type with a petcock for filling it with air, water or whatever else was handy.

Two significant advances in pneumatic tire construction came into being with the advent of the "wired tire" and, later, the clincher tire-rim principle.

The first of these, the so-called "wired tire", used the wire bead which was an endless wire ring embedded in an enlarged edge of the tire casing. This was invented by Charles Kingston Welch, an engineer from Coventry, England, who secured a British patent for such a tire on bicycle (Pat. No. 14,563) and a U.S. patent for a similar one on carriages (Pat. No. 512,594). Welch devised a drop-center rim to hold his tire on a bicycle, but the carriage tire was set over the steel tire of the carriage wheel and held on by the lips of the tire itself through which ran the wire bead.

At the same time, Alexander T. Brown and George F. Stillman in the United States were granted a patent for a wire-reinforced beaded tire (U.S. Pat. no. 488,494). Using a metal cycle rim, the tires were held in place by grooves in the rim which were lower than the edges of the rim, but were above the bottom of the center of the rim (the hollow, or drop-center, rim). The patent held by these two Buffalo, New York, engineers was eventually acquired by the Dunlop Company as was the Welch patent.

An American, living in Edinburgh, Scotland, came along with the second major innovation in the pneumatic tire industry. William Erskine Bartlett is credited with the invention of the clincher tire, which was put to good use in the automobile industry for many a year. Essentially the concept was this. Combining tire designs with thickened edges and rims, or felloes, in the shape "of metal troughs with edges inclined towards each other," Bartlett modified his tires several times before he arrived at a true clincher-type tire. Since he was the Managing Director of the North British Rubber Company,

the patents became its property, although the Dunlop Company was able to acquire them in 1903, at a reported price of 120,000 pounds. The price was closer to 200,000 pounds ($1,000,000).

The method by which a tire was attached to a rim appeared to be the focus of many of the patents of 1891 and 1892. A New York City inventor, E. R. DeWolfe, used "thin spring metal having beaded wire edges" which he molded into the base of the outer casing. While this appears to be another in a long series of metal attachments, it is important inasmuch as it helps establish the principle of embedding wires in rubber tires, a fabrication technique which would lose favor for some time, only to be reintroduced as a permanent part of the tire structure itself.

More important immediately was the patent of Thomas B. Jeffery. Jeffery was the other half of Gormully and Jeffery, bicycle manufacturers from Chicago, Illinois. Jeffery seems to be the experimenter in the corporation and was the holder of a number of U.S. Patents concerning tire construction. The patent granted to him in 1891 was the more significant as it dealt with the shaping of a tire which had hooked edges which were supposed to engage a matching set of hooks into which the rims were fashioned. In 1892, a patent of his, separated the hooks from the rim (Pat. No. 466,565). Pat. No. 466,789, granted to Mr. Jeffery just a week later on January 12, 1892, had

"exterior projecting beads or spurs" which fitted into an inward-curving rim. Again, the precedent was being set for embedded metals. In the vast history of inventions and patents, the Jeffery patents were not protected in France; hence, at an opportune moment, Michelin et Cie entered the burgeoning bicycle tire market.

Naturally, there were many other interesting ideas and inventions, many of which received patent status. Alexander T. Brown was given a patent for using a metal band which encircled the rim and which was moved by a special tool to engage the beads (Pat. No. 474,589).

George F. Stillman, who worked with Mr. Brown, received a patent in his own right in 1893 for a tire in which the air pressure activated a hinged spring-clamp which forced the beads of the tire against the rim itself (Pat. No. 495,277).

Alexander Straus, from New York City, received Pat. No. 474,423 for utilizing the linen or canvas outer casing to form a hollow tube through which ran the wire bead to hold the tire in place.

By comparing the dates of the various patents granted for pneumatic tires with those granted for solid rubber tires we can see that the development of both types was concomitant. Moreover, there developed a rather lively rivalry between manufacturers of the different types. Promotionals for one type loudly sang its praises over

Rubber Tires

the over, and vice versa.

A case in point might be that of the "Victor" tire. In the early fall of 1890, A. H. Overman received a patent for a cushion tire (solid rubber tire) which was arched in shape and contained no fabric. It was promoted as a substitute to the pneumatic tire and marketed on the Victor bicycle which was Mr. Overman's company. In attempts to achieve dominance in the market place he "had his buggy fitted with ball bearings and the Victor arch or cushion tire which he was strongly advocating, and with which he was trying to stem the tide of favor sweeping toward the pneumatic", (The Automobile, August 24, 1916). In so doing, Mr. Overman secured for himself a place in carriage transportation history for being one of the first to use rubber tires on his carriage.

At this point, the question arises regarding just how the pneumatic bicycle tires got here in the first place. The answer seems to lie with the St. Georges' Engineering Company's shipment of three "New Rapids" bicycles to one Samuel T. Clark of Baltimore, Maryland, in 1889. In early 1890, the Sweeting Cycle Company of Philadelphia got a shipment of bicycles from England and they had pneumatic tires on them. It was soon perceived that, in spite of the derision heaped upon them earlier, these pneumatics raised eyebrows when used at cycling meets and races. Cyclists using pneumatic tires were able to attain faster speeds.

It was only a step from the bicycle to the racing sulky; this appears to be the way pneumatics got into the carriage world, and it all seemed to happen at once. George R. Bidwell secured the patents of A. W. Thomas, granted in June 1888, and set up the first pneumatic tire manufacturing company in the United States. It was located at Sixty-sixth Street in New York City, near Columbus Avenue. The first orders were given to the New Jersey Car Spring and Rubber Company (Jersey City, N.J.) and to the B. F. Goodrich Company (Akron, Ohio) in March or April of 1891. In the Spring of 1892, George Bidwell "fitted pneumatic tires to two Stivers runabouts and drove one of these about the city," (Ibid).

In July, 1892, Sterling Elliot sent a sulky with ball bearing wheels and pneumatic tires to the racing driver, Bud Doble, hoping that Doble would use them in the upcoming Blue Ribbon

meet at Detroit. Doble could not be bothered with the new contraption and half-contemptuously loaned it to Ed Geer who had been envisioning just that sort of thing and had asked for it. Greeted at first by jeers from the race goers, things began to change quite a bit when everyone began to notice that Geer and his drivers were achieving faster records on the track, by 2 to 4 seconds. Sulky makers were inundated with requests for the new sulky, but makers such as Frazier and Caffrey and others could not supply the demand. Drivers even used bicycle wheels on existing sulkies in their enthusiasm to try the new-fangled pneumatics. (Cfr. The Carriage Journal, Vol. 15, No. 2, p. 283).

Elliot assigned his patent for a pneumatic-tired sulky wheel to the Hickory Wheel Company. This patent (Pat. No. 494,113, granted 1893) was declared invalid by the United States Circuit Court in Chicago (1900), upheld by the Supreme Court the same year. The courts held that the tires were no different than bicycle tires and that they were on a different type of vehicle made no difference, i.e. "did not give them the elements of novelty".

The Sercombe-Bolte tire had a flat metal edge in it, patented in 1892, and it engaged a rim whose edges hooked outward. Not a popular tire or mechanism or attachment, it was one of the early tires to be used on sulkies to a great extent. It was abandoned because it collected dirt badly, was hard to put on, pinched the air tube, and the rim tended to cut the tire.

All of these inventors, as much as we would like to assign the premier makership place to them, seem to have been superceded by Robert William Thomson, about whom we have spoken earlier. It seems that someone had a carriage in New York City in 1847 which was equipped with his tires. Carriages which had Thomson tires "were offered by sale by one or more prominent London carriage houses." In Scientific American, 1847, mention was made of that carriage which had appeared in New York. This gives Thomson claim to have been the one to use pneumatic tires in the United States first, albeit by a third party.

That the pneumatic tires were eminently suited for carriages was a matter which seems to have followed naturally from their very invention. H. H. Mulliner of Birmingham, England, writing in Live Stock Journal

Almanac, c. 1900, says; ". . . the result convinced me that (if the risk of puncturing was not great) it would only be a matter of time for all pleasure carriages to have them." In the same article, he writes:

Most people have ridden in carriages with solid rubber tyres and know their advantages, but I assert that the advantages of pneumatic tyres over solid tyres are infinitely greater than the advantages of solid rubber tyres over the ordinary iron ones . . . pneumatic tyres will, I believe, wear just as long over the very roughest of country roads as they will over the wood pavement in London.

Live Stock Journal Almanac, c. 1900, p. 54

One of the greatest advantages for urban dwellers who ventured to drive carriages along the city streets concerned the ease with which the pneumatic tire went over the tram lines. Mr. Mulliner observes further:

A carriage with the ordinary iron or solid rubber tyre rolls for a time in the lines instead of following the course of the horse, and then swerves sideways, but with pneumatic tyres this is impossible. This is an enormous advantage in large towns where the streets are intersected with tram lines, as collisions often occur owing to the wheels of vehicles going in opposite directions being held in the lines.

Ibid, p. 54

The tires which were being used on carriages were specifically made for carriage use. They consisted of three parts: an endless rubber tube called the "chamber", in which air is compressed by means of a pump; an envelope composed of canvas provided with a coating of rubber which prevents abnormal expansion of the chamber, and the felly, or rim upon which it is set. All of this might seem quite ordinary and taken merely for granted, given our familiarity with automobile tires. It was not so in 1891. It owed its great advance to the work of P. W. Tillinghast.

Among the many tires brought out, the 'single tube' tire, patented by Pardon W. Tillinghast, May 23, 1893, and first introduced by Colonel Albert A. Pope on his 'Columbia' bicycles, became in time more popular than any other; in fact it grew to be the standard type for cycles in the United States. The same principle had been worked out in England by A. Boothroyd, who failed to take out a patent, so that the invention

THE CARRIAGE JOURNAL

Rubber Tires

became public property in that country.... Single tube tires have been used extensively on sulkies and light carriages in America....

Pearson, Rubber Tires and All about Them,

New York: The India Rubber Publishing Company,

c. 1906, p. 39

It was Tillinghast's invention which made possible the quick acceptance of the pneumatic tire.

Before saying something about the specific companies which made such tires or the carriage companies that used them, note should be made of the process by which such tires were made.

In order to make an envelope, the workman passes around a circular bronze form ... a band of the gummed canvas ... and cements the two ends together. this first band is wider than the form, since the workman is to place in its upturned edges the lateral projections of very hard rubber that serve to hold the envelope in the groove of the rim. He next glues in succession upon this backing a band of rubber and a band of canvas, and finished with a band of rubber called a 'crescent,' and which is thicker in the middle than at the edges. Pneumatic tires for carriages have four thicknesses of fabric.

Scientific American Supplement, No. 1271

May 12, 1900. p. 20368

The construction of the air chamber was done differently.

The sheet of rubber engages with the screw of a drawing machine, which kneads it and discharges it, through a round aperture, in the form of a seamless tube. It is afterward placed upon a rod and covered with canvas to prevent its distortion, and surrounded with talc and placed in a steam-stove. Upon making its exit from the latter, it is tested.

Ibid, p. 20368

The manufacture of pneumatic tires was the provenance of both those companies which made bicycle tires as well as companies specifically making carriage tires alone. The American Carriage Directory for 1899 lists eleven makers under the title "Rubber Tire Mfrs.":

American Rubber Tire Co. (New York City)

Beck and Corbitt Iron Co. (St. Louis, Mo.)

Chase, L. C. and Co. (Boston, Mass.)

Kelly, Maus and Co. (Chicago, Ill.)

Kendall Rubber Tire Co. (Providence, R.I.)

Kimbark, S. D. (Chicago, Ill.)

Lockwood and Brown (Amesbury, Mass.)

Meeker Mfg. Co. (Dayton, Ohio)

Newton Rubber Works (Newton Upper Falls, Mass.)

Rubber Tire Wheel Co., The (Springfield, Ohio)

Victor Rubber Tire Co., The (Springfield, Ohio)

These appear to be the most prominent makers although B. F. Goodrich, Goodyear and other tire companies did make and market pneumatic tires for carriages.

The companies which limited production to specific vehicles which had rubber tires were the sulky makers and the makers of speed wagons and matinee wagons, such as George Werner from Buffalo, New York. Well-known carriage makers could and would supply carriages with pneumatic tires. There is a Governess Cart at Morven Park which has them and a landau as well, this last carriage made by "Ferd. French and Co." Those carriages which were to be used on paved city streets were available with such tires. It appears, though, that there are not a lot of surviving examples of such carriages. Many do exist with solid rubber, or cushion rubber, tires. Those with pneumatic tires are decidedly less numerous.

In the Twelfth Census of the United States, 1900, Vol. X, it was noted that 1000 patents had been granted for the subclass "pneumatic tires", certainly an indication of the interest in their continued use and betterment. The reason that they did not receive wider audience was only that the carriage era itself was a threatened market and the story of the pneumatic tire from the turn of the century on is more the story of, first, the bicycle, and, then, the automobile. In spite of the fact that a few companies tried to make a specialty of a buggy which had wire wheels and pneumatic tires, such as the Bailey Bike Wagon Company, the use of the pneumatic tire on carriages remained minimal. Those who wanted rubber tires on their carriages chose the solid, cushion tire, and the countryman clung to the iron or steel tire.

The aspirations of man for calmer and quieter enjoyment and undisturbed association with his thoughts may be a sort of half-poetic way of stating a hard fact in favor of the rubber tire, but such is the nature of the hidden force and attraction that is designed to make the rubber tire on vehicles the inevitable and irresistible accompaniment of wheel constriction in the carriage shop of the future.

Carriage and Wagon Builders: Association of Philadelphia

Official Souvenir Catalogue, Third Annual exhibition

March 28–April 2, 1898.

THE CARRIAGE JOURNAL

Rubber Tires and the Carriage Industry
Part 111 — The Contemporary Scene
Ken Wheeling

The "carriage shop of the future" envisioned at the Third Annual Exhibition of the Carriage and Wagon Builders' Association of Philadelphia, and referred to at the close of Part 11, never materialized, in spite of the fact that the catalogue of the exhibit predicted it in effusive terms.

It is coming fast. We are tired of the discomforts of steel. The world is rushing cushionward, springward; away from hard mechanical motion, away from the sense of means which produces the sense of enjoyment in us, and towards those means and facilities which hide the mechanism and help us to forget the force and wear and pull of parts that are requisite for the mechanical part of our esthetic enjoyment which the rubber tired vehicle is destined to bring us in increasing measure.

Official Souvenir Catalogue, Third Annual Exhibition, Carriage and Wagon Builder's Association of Philadelphia, March 28 — April 2, 1898.

Little did anyone realize that the carriage would soon be eclipsed by the automobile, and that the rubber tire industry would become dependent upon the gas guzzling behemoths. Perhaps, in retrospect, we should recognize that the automobile depended upon the success of the rubber tire.

Nonetheless, all discussions about carriage wheels equipped with rubber tires became superfluous, and the carriage industry inexorably waned into almost non-existence. Here and there, religious communities kept their rigorous adherence to the horse and buggy, and an occasional farmer or some other countryman kept a carriage for nostalgia's sake. By and large, fine carriages were confined to private estates or the show ring, and in ever decreasing numbers.

The resurgence of the "carriage industry" was a result of a renewed interest in carriages on the part of collectors, restorers, horse lovers, and others. Spurred on by associations such as The Carriage Association of America, the American Driving Society, and other groups, old shops sprang to life and new shops opened and a thriving mini-industry emerged. An account of that renewal need not detain us here. Suffice it to say that the desire to restore old rubber-tired wheels for actual use, and the desire to build new wheels, affected the rubber tire industry and brought a new demand for rubber tires.

For a short time old supplies were available from the regular brand name dealers, but it was found that the rolls of rubber which were in storage were not up to standard. The rubber had lost its ability to creep and that was essential to the success of rubber tiring. It is essential to realize that the tire moves ever so gradually around the rim of the wheel. This is due to the compression of the rubber when the wheel itself becomes parallel to the road surface. This is most evident when an old rubber-tired wheel is put into use. Gradually a space appears between the ends of the rubber tire and the gap increases with each successive use. The wires holding the tire are exposed to the elements, become rusty and eventually break, usually when the driver is just about "five miles from home."

Large commercial tire manufacturers were not really interested in supplying the small amount of rubber tire material required for wheelwrights and restorers of horse-drawn vehicles. Small local rubber companies such as the Jasper Rubber Company, in Indiana, had most of the dies and could produce enough to satisfy the demand. Large parts dealers in the Amish country carry the rubber tire material in rolls, and in most sizes.

In the United States, there was a decided preference for the rubber tire which had the wire running through it; while in England and Europe, preferences focused on the clincher-type. This difference in preferences has affected the availability of materials with which to continue the rubber tire trade and has determined the course of that market in several respects.

Before proceeding to a discussion of that current market, it is necessary to understand the process of applying the tire to the channel. The process has determined which machinery would be salvaged and put into use, which would be "manufactured" again, and what mechanisms had to be put into place in order to insure a steady supply of channel iron, rubber tires, wire, and the other requisites for doing the work.

Many companies had patented the machinery for applying rubber tires, and some were more popular than others. The Goodyear Style "D" machine was a widely accepted model, one which essentially demonstrates the manner by which the rubber tire was applied. The process was as follows:

A piece of rubber was cut to its proper length, a length that was predetermined from a tire schedule, and the wires were inserted into it. These wires were at least one foot longer than the rubber. The operator then clamped one end of the wires in a "double clamp" and then proceeded to clamp the end(s) in the right-hand tire machine jaw, with the bottom of the tire upward. The wires were scarfed, i.e., cut and filed to create a lap joint. The wires were then released, the tire turned right side up and reset with the scarfed sides down. The rubber tire was then passed around the wheel, "threading the free ends of the wire" into the left-hand jaw, adjusting the scarfed ends on the left side exactly the same distance as the right side. The tire was then drawn up to the channel. At this critical juncture, the rubber was compressed "backward" along the wire, leaving the bare ends of the wire free. A paste was applied to the channel to

Rubber Tires

facilitate sliding the rubber back into place, and then the wires were brazed together. After the joint cooled down, a thin slice of rubber was cut off each bare end to insure "new" rubber and a joint cement was applied. The tire was then put into the Joint Closer and the rubber forced back along the channel and into its proper position, where the ends were put together and cemented.

While we may set this process down in abbreviated and simplistic terms, it took an incredible amount of experience to get this practice correctly. George Isles, in *The Restoration of Carriages*, comments, "It takes lots of practice and years of experience to judge the expansion and contraction of the channel iron and the rubber tire and a novice is not likely to get this right for some time. So, installing your own rubber tires is likely to be an exercise in frustration." (pp. 90—91).

The machine favored by contemporary wheelwrights is one made by the Sweet Rubber Tire Company around the turn of the century. The major reason for this choice is the ability of the machine to engage the two ends of the inserted wire in opposite jaws and draw the whole tire on, moving upwards along each side simultaneously. This appears to mount the tire more evenly and the wheelwright does not have to work the tire back by compression on one side only. The Witmer Coach Shop has had new machines made using the parts of an old machine for the molds. This they market to the trade. Several of the wheelwrights with whom I spoke expressed satisfaction with this particular machine and indeed prefer it.

At the present time, solid rubber tires are being manufactured in Indiana and Ohio. The major source for this is through a shop such as Witmer's which owns the dies that are loaned to such small commercial rubber manufacturers. Only the two-wire rubber tires and the single-wire cushion tires are available and no one is currently making the three-wire type which was usually reserved for the larger size tires such as the three inch, three and a half inch and the four inch tires. An inspection of the current catalog of the Witmer Coach Shop reveals that three types of rubber tires are being offered: the N.S. Round Top in five sizes, the Flat Top in six sizes, and the Old Style in five sizes. The N.S. Round Top are similar to the old "wing rubber" tire of the Goodyear Company, and covers the top of the channel. The Old Style has a chamfered side which fits the channel but does not cover the rims of the channel. The round rubber "cushion tires" are available in eight sizes.

The side wire type is not available in the United States but can be had in Europe where it is more popular. Mr. Witmer told me that there is a shift in this as the Europeans are finding that the competition vehicles which are fitted with this type of tire are experiencing difficulties. Those wheels equipped with such a tire seem to lose the tire under the stress of some of the movements required of such competitions.

Both the rubber tires, and the wires which bind them, must be ordered in large quantities. The wire comes from Seneca Wire in Seneca, Ohio, and is very similar to the old wire styles. The rubber presents a different problem in that it cannot be kept too long in its rolls before being used. After eight to ten years the rubber begins to crack, something that doesn't really affect use at that age, but which does affect saleability. The only real problem that seems to affect the tires after they have been applied arises when owners allow the vehicles to sit too long in one spot. A flat spot occurs and this is extremely difficult to overcome. Hence, the long tendered advice to carriages owners is to put their vehicles up on blocks when they are going to be stored for long periods of time.

The channel iron which holds the rubber tire is not being currently made in the United States and must be imported from Europe, mainly from Belgium. Again, large amounts of this product must be secured at the same time to make it financially possible to get it.

Pneumatic tires are also in demand, for show buggies, pleasure carts, and racing sulkies. The two major manufacturers of such carriages, The Houghton Sulky Company in Marion, Ohio, and the Jerald Sulky Company in Waterloo, Iowa, are experiencing some degree of difficulty in obtaining the tires they really want and tires that are up to the specifications they prefer. At the present time, there are no pneumatic tires being made in the United States or Canada, and they must be imported from abroad.

The tire currently used by the Houghton Sulky Company is one marketed by Kenda, which is made in Taiwan. It comes in four sizes, 16 inch, 20 inch, 24 inch, and 26 inch. Mark Bauer would prefer to see a better quality tire than this, but none is available to him now. The rims for the tires used at Houghton are made by the Sun Metal Products Company in Warsaw, Indiana. Houghton also uses a hard rubber tire on carriage wheels, one obtained from Holmes Wheels. Kenda is also supplying a puncture-proof tube for the pneumatics and this seems to be very serviceable at the present moment.

The people at Jerald Sulky Company have solved the problem of pneumatic tires in a different manner. Their tires are made on a special mold, with straight treads, and appear to be the only straight-treaded tire currently being made. Jerald owns the molds from which these tires are made, and these molds are copied from patterns supplied by Jerald. The tires are made in two sizes: a 24" (diameter) x 2.125 (width) and a 26" x 2.125. They are essentially bicycle tires although they differ a great deal from the tires which are specifically made for a bicycle. These tires intended for use on horsedrawn vehicles contain a different type of nylon and more of it. This means that there are more threads per inch in these particular tires. This has led to a bit of a marketing problem since some dealers realize that they can supply replacement tires with a greater profit margin than the ones obtained from Jerald. Such replacements are not of the same caliber as the original wheels. In addition to using these pneumatic tires on show buggies and pleasure carts, they are also being used on a training cart which is being marketed for standardbreds.

Jerald also markets a racing sulky, but uses an entirely different pneumatic tire on this vehicle, a tire that is currently being made by the I.R.C. Company in Japan, to its own patterns and specifications.

Bill Card, at Jerald, expressed some frustration with the availability of pneumatic tires and the precarious nature of the supply. Jerald was able to secure its tires from the Carlisle Tire and Rubber Company in Carlisle, Pennsylvania, up to about four years ago. This was the last company to make bicycle tires in the United States. Due to the pressure from foreign competition, it ceased production abruptly and its customers were forced to search elsewhere for the tires they needed.

Asked what problems owners could encounter from the pneumatic tires currently being produced, Mr. Card explained that horse people never seem to carry an air pump as a standard part of their equipment and habitually run the tires half-flat. Not only does this produce a drag which inhibits the full draft, such under-inflation causes rim cuts which eventually destroy the tire.

It is difficult to assess the scope of the rubber tire industry in terms of numbers of rubber-tired wheels produced. Witmer's Coach Shop estimated that they produced about 900 sets of four per year, and that included pushcart wheels; The George Daniels Wagon Factory in Rowley, Massachusetts puts the figure there at about 500 sets of four, exclusive of pushcart wheels. A census of the industry would be well worth taking to gain a proper perspective of just how many wheels with rubber tires are being produced.

As we look back over the tumultuous rise of the rubber tire industry, the involved court battles over patent rights, the inventions which multiplied at an expansive rate each year, and the resurgence of a demand during this current "horse-and-buggy" era, we can reflect about the long way it all came since caoutchouc first intrigued the Spaniards in 1525.

In conclusion, we might remember with some kind of a "harump" the denunciations of John Taylor, the Water Poet, who published in 1623 *The World Runnes on Wheels*. He "warned that those who would ride in them (hell-cart coaches) would be tost, tumbled, rumbled and jumbled." (CARRIAGE JOURNAL, 1,4, p. 101). At least, thanks to John Thomson and his successors, the tossing and jumbling has been greatly eased and the body politic has careened along in heedless flight, and "runnes on wheels", and those wheels have been cushioned by the addition of a rubber tire in one form or another. I do think, however, that we might agree with Taylor when he says: "the chained ensnared world doth follow fast."

The author wishes to thank Susan Green of Orwigsburg, Pa., for collecting a great amount of the research material for this article in the C.A.A. files, and also to thank Tom Ryder of England.

The Carriage Journal, Vol.28, No. 1, Summer 1990

VIII
PATENT WHEELS

193

PATENT WHEELS

This chapter begins with an article that gives a brief history of patent wheels. Of particular interest is its discussion of the patent lawsuit brought by James D. Sarven against Elihu Hall & Co. for patent infringement of hub design. Although there appears to be little similarity between the Sarven and Warner designs, the court found in favor of the plaintiff, ruling that all wheels with a "wooden hub, into which spokes entered by a mortise and tenon, having any metallic support for the spokes outside of the hub" were infringements of the Sarven patent. This ruling signalled the demise of Elihu Hall & Co., the manufacturer of the Warner Patent Wheel. The Sarven Wheel, patented in June 1857, became the first, and ultimately the most popular, of what was to become a deluge of different types of wheel hub designs.

In reading the articles that follow, it is clear that the Sarven patent was not immediately well-received by those in the carriage industry. Once again, as with most technological developments occurring in the industry during the latter part of the 19th century, it took some time for the merits of the wheel to be recognized by most wheelmakers. Throughout this period, as the articles in the "Hub" chapter relate, many wheelmakers thought that they had other, superior methods for making a stronger and better wheel hub.

COACH-MAKERS'
INTERNATIONAL JOURNAL.

JUNE, 1872.

The Sarven Patent Wheel.

AN IMPORTANT DECISION IN ITS FAVOR.

On April 23rd, 1872, the United States District Court, sat in New Haven, Conn., and in the case of James D. Sarven, *vs.* Elihu Hall & Co., which was a bill in equity, to restrain the defendants from using certain improvements patented to the plaintiff for making wheels. Judge Woodruff read an opinion deciding in favor of the plaintiff, with an order for an injunction.

"The New Haven Wheel Co. appear as one of the principal plaintiffs against Elihu Hall & Co., who are manufacturing the so-called Warner patent. The decision of the court sustained the Sarven patent as against many other so-called patent wheels, and was therefore a test case. All infringing patents are therefore enjoined from further manufacturing. For a considerable time the plaintiffs have been injured by a large number of infringements which the success of the wheel has induced, some twenty-five or thirty have sprung up in the course of three or four years. The matter was such an important one that a special term of the United States District Court, Judges Woodruff and Shipman sitting, was held in Hartford last December to try the case, and the trial excited considerable attention, on account of the extensive interests involved. The ablest patent lawyers in the country, including Judge Fisher, late Commissioner of Patents, Keller and Blake of New York, Thurston of Providence, Beach, Ingersoll and Earle of New Haven, were engaged. Such experts as the Rennicks and Treadwell of New York, and Waters of Boston, were also employed. The case has been a long one, but it is probable that the decision given will be final."

By private letter, we learn that the court has given E. Hall & Co. until September next to close up and has appointed a master to receive statement of all wheels made by them, assess damages—and report same at next term of the court. E. Hall & Co., go under bonds until September, when the injunction takes effect. It is expected that all other infringements will be enjoined immediately.

THE HISTORY OF PATENT WHEELS.

AND HOW THE MATTER NOW STANDS.

THIS matter being one of considerable importance, and having called forth many inquiries from correspondents, we have taken some pains to get all the information possible upon this subject. In order to understand it more clearly, it may be well for us to give a short history of patent wheels. It had been long known and felt that the weakest part of a wooden wheel was at the hub. In order to give the wheels sufficient strength, a large hub was necessary, which was thought clumsy upon a light carriage. To overcome this defect, metal hubs of various kinds were invented and tried ; but the large majority of them proved failures. About the year 1856, J. D. Sarven, then carrying on the carriage business in Columbia, Tenn., conceived the idea of making a strong but light wheel, by using a very light wooden-mortised hub. Instead of staggering the spokes to give them strength, as had always been done on light wheels, he mitred the shoulders, so as to form a solid arch on the outside of the hub, and, in order to give lateral support to the spokes forming this arch, he brought a metallic flange to bear on each side, and connected them together by rivets or bolts ; thus making a strong wheel, and yet retaining the elasticity of the one made entirely of wood.

Upon this wheel he obtained a patent. Although a good wheel, he had great difficulty in its introduction. Carriage-makers doubted its construction, and opposed it as an innovation, so that several years elapsed before it came into use, and its merits were tested. Mr. Sarven, failing to introduce it to any extent, gave to Woodburn & Scott, of St. Louis, and the New-Haven Wheel Company, the exclusive right to manufacture the wheel in the United States. They found the same difficulty that all the great patents have met with in their introduction : the opposition of interest and prejudice, in their most determined forms. But after several years of persistent effort, they partially succeeded in introducing the wheel. The present lessees of the Sarven wheel, the Woodburn Sarven Wheel Company, of Indianapolis ; the Royer Wheel Company, of Cincinnati ; and the New-Haven Wheel Company, of New-Haven, have been using the greatest care in sustaining the character of the wheel by using nothing but choice second-growth stock, and having that thoroughly seasoned and carefully made, and by this means they have aided to give the Sarven wheel its present popularity.

continued next page

Patent Wheels

The patent expiring June 8th, 1871, was extended for seven years. As soon as the patent became valuable, the ever-inventive genius of carriage-makers tried to improve (perhaps they thought to evade) the patent, and a large number of patents, following each other in rapid succession, were issued. Among these were the Warner, Thresher, Whelan, and many others. Mr. Sarven, and those interested with him, believing that these wheels embodied the essential principles of the Sarven patent, and that they were therefore an infringement, brought suit against the Wallingford Wheel Company, of Wallingford, Ct. This was made a test case, and was brought fairly before the courts. For one cause and another the case was put off several terms of court, until Judges Woodruff and Shipman, of the United States District Court, appointed a special term to try the case. The case being one of great interest, and involving a large amount of money, the most eminent legal talent in the United States was employed on both sides. The judges decided that all wheels with a *wooden hub, into which spokes entered by a mortise and tenon, having any metallic support for the spokes outside of the hub*, were infringements of the Sarven patent—a decision which makes several of the other patent wheels infringements on the Sarven wheel, and which compels the manufacturers of such wheels to stop manufacturing, unless they can make some arrangements with the Sarven interest.

The character of the judges in the above patent law case would seem to make the case a decisive one.

We have had many inquiries as to the *liability for damages* on the part of parties using wheels which are decided to be infringements on a patent. We understand the law to be this: When an *article* is patented, the patentee can attach the article wherever he finds it. This, then, according to the letter of the law, seems to make not only the man who manufactures it, the merchant who sells it, the carriage-maker who uses it, but the person who buys it, *all liable* for damages. The patentee can make his own selection as to which party he makes pay. Congress has of late years given increased stringency and severity to the laws affecting the infringement of patents, which, of course, is right. Property in a patent, if it be legal and straight, ought to have the same protection as any other property. It is a striking fact that, for lack of being thus protected, the majority of the great inventors have died poor, while other parties, without their brains, have taken their inventions and made fortunes.

The Hub, Sep 1872

PATENT OR WOOD-HUB WHEELS.

THIS is a question which directly interests the carriage-makers of the present day. Patent wheels became very popular within a few years after their introduction, so much so that many carriage-builders dispensed with wood-hub wheels entirely. But are they equal to a well made wood-hub wheel? Let us consider the construction of each.

First, the patent wheel. At the hub we find the spokes driven into an iron band, and resting on the end on the wood of the hub, but the mortice in the iron being tapering, the spoke is as firm as though it were set into a solid block of iron, and when it receives a blow at the felloes, it repels the shock with so much force that it brooms the spoke, or drives it into the felloe, even if it rests well on the tire. Then the tire becomes loose, or the wheel "rim bound." This shows that tires can not be kept tight so long on patent as on wood-hub wheels, the wood-hub allowing a little spring, which obviates trouble at the felloe. Next, the spokes in patent wheels are in a row, while in the wood-hub they are generally dodging. The latter makes much the stiffest wheel. An inch spoke, with half an inch dodge at hub, makes nearly as strong a wheel on side strain as though the spokes were all one and a half inch. at the base. The back spoke, with three-quarters inch dish, acts as a brace, preventing the wheel from turning wrong side out, which is a great trouble with patent wheels. Then too, as patent wheels have their spokes in a row, when lateral strain comes the wheel easily turns wrong side out, and often slivers the spokes at the hub. It will yield much more readily than a dodged spoke wheel, no matter how much dish it has. And I have sometimes thought that the more dish a patent wheel has, the more liable the spokes are to sliver, especially with very hard usage, for when there was enough strain to turn the wheel wrong side out, the tire would have a tendency to hold it dished backward as much as it was front, and to have the strain continue, it would be more liable to splinter the spokes at the hub than if there were less dish. This slipping backward and forward is what causes spokes to break at the tenons in the felloes, as the rim is held firm by the tire in the same plane. There are many patented devices for holding the spokes at the hub, but in my opinion they all have the above described difficulty, viz: they hold the spoke too firm endwise.   M. S. STOTLER.
THERESA, N. Y.

The Hub, Mar 1878

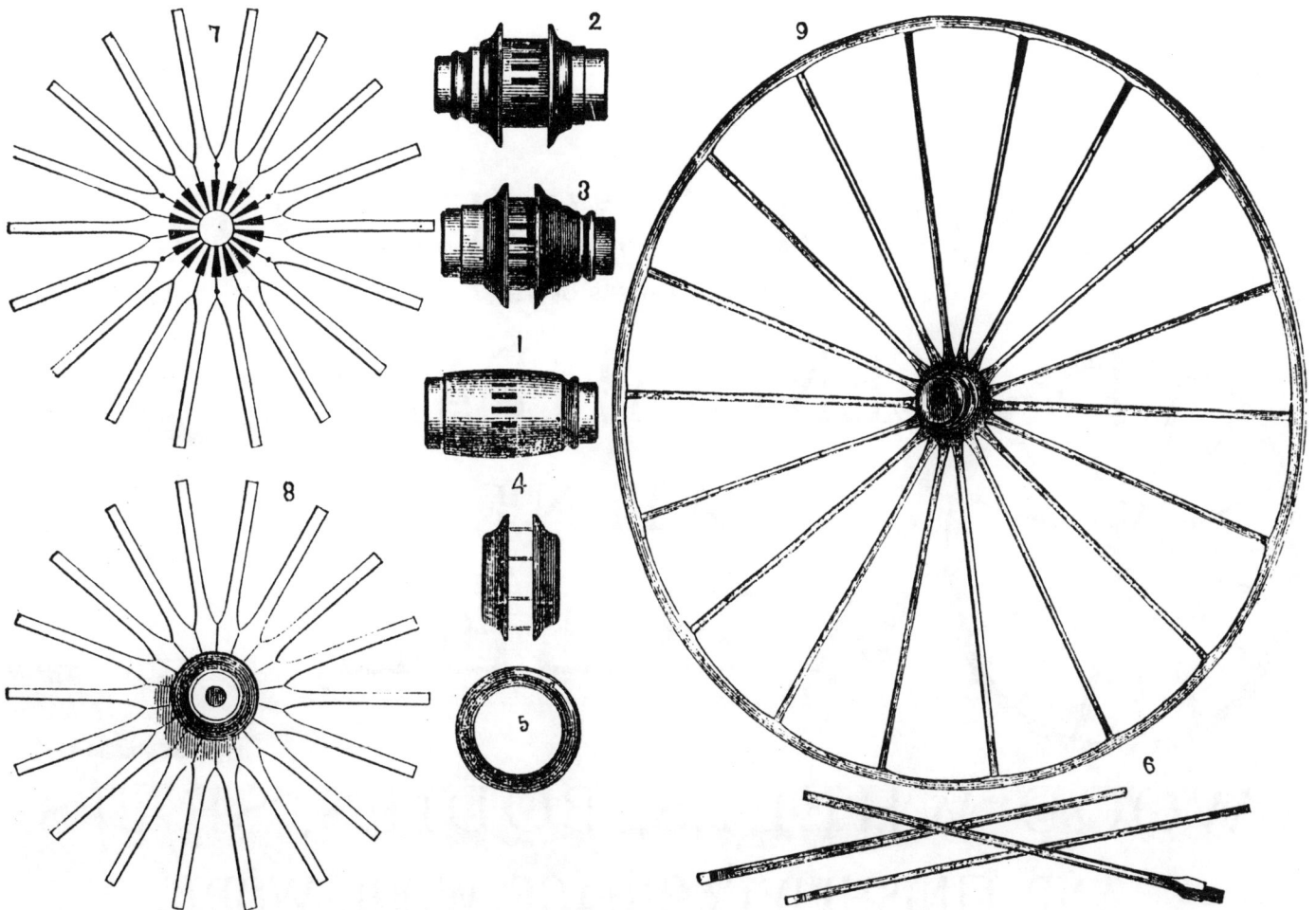
Coach-makers' Int'l. Journal, Mar 1873

Coach-makers' Int'l. Journal, Mar 1873

THE PALMER PATENT WHEEL

WITH MALLEABLE IRON HUB,

Combining every Requisite of Simplicity of Structure with the ultimatum of Strength, Durability and Beauty,

PATENTED JANUARY 28th, 1873.

MANUFACTURED BY THE

Davis Wheel and Palmer Fork Company,

SOLE ASSIGNEES,

109 and 111 West Broadway, New-York.

Fig.1

FIG.2

IN OFFERING this wheel to the public, we do it with the full assurance that nothing equal to it for beauty, simplicity of structure, strength and durability combined, has ever before been manufactured. It is constructed with the approved standard number of fourteen spokes, wedge-shaped and dove-tailed at the centre between the clamping-plates, and with a hub of the most delicate and beautiful form, created from light malleable iron shells, presenting an appearance of symmetrical proportions never before attained in any other wheel.

Fig. *1* represents a front view of our light wheels for road wagons and buggies, with a hub of only about three inches in diameter.

Fig. *2* presents a view of the central portion of the wheel with the back shell and moon-plate nut removed *A* is the axle-box passed through the tube *S*. *B* the spokes. *G* the front shell of the hub in position. *D* the back shell of the hub. *E* the moon-plate nut, which when in use, being passed over the axle like the ordinary moon-plate, and with right and left threads screwed into the inside of the rear end of the back shell presses against the shoulder of the axle let into the shoulder *M* of the box, thus holding the wheel firmly to the axle. *R* is the back clamping-plate slipped over the tube *S* and pressed against the spokes *B* by the nut *K* screwed on said tube.' *O* is the lug on the box to keep it from turning in the back shell, being held thereto by a slot into which it is fitted. *L* represents the border of the clamping-plate outside of the recess into which the shells are fitted. *P* shows the corrugations of the nut *K* through which the lug *N* of the back shell passes in penetrating the hole in the clamping-plate and thereby locking said nut.

Fig. 3 shows the construction of the tube *S* as cast fast to the front clamping-plate *C*, the face of which is roughened or machine-turned for the double purpose of making it true and giving a hold to the spoke.

The back clamping-plate *R* is represented in position to be passed over the tube *S* and held there against the spokes by the pressure of the nut *K* screwed on the thread *T* of the tube and extending the force of said pressure to the outer diameter of the corrugations *P* which, when up to place, is held from jarring loose by the lug *N* on the back shell *D* as shown in Fig. *2*.

Fig. 4 gives a front view of the inner portion of the wheel with the front shell, box-nut and oil-cup removed. *A* is the box with the front end closed. *C* is the front clamping-plate with a portion cut away to show the manner in which the spokes *B* are inserted wedge-shaped and dove-tailed between the plates. *D* is the back shell of the hub. *E* is the moon-plate nut. *F* the box-nut which screwed on the thread *X* draws the back shell *D* firmly against the back clamping-plate *R*. *H* is the oil-cup nut which screwed on the thread *J* on the end of the box *A*, and inside of the band or front end of the front shell *G* gives a beautiful finish and serves in unison with the nut *F* to relieve the strain on the tube-nut *K*, and holds the shell *G* closely against the plate *C*.

FIG.3

FIG.4

By a simple device not shown in the cut the box-nut *F* and the oil-cup nut *H* are absolutely secured against jarring loose.

The advantages this wheel possesses are very apparent.

1st. The centre contains all the wood it is possible to have for strength—the spokes being of full size where they enter the hub.

2d. The spokes being wedge-shaped and dove-tailed, and brought to place by great pressure and there firmly held, act separately as the key-stone of an arch, and can neither be drawn out of the clamping-plates or driven in on the box.

3d. The dish of the wheel is made positive by its peculiar construction.

4th. Any shrinkage occurring cross-wise of the spokes can be taken up by a slight turn on the single nut on the tube.

5th. A broken spoke can be replaced without taking off the tire or felloe.

6th. Any part of the wheel giving out can be replaced by a new one, as all the parts are made interchangeable, thus enabling the user to rebuild his wheels from time to time as may be required.

7th. The hubs can be left off till the tire is set and the wheels painted, when they can be put on and adjusted in a few minutes.

8th. The wheels can be trued by simply putting a thin bit of paper or shaving between the shells and clamping-plates, or by slightly filing off one side of the back shell where it presses against the clamping-plate.

9th. This wheel held to the axle by the moon-plate nut, and with the axle-box closed at the end, can be applied to vehicles of all sizes, from the lightest sulky to the heaviest truck, obviating, by its construction, the objection of lubricating matter (always unsightly) oozing from the end of the box, at the same time giving grace and beauty to the front end of the hub, and doing away with what has ever been regarded by al. coachmen as a nuisance—to wit; the loosening of three nuts to each wheel of a carriage when oiling became necessary. Duplicates of any part of these wheels are always on hand and furnished at the shortest notice, and none but the best materials are used in their construction. Quotations of price, per set, include boxes and axles, as the box being part of the hub goes with them in all cases.

SEND FOR PRICE-LIST.

The Hub, May 1873

OLDS'
PATENT WHEEL.

GUARANTEED THE BEST PATENT WHEEL MADE.

The following cuts represent sectional views of the Olds' Patent Wheel:

FIG. 1.

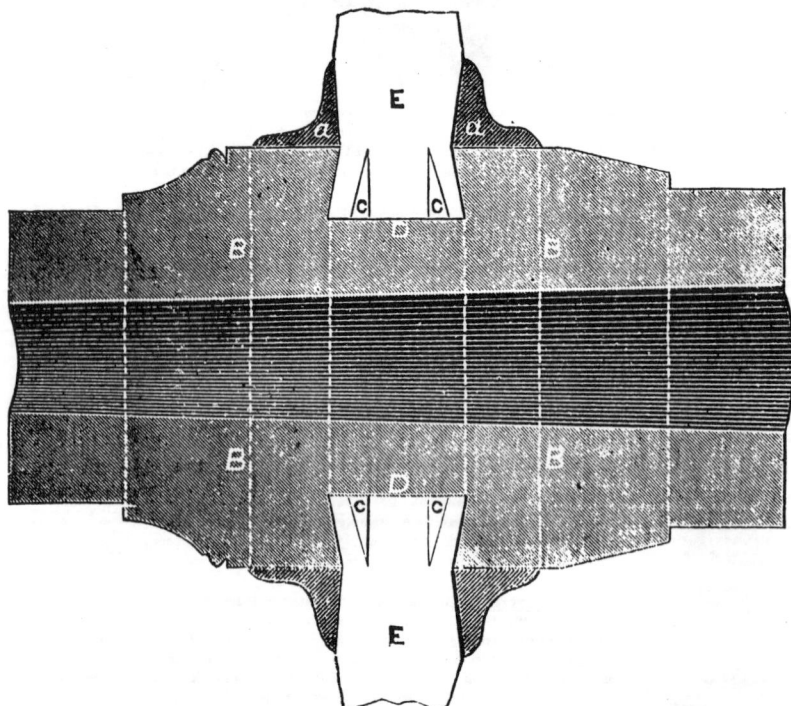

FIG. 2.

Fig. 1, is a sectional view of the wheel in a plane passing through the transverse axis of the hub and the longitudinal axis of the spokes.

Fig. 2, is a sectional view through the longitudinal axis of the hub, bisecting the spokes.

We wish to call your attention to the manner of securing spokes to wood part of hub. The iron band of hub being same as band of the Warner Wheel, formerly manufactured by us, excepting the partitions between spokes being brought to a sharp edge at the bottom or exterior of wood hub, thus enabling us to miter the spokes together in the groove of wood hub, forming a solid arch inside of the groove in wood hub. By referring to *Fig.* 2 you will see we turn the groove larger at bottom than top, thus forming a dovetail groove. Before driving the spokes, we slot their ends and insert wedges. As the spoke is being driven, the base of wedges come in contact with bottom of groove, and are forced into ends of spokes, spreading their ends and forming a perfect dovetail in the groove. You will readily see the impossibility of the spoke getting loose in the hub, the tendency being to get tighter when wrenched, giving the OLDS' WHEELS an advantage over any patent wheel ever invented, to stand the lateral strain when the tire may be loose without loosening the spoke in hub. It also possesses the advantage of the impossibility of grease getting to the spoke in case of cracked or porous boxes, also the greater facility in setting boxes, there being no ends of spokes to cut off, thereby making more perfect fits of boxes in hubs.

We have tested this wheel thoroughly in comparison with the Warner Wheel, and find it superior in spoke-holding properties of the hub.

We hereby guarantee the OLDS' WHEEL fully equal to the Warner, and superior to any patent wheel in the market. The external appearance of our wheel is exactly the same as the Warner Wheels. formerly manufactured by us.

In offering our new wheel to our customers, WE KNOW they will be as well satisfied as they have been with the Warner.

Hereafter, orders for patent wheels will be filled with the OLDS' PATENT WHEEL.

Price list furnished on application. Address

N. G. OLDS & SONS,
FORT WAYNE, IND.

The Carriage Monthly Advertiser, 1874-75

The Carriage Monthly Advertiser, 1874-75

H. O. YOUNG, President.   GEO. L. ROUSE, Vice-Pres. and Sup't.

ROYER WHEEL COMPANY,
NOS. 338 TO 358 WEST THIRD ST.
CINCINNATI, OHIO,
MANUFACTURERS OF

The Old Reliable Sarven Patent Wheel, Stoddard Patent and Plain Wheels.

SARVEN'S PATENT WHEELS.
The Best Wheel Made.

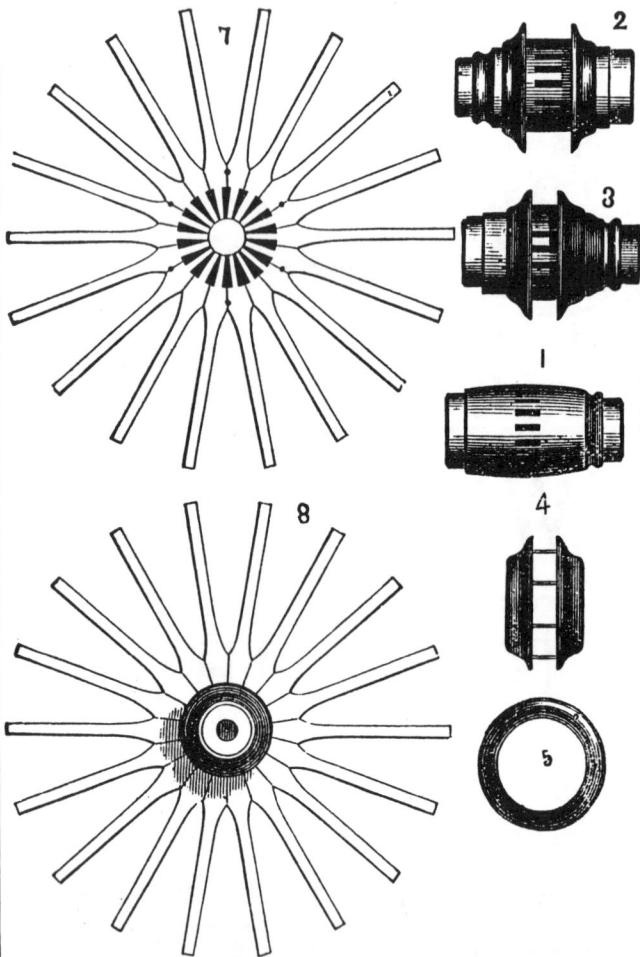

All Sarven wheels have the spokes in contact at the circumference of the flanges, and rivets from flange to flange through the arch of the spokes. Many shoddy wheels resemble the Sarven in this respect that are as worthless as those without the resemblance, but none are genuine unless as described.

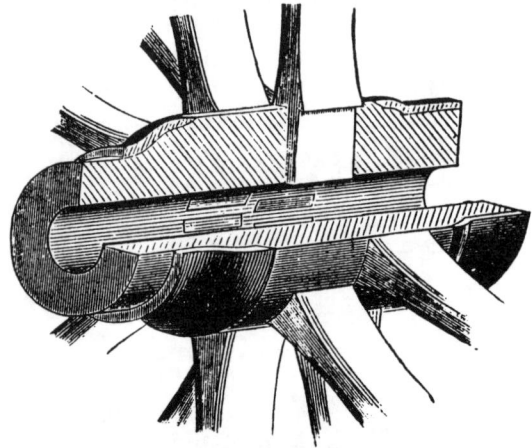

THE STODDARD WHEEL.

The favor with which the Stoddard Wheel has been received by carriage-makers, and that portion of the public using light sporting and road wagons, requiring a wheel with a hub from 2¾ to 4¾ inches in diameter, and of flexibility equal to that possessed by the ordinary wooden hub, with greater strength, has exceeded our most sanguine expectations, and gives us greater confidence that it will meet the requirements not hitherto obtained in any other make or device.

This end is secured by banding the hub, point and butt, with light neat malleable iron bands, which are heated and pressed on by our 70-ton hydraulic jack. The inner rim (next the spokes) of each band is beveled inside, the hub shaped to fit the bevel, then the bands are pressed on warm, fitting as tightly as the best iron will withstand, thus securely holding down the growth layers of the wood. The center part which is not covered by the iron is then neatly turned down to the outer circumference of the bands, leaving the outside neat and stylish, the iron not projecting above, but forming part of the outline of the hub. We then mortise as usual, except that we *INVARIABLY* use only **14 spokes** in the lighter wheels, the mortises are larger than usual, which enables us to use a good heavy tenon on the spoke. The hubs being well seasoned and banded, the spokes can be driven as tight as desired, a result not always obtained in the light wooden hub wheel as ordinarily made. We claim this wheel to be stronger than the old style wheel and more durable, that there is no possibility of the hub splitting or checking by ordinary exposure, and on examination it will be pronounced as light and stylish as any made.

Made with full stagger, three-quarter or one-half as desired, but three-quarter stagger preferred.

ROYER WHEEL CO.,
CINCINNATI, OHIO.

EXPLANATION OF DRAWINGS OF SARVEN'S PATENT WHEEL.

No. 1. Mortised Hub, of Selected Timber.
No. 2. Hub, with the Iron Flanges partly forced on, ready to receive the Spokes.
No. 3. Hub, showing the position to which the Flanges are forced after the Spokes have been driven.
No. 4. Side view of Flanges, as riveted together in finished Wheel.
No. 5. Front view of Flange.
No. 7. Front end view of the Wheel with the end of Hub cut away to show the Arch made by the Spokes, and also the Tenons of Spokes.
No. 8. Front view of the Wheel before pressing on the Flanges.

OUR MATERIAL IS ALL OF THE BEST OHIO SECOND-GROWTH HICKORY, OAK, ELM AND ASH.

Branch House, styled "New York Sarven Wheel Co.," at 83 Bowery, New York City.

The Carriage Monthly Advertiser, Feb 1877

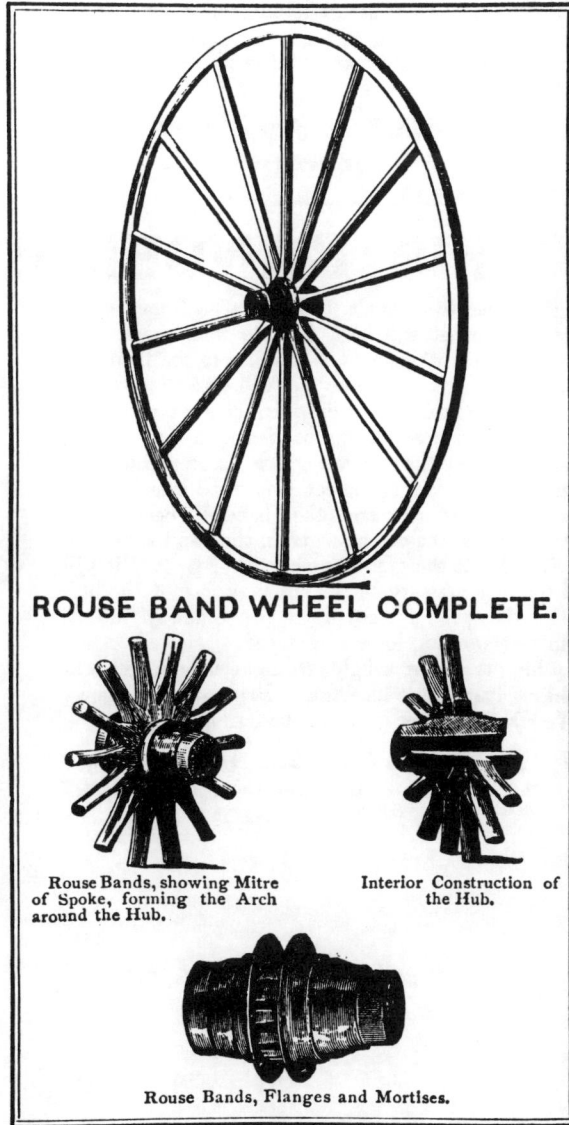

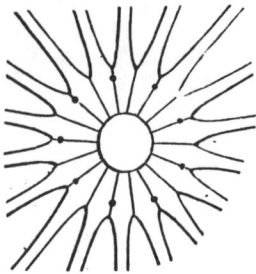

CORRIS' PATENT WHEEL.

We would call attention to our patent wheel, an engraving of which we here show. A large and extended experience of many years enables us to confidently assert that there is not the least doubt of its superior strength, durability and lightness; indeed, this must be apparent to any person examining the manner of its construction.

We furnish spokes and other parts of the wheel, for repairing purposes, at reasonable prices.

WM. CORRIS & CO., Sole Manufacturers,

ROCHESTER, N. Y.

First premiums at State and County Fairs.

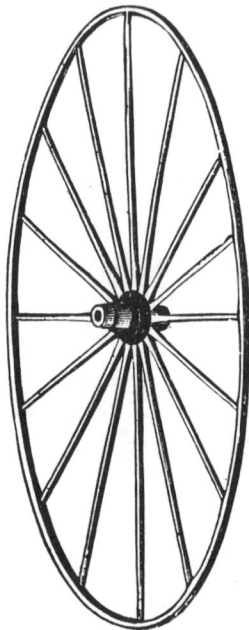

Wm. Corris' Patent, Jan. 6, 1874.

CORRIS' BANDED-HUB WHEEL.

We offer this wheel to the trade, being satisfied from its success the past year that it will be a favorite with them. We present a wheel, which, whilst it is not subject to the liabilities of failure of the ordinary wood hub, yet will give the required amount of strength, elasticity and lightness of appearance.

The hubs being well seasoned and banded with rings of best iron, enables us to give the spokes a much greater side drive than is possible to do safely in the common wood hub.

We consider it superior to any other banded wheel in the market. As shown in the cut here given, the band is oval on inside and pressed into the groove prepared for it; and it will be observed that the pressure on the outer surface of the hub is so great that the spoke is really dove-tailed sideways into it, and all liability of working loose is removed.

Parties wishing to obtain a light, strong, elastic wheel, will find it to their advantage to use this. Price same as flange wheels. We make no other qualities than xx and xxx.

WM. CORRIS & CO., Sole Manufacturers,
ROCHESTER, N. Y.

CORRIS' PATENT WHEEL.

CORRIS' BANDED-HUB WHEEL.

The Carriage Monthly Advertiser, Feb 1877

Patent Wheels

The Rubber Cushioned Axle,

(PATENTED IN THE UNITED STATES, CANADA AND EUROPE,)

Is now manufactured under our licenses by the following well-known firms:

The Tomlinson Spring Co., Newark, N. J. | Messrs. S. Rogers & Co., Stanfordville, N. Y.
Messrs. Sheldon & Co., Auburn, N. Y. | " D. Dalzell & Sons, South Egremont, Mass.
" Ives & Miller, New Haven, Conn. | The Spring Perch Co., Bridgeport, Conn.

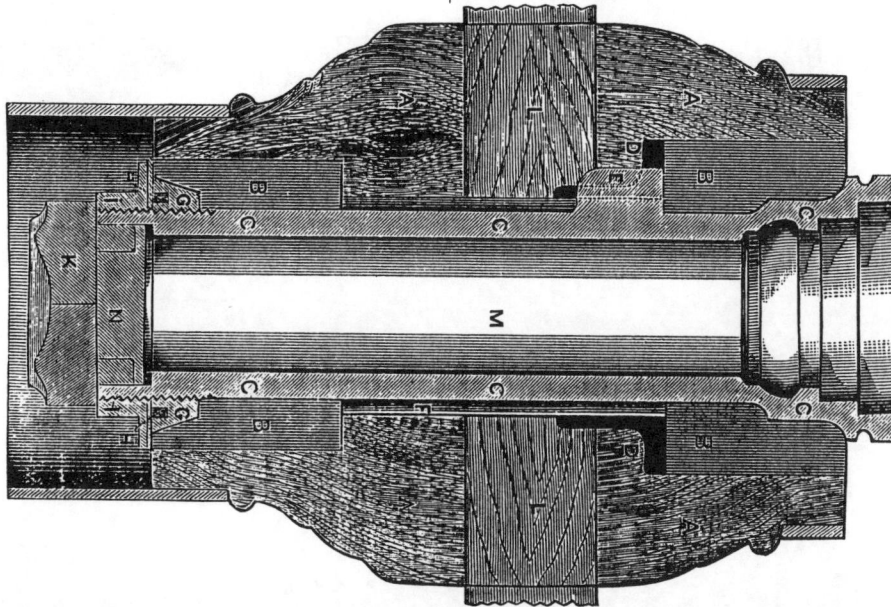

Fig. 1. Half section of Hub, with Rubber Cushions. *B,* Rubber Cushions ; *D,* Slotted Retaining Sleeve ; *E,* Spur or Fin on Axle Box. *D,* Compression Nut.

Fig. 2. Section of Axle Box, showing position of Slotted Retaining Sleeve.

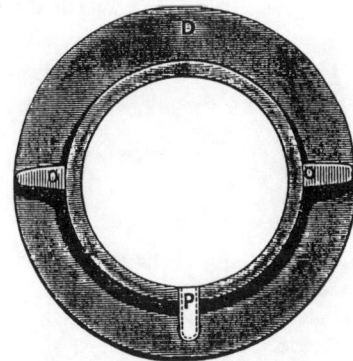

Fig. 3. Slotted Retaining Sleeve.

To either of the above mentioned houses, orders may be addressed for our improvement upon any style of axle in general use, except the "three-bolt mail." Having made arrangements for the general introduction of our CUSHIONED AXLE to the Carriage Trade—its merits being fully established during the past two years upon upward of fifty vehicles, both light and heavy, in constant daily use over the rough pavements of New York, we beg to present briefly its salient features. In a hub of any ordinary size or form are embedded the ELASTIC RUBBER CUSHIONS, firmly held in their place by our PATENT DEVICES, both simple and effective, securing for the first time in the construction of carriages, the same results achieved by the application of RUBBER CUSHIONS or SPRINGS upon railway cars.

Vibration, jolting, pounding, noise, and the *crystallization* of *iron* and *steel* are greatly diminished, if not wholly avoided, and the *strain* and *shock* upon all parts of the vehicle when passing over rough roads, pavements and rail tracks, taken up and born by the CUSHIONS, thus acquiring to an extent never before reached the great *desiderata,*

SAFETY, COMFORT AND ECONOMY

in the use of all wheeled vehicles, whether for passengers or freight.

The CUSHIONS cannot be displaced by use or worn out, and being *tightly fitted,* effectually prevent *oil or grease* from penetrating the hub.

The adoption of the CUSHIONED AXLE by many of the leading *carriage-builders* in this city and elsewhere, under whose supervision abundant tests have been made, together with the *unreserved endorsement* by ALL the many owners of carriages to which it has been applied, warrant us in offering to the *trade* our RUBBER-CUSHIONED AXLE, not as an untried or doubtful experiment, but as a *perfected, valuable improvement.*

THE RUBBER CUSHIONED AXLE CO.,

BROADWAY, SEVENTH AVENUE AND FORTY-THIRD STREET,

NEW YORK.

B. F. BRITTON, President.
J. B. SAMMIS, Secretary and Treasurer.
G. W. HAYES, Superintendent.

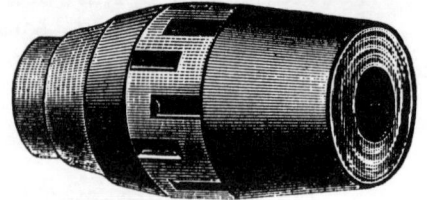

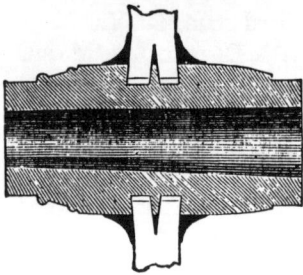

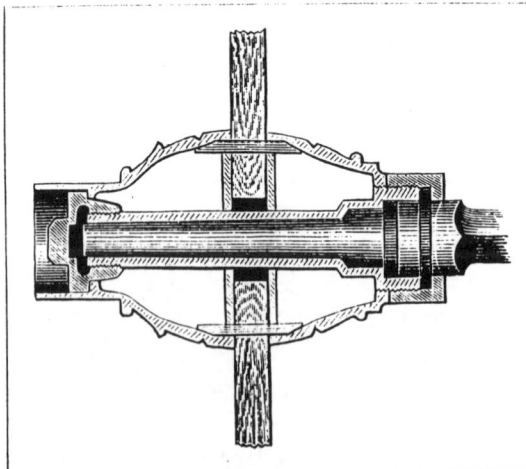

WARNER PATENT WHEELS,

—THE BEST WHEEL EVER MADE.—

The following Manufacturers are Licensees of this Patent, and are supplying the Wheels to a large and rapidly increasing trade.

HOOPES, BRO. & DARLINGTON, Limited,   West Chester, Pa.
PHINEAS JONES & CO., - - - Newark, N. J.
WM. STEVELY & CO., - - - - " "
WALLINGFORD WHEEL CO., - Wallingford, Conn.
WM. B. CLARK, - - - Waterloo, N. Y.
FOSTER, HOWE & CLEARY, - - Merrimac, Mass.
R. F. BRIGGS & CO., - - - Amesbury, Mass.
LOCKE & JEWELL, - - - " "
ROBERT ROWE & SONS, - - Brentwood, N. H.
ALMON WARNER, - - - Belvidere, N. J.
SPRINGER, MORLEY & GAUSE, - - Wilmington, Del.
CINCINNATI WHEEL CO., - - Cincinnati, Ohio.
ROBINSON WAGON CO., - - " "
I. & J. LAFORGE, - - - Rahway, N. J.
THE WHEEL AND WOOD BENDING CO., Bridgeport, Conn.
THE OLDS WHEEL CO., - - " "
THE DANN BROS. & CO., - - New Haven, Conn.

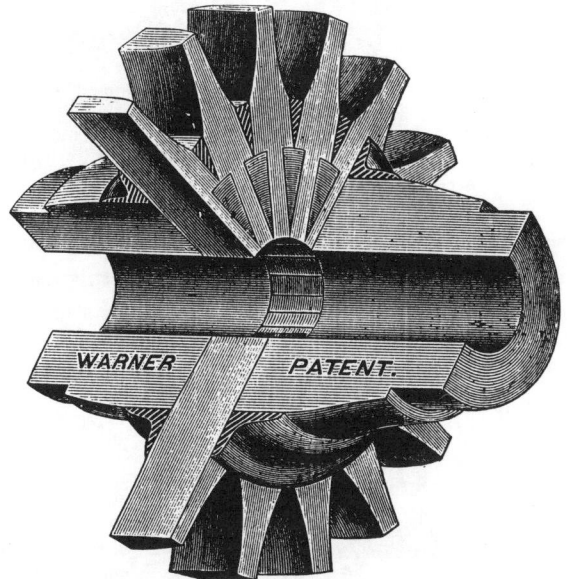

It is the only wheel having spokes driven THROUGH mortises in a **SOLID METAL FLANGE**, into mortises in a wooden center.

It is **STRONGER** than any other.

It is **MORE DURABLE** than any other.

It is as **EASILY REPAIRED** as a plain wheel.

It is **LIGHT AND STYLISH.**

If properly made, the spokes will NEVER BECOME LOOSE IN THE HUB.

A few more Licenses will be granted to suitable parties.

All persons infringing the Patent, will be held to the strictest accountability.

For Information, address

B. A. TREAT, (Attorney for E. Hall & Co., Owners,)

WALLINGFORD, CONN.

The Carriage Monthly Advertiser, Nov 1880

SILVESTER PATENT TIRE

TIRE
FELLOE
SPOKE SPOKE

IS A
UNIVERSAL
FELLOE-CLAMP,

THE **OPEN-HEARTH STEEL** I manufacture this Tire from is warranted to me to stand all the strain I claim for it. It is specially made for this use. It is tough and cannot be broken, but will bend cold in any shape, as shown by the engraving. It will stand a strain of 60,000 pounds to the square inch.

The number of Tire represents the number eighths of an inch the felloe is thick.

PRICE-LIST.
Per Set, subject to changes without notice.
Steel only.

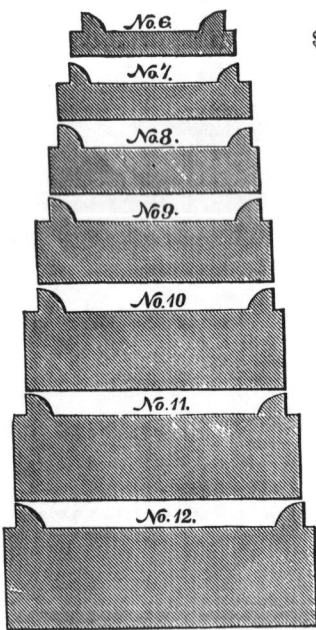

No.6 — $1.75
No.7 — 1.95
No.8 — 2.95
No.9 — 4.15
No.10 — 5.55
No.11 — 7.10
No.12 — 8.90

All sizes above No. 12 are made of Steel or Iron to seven inches in width and any thickness as desired. No. 13 is made of Iron, for arm wagons, one and one-half inch felloe.

A few of the Advantages this Tire has:

This Tire is a positive and absolute protection to the felloe. The vertical flanges enclose the felloe in such a manner as to prevent the tire from coming off; render the use of screws and bolts unnecessary. The lateral flanges which project outward beyond the felloes, protect the felloes from damage by *rail road tracks, curb-stones, rocks, etc.*, which cannot wear the paint off or destroy the felloe; both felloe and tire are benefitted by being left without bolt-holes. The felloe cannot split, as the flanges bind it together. The tire cannot ship off if the felloe shrinks, and does not need resetting one-half as often by reason of the felloe being constantly protected and in place. The wheel is strengthened sideways by the flanges. A wheel set with this tire will outlast three wheels with common bar-iron hoop tire by its protection. These flanges also strengthen the tire and felloe, and prevent the felloe bending inwards between spokes. It gives the tire a light appearance. A part of the edge is level with and painted like the felloe.

It is a great value to a wheel, as the wheel cannot get out of shape, unless there is a tremendous strain, sufficient to bend both tire and felloe edgewise. To make a straight wheel, put this *straight* tire on, and the wheel must be straight and remain so, as it has the strength of the tire edgewise combined with that of the felloe to hold it in position.

This Tire has great advantages over the common bar-iron hoop tire when necessary to reset it, as there are no bolts or screws to take out and replace, and the felloes are not ruined by boring four, eight or twelve extra holes at every resetting; the old holes fill with mud and water, and rot out the felloes. The expense of new tire-bolts is unnecessary.

This Tire cannot be broken; neither frost nor rock can break it. The metal is made specially to meet these requirements.

This Tire can be set by any blacksmith in one-half the time of the common bar-iron hoop tire, by being welded, then *expanded* by *heat*, so the tire easily slips over the *felloe*. When cold, the job is complete. No bolts are used or required. The tire is much stronger without them. The felloe is much stronger without them. The steel welds like soft iron. In resetting there are no bolts to take out or replace.

A little router like a spoke-shave prepares the felloe for the tire. To make the flanges, use a top-swedge in welding.

I am the sole owner and manufacturer of this Patent Tire in the UNITED STATES, ENGLAND AND IRELAND, and am prepared to furnish daily 160,000 lbs. I offer these tires for sale, for cash, and solicit patronage. Parties keeping this tire in stock, or purchasing large quantities for consumption or sale, I shall take pleasure in advertising them, and ask them to advertise this tire in their catalogues and otherwise for our mutual benefit. If engravings are wanted for this purpose, they will be furnished on application. Orders and correspondence solicited.

BENT COLD

C. B. CLARKE,
719 Locust Street,   ST. LOUIS, MO.

Common Tire after Two Years use.
Felloe and Spoke Ruined.
Entire New Rim, Hub, Spokes, Tire, Felloes, Bolts and Painting a Necessity.

Patent Tire after Two Years' Use.
Felloe and Spokes Perfect.
Entire Wheel Good as New.

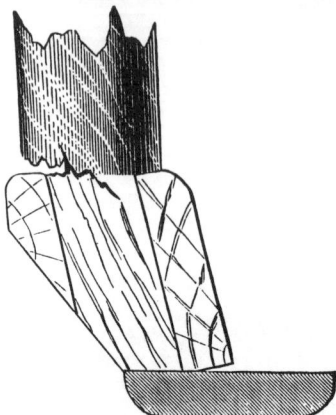

Price List of Plain or Patent Wheels, with Silvester Patent Tire on:
Free on board cars in St. Louis, Baltimore, Boston, Buffalo, Charleston, Chicago, Cincinnati, Cleveland, Galveston, Louisville, Memphis, Milwaukee, Mobile, New Orleans, New York, Philadelphia, Providence.

Width of Tire.	A	B	C	D		Width of Tire.	A	B	C	D	
No. 7.—	$19.66	$17.15	$13.85	$13.00	per set.	No. 10.—	$29.15	$27.00	$23.60	$19.75	per set.
No. 8.—	20.70	18.15	14.85	14.00	"	No. 11.—	34.75	30.35	25.95	23.65	"
No. 9.—	22.95	19.45	16.10	15.25	"	No. 12.—	40.35	35.90	34.80	32.55	"

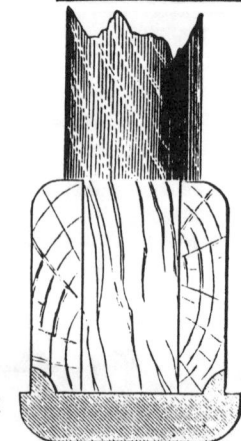

The Carriage Monthly Advertiser, July 1884

Patent Wheels

The Dayton Patent Compressed Band Hub and Tenon Wheel,

MANUFACTURED BY

PINNEO & DANIELS,

DAYTON, OHIO.

Cut No. 1 represents a piece of the best Imported Norway Iron, cut and coiled ready to weld.

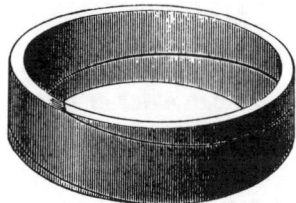

No. 1.

Cut No. 2 represents a Band after it is welded and rolled. These Bands, having a continuous weld, as shown in cut No. 1, cannot slip, as sometimes happens in short lapwelded Bands. Our Bands are made with a lip, as shown on the upper side of cut No. 2. We manufacture all our Bands.

Cut No. 3 represents a Block with the ends turned to the finishing size, but the center left ⅜ of an inch larger. The end of the Block to the left is just as all of our Blocks are turned ready for the Bands. The

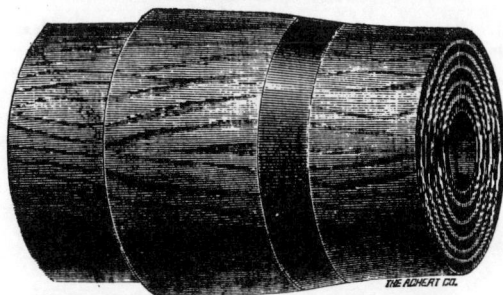

No. 3.

end to the right has a Band seated, showing the amount of compression both Block and Band receive. It will be seen that we do not weaken our Hub by cutting grooves for the Bands to rest in, but strengthen the entire surface by applying the Band over all and compressing them into the wood flush This can only be done by using our Patent Band, with the overhanging lip, which extends out over the shoulder formed by the enlarged center.

Cut No. 4 represents a finished Hub with a longitudinal section removed, showing a cross-section of the Bands as they appear after the Hub is finished. It will be seen, by the highly condensed condition of the wood beneath and adjacent to the Bands, that no grooves have been cut to receive them, but that they are pressed bodily into

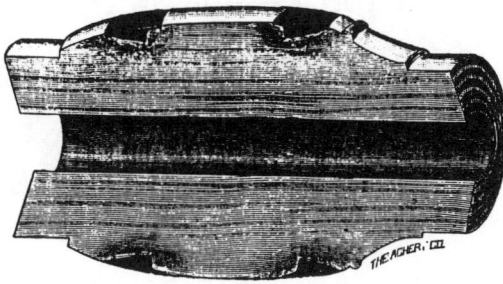

No. 4.

the wood. This not only precludes the possibility of the wood escaping from beneath them, but it also prevents the Bands from ever becoming loose.

No. 5.

Cut No. 5 represents the large end of a Spoke, with a groove cut in both sides of the Tenon next to the shoulder, making the thickness of the Tenon where the groove is cut just as we want it when ready to drive, he lower end of the Tenon being about one-eighth of an inch thicker This we reduce by compression to the desired thickness.

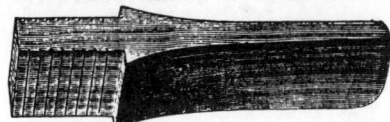

No. 6.

Cut No. 6 represents the same Spoke after the Tenon has been compressed and scarfed ready to drive.

Cut No. 7 represents a lateral section of our Hub with the Spokes driven. We cover the walls of the Mortises and the compressed end of the Tenon with hot glue before driving them. Tenons driven in this manner after having been compressed will expand as nearly as possible to their original size and shape, as shown in cut No. 5, and in so doing they imbed themselves so firmly into the walls of the Mortises below the surface that the Hub and Tenon are inseparable, as shown in this cut.

In summing up, we present you with a Wheel with the Hub reduced by compression three-eighths of an inch in diameter, with its surface pores closed and solidified, and the whole Hub made much stronger and more durable than it is possible for wood to be in its natural state, and we hold it in this position by lipped Bands that will not

break or allow any of the fibers to escape from beneath them ; neither can any of the wood rise above them.

No. 2.

We compress all our Tenons below the groove, and they invariably expand beneath the surface of the Hub after they are driven, just as they show in cut, making it impossible for them ever to become loose This method of constructing a Band Hub Wheel overcomes all objections both to a Plain Wood Hub Wheel, which are that hubs sometimes crack and spokes work loose in the hub, and to a Band Hub Wheel, which

No. 7.

are that the bands become loose, or the wood shrinks away from the bands, or rises above them, in either case cracking the paint and varnish around them.

These Bands, being seated in the Hub without grooves, will never become loose, and with the surface pores of the Hub closed and firmly held by the Bands, make the Hub impervious to moisture, and thereby prevent expansion and contraction, which are the cause of paint and varnish cracking round the bands after the work is finished.

The Carriage Monthly Advertiser, May 1885

The Carriage Monthly, Nov 1894

211

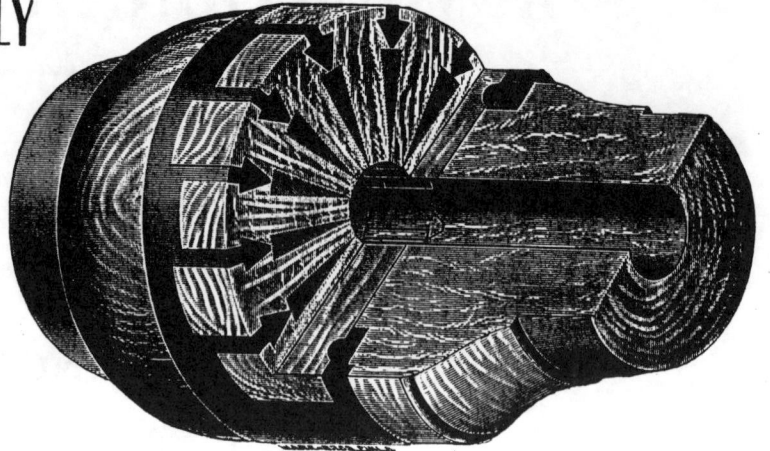

The Carriage Monthly, July 1901

212

IX
WHEEL REPAIR

WHEEL REPAIR

It is when a wheel comes into the shop for repair that the strengths and weaknesses of wooden wheel design and construction become apparent. The repair process, with its trial and error approach, generated many of the improvements in wheelmaking.

For example, in the first section in this chapter is an article describing a blacksmith-made tool that was used to tighten spokes in a wheel without removing the wheel from the vehicle or unhitching the horse. The section on "Repairing Sarven Patent Wheels," p.220, contains several articles that discuss the unique construction of this type of wheel, which requires special handling. (The company issued instructions explaining that the felloe joint must be tight, the spoke tenons must not project past the felloes, and there must be a minimum of 5/8" dish in the wheel.) The "Tire" section contains articles ranging from advice on dealing with the tendency of tire bolts to turn, to a description of tire pullers for removing the iron tire from the wheel. The "Shop Helps" section contains hints and ideas on making tools for use in the wagon shop, including plans for a rack that holds rim and hubs.

Wheel Repair

SHRINKING OF SPOKES IN THE HUB.

NEW-YORK, May 10, 1879.

To THE EDITOR.—DEAR SIR: There is a carriage that was built by Million, Guiet & Co., now in the city, which is a fine job in other respects, but the spokes of the wheels have shrunk in the hub. I would like to know from some of your wheel correspondents what would be the best method to secure the spokes in the hub, and oblige yours, D.

ANSWER.—In answer to D.'s query, we give the means used in a similar case which, under very unfavorable circumstances, proved to be a permanent cure.

The spokes were removed from the hubs, each one marked, insuring return to the mortise to which it was originally fitted. Upon examination of these mortises, it was found that the grease had worked through the ends of the hub, also along the tops of the boxes, and of course into the mortises. As a preliminary, these were treated to a strong solution of soda, boiling hot, the application being made with a small swab, care being taken not to saturate the entire hub; this was done to neutralize the grease. The tenons of the spokes were also dipped into the soda solution, and placed near the oven to dry. Upon callipering the mortises it was found that the surging or swaying of the wheels while "running loose" had enlarged the mortises at the bottom; this enlargement being measured, wedges were prepared which were long enough to reach from the box nearly to the surface of the hub. To allow size enough for these to resist the pressure, the backs of the mortises were pared away until room was gained to admit a wedge measuring $\frac{5}{32}$ds of an inch at the thickest part, tapering to a razor edge at a point near the surface of the hub. These wedges were securely cemented against the back of the mortises, the base resting down upon the box, to avoid being forced out of place when the spokes were entered. This cement was composed of the following ingredients: 6 oz. white frost glue, 2 oz. bonnet glue, $\frac{1}{2}$ oz. American isinglass. These were soaked and melted separately, being mixed together while hot; and to this was added a quantity of pure pulverized white-lead, dry, enough to make a moderately thick cement, possessing the properties of drying quickly and not being easily affected by grease.

After cementing the wedges, the hubs were laid away to dry. The spokes were found to need tightening sidewise, also. To accomplish this, pieces were cut from the thinnest size of "woodbury duck," or sailcloth (new); the ends of the tenons were slightly rounded, and the duck or canvas secured on each side of the tenons, passing over the ends but not reaching quite to the shoulders of the spokes; a firm grip of the vise after cementing insured a tight adhesion. When this canvas became dry the re-driving began; the wheels, having the bands upon them, had to be fastened upon the rimming stand, a driving-set or gauge being used to regulate the "dish," which was the same as in the first instance. A tough hickory block was bored to pass over the rim tenons neatly, and rest upon the shoulders cut for the rims; this block received the blows of the mallet, and each spoke was tried sidewise and made to fit the same as in new work, while the callipers showed the mash to be $\frac{1}{8}$ in. at bottom of mortise, decreasing to very slight pressure at surface of hub. The cement was used both in the mortise and upon the whole surface of the tenon, and very hot, requiring rapid driving to prevent setting.

This is not the quickest way of repairing, but for any class of work warranting such an expenditure of time, it will be found satisfactory in result.

There is an article made from lignose or wood fiber, in sheets of different thicknesses, that merits a trial for the purposes mentioned above, but we think it is best to recommend that which has already been proven.

DUBOIS.

The Hub, Aug 1879

AN IDEA FOR FASTENING LOOSE SPOKES.

REPAIRS done while you wait, are very popular in coach making, as well as in other trades. The repairer is often required to tighten a loose spoke of a wheel immediately, as the owner can spare the vehicle only for a very short time, especially if it be a delivery wagon or market cart, or even a professional gentleman's brougham or buggy.

To take the tire off, put in a new spoke, and contract the

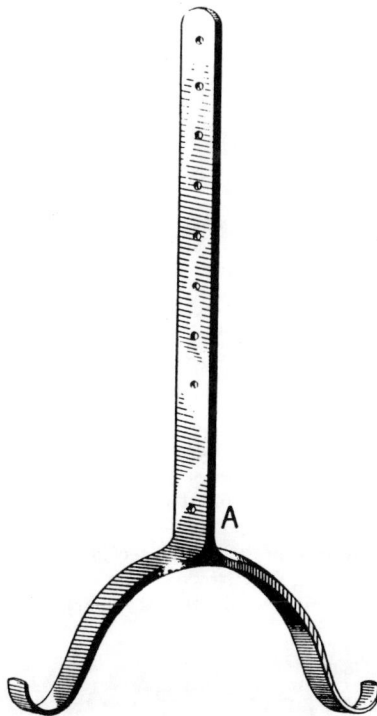

FIG. I. FIG. 2.

tire means time, and leads to painting the rim. The old fashioned method for a quick job was to fasten the spoke with spoke-plates, which is often practiced to this day in small wheelwright shops. This does not make a neat job, and damages the paint. A neat, quick and simple method of spoke tight-

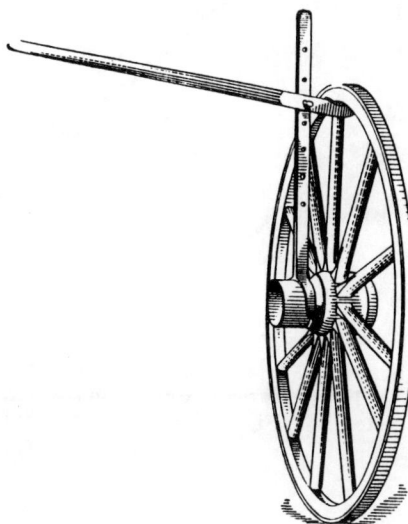

FIG. 4.

continued next page

ening, as witnessed in a London coach maker's shop, is described herewith, which with the accompanying sketches of the tools used, will make it more plain to the reader.

The wheel was not taken off, neither was the horse taken out of the shafts. The wheel was turned with the loose spoke perpendicular to the ground, then the saw was run between the shoulder of the spoke and the felloe, to make a level bearing to allow of a small piece of leather being drawn between. Then the setter's stand, Fig. 1, was placed on the end of the hub and the lever bar, Fig. 2, was placed with one end under the felloe and resting on a pin in the upright stand which forms a fulcrum. The rim was raised by it sufficiently to admit of the

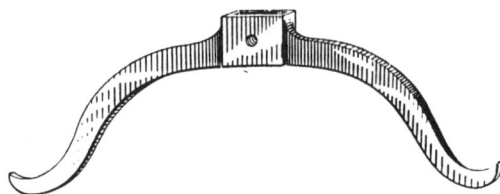

FIG. 3.

leather being placed between the two bearings, the lever was then released and a hole drilled in the tire to admit of a long chisel-pointed rivet or screw, a hole was then bored through the broken tang into the spoke and the rivet or screw was driven home. This made a neat and strong job and was done in a short time. Fig 3 represents another section of the stand, which is fitted to Fig. 1 at A with caulkings and secured by a thumb screw. The smith made this tool, which is

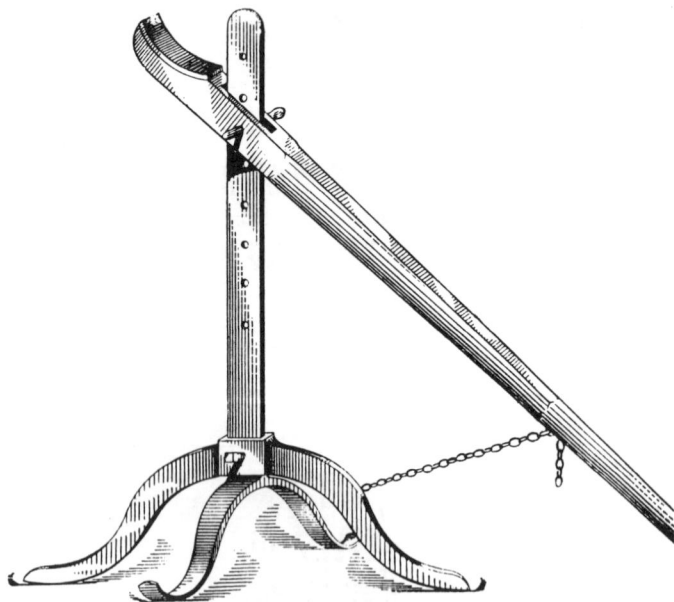

FIG. 5.

in three pieces, and can be be put together easily, fitting it to serve the purpose of an ordinary setter. Fig. 4 shows two parts in use as a spoke setter, and Fig. 5 shows all parts complete as an ordinary setter.

The Hub, Jan 1895

CARRIAGE AND WAGON REPAIRS.

Suggestions, Directions and Advice from a Practical Worker.

XXXIII.

BY A VETERAN.

The hub in the farm wagon wheel often wears down under the shoulder of a spoke and by letting in water the spoke tenon rots, as well as the hub itself. This is due to the soft spots so common in wood, particularly when, like the hub, the outer or sap wood is the part that is the most exposed. If the battering down of the shoulder is discovered while the hub is comparatively new, the repair can be made quickly and cheaply. The first thing to do is to clean out the dirt that has collected at and around shoulder, using a thin blade so as not to increase the trouble. Then, if possible, work in keg white lead; if it is about the consistency of workable putty all the better. Then prepare two iron wedges

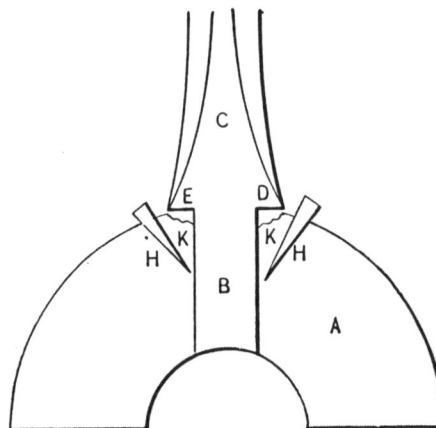

Fig. 172.—Spoke and Mortise Shoulders Forced Up by Iron Wedges.

the full width of the spoke, making it thin or blunt, as your judgment may dictate. The office of these wedges is to raise the battered wood up to the shoulder of the spoke. Fig. 172 shows a cross section of a hub with one tenon in place, and the opening between the shoulder of the spoke and battered wood around the mortise, somewhat exaggerated, to more clearly show the injury that is to be repaired, A, the hub; B, the tenon; C, the spoke; D and E, the damaged shoulders. As will be seen, the thickness of the wedges must depend upon the spaces at the openings D, E. These wedges should be sharp enough to allow for their being driven in without the use of any other appliances. It is well to say here that it will seldom require a wedge to be more than 3/4 of an inch long and 1/8 of an inch thick on the top. Burr the outer edges with a burr punch in such a way as to set the burrs out at the top 1-32 of an inch; four or five burrs will be sufficient on each side. When ready to drive set the wedge at an angle, as shown by H, starting it about 1-16 of an inch from the outside of the shoulder of the spoke, and drive in until the top edge comes flush with the outside of the hub. The effect will be to force the triangles, K, up to and hard against the shoulders of the spoke and completely close the opening, and if the wedges are of the right thickness the wood will be made compact and hard and it will give a solid and

continued next page

Wheel Repair

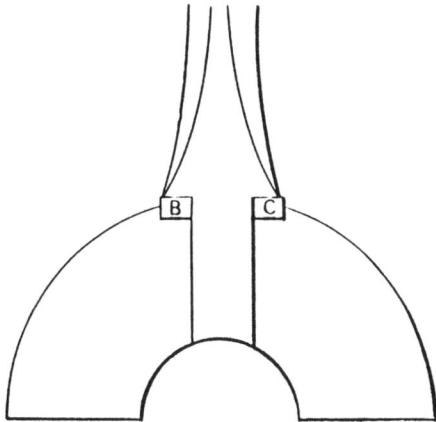

Fig. 173.—Shoulders and Mortise Repaired by Insertion of Solid Blocks.

water-tight shoulder. If the burrs are properly cut the iron wedges will not jar out and a permanent repair will be effected. Wood wedges will not answer, as they cannot be driven in without first making an opening, and they will become loose by the jar that is put upon them.

If the hub is badly battered, as shown by Fig. 173, the wedges will not suffice, and other means must be resorted to. The job is an easy one if the tire has been removed, as the felloes can be removed and the spoke drawn from the hub. When the tire is not removed the work is more troublesome, but the work to be done is the same in both cases. The old wood must be cut away until a solid shoulder is reached, as shown by square openings, B and C, the boxed out space being the full

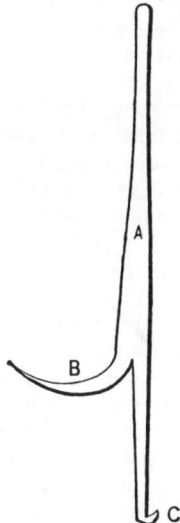

Fig. 174.—Puller for Removing Old Tenons from Hub Mortises.

width of the spoke and deep and wide enough to remove all the battered wood. If the spoke is in place it will be necessary to spring the felloe, the same as when springing in a spoke, and when sufficient opening is had secure the lever so that the spoke will remain up above the surface of the wood. Cut the wood out true and slightly level the ends and side, so that the wood filling will rest against the hub wood, and make a lighter joint than it would be possible to make if the sides were cut square. In filling in these pieces under the shoulders of the spoke it will be necessary to cut the trench, the full depth and width, through to the back end of the hub. Make the filling pieces of good hard wood and set them with the edges of the layers up. They should be fitted so tight as to require good heavy blows of the hammer

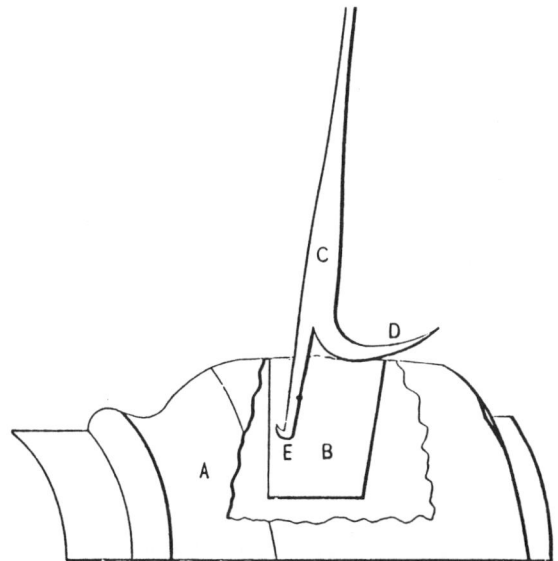

Fig. 175.—Shows the Puller with Hook End Buried in the Tenon.

to drive them in place. It will be well to clamp them down, so that when being driven in the filler will not raise from the bottom of the trench. Use plenty of keg white lead when driving them in place; the amount that is not forced out will be sufficient to keep out the water. When the fillers are in place release the lever and strike a sharp blow upon the tire with a heavy hammer directly over the end of the spoke. This will set the shoulder down snugly upon the fillers and make a repair that will be durable. After the spoke is set down clean off all surface wood of the fillers that may project above the surface of the hub, set in two or three nails through the fillers back of the spoke to prevent their working up. If well done the job will not only prove satisfactory, but it will leave a smooth finish and will not show after the wheel has been painted.

A broken spoke, one that has broken off close to the shoulder at the hub, if the tenon is tight in the hub, will often cause annoyance and work to remove it. The easiest and best way is to bore as many holes lengthwise in tenon as possible with an auger that is enough smaller than the mortise, to insure not cutting into the sides of the mortise when boring the tenon; then with a lever made with a heel, something like that of a nail puller, and a hooked end that will enter the hole in the tenon, force the hooked end down into the hole until the heel rests upon the hub, draw the lever over and almost any tenon can be drawn out by a single operation. Fig. 174 shows this lever, which can be made by any blacksmith, and it will pay its cost in a very short time. The handle of the lever, A, should be about 2 feet long and of 1¼ x 5-inch iron. The heel, B, should be slightly curved and flattened out so as to present an inch surface at the face or bearing. The hook should be shaped so as to bite the wood by the slightest movement of the lever. In making, split the end of the bar about 2 inches, turn back one side to form the heel, leaving the other side nearly straight except at the lower end, where the hook, C, is turned up. With a lever of that kind great power may be had, power sufficient to draw the tightest tenon after it having been bored as we have directed. This will be found a much simpler method of drawing old tenons, no matter whether in a hub or some other part of the wagon, and will not consume one-half the time required to bore and mortise out as is common. The lever is shown in place in a mortise in the cross section, Fig. 175, A, the hub; B, the bored tenon; C, the lever handle; D, the heel; E, the hook, and it requires

continued next page

Wheel Repair

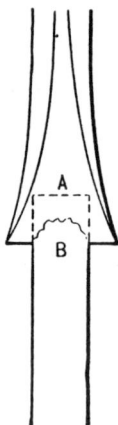

Fig. 176.—Repair of Face of Spoke.

but a glance to determine how practical and useful this simple tool is and how little time will be required for the blacksmith to make one.

In times past, and, for ought we know, there are those who even now follow the old method, the custom was if a spoke got loose in the hub and the mortise was badly battered on the front and back, the tenon was wrapped in wet canvas and the spoke was then driven in, the canvas being depended upon to fill the vacant space caused by the deficiency of wood. This seemed for the time to insure a tight spoke, but a few months' use served to show its weakness, as the canvas wore out and the condition of the tenon and mortise was worse than it was before the canvas repair was made. There is but one proper way to make this repair, that is to use wood to supply the place of the wood that has been worn away. This wood must be attached to the tenon or to the mortise. Our preference is to face up the back and front edges of the tenon, then clean out the mortise, cutting away all the bruised wood, testing it with a spoke tenon gauge to see that the face and back are cut so as to hold the spoke to the position that it originally

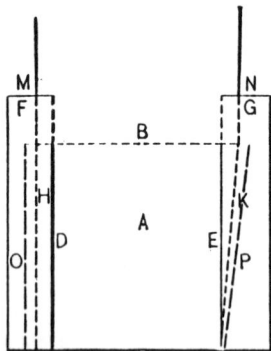

Fig. 177.—Side View Showing Block and New Lines for Tenon Front and Back.

occupied. Then mark the draft on the back edge of the tenon, being sure to have the front edge perfectly true with the front edge of the spoke. Give about the same draft to the tenon as would be given were the spoke and hub new; ⅛ of an inch is enough for a tenon 1½ to 2¼ inches long, having all the bruised wood cut away from both the tenon and mortise the exact size to make the tenon after the new wood has been glued on. Select

good hard wood of the same kind as that of the spoke. If the spoke is badly bruised, so much so as to damage it above the shoulder, as shown by Fig. 175, as shown by jogged lines, A, Fig. 176, and the latter are solid, cut out of the face of the spoke, as shown by the dotted lines, B, as deep as may be necessary to reach the solid wood, doing this on both edges if necessary; then glue on strips, as shown by Fig. 177, which shows the side of the spoke with the shoulder cut away on one side that the position of the pieces glued on may be shown to a better advantage than if the spoke was shown complete. A, the side of the ten; the dotted line, B, shows the line of the shoulder; D and E, the face and back of the tenon after the bruised wood has been cut away. This is somewhat exaggerated in order to make the operation clear. The sections F and G show the blocks glued to the tenon with the ends cut up on the spoke above the shoulders, as shown by B, Fig. 176. Glue the blocks firmly, and when the glue is hard face off the front to the dotted line, H, which line is straight with the face of the spoke above it. Then dress off the back, cutting the tenon to the dotted line, K, which gives the draft for the tenon. Clean off the top of block, G, flush with the back of the spoke. The pieces, F and G, must be well fitted in at the tops, and when the clamps are tightened tap the bottom end with a hammer to insure their being up tight at the tops, M and N. After the blocks, F and G, are properly cleaned off drive in two or three 1-inch wire nails to insure the wood keeping its place when driving the spoke in. By the dotted lines, H and K, we have shown the tenon as it would be if the mortise has not been cut away beyond its natural lines. When this is the case the front and back edges of the tenons must be trimmed, as indicated by the dash lines, O and P. The amount of this projection can easily be determined by the spoke gauge, which will give the full width of the mortise.

Repairs of the character we have mentioned in this article are such as are warranted by the general good condition of the other portions of the wheel. When the whole wheel needs repair—that is, when every spoke is loose and the hub badly conditioned throughout—the best "repair" is a new wheel. There was a time when extensive repairs were warranted, but that time is past, as at the present time new wheels can be purchased at a price that makes the new cheaper than the repaired. It was not so when the wagon maker bored and mortised his hubs, hewed from the rough and rounded his spokes, but now if the wagon maker prefers he can purchase hubs, bored and mortised, spoke turned and tenoned at one end and short felloes sawed, ready for boring. Then, too, in regard to the repairs mentioned, they apply entirely to heavy wheels; light ones will require different treatment.

The Blacksmith & Wheelwright, Feb 1903

Wheel Repair

Some Practical Hints on Wagon Repair Work.

W. G. BRECKON.

The following hints have been gleaned from an experience extending over quite a number of years and are given to the craft in the hope that they will prove of value and benefit.

If when setting the tire on a light wheel, you should pull it to dish, stretch the tire cold by pounding on the edge with a fuller.

To tighten the box in a wheel do not wrap the box with canvas and drive in. It is sure to get loose again. Take a piece of pine board and split off a lot of very thin wedges. Place the box in the hub, get it centered, and fill up tightly between the box and the hub with the pine wedges. When you have driven in all the pine you can, get some hard wood wedges and wedge, as you would on a new job.

Sometimes a wheel comes to the shop with the tire spiked on. To drive the cold chisel between the rim and the tire is rather hard on the former. Find where the spike is, center punch it, put the wheel under the drill and bore the spike out.

If when taking a weld on large iron the fire is dirty, take some salt and throw into it. It will clean your fire and make it burn brightly.

When replacing an old rim with a new one don't drive the old rim off, as there is danger of disturbing the spokes in the hub. It is better to split the old rim off by driving wedges into the face of the rim, as shown at B.

Light delivery wagon wheels should be riveted each side of the spoke in the rim (C in the engraving) to prevent the tenon of the spoke from splitting it. Place the wheel on the wheel horse, face up, mark where the rivets come and bore with a screw bit, which the rivets will follow tight, and drive your rivets in. Take the wheel off the horse and screw it face down on the iron tire platform, place the burrs on, cut the rivets the right length and rivet them. Proceed with the other wheels in the same manner.

A wheel two feet six inches high, with a four-inch tire, is sometimes a troublesome thing to rim. By putting on rims two inches higher than the wheels you will overcome a lot of that trouble, as a bent rim will pull down easier than it will pull open.

Suppose that a wheel with a four-inch tire comes to the shop to have a couple of new spokes put in. The tire is tight on the rim and you do not wish

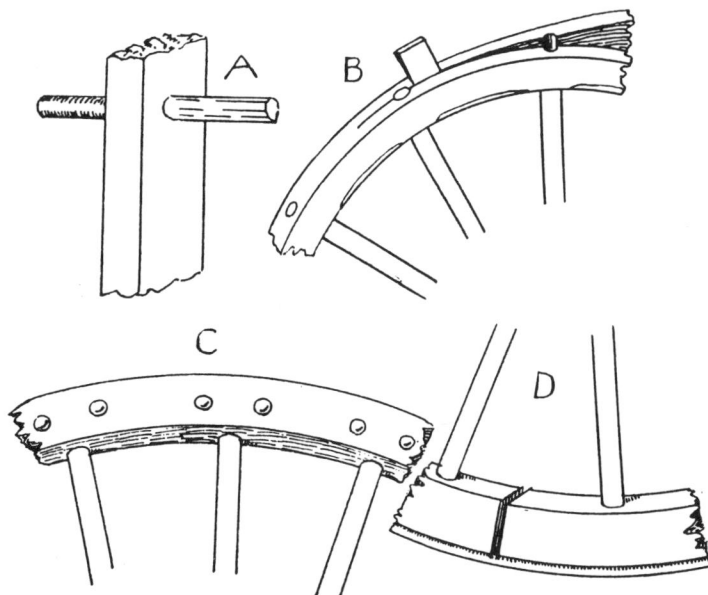

SOME PRACTICAL HINTS ON WAGON REPAIR WORK

to injure the wheel. Just take an old, coarse saw and saw a piece out of the joint, D, strike a few sharp blows on the sole of the tire and you will find that your tire will come off easily.

When the bands on a hub get loose do not take them off and cut and weld them; it is unnecessary. Make a lot of thin wooden wedges. Then take a thin iron chisel and drive into the hub pretty close to the band, pull out and drive one of the wedges in its place. Do this all around the hub and the band will be as tight as necessary.

The first point in making wheels is to select good material. The second is to have the wheel proportion proper. Be sure that your hub is large enough; it should be of such a size that it will allow the spoke to be in the hub as deep as it is wide, and to allow a space of one eighth inch between the spoke and the box. If the spokes touch the box they will soon become loose.

Always use a dodged hub. It is the best because the dodged spoke acts as a brace, and also the hub is not cut away so much in one place as when mortised straight. Drive the spokes so that the wheel will have one eighth dish. When the tire is set the wheel should not pull to dish. If it does, the spokes are bent or they have moved in the hub. Be very careful in mortising the hub. First, get the feet of the spokes all on taper, about one eighth inch for heavy wheels and one sixteenth for light ones. When the mortise is made it should be exactly the same shape as the foot of the spokes, but one eighth inch smaller endways, so that it will drive tight. Of

course, you have to be guided by the material in the hub as to how tight you set the spokes to drive. Take a hammer of suitable weight, dip the spoke in some very thin glue and drive it into the mortise. Do not drive with a wooden mallet, because you cannot drive a spoke as tight with it. Do not set the spoke too tight, for when it is driven it will cause a shoulder to form on the back edge, which will allow the wheel to pull to dish and the spokes will not remain tight. The next move is to cut the spokes off the required length.

When cutting the tenon on the spoke do not cut it straight with the spoke, for if you do when the rim is put on the joints will be hollow. Always cut the tenon a little back of, but straight across, the wheel. This will cause the joints to be high, so that when the wheel is screwed down on the platform it will bring the face of the wheel level. When the tenon is cut on the spoke, take a chisel and cut a shaving off the top and bottom. This will prevent it from splitting the rim. There should be a wedge driven in the tenon after the rim is on, to get it up to the shoulder of the spokes so as to enable the workman to cut the correct joint.

The American Blacksmith, June 1909

Wheel Repair

HOW TO FORCE SPOKES FROM A SARVEN HUB.

MR. EDITOR :—I am a jobber, and have a good many Sarven wheels to repair, which give me no small amount of trouble, it being impossible to pull out, pry out, or coax out a spoke tenon from the hub. Invariably they break off, and then I am obliged to dig them out the best I can. Is there any way, by fair means or foul, that a man can get out one out of fifty spokes without breaking off the tenon ?

By setting me in the way of accomplishing this feat, you will be entitled to my blessings, while greater favors will be paid for in corresponding proportions. Please answer, if you can, through *The Hub*, and oblige,

A CHICAGO JOBBER.

ANSWER.—Mr. Tom Sykes, who has had some experience in repairing Sarven wheels, sends *The Hub* the following reply to the above inquiry, prefaced with the statement that "Chicago Jobber," upon following his suggestions, will be enabled to remove without breaking, not only one out of fifty spokes, but, proper care being observed, forty-nine spokes out of fifty.

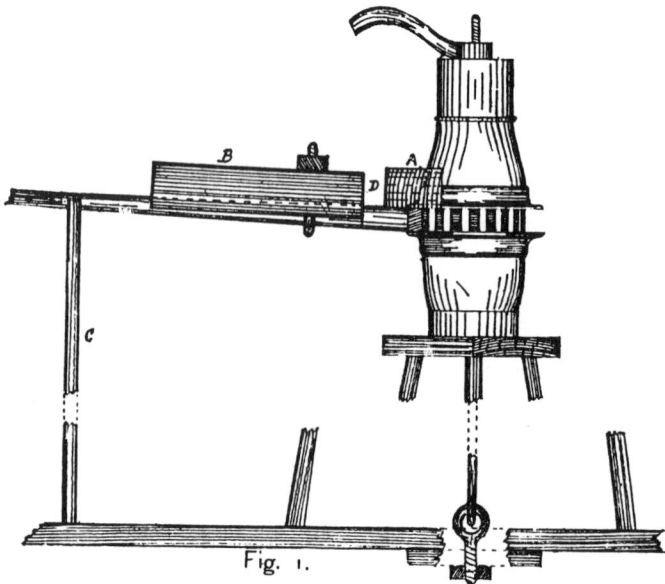

Fig. 1.

In Fig. 1 is shown a Sarven hub placed upon a wheel-horse or bench, securely fastened to the floor in the manner usually followed. One spoke remains in the hub, and is in proper position to receive the wedge by the aid of which the tenon to the spoke will be forced from the mortise in the hub.

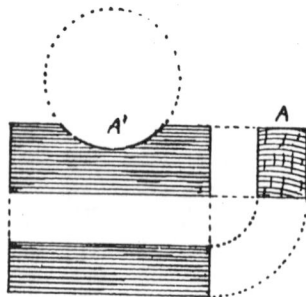

Fig. 2.

Fig. 2 presents three views of the piece A, Fig. 1. This is a block of hard wood, 12 in. long, 4 × 5 in. The side A is concaved to fit against the hub as shown at A', the dotted line representing the circumference of the hub.

In Fig. 3, A presents the side, B the end, and C the underside view of piece B, Fig. 1. This piece is of hard wood, about two feet long,

3 × 4 in. Through the center of the under-side runs a groove, worked out to fit the front edge of a spoke, as shown at B, B', with the upper side studded with short spikes.

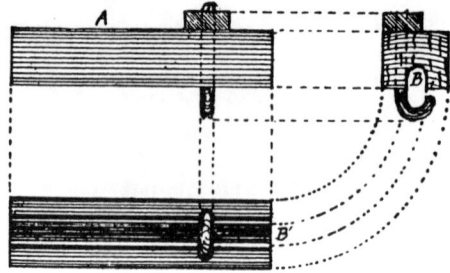

Fig. 3.

Fig. 4 is a larger view of this section, and the spikes can here be more plainly seen. Fig. 4 also shows more plainly the hook that clasps the under side of the spoke. This hook should be made of tough, hard half-inch iron, with the hook on the lower end, and a substantial thread cut on the upper end ; and it should be placed in the piece A, Fig. 3, about four inches from the end, as shown in the cut.

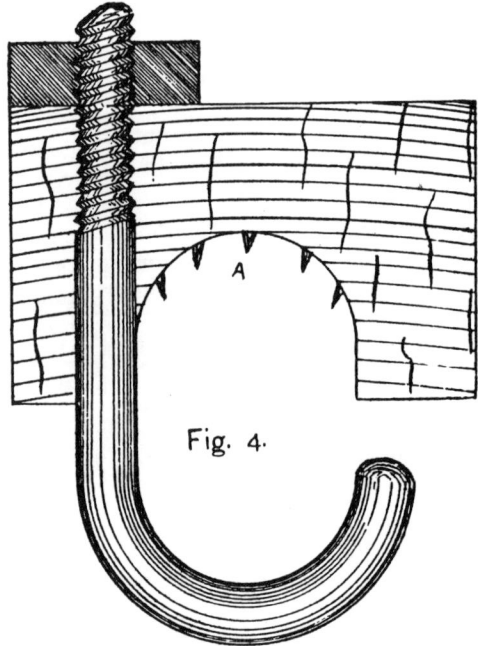

Fig. 4.

Fig. 5 is the wedge, 2 feet long, 3 × 2½ in. at the end A, and tapered as shown. It will be necessary to use for the wedge straight-grained wood, and the toughest wood that can be procured.

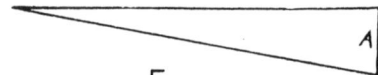

Fig. 5.

Place the wheel, and secure it upon the horse. Put near the end of the spoke the standard C, Fig. 1. Place the pieces A and B in position, leaving the opening between the two blocks, of 1 inch at D. Tighten the hook, and secure still more with thumb-screw to the spoke to the block. Coat the sides of the wedge with tallow, and insert the end in the opening D, and drive in the wedge.

The Hub, Dec 1882

220

Wheel Repair

REPAIRING SARVEN PATENT WHEELS.

———

ATLANTA, GA., Sept. 8, 1881.

EDITOR OF THE HUB—DEAR SIR : I wish to communicate through *The Hub* an idea, new to me, which I have just carried out successfully. It may be that others may have done the same before, but if so I have never heard of it.

It is a very generally expressed opinion that if the flanges become loose on the hubs and the spokes loose in the hubs, there is no remedy or way to fix them to make a good job. I recently had a set of wheels as described, and, at first thought, I was inclined to replace them with a new set ; but the spokes, rims and tires were all good, and all fitted, so I hesitated, and finally concluded to repair them ; which I did in the following manner :

I first took off the tires, cut the rivets in flanges and knocked them off the hubs. I then drew the spokes, canvassed the tenons, and drove them back with glue. Then I got a piece of tin, a little wider than the bearing of the flange on the hub, and banded the hub with thin tin, and nailed it on to keep it from slipping. Then I heated the flanges, to expand them a little, and drove them on, cooling them with water. I then put in the rivets, set the tire, and after testing the wheels with hard usage I find that they have never moved.

Hoping this may be of service to the craft,

I am yours truly,   N. C. SPENCE.

The Hub, Oct 1881

Replacing Spokes in a Sarven Wheel.

MILL VILLAGE, NOVA SCOTIA, Nov. 28, 1889.

MR. EDITOR :—I was asked the other day the proper way to put in a spoke or re-spoke a Sarven wheel. Would you please answer this question through your columns.   Yours,   C. A.

In putting one or two spokes in a Sarven wheel, remove the rims, cut those bolts going through the broken spokes, remove the spokes to be replaced. Cut the new spokes in size, and specially so at the square end the same as those removed, as they must fit exactly. Drive them in, and take care that they stand in the right position, neither back nor forward. Replace the bolts and the rest is the same as on other wheels. To respoke the whole wheel write to a manufacturer of Sarven wheels, state number and size for Sarven wheel spokes, if for front or back wheel, or outside diameter of wheel, which is better, as it does not pay to make them by hand.

The Carriage Monthly, Dec 1889

Repairing Sarven Patent Wheels.

E. A. BARNARD.

If you have one or more spokes to put in, take out all the rivets necessary, replace your spokes, set the tire and put your rivets in: If, however, you have all the spokes to put in, take out every other rivet, or in other words 4 rivets. Take out all the spokes, being careful to save 2 spokes for a pattern. Now take 2 new spokes, fit them exactly like the old ones, making them 1-32 of an inch thicker. Now put the 2 spokes in the vice and bore a hole in the center of the spokes, like in the old ones, use good glue and drive one on each side of a rivet. Do the same with the other rivets, until you have 8 spokes in pairs around the wheel. Then fit and drive the other 8 spokes in pairs and let the wheel stand over night, or until glue has set. Then put on the felloe and replace tire bolts and rivets and you will have a first-class job. It is expensive, however, to respoke a wheel all new. If customer does not feel like paying for that kind of a job, I try and find an old patent wheel, cut down to size, put on felloe and tire, and I get a good job and a good price for the job.

The American Blacksmith, Feb 1906

How to Fit Spokes for Sarven Hubs.

W. A. SHORT.

Cut all the rivets, then take out all the old spokes. After this is done straighten up the flanges in good shape, before driving any spokes. If there are any old spokes that can be used as a pattern I take one and fit all the spokes just like it. If I cannot get an old spoke, then I make a pattern thus: Draw a circle the same size as the hub where the shoulder of the spokes rests. Then measure your spoke from the shoulder to the widest part of the spoke as shown at A. Now supposing you have laid out a 3-inch circle, draw another circle outside of the first one, and as much larger as the length of A. Now divide the circle into 16 equal parts, and make your pattern the exact size as shown at B. This gives the face of spoke. Now make each spoke the exact size and shape of this pattern, and make the face of the spoke straight. Slope the back of the spoke about $\frac{1}{8}$ of an inch from shoulder to point. I use a spoke large enough to fill the flange full. After all the spokes have been fitted I then drive them as per numbers in the diagram; number two opposite number one and so on until eight are in. Then I fill each vacant place exactly, and am very careful not to let one spoke force another out or sideways. I use hot glue in driving, and after the spokes have all been driven in, I put a bolt through the hub from end to end, and tighten up the bolt with a large wrench, which brings the flanges up tight on the spokes, if they have been forced apart in the least. I now drive each spoke again, and then bore for the rivets. Use a $\frac{7}{16}$-inch bit for a $\frac{3}{32}$-inch rivet, and so on so that the rivet will be tight in the hole, and not bend in riveting. Cut the rivets $\frac{1}{4}$-inch longer than distance through hub, and use a rivet set to make a nice job.

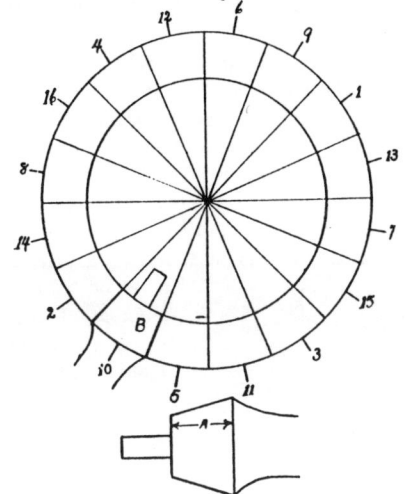

The American Blacksmith, May 1907

A Good Tire-Drag.

If there is anything with which we are displeased, it is the tearing to pieces with the tire-drag, in drawing on tires. While we were engaged practically at the business, we often noticed the tread of the rim jagged and torn by the indentations of the tire-drag. While such action has caused us many a bitter feeling, we are pleased at present in being able to present to our fellow craftsmen—while it may not be new—a very simple and effective tire-drag.

The drag is made as per *Fig.* 1. We first take a piece of good tough hickory, about two inches wide by one inch thick. Form the lever *C*, by rounding, leave the portion *D* the full size; then mortise out in the end left square, as shown at *B;* into this mortise *B*, we insert the drag hook (*Fig.* 2), which is held in position by means of a bolt passing through the hole at *A*.

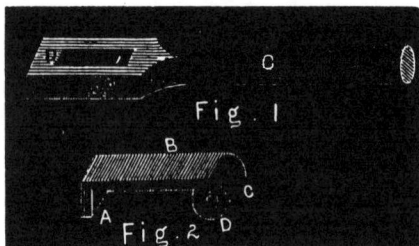

The drag hook (*Fig.* 2), is made of iron three-quarter inch wide by one-half inch thick, and should be of some solid or dense iron. First form the eye *C*, with the hole *D*, for the insertion of the securing bolt. Next place the hook in the recess, and mark off on the under side of *B*, at the end of *D*, (*Fig.* 1) for turning down the hook *A*, and leave the same about three-quarters of an inch long. The lever being made of wood prevents any marring of the rim of the wheel. We are confident this utensil will be gladly hailed by our brother tire setters.

The Carriage Monthly, Nov 1874

DESTROYING RIMS BY REASON OF BUR ON THE TIRES.

In our correspondence-box, we find one letter asking:

" I have experienced much trouble by having the rims of wheels torn at back portion by the bur on the inner surface of the tire. Can it be prevented? Please answer at my expense."

Editor's Answer.—We had thought that this old trouble had been done away with years ago, and for that reason we have not thought the subject worth mentioning before. We will now explain the cause, and prescribe the remedy.

The cause of the extreme amount of bur on the inner surface of the tire is the result of using an improperly shaped drill; which, after having passed through the inner surface of the tire

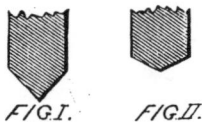

with its point, ceases direct cutting, and commences forcing the shell of metal into the softer material which the tire encircles. The drill is too pointed, and, instead of cutting through the crust of the inner surface, punctures it in a rough, jagged manner, thus causing the formation of the bur above mentioned. Fig. 1 represents in outline the shape of the drill which does the puncturing. Fig. 2 represents the drill which is best suited to cut a clean hole.

To bore through iron without leaving some little bur, requires what is termed a "center drill," which would not answer for boring tires, as it cuts out the inner surface in one piece, which would not allow of boring the wood, and if forced would result in splitting the rim. Use No. 2 drill, as per sketch, and you will be relieved of further trouble.

The Hub, Dec 1874

Piecing Tires.

From various causes, sections of tires have to be removed and new pieces inserted, which, in the main, is a perplexing job; therefore, with a view to making the task an easy one, we present the above illustration with the following explanation. *A*, *A*, represent sections of the intact portion of the tire; *B*, the piece to be inserted; *C*, and *D*, upper and lower sections of

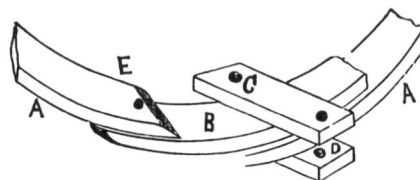

clamp securing section of tire and piece, the clamp consisting of two ordinary clip bars, and secured with bolts. To proceed, we first scaff the tire at one end, and next select a proper sized piece, and secure both with a rivet, as at *E*, and then secure the other end of tire and piece with the clamp. We then weld at *E*, and measure with traveler the tire, cut off scaff, and weld and fit. We have seen the above method employed many times, and from its simplicity and utility, cheerfully recommend its use in piecing tires.

The Carriage Monthly, Oct 1877

HOW TO REMOVE A TIGHT TIRE WITHOUT INJURY TO THE WHEEL.

———

Cincinnati, O., Sept. 23.

Editor of The Hub—Dear Sir: Will you have the kindness to add another chapter to your already numerous and valuable " How Papers," on the following subject. Occasionally we have some of our buggies sent to us, to replace a spoke or so which has become broken by accident. We experience much difficulty in getting the tire off without injuring the rim. We prefer not to cut the tire if we can remove it readily without. Please give us a receipt for the ailment, and oblige,   M.

Answer.—Our receipt for this trouble is a simple one. If the tire is heavy, make two or more pieces of iron, as per accompanying illustration, three-fourths of an inch thick and somewhat narrower than the tire.

Heat one of these pieces to a red heat, and lay it on the floor, concave side up; and on this hot iron place the tire, moving the wheel the length of the iron, at intervals of ten seconds. While one piece is being thus used, let another, or two other pieces, remain in the fire so as to replace the first one when it becomes cold. A few minutes will serve to expand the tire, when it may be removed with a tight pressure without injury to the rim.

Light tires may be removed still more readily, by placing the hub against a vertical post, and placing a piece of thin sheet-iron next the tire, to which hold a paint-burner, and cause the wheel to revolve rapidly. This requires but a few moments.

Neither of the above methods will injure the paint of the rim, providing ordinary care is observed.   N.

The Hub, Nov 1884

Wheel Repair

SIMPLE DEVICE FOR DRAWING RUSTY SCREWS FROM RIMS.

ANNOYANCE is often experienced in drawing old screws from rims after having worn away and rusted; but much of this can be obviated by use of the simple device represented in the accompanying cut, where A represents the tire; B the rim, with the screw inserted, the device being shown in the act of withdrawing the screw; and C a concave piece of iron, covered with leather, to prevent scratching the paint, and adjustable to almost any rim by turning the square-head screw.

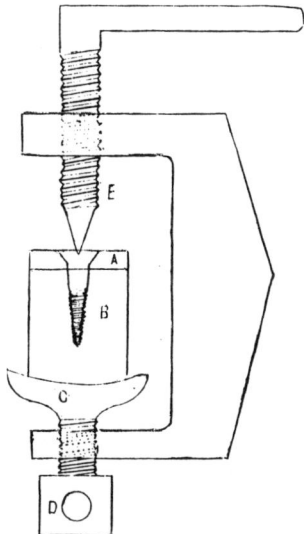

Clean out the slot in the screw-head as well as possible before adjusting the screw-driver E, which should be of hard and tenacious steel. The thread on the screw-driver should be No. 10, in order that the screw may not crowd it in turning out. The handle of the screw-driver need not be more than 6 in. long.

I have used the above-described arrangement for a good many years, and find it very successful in drawing at least 90 per cent. of old screws and I now present the idea to readers of *The Hub* for what it is worth.

When using the apparatus, tighten up the lower screw before turning the screw-driver, as one works independently of the other.

The Hub, Nov 1884   R. H. LEE.

HOW TO HOLD TIRE-BOLTS FAST.

WILMINGTON, O., Dec. 18, 1886.

EDITOR OF THE HUB—DEAR SIR: I am a constant reader of *The Hub*, and I often see inquiries regarding certain ways of doing work, and feel deeply interested in some of them. I see in the December number that some one by the name of P. T. S. wants to know how to prevent old tire-bolts from turning when removing the nut, and, as I have a less expensive way than that expressed in the answer, I thought I would send you a brief cut of the little tool, which any blacksmith can make, and if you deem it worth giving to the trade, I shall feel greatly repaid for my slight trouble.

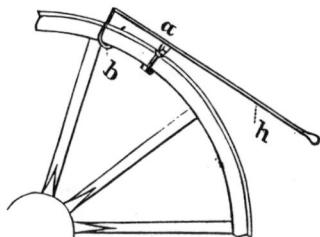

You can perhaps see my idea from this sketch of the wheel: *h* is the handle; *b* is the end turned down to clasp the rim; *a* represents a steel prong about 1½ in. in length, and 2½ in. from the end turned down; this prong is sharpened like a chisel, and rests on the head of the bolt. With the part turned down clasped over the rim, just press against the handle with your side and then you can turn any bolt. I have never found one that it would not hold.

Wishing *The Hub* the success in the future it has had in the past, I am, as ever, its faithful reader,   JAMES H. HART.

The Hub, Jan 1887

A SIMPLE MACHINE FOR REMOVING TIRES.

(See three Illustrations accompanying.)

EDITOR OF THE HUB—DEAR SIR: The question is often asked: "Can you give me a device for removing tires from painted wheels without injury to the paint?" To begin with, a painted wheel should never be struck with a hammer or mallet—every one will agree with me there. To do away with this a machine must be used. I have designed such a one and tested its merits for nearly a year. It may be made by

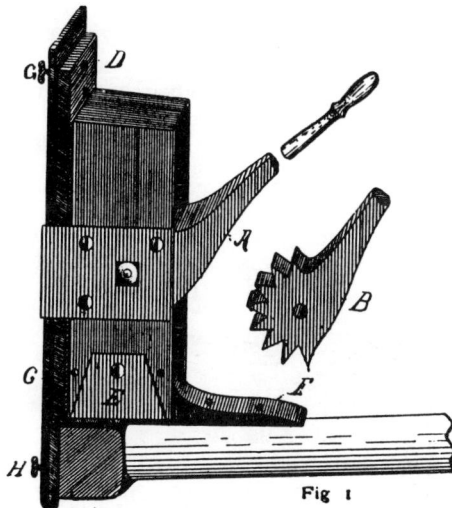

Fig. 1

any mechanic. It is designed for the benefit of the readers of *The Hub*, and I have numerous other devices which I propose to give them in future issues of this journal. Fig. 1 shows the machine in complete working order, 4½ in. high, 1¼ in. tire steel. This is what holds the lever and foot. A is the lever, 17 in. long and ⅝ in. thick, made of soft steel. B is the gear of the lever with ¼-in. cogs sawed in with an iron saw, and chipped out with a narrow chisel. C is the tire remover. D is the connecting part of C with cogs; this piece is to be ½ in. thick and cogged to fit the lever cogs. E shows the heel of the machine, covered with leather to prevent scratching the paint, fastened on with a small bolt. F, the foot of the machine, which is also covered

Fig. 2

Fig. 3

with leather, is riveted in through the uprights. This foot rests on the rim and spokes when removing tire. G shows the screw, which adjusts to different widths of tire. H shows the set-screw for different thicknesses of tire. Fig. 2, J is the back view of D. Fig. 3 shows back view of C, which makes the little machine complete. There should be two machines in a shop where heavy and light wheels are used. This size should be used for nothing over 1¼-in. tire, as the cog-wheel and lever are only 1¼ in. across. Yours truly,   D. E. B , Merrimac, Mass.

The Hub, Apr 1889

Smith Shop

PUTTING NEW RIVETS IN AN OLD WHEEL.

(See Illustrations Accompanying.)

THE blacksmith is often called upon to either tighten the rivets in a patent wheel or put in new ones, and to do either successfully he is compelled to remove the tire, because of the want

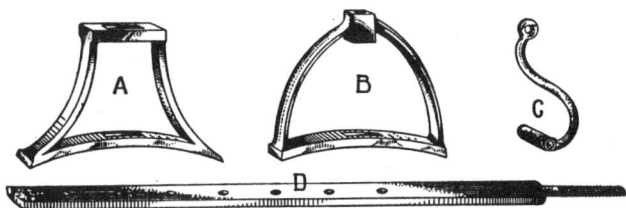

FIG. I.

of suitable appliances, and hours of valuable time are lost by neglecting to provide accessories for facilitating labor. This is the case in small shops, in particular the very places where the most complaints are made, of not being able to compete with the big houses. If a tithe of the time wasted in making sense-

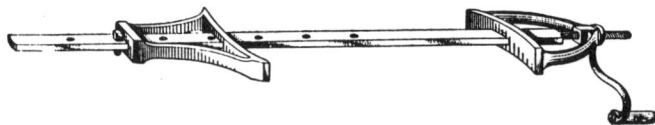

FIG. 2.

less complaints was spent in perfecting details in connection with shop appliances, the repair shop would be far more profitable than it is, but as it is so much easier to complain than to work, the air is full of fault-finding and the earnings small.

FIG. 3.

The tool here shown is one that would be handy in any shop. It is made in several parts, and is virtually a clamp specially fitted up for repairing wheels. Fig. 1 shows the parts detached, A the slide piece, B, the saddle, C, the crank and nut, D, the bar.

Fig. 2 shows the clamp together and ready for use, and Fig. 3 the clamp in position on the wheel.

If the spokes are very loose, it is better to take out the old rivets one at a time, and to do this take a body clamp, as per Fig. 2, place it on the wheel, as in Fig. 3, remove the old rivets, and tighten the screw so as to draw the spokes down solid; then put in the rivets, and be careful that they do not bend, for if they do they are spoiled; have a tool to fit in the anvil for the head of the rivet to rest in, and have a set tool for the back. Wheels treated this way have lasted for years, where if the rivets had simply been tightened they would have soon worked loose again.

The way a patent wheel is constructed it has very little of the wood left at the tenons, and the flange and rivets have to do the work. The light wheels are worse than the heavy ones about working loose, for they have less surface, and get more rough treatment in proportion than the heavy wheels.

The Hub, Jan 1890

The Old Fashioned Tire Trestle.

EDITOR OF THE CARRIAGE MONTHLY.—I wish to show the readers of the CARRIAGE MONTHLY one of the old fashioned tire trestles, as they were made by our grandfathers. They are still in existence in small shops at the present time, and are one of the necessary fixtures for setting tires. I will not go into the details of how to set tires, but how to make one of these trestles.

The illustration *A* represents one of the most simple tire setting trestles that can be made; *B* is the press screw, to hold the wheel in position. In this case, the nut guiding the thread is at the bottom of the trestle, but this can be reversed, the rod hooked into the bottom of the trestles with a large nut, having a surface of about 4 inches, and handles on each side. This is more convenient than a block of about 6 inches in diameter, with a smaller nut above the block. The length of this piece is about 6 feet; but for lighter work, where the diameter of the wheels does not exceed over 50 inches, the pieces are sufficiently long when 4 feet 6 inches.

For heavy work, the pieces are 3 by 9 or 12 inches, and about 6 feet long, while for light wheels, they are 2 by 7 or 8 inches. *C* and *D* illustrate one of the pieces, and how they are lapped together. *D*, one of the iron plates that cover the face of the trestle, bolted with 3/8-inch bolts through the four centers, and the rest with 1½-inch No. 16 screws. The iron plates for a light trestle, are 3/8 by 2 inches, and for a heavier one, ½ by 3 inches, and 2 inches long. No. 18 or 20 screws should be used.

The Carriage Monthly, Nov 1891

MARTIN'S COMBINED RIM, WRENCH AND BOLT CUTTER.

THIS machine is entirely new in design, construction and application. The need of such a machine is apparent to everyone, and its application will be seen and appreciated at sight.

Mr. Martin, the inventor and patentee, is a thoroughly practical and reliable blacksmith, with something over thirty years' continuous work at the forge. He states that he has searched the carriage and blacksmiths' journals for years, hoping to procure such a machine for his own use. The constant need of a machine of this character in the shop of a practical smith and mechanic, prompted its invention.

No doubt every carriage builder that makes any number of new jobs in a year, and every blacksmith that has large numbers of tires to reset during dry seasons, has longed for a machine that would expedite and facilitate the slow, laborious and tiresome work of putting in and taking out tire bolts. Especially is a machine or other appliance needed to aid in removing the nut when it is rusted fast and the bolt worn loose in the rim.

This machine should be securely lagged to the floor, when it serves as a wheel horse. The wheel is held firmly to place by the screw clamp as shown, when bolt holes may be bored in the rim and other repair work done.

In the use of the rim wrench and bolt clipper on new work, the bolts are placed in the rim and the nuts started in the usual way. The rim of the wheel is then placed against the front of the wrench " A," and held in position by the foot lever and chain, as shown, and the nuts are run on and tightened in less time than it takes to describe the operation.

The wrench is hinged or pivoted in the frame at the point where the bolt is seen, and is dropped or swung down out of the way when the bolt cutter is in use.

The bolt clipper " B," is one of the principal and desirable features of this machine. After the nuts are set, the wrench is dropped as described and the bolt ends are clipped in less than half the time with a hand clipper, and with less labor, as the operation is performed by a single motion of the foot on the treadle, as shown, leaving the hands free to handle the wheel.

The use of the machine on repair work is the same, with the addition of taking out the old bolts.

If the bolts are not worn loose in the rim so they "turn," the nuts are rapidly run off by applying the wrench as described. If they are loose and nuts rusted fast, as is so often the case, it is simply necessary to set the bolt holder up, and screw the square centre or chisel tight against the head, and the nut is easily displaced. As the nuts come off they pass through the hollow die and drop into the box on the frame, as shown, where they may be had if it is desirable to use the old bolts again.

This box is divided into apartments, and serves also as a receptacle for new bolts and such tools as are required about work of this kind.

An experienced blacksmith states that he once had an arrangement with a liveryman to reset his buggy tires at $2.00 per set and do all the work, or $1.00 per set if the liveryman would take out and put in the bolts.

Why not procure one of these machines and save the other dollar?

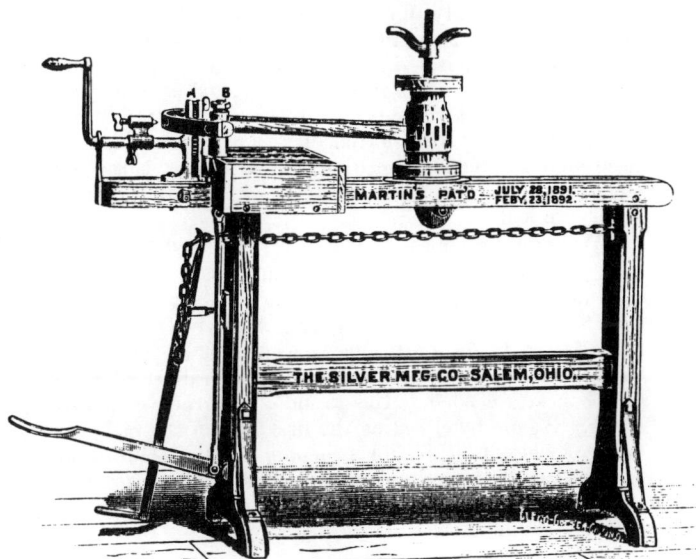

COMBINED RIM, WRENCH AND BOLT CUTTER.

A few testimonials following from prominent and experienced carriage builders are worthy of note.

TESTIMONIALS.

C. P. Kimball & Co., Carriage Builders,
Wabash Ave. and Harrison St.,　　　Chicago Dec. 15th, 1891.
Mr. Marshall Martin.

Dear Sir :—We noticed in THE HUB a short time since, that you had an arrangement for turning on tire bolts, and for cutting off the end of the bolt. We have seen something of this sort elsewhere, and if you have the machine perfected, would like to give it a trial, as we believe that such a machine properly built would save a great deal of time in the smith shop. Hoping to hear from you in the matter, we remain,

Very truly yours,　　C. P. KIMBALL & Co.
The Silver Manufacturing Co., Salem, O.

The Hub, June 1892

TIRE PULLERS.

BLACKSMITHS find it difficult to remove a tire from a wheel without injuring the tire or the rim. Very often, particularly in busy times, after repeated efforts, they are finally obliged to cut the tires. This necessitates rewelding—a loss of time, fuel and patience.

The Smith Giant Tire Puller, which we illustrate, will therefore be appreciated by every blacksmith. It has a powerful leverage, capable of grasping and removing any tire that is in condition to be reset, without the slightest injury to the wheel, and, with proper care, without marring the paint on new work. It can also be used for truing tire.

This tool is offered to the trade by S. D. Kimbark, Michigan-ave. and Lake-st., Chicago, Ill., at a very reasonable price. Although a new article, it has a large sale, and pleases all who use it. It is as useful as a tire upsetter, tire bender or tire bolter; it belongs to that family of machines; indeed, in practice it comes before them. This is the season when all such machines are in daily use in the shop. Mr. Kimbark can supply promptly any of them that may be wanted.

The Hub, July 1898

Wheel Shop.

RACK FOR RIMS AND HUBS.

WHEN down in Camden, New Jersey, a few days since, we visited the carriage shop of our friend, Charles S. Caffrey, and saw there the nicest arrangement for preserving bent rims in a proper shape until used, or for storing hubs, that it has ever been our fortune to inspect, being one of his own invention. At our request Mr. Caffrey has very kindly furnished us with the sketch from which we have had the accompanying engraving made. The entire figure is supposed to represent the floor of one of the stories—in this instance one of three—wide enough to take in several tiers of rims, arranged one above another—a bird's-eye-view of which is here presented—and still allow them to project one inch over the edges in the center at J. This is done so that they may be conveniently handled when required for use. The length of the rack may be extended to meet the requirements of any manufacturer, as will readily be seen.

The lines A and A define the outside edges of one of the horizontal floors—one of three—B and B being two imaginary lines in dots to show where the forms or standards c are to be placed for securing the ends of the rims. The width between the lines A and B is four inches, showing that the standards or pillars supporting the floors must be placed four inches from the outsides respectively. These standards answering also as forms for the ends of the rims at d and running laternly with the floors, are made of pine boards, fourteen inches wide and one inch in

thickness; with another at e standing at right angles with the first. These two —d and e—combined serve as pillars or standards for supporting the whole row or tiers of floors. This rack may be carried to any height — in this instance three stories — three feet apart, and one above the other. The width between the standard e is four inches shorter than the rims to be placed there are in diameter, in order to preserve the original shape, and also to secure them from any liability of straightening. For instance, the distance from e and e, each respectively, is four feet. This is designed to hold a four-feet four-inch rim, or rims, contracting them four inches at the points. It will readily be seen that, by this arrangement, there is not only the holding of the rims securely, as they lie piled one upon another, but they are also preserved in the shape in which they come from the bender's hand originally. Indeed, Mr. Caffrey assures us that he can take a badly formed rim and actually put it in good shape by placing it in his rack for a few days. A rack of this description, say nine feet high—three stories —two feet six inches wide and some twelve feet long, omitting the hub rack D, will hold about seventy-five sets of bent rims, greatly economizing room in cities where such is a desirable object. In addition to this, a single rim, when wanted, can be laid hold of with the most perfect ease. The foregoing description applies to the "institution" solely as a rack intended only for storing rims. We will now change it a little, so as to make it a rack for storing hubs.

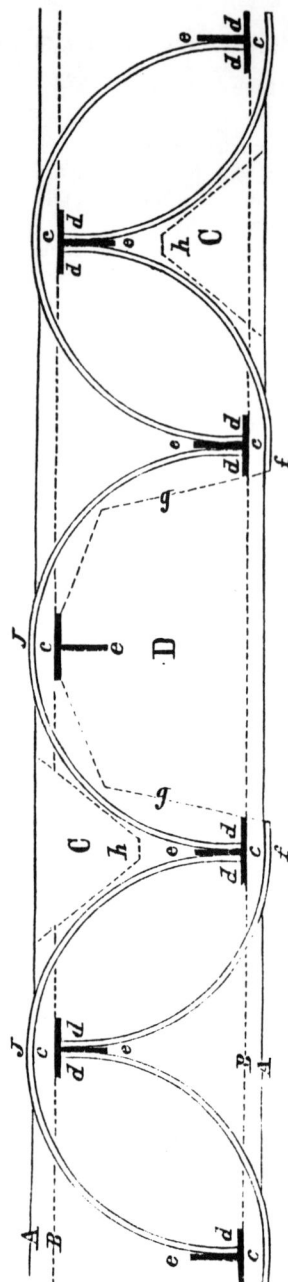

For this purpose—if much room is needed—we now cut in halves two tiers of rims at f and f, and use the recess D thus provided for a hub rack in the same frame. This is represented by the dotted lines g and g marking the location for a vertical position which forms a closet for the hubs. By this arrangement we economize space otherwise lost.

We would suggest here a little improvement of our own. Suppose we restore the cut-off rims at f and f and use the space left by the rims—as for example at C—and then form different hub racks, as indicated by the dotted lines h and h, for partitions surrounding it, where any such space occurs, by multiplying them in different parts of the rack, if needed. This would still further economize room. Again, by elevating the first floor of the rack, say twelve inches, facilities for sweeping out the shop would be afforded, and thereby cleanliness and order secured—a matter too often neglected in the workshops of most carriage manufactories.

BOX-SETTING MACHINE.

WE have received from Messrs. P. A. Fisher & Son, of Beardstown, Illinois, the plan of a Box-Setting Machine, with a description of the same, which they say they do not intend to patent, and therefore offer it freely to the trade, although it is one of their own invention.

In the drawing, the A's represent four base timbers, the B's the upright posts, and the C's the braces to stiffen them. D represents a shaft, with a heel or flange E, and F the arms attached to flange E, with slots in the ends for a thumb-screw, to which is added a hook at the end by which to clamp a wheel; and G shows the shaft or handle of the boring tool. We use the extension bit for

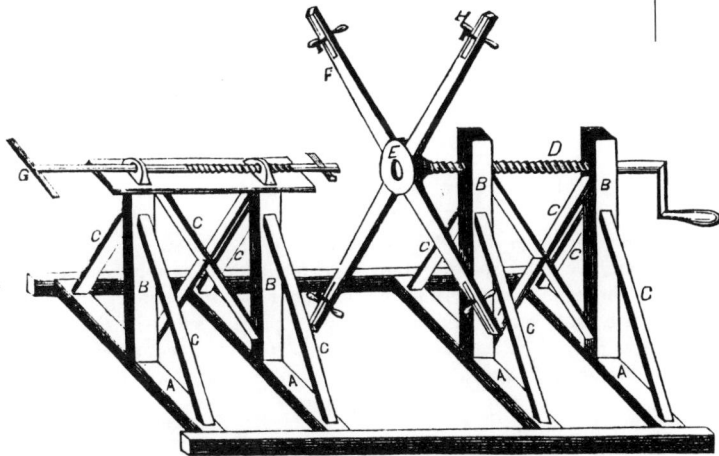

FISHER'S BOX-SETTING MACHINE.

boring. The flange E has an apperture for the hub of the wheel to set in. The hole in the hub being first plugged and the center found with the compasses, the wheel is afterward placed in the chuck and fastened with the thumb-screw H, and made true. This done, the wheel is next put in motion with the tool G to the centre, and the boring performed. We have bored hubs so correctly, that when the boxes were set in the wheels they were so true that they needed no wedging afterward. The machines that set boxes in the old way are not correct. The hub sometimes verges enough to throw the wheel two inches out of true, while ours sets by the verge which must be correct.

The Coach-makers' Magazine, Nov 1866

To Prevent Boxes from Working in the Hub.

NEW RUTLAND, ILL., July, 1875.

MR. EDITOR:—Seeing in the July number of the MONTHLY your reply to W. M. M. on preventing boxes from becoming loose, I am tempted to give my plan, which is as follows: After fitting the box in the hub, as advised in your article, mix a kind of paste composed of yellow ocher and glue, just thick enough to cover the box nicely. Spread this on the box, and drive it (the box) into the hub, and my word for it you will never be troubled with boxes becoming loose, as the paste will become almost as hard as steel.   TIMOTHY TUGMUTTON.

The Carriage Monthly, Aug 1875

Wheel Pit.

We present to our readers an illustration of a wheel pit, which our agent sketched at the shop of H. G. Crum, Tiffin, Ohio, and which we judge will be of interest to a large number of our readers. This pit is easily made. It consists of a base piece A, 2½ inches thick by 7 inches wide, and about 2 feet long. It is made heavy, so as to be steady; any piece of waste plank will do. Two upright pieces are framed to the base, and of sufficient length to support the highest wheel without the spokes coming in contact with the bottom. The piece B should be made of oak or very close-grained ash, in size about 1½ by 5 inches, and hollowed out on the top, as shown. To hold the hub in position,

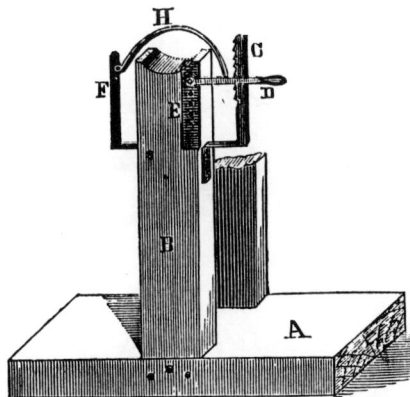

we make use of irons, having them the same on both uprights. C, F, are two braces, at which to fasten the lever D and band H. C is made with teeth, as shown, while F has seven holes, into which a pin passes for holding one end of band H. The lever D has a projection or lip on the inner side, which catches underneath the teeth when holding the band H down over the hub. The band H has a hook formed on the end, which hooks to a pin riveted to the lever D. When the hub is placed in the pit, the band H is pressed down over its end and hooks to the lever, and the lever is then forced down hard by the hand, and secured by the brace C. An iron plate (as at E) is let into the piece B, and which has a number of holes, into which the lever D is secured by a pin; these holes in the braces and plate are to regulate the band and lever, so that any size hub can be placed into the pit.

We have only shown a part of the back upright, as it is made the same as the forward upright piece.

The Carriage Monthly, Aug 1875

Wheel Repair

SIMPLE MODE OF GETTING DIAMETER OF WHEEL.

A, IN accompanying drawing, is a piece of whitewood, straight on front, with half-circle at one end, spanning the hub, as represented in the drawing.

In getting the diameter of wheel, measure from the center of the hub to tread of spoke.

In getting the length of spoke, set the two points F F at equal distances from the hub, marking any desired length on spoke;

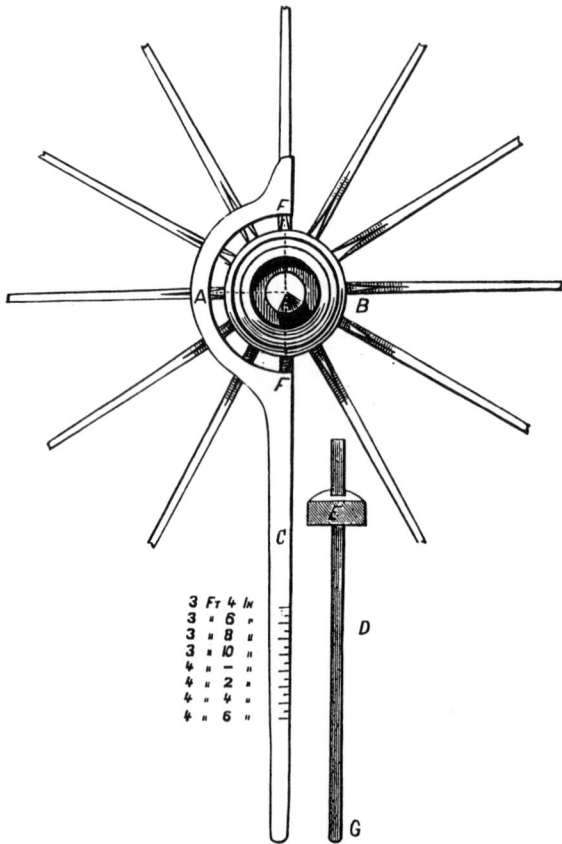

then the gauge D, with spur in the head, E, is set to the mark on the spoke, thus gauging all around.

I do not know that I have made my explanation very plain, but the contrivance is a simple one, which I hope the sketches accompanying will be able to show, even if my letter does not.

DAVID NORTON.

The Hub, Nov 1875

TRUEING GAUGE.

A VERY simple and reliable gauge for trueing axle-boxes may be made in the following manner.

Let the tapered spindle A, Fig. 1, be of proper size to move easily within the box; the crane B of sufficient length to reach from the hub

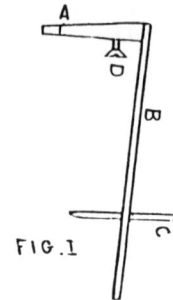

FIG.1

or spindle to the felloe of the largest wheel used. Fasten securely the crane to the spindle at a slight inclination, as represented in Fig. 1. Through the crane B make several holes ⅜ in. diameter, in one of which the pin C is placed. In this spindle place a screw D, allowing the head to project as represented, forming, as is intended, a rest for the spindle upon the end of the box. In use the gauge will appear as in Figs.

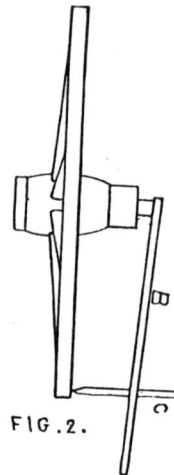

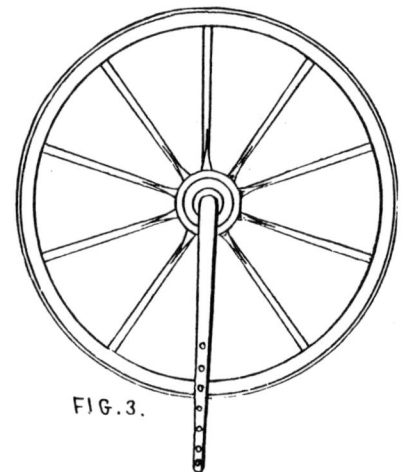

FIG.2.   FIG.3.

2 and 3. The pin being adjusted in the proper hole, and fastened at a required length to come in contact with the face of the felloe, the screw in the spindle will keep it at a proper height, while the weight of the arm of the crane will cause the spindle to press against the side of the box when in use.

TOM SYKES.

The Hub, Oct 1883

Wheel Repair

IMPROVED DEVICE FOR TRUING AXLE-BOXES.

In the October *Hub* a contrivance was illustrated for truing axle-boxes, which is a very old and clumsy device. I have a lively recollection of using just such a rig myself for awhile,—just long enough to make a better one, of which I now present a drawing.

This is very easily constructed, and is made exactly like the old one illustrated by Mr. Sykes, except as to the row of holes to accommodate the height of wheel, and the wooden pin which runs against the tire. Instead of the row of holes I use a long slot and sliding block, and instead of the pin, a long screw is introduced.

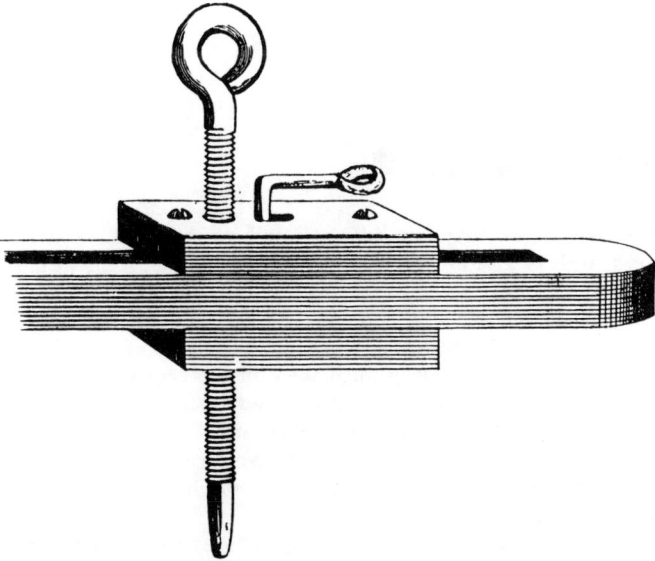

Fig. 2 is the top piece, showing holes for a screw or pin, and for the round-head screws which hold the block together. These screws go through the middle piece and into the bottom cap. This figure also shows the slot in which the lever moves.

Fig. 3 shows the middle block, sliding bolt, and spiral spring. The block is $\frac{1}{2}$ in. square and $2\frac{1}{2}$ in. long, and the bolt is $\frac{1}{2} \times \frac{5}{16}$ in., with $\frac{3}{16}$ in. tail, as shown. This leaves $\frac{3}{16}$ in. of wood under the bolt, through

which a hole is made to correspond with the hole in the bolt through which the lever passes. This hole in the wood is the bottom bearing for the lever. The hole in the bolt must be filed out a little oblong on the bottom, to allow free action of the lever.

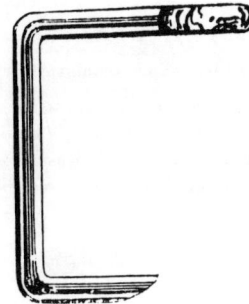

FIG. 1.

Fig. 1 shows the device complete. Near the end of the arm or crane, the block is held in position by its friction on top and bottom of the arm, and it can be made more or less tight by tension on the screws near the end of the block.

The screw is held in place by a spiral spring, and a bolt concealed within the middle section of the block, plainly shown in Fig. 3. This bolt has a thread in the end, to correspond with the thread of the screw into which it is forced by the spring.

If it is wished to raise or lower the screw more than a thread or two, by pressing on the thumb-piece or lever the sliding-bolt is drawn back and the screw released, and it will then move up and down at will. For a shorter distance, simply turn the screw.

The sliding block is made of three pieces of hickory, two of which are shown in Figs. 2 and 3, and the other is simply a bottom or cap of the same size as Fig. 2.

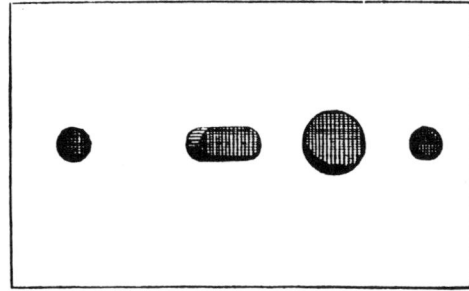

FIG. 4.

Fig. 4 shows the shape of the lever. It is made of $\frac{1}{8}$ in. brass or iron wire. A small recess is cut in the top of the bottom cap, to receive the lower elbow of this lever, which projects through the middle block. This prevents the lever from turning. The arm or trammel is made of $1\frac{1}{2} \times \frac{1}{2}$ in. ash. The slot is $\frac{1}{2}$ in., and of length to suit the maker.

I made one of these gauges in 1874, and it has been in constant use ever since. The best way to try a wheel is to place it on the axle-arm.

You will observe that, in Mr. Sykes's drawing, the gauge is used on the wrong end of the hub. It should be on the back of the hub.

GEO. W. KERR,
BRIDGEPORT, CT.   Foreman Bridgeport Cart Co.

The Hub, Jan 1884

FIG. 2.

Wheel Repair

A Simple Way to Make a Box for Tapering Tenons on Spokes.

MR. EDITOR :—The following I consider a good and simple way to make a box for tapering tenons on spokes : Take a piece of 1-inch thick ash, 2 inches wide and 3 feet long, two pieces of ¾-inch ash, 30 inches long, tapering from 2 inches wide at the narrow end to 3¾ inches at the broad end; rabbet down ¼ inch on top on the inside so that your jack plane will work freely, and screw the two tapering pieces against the first-mentioned piece on each end, leaving the bottom piece extend past the side pieces 4 inches on the back, or wide side, and 2 inches on the front. You now have a box ¾ inch on small end and 2⅛ inches on wide end, and you can taper any spoke from ¾ to 2 inches wide, as illustrated in *Fig. 1.* *A* represents the

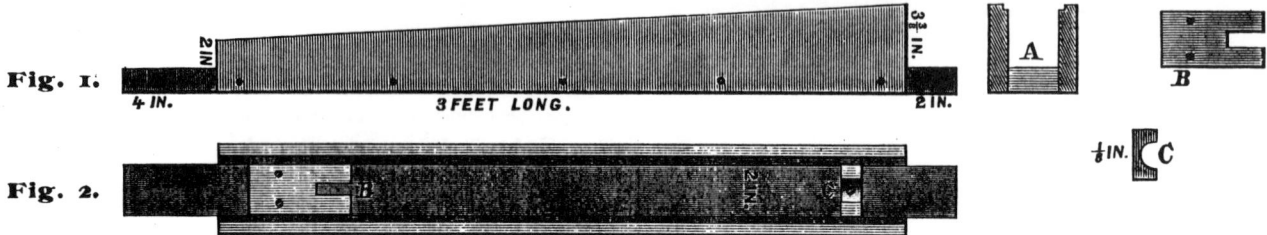

Fig. 1. 4 IN. 3 FEET LONG. 2 IN.

Fig. 2.

back view of the box; take block *B* and screw it to the box at *B*, *Fig.* 2, with two screws to whatever size spoke you want to taper; then screw block *C* to *C*, *Fig.* 2, from 12 to 18 inches from block *B*, so that when you put your tenon of the spoke in the slot in the block *B*, the body of the spoke will rest in the hollow of the block *C*. Block *C* should be about ⅛ inch thick in the center, so as to have your spoke rest on the end of the tenon and block *C*. It is necessary to have block *B* as a stop for the spokes, and *C* to raise the spokes, as sometimes you will find spokes that have sprung backward a little and yet not enough to make them useless, and by using both blocks you will have no trouble in tapering them.

Yours truly,   D. A. FISTHYLER.

The Carriage Monthly, June 1886

SPOKE TRESTLE.

(See Illustration accompanying.)

A FEW days ago, in making one of our usual about-town trips, and while in the wheel-room of one of our many factories, we saw a novelty which gave us considerable pleasure. It was a spoke trestle. Its duty was to carry or hold the spokes during their various manipulations. It was made with four corner posts of which A A A A, are the bottom sec-

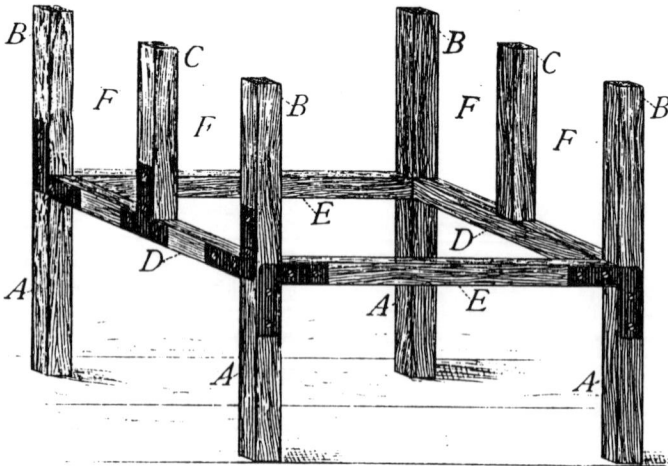

tions; B B B B, the top; D D, the end pieces framed in the corner posts; E E, the side pieces, also framed in the corner posts; C C, two upright posts framed centrally in the end pieces, D D, and giving the spaces F F F F, which keep the spokes separate. The corner posts and end pieces were secured with corner plates as shown in cut, made of 1¼ in. band-iron.

The Hub, June 1889

Hub Borer.

SPRINGFIELD, MASS., Jan. 31st , 1889.

MR. EDITOR :—I would like to ask you to let me know where I could buy a hub reamer, or what I would call an old fashioned hub borer. If you don't know, perhaps some of your subscribers would be kind enough to let me know.

Respectfully yours,   C. E. M.

These hub borers can be bought in hardware stores where they make specialties of coachmakers' tools, but this is only the case in large cities, and inconveniently out of the way for most of those who would want something of this kind. For those who cannot obtain a hub borer, give the blacksmith a piece of cutting steel ⅛-inch thick by 1½ inches wide, taper the front end to 1 inch wide, draw out a sharp edge, flatten the upper end to ¾ inch wide by bending it over to give strength to the end where the handle is put on. Now bend it the size and taper wanted, sharpened well, and you will have a hub borer the size you desire and will not cost much more than it cost in the hardware store and express charges.

The Carriage Monthly, June 1889

230

Wheel Repair

A Machine For Filling Wagon Wheels.

D. M. LOVE.

Now that spring is here and wheel work is commencing to come in, perhaps some of the craft would like a little help on filling wagon wheels. In the country shops, where all kinds of repairing is done, we have to economize on room as much as possible. We cannot have a place fitted up expressly for filling wheels as they have in large shops. Many of us get along without much of any place, although we must have a wheel jack on which to screw wheels while putting on rims. This jack can also be used for filling wheels if we have something to hold the spokes while driving.

The accompanying engraving shows an arrangement I have used for some time. I give it here, hoping it will be of use to some of the brothers. Take a plank 1½ by 8 inches by 2 feet, and taper one end down to three or four inches. Bore a hole in the wide end for the wheel bolt and place on top of jack. Bore a small hole in one corner for a pin to hold the plank from twisting around. Fasten a leg under the free end of plank, to hold it level with the jack. Now measure 18 inches from the wheel bolt and place a ⅝-inch T bolt with a long thread, so it can be screwed up or down to fit the wheel. While driving, the spoke will rest on the bolt. Now put an iron plate on the side of the jack with 2 or 3 notches in it to hold the end of the treadle. Run a rod from treadle up through the plank to hook over the spoke and hold it down on to the T bolt while driving. Then make a cone shaped block, to fit into a wheel, with a hole through its center for the wheel bolt. Then when hub is screwed down it has to turn true on the bolt. Two or three sizes of these blocks will fit all wheels.

A MACHINE FOR FILLING WAGON WHEELS

A Tool for Pulling Tenons.—Take a lag screw and weld it on a rod about as long as the spokes. Then weld a rod cross the end for a handle. Now take an old skein and slip it over the rod. Bore a hole in the tenon and turn the lag screw in tight. Now drive back on the skein. This is the best tool to pull broken spokes that I have ever seen. It may be what Louie Gillmann, Jr. of Tennessee is looking for.

G. S. O., Missouri.

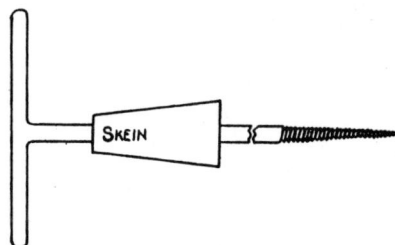

A TOOL FOR PULLING TENONS

The American Blacksmith, Feb 1906

The American Blacksmith, May 1912

Several Practical Helps for the Wagon Shop

Labor and Time Saving Tools and Devices Easily Made in the Shop

J. H. HAYDEN

IT isn't necessary, by any means, for the wagon and carriage repairman to put up with makeshift devices in his shop simply because he doesn't feel able to afford the modern machines now found on the market. There are so many machines and tools that can be made right in the shop that there is hardly any excuse for the wagon repairman to handicap himself with old-time, obsolete, makeshifts.

For example, take the old, wobbly wheel bench upon which the wheel wiggled and wobbled despite the efforts of the workmen with the added help of various sized bolts, nuts, wedges and what not. This old-time, temper strainer wasted more of the vehicle worker's time than a modern wheel bench would require for all the operations that can be done on it. Compare the old-time wheel holder with the device suggested in the engraving, Fig. 1.

This wheel support can be easily and quickly made from material readily found in the average general shop. The base of the wheel support proper is a large pipe flange. This is fitted with a piece of pipe of proper diameter and length and the top end of the pipe is fitted with a coupling. The bolt or threaded rod to go through the wheel hub is now fitted to go into the coupling at the top end of the standard. This may be done in several ways. After the coupling is run onto the pipe end a disk may be dropped into the coupling and onto the end of the pipe. The end of the bolt, B, in the engraving is then fitted with a nut or flange or if already fitted with a head so much the better. Now place the bolt, B, on the disk in the coupling and carefully center it and square it up plumb and straight. Then pour lead around it until it is held tightly in the end of the pipe coupling. Another method would be to center the bolt, B, in a pipe nipple to fit the coupling, pour lead

into the nipple to hold it in and then screw it into the coupling.

The other parts of the wheel holder are easily made and followed from the engraving, Fig. 1. The cones, marked C, C, are of wood, with a hole bored through the center of each so that they can be put on the rod, B. One of these cones is put on the rod permanently while the other one is removable for placing on the rod after the wheel has been placed on the stand. The second cone has a disk of tin or sheet metal fastened to its upper flat end so as to protect it from the nut, N, when this is screwed down on it to hold the wheel.

The nut, N, as will be seen, is fitted with a handle so as to be easily and quickly operated in fastening the wheel on the standard. The piece marked P, is a piece of heavy plank supported on one end by the pipe stand and at the other end by the support, S, which is made or ordinary pipe fittings as shown. The spoke holder, H, is a threaded rod with a Y forged on one end. The threaded end is allowed to turn down through a flange to allow for

different sized wheels. The rod, R, has a hook on its upper end and is operated by the foot lever, L, which is fastened as shown. To facilitate the movement of the rod, R, a coil spring, X, is placed on this rod. It is held in place by a key above it.

This device can be used for doing almost all wheel operations. The spoke holder, H, with the rod, R, will enable the workman to drive spoke accurately and easily. Then, too, this device holds the wheel steady while tenons are cut and during rimming operations.

A simple guage for use in connection with the wheel stand, is shown in Fig. 2. This spoke guage, placed on the rod, R, and held between the upper cone, C, and the nut, N, will enable the workman to accurately guage his spokes. The guage is cut out of a good piece of oak and trimmed down as shown. The hole at H is reinforced with a large washer or disk of sheet metal to save wear and to prevent the nut, N, of the wheel stand from cutting into the wood. The arm, A, is made to fit the long piece quite snugly so as to allow no free motion when the guage is set.

This guage arm is held in place by a thumb screw. The markings on the long arm of the guage enable the user of the guage to quickly adjust the small arm.

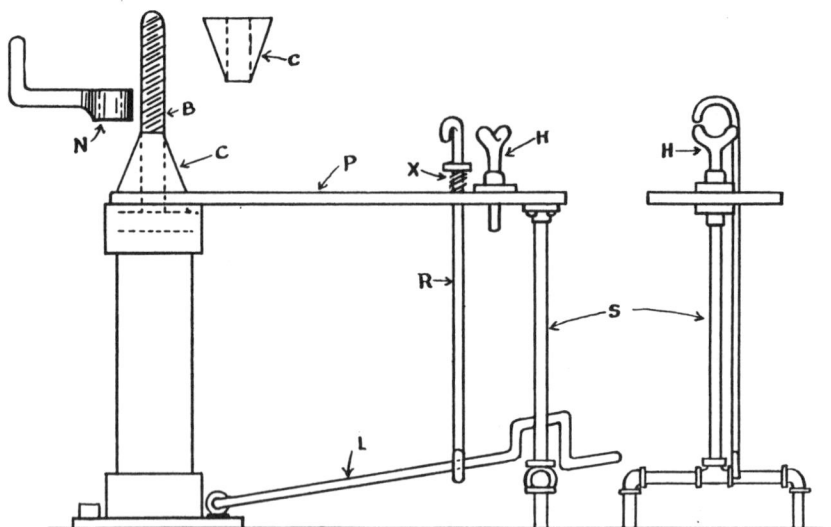

FIG. 1.—A MODERN WHEEL HOLDING MACHINE

The American Blacksmith, Feb 1916

X
TABLES AND CHARTS

233

TABLES AND CHARTS

This chapter represents the industry's attempt to standardize certain aspects of the wheelmaking process (e.g., dish of wheels, draught of tires). As noted in the text accompanying the charts and tables in this chapter, however, not everyone involved in wheelmaking was in agreement as to the wheel proportions and measurements given. Furthermore, many of the figures varied depending upon use factors. Still, these tables and charts, and the measurements contained in them, are as useful for comparison purposes today as they were a century ago. For example, one of the tables, p. 235, includes the distance between spokes for 3' and 5' wheels, while the Gribbon's chart on p.237 gives the amount of dodge or stagger of hub mortises, and the amount of dish for various types of vehicle wheels. These measurements can be used in wheelmaking today.

CIRCUMFERENCE OF WHEELS AND DISTANCE BETWEEN SPOKES.

MR. HENRY RICHARDSON, of York, England, has sent us the following table, showing the circumference of wheels from 3 to 5 feet diameter, and the distance of the spokes in the felloes :

Diameter.		Circumference.		No. of Spokes.	Distance in part in felloe.
Feet.	Inches.	Feet.	Inches.		
3	0	9	5	12	9½
3	1	9	8¼	12	9 8-9
3	2	9	11¼	12	10
3	3	10	2½	14	8¾
*3	4	10	5¾	14	*9
3	5	10	8¾	14	9¼
*3	6	11	0	14	*9½
3	7	11	3	14	9 9-14 15 9
3	8	11	6¼	14	9 6-7 15 9 1-5
3	9	11	9¼	14	10 1-11
*3	10	12	0¼	16	*9
3	11	12	3½	16	9¼
*4	0	12	6¾	16	*9½
4	1	12	10	16	9¾
4	2	13	1¼	16	9 13-16
4	3	13	4¼	16	10
4	4	13	7¼	16	10 1-5
4	5	13	10¼	16	10⅛
4	6	14	1½	16	10⅜
4	7	14	5	16	10 4-5
*4	8	14	8	16	*11
*4	9	14	11	18	*10
4	10	15	2	18	10 1-5
4	11	15	5	18	10⅛
5	0	15	8½	18	10½

"The first four, marked with stars, will be found the most useful sizes for four-wheeled carriages, and the last two for two-wheeled ditto." What have our American wheel-makers to say of these figures?

The Hub, Nov 1875

Tables and Charts

VALUABLE WHEEL TABLES.

Table No. 1: By H. M. DuBois.

(See List below on this page.)

WE are in constant receipt of letters inquiring our opinion as to suitable dimensions and proportions for wheels adapted to certain classes of vehicles. · Fragmentary answers to fragmentary questions of this class are always more or less unsatisfactory, and oftentimes even misleading. We consequently take special pleasure in presenting to our subscribers the accompanying list of wheel dimensions, carefully condensed and arranged in a form convenient for ready reference, which has kindly been prepared for us by Mr. H. M. DuBois, of Philadelphia, the well-known wheel-maker; and this will be followed in our next number by a second list of the same character, but longer and more detailed, prepared for us by Mr. John D. Gribbon, of New-York, principal of the Class in Carriage Drafting and Construction.

These valuable tables not only contain full answers to a multitude of questions, both asked and unasked, but afford the reader an opportunity to compare the opinions of two competent authorities; and they well deserve being closely studied and then carefully preserved for reference.

TABLE No. I.

DuBois Table of Comparative Proportions of Carriage Wheels,

Based on approximate carrying capacity of mathematically proportioned Spokes, and representative style of Vehicle for which they are popularly used.

Prepared for The Hub by H. M. DuBois, of Philadelphia.

*Lbs.	Hubs.	Spokes.	Number of Spokes.		Rims.	Tread.	Height.	Style of Vehicle for which the proportions given are suitable.	Approximate weight of finished job.
180	3¼ × 6 in.	⅞ in.		16	15-16 in.	¾ in.	4 ft. 8 in.	Track Sulky.	70 lbs.
275	3 × 5 in.	⅞ in. light.	14	14	⅞ in. light.	¾ in. full.	3 ft. 9 in. and 3 ft. 11 in.	One-man No-top Wagon.	100 "
350	3⅜ × 6 in.	⅞ in.	14	14	⅞ in. full.	⅞ in. light.	3 ft. 10 in. and 4 ft.	Two-men No-top Wagon.	190 "
425	3¼ × 6 in.	15-16 in.	14	14	1 in. light.	⅞ in.	3 ft. 9 in. and 4 ft.	Top Wagon, Side-bar.	225 "
500	3⅜ × 6 in.	1 in.	14	14	1 in.	⅞ in.	3 ft. 10 in. and 4 ft.	Top Wagon, Elliptic spring.	240 "
600	3½ × 6 in.	1 1-16 in.	14	14	1 1-16 in.	⅞ in.	3 ft. 8 in. and 3 ft. 11 in.	Top Wagon, Elliptic spring.	290 "
800	4 × 7 in.	1⅛ in.	14	14	1 3-16 in.	1 in.	3 ft. and 3 ft. 10 in.	Extension-top Phaeton.	550 "
800	4⅛ × 6 in.	1⅛ in.	12	12	1⅛ in.	1 in.	2 ft. 6 in. and 3 ft. 4 in.	Light Cabriolet Phaeton.	500 "
900	4¼ × 7 in.	1 3-16 in.	12	14	1¼ in.	1 in.	3 ft. 3 in. and 4 ft.	Extension-top Phaeton.	600 "
1000	4¼ × 7 in.	1¼ in.	14	14	1¼ in.	1 in.	3 ft. 8 in. and 4 ft. 2 in.	Physicians' Phaeton.	350 "
1000	4¼ × 7 in.	1¼ in.	12	14	1¼ in.	1⅛ in.	3 ft. 4 in. and 4 ft.	Four-passenger Phaeton.	700 "
1500	5 × 7 in.	1⅜ in.	12	14	1½ in.	1¼ in.	3 ft. 3 in. and 4 ft.	Six-passenger Phaeton.	900 "
1900	5½ × 7 in.	1½ in.	12	14	1⅝ in.	1¼ in.	3 ft. 2 in. and 3 ft. 9 in.	Light Barouche.	1100 "
1900	5⅝ × 7¾ in.	1½ in. heavy.	12	14	1⅝ in.	1¼ in.	3 ft. 2 in. and 4 ft.	Landau.	1400 "
2400	5¾ × 8 in.	1⅝ in.	12	14	1¾ in.	1⅜ in.	3 ft. 3 in. and 4 ft. 1 in.	Clarence.	1400 "
3800	7 × 9 in.	1¾ in.	12	14	2 in.	1½ in.	3 ft. 2 in. and 4 ft.	Landau.	1700 "
3000	7¼ × 9 in.	1⅞ in.		14	2 in.	1½ in.	4 ft. 4 in.	Dog-cart, 2 wheels.	

Left margin (rotated): • Approximate carrying capacity of Wheels in proportion to Spokes. Exclusive of weight of finished job.

NOTES.—1. The carrying capacity of wheels depends to a great extent upon their *height* and the general condition of the roads over which they travel; the approximate figures here given refer to a standard of proportioned spokes accurately tested, and the weights given are from these tests—which are within safe limits—in ordinary cases.

2. Height of wheels can never be absolutely given in tables, these heights depending upon the style of vehicle to be fitted. Heights here given are those oftenest used in regular orders.

3. In no case should an axle be used whose box or bush requires removal of wood from center of hub reducing the length of spoke tenons to *less* than the width of spoke at the shoulder, where it rests upon the hub.

The Hub, Apr 1882

GRIBBON'S WHEEL CHART,

SHOWING

Standard Proportions of Light and Heavy Wheels.

Prepared by John D. Gribbon, of New-York,

PRINCIPAL OF THE CLASS IN CARRIAGE DRAFTING AND CONSTRUCTION.

Style of Vehicle.	Approx. weight when finished.	Height of Wheels.		Hubs.	Spokes.	Number.	Dodge.	Rims.	Dish.	Tire.
	Lbs.	ft. in.	ft. in.	Inches.	Inches.		Inches.	Inches.	Inches.	Inches.
Sulky	90	4 8		3½ × 6	1	16	1	1	None.	¾ × ⅛
Sulky	70	4 8		3¼ × 6	⅞	16	⅞	⅞	None.	¾ × ⅛
Sulky	50	4 6		3 × 6	¾	16	¾	¾ 1-16	None.	⅝ × 3-32
Dog-Cart, two-wheel	....	4 6		6½ × 9 4⅛ × 5¼	1⅞ × 1 5-16 1½ × 1¼	14	⅜	2	¾	1½ × ½
Dog-Cart, two-wheel	....	4 7		6⅝ × 9 4¼ × 5⅜	1⅞ × 1 5-16	14	None.	2	¾	1½ × ½
Dog-Cart, two-wheel	....	4 8		6 × 9 4¼ × 5	1¾ × 1¼	14	None.	1¾	¾	1¼ × 7-16
Dog-Cart, two-wheel	....	4 8		6 × 8½ 4⅛ × 5	1¾ × 1 5-16	14	⅜	2	¾	1½ × ⅜
Dog-Cart, two-wheel	....	4 8		6 × 9 4¼ × 5	1⅞ × 1 5-16	14	⅜	1⅞	¾	1½ × ½
Dog-Cart, two-wheel	....	4 9		7½ × 9 5¼ × 6	2 × 1 5-16	14	None.	2⅛	¾	1½ × ½
Dog-Cart, two-wheel	....	4 9		6 × 9½ 4¼ × 5	1⅞ × 1¼	16	None.	2	¾	1¾ × ½
Dog-Cart, two-wheel	....	4 10		7¼ × 8 5¼ × 6½	1¾ × 1¼	14	None.	2	¾	1½ × ½
Dog-Cart, two-wheel	....	4 8		8 × 9 5¾ × 6½	2 × 1⅜	16	None.	2¼	¾	1¾ × ½
Dog-Cart, two-wheel	....	4 10		5½ × 8 4 × 5	1½ × 1	14	⅜	1⅝	⅜	1¼ × ⅜
Dog-Cart, two-wheel	....	5		6½ × 9 4¼ × 5	2 × 1 5-16	16	⅜	2⅛	¾	1½ × ½
Dog-Cart, two-wheel	....	5		5½ × 7½	1½ × 1	16	⅜	1½	½	1 × 5-16
Dog-Cart, two-wheel	....	4 6		4 × 6½	1¼ × ⅞	14	⅜	1¼	½	1 × ¼
No-top Wagon, one-man, full springs	85	3 10	3 11	2½ × 5	¾	12 12	¼	⅝	1-16	⅝ × 1-16
No-top Wagon, two men, full springs	175	3 11	4 1	3¼ × 6	⅞	14 14	¼	⅞	1-16	¾ × 3-16
No-top Wagon, half-springs, two men	190	3 11	4 1	3¼ × 6	⅞	14 14	¼	⅞	1-16	¾ × 3-16
No-top Wagon, half-springs, two men	150	3 11	4 1	3⅛ × 6	¾ 1-16	14 14	¼	⅞	1-16	¾ × 5-32
No-top Wagon, half-springs, two men	145	3 11	4 1	3⅛ × 6	¾ 1-16	14 14	¼	⅞	1-16	¾ × 5-32
No-top Wagon, half-springs, two men	115	3 10	4 0	3 × 6	¾	14 14	¼	⅞	1-16	¾ × ⅛
No-top Wagon, half-springs, two men	100	3 10	4 0	2⅝ × 5½	¾	12 14	¼	¾	1-16	⅝ × 3-32
Top Wagon, elliptic springs	290	3 8	3 11	3⅝ × 6	1 1-16	14 14	¼	1 1-16	1-16	⅞ × 3-16
Top Wagon, elliptic springs	260	3 8	3 11	3¼ × 6	1 1-16	14 14	¼	1 1-16	1-16	¾ × 3-16
Top Wagon, elliptic springs	240	3 8	3 11	3¼ × 6	1	14 14	¼	1	1-16	¾ × 3-16
Top Wagon, elliptic springs	215	3 9	4 0	3⅛ × 6	⅞ 1-16	14 14	¼	⅞ 1-16	1-16	¾ × ⅛
Top Wagon, half-springs and side-bars	250	3 10	4 1	3⅜ × 6	1	14 14	¼	1 1-16	1-16	⅞ × 3-16
Top Wagon, half-springs and side-bars	195	3 10	4 1	3⅛ × 6	⅞	14 14	¼	⅞	1-16	¾ × 5-32
Top Wagon, two men, full-springs	230	3 11	4 1	3⅜ × 6	1	14 14	¼	1	1-16	¾ × 3-16
Pony Phaeton	....	2 6	3 4	4½ 4⅝ × 6½	1⅛ × ¾ 15-16 × 13-16	10 12	¼	1¼	¼	1 × ¼
Physicians' Phaeton	350	3 8	4 2	4¼ × 7	1⅜	12 14	¼	1⅜	¼	1⅛ × ¼
Physicians' Phaeton	....	3 8	4 2	4⅛ × 7	1¼	14 14	¼	1 5-16	¼	1⅛ × ¼
Light Cab Phaeton	500	2 6	3 4	4¼ × 6	1⅛	12 12	¼	1⅛	¼	1 × 3-16
Light Cab Phaeton	....	2 10	3 6	4¼ × 6½	1 1-16	14 14	¼	1⅛	¼	1 × 3-16
Light Cab Phaeton	....	2 11	3 10	4 × 7	1¼	12 14	¼	1¼	¼	1 × 3-16
Light Cab Phaeton	....	2 11	3 9	4¼ × 7	1 5-16	10 12	¼	1 5-16	¼	1⅛ × ¼
Extension-top Phaeton	600	3 0	3 10	4 × 7	1⅛	12 14	¼	1 3-16	¼	1 × 3-16
Extension-top Phaeton	....	3 3	4 0	4⅛ × 7	1 3-16	12 14	¼	1¼	¼	1 × ¼
Extension-top Phaeton	....	3 4	4 1	4¼ × 7	1¼	12 14	¼	1¼	¼	1 × ¼
Four-passenger Phaeton	700	3 4	4 0	4¼ × 7	1¼	12 14	¼	1¼	¼	1⅛ × ¼
Four-passenger Phaeton, single top	....	3 10	4 1	4 × 7	1⅛	16	¼	1¼	¼	1 × ¼
Four-passenger Phaeton, no top	....	3 10	4 1	3¾ × 7	1, scant.*	14	¼	1⅛	¼	1 × 3-16
Four-passenger Phaeton, turn-out seat	....	3 6	4 0	3¾ × 7	1⅛	14	¼	1¼	¼	1 × 3-16
Six-passenger Phaeton	900	3 3	4 0	5 × 7½	1⅜	12 14	¼	1½, scant *	¼	1¼ × ¼
Six-passenger Phaeton	....	3 4	4 1	5 × 7½	1½, scant.*	12 14	¼	1½	¼	1¼ × 5-16
Coupé-Rockaway	960	3 2	4 1	5 5¼ × 7½	1½, light.	12 14	¼	1½	½	1¼ × ⅜
Six-passenger Rockaway	....	3 4	4 1	5 5¼ × 7½	1½, heavy.	12 14	¼	1⅝	½	1¼ × ¼
Six-passenger Phaeton	900	3 3	4 0	4⅞ 5 × 7½	1⅜, full.	12 14	¼	1½	⅜	1¼ × ¼
Four-passenger Coupé-Rockaway	....	3 2	4 1	4⅞ 5 × 7½	1⅜ × 15-16	12 14	¼	1½	⅜	1¼ × 5-16
Four-passenger Rockaway, curtain quarters	....	3 7	4 0	4½ 7	1¼ × ⅞	14 14	¼	1¼	⅛	1⅛ × ¼
T-Cart	....	3 0	3 11	4¼ 4½ × 7	1¼	10 12	⅜	1¼	¼	1⅛ × ¼
T-Cart	....	3 2	4 1	4½ 4¾ × 7	1⅜	10 12	⅜	1½, light.	⅜	1⅛ × ¼
T-Cart	....	3 3	4 2	4½ 4½ × 7	1⅜	12 14	None.	⅜, full.	⅜	1⅛ × ¼
T-Cart	....	3 3	4 2	4½ 4¾ × 7	1½	10 12	None.	1½, full.	½	1⅛ × ¼
Landau	1,770	3 3	4 1	7 × 9 7⅜ × 9	1¾ × 1 3-16, 1½ × 1⅛, full.	10 12	¼	2½, sawed.	1⅛	1½ × ½
Landau	1,450	3 3	4 1	6 × 8 6¼ × 8	1⅝ × 1⅛, 1⅝ × 1 1-16, full.	10 12	¼	1¾, bent.	⅞	1⅜ × 7-16
Landau	1,400	3 2	4 1	5¾ × 7¾ 6 × 8	1⅝ × 1⅛, 1⅝ × 1 1-16, light.	12 14	¼	1⅝	⅞	1¼ × 5-16
Landau	1,300	3 3	4 1	5¾ 5⅝ × 8	1⅝ × 1 1-16, 1 5-16 × 1 1-16	12 14	¼	1⅝	⅝	1⅜ × ⅜
Landau	1,300	3 3	4 0	5½ × 8 4¾ × 8	1½	10 12	¼	1⅝	⅝	1¼ × ⅜
Medium Size Landaulet	1,250	3 3	4 0	5¾ 6 × 7½	1⅝	10 12	¼	1¾	¾	1⅜ × ⅜
Small Size Landaulet	900	3 1	3 8	5¼ × 7½ 5½ × 7½	1½	10 12	¼	1⅝	⅝	1¼ × 5-16
Coupélet	....	3 2	3 11	5 5¼ × 7½	1⅜	10 12	¼	1½	½	1¼ × ¼
Full Clarence	1,300	3 3	4 1	5½ 5⅝ × 8	1⅝ × 1⅛ 1⅜ × 1 1-16	12 14	¼	1¾	1	1⅜ × ⅜
Medium Clarence	....	3 4	4 1	5⅝ 5¾ × 8	1¾ × 1 3-16 1½ × 1⅛	10 12	¼	1¾	1	1⅜ × ⅜ 7-16
Three-quarter Clarence	1,100	3 4	4 0	5¼ 5⅜ × 7½	1½ × 1 1¼ × 15-16	10 12	¼	1⅝, light.	¾	1⅛ × ¼ 5-16
Coach, close-quarter	1,480	3 4	4 2	6 6¼ × 8½	1¾ × 1 3-16 1½ × 1⅛	12 14	¼	1⅞	¾	1½ × ⅜ 7-16
Coach, curtain quarters	1,105	3 4	4 1	5⅛ 1¼ × 7½	1½	12 14	¼	1½	½	1⅛ × ¼
English Brougham*	1,000	3 0	3 8	5½ × 7½	1⅝ × 1⅛ 1⅜ × 1 1-16	10 12	¼	1⅝	1	1¼ × ⅜
English Brougham	....	3 0	3 8	6¼ 6¾ × 7½	1⅝	10 12	None.	1¾	½	1⅜ × ⅝
Light Coupé †	700	3 0	3 6	5¼ 5½ × 7½	1⅜ × ⅞ 1¼ × ⅞, light.	10 12	¼	1⅜	½	1⅛ × ¼
Light Barouche	1,040	3 2	3 9	5½ 5¾ × 7½	1½ × 1 1¼ × 15-16	10 12	¼	1⅝	½	1¼ × ¼
Heavy Barouche	....	3 3 or 4	4 1	5½ 5¾ × 8	1¾ × 1 3-16 1½ × 1⅛	10-12 12-14	¼	1¾	1	1⅜ × 5-16 ⅜

continued next page

Tables and Charts

Gribbons Wheel Chart
(continued from previous page)

Eight-spring Victoria	1,100	3	1	3 10	5¼	6 ×7¼	1¼	1½, full.	10	12	¼	1⅞	1	1⅜×⅜
Victoria	850	3	1	3 10	5¼	½×7½		1½, full.	10	12	¼	1¾, light.	¼	1⅜×5-16
Light Victoria		2 11		3 10	4⅞	¼×6½	1¼×⅞	1⅜×13-16	10	12	¼	1¼	¾	1 3-16×¼
Light Victoria		3 0		3 11	4¼	4⅛×7	1¼, light.		12	14	¼	1⅜	⅜	1⅜×¼
English Victoria	800	2 7		3 6	5¼	5½×7½	1½, heavy.		10	12	None.	1¼	¾	1⅜×⅜
French Victoria		2 11		3 8	5	5¼×7	1⅜, light.		10	12	None.	1½	½	1¼×¼
Victoria, extra heavy		2 11		3 8	5¼	5½×7	1½		10	12	None.	1⅝	½	1¼×¼

NOTES.—1. Heights given are not arbitrary, being governed by the conditions of the job to be got up.

* In some cases the above sizes have been much increased; the spokes, for instance, in a certain job have been made two inches, and the other parts in proportion, according to the fancy of the builder or customer.

† These sizes are increased up to 1⅝ inch spoke, according to the size and weight of the coupé, the swell-front coupé sometimes requiring the latter size, but seldom.

The Hub, May 1882

This letter refers to above chart

WHEELS ADAPTED FOR COUNTRY ROADS.

FLORIN, LAN. CO., PA., June 5th, 1882.

EDITOR OF THE HUB—DEAR SIR : I have been reading your *Hub* for three years past, and from the reading of it I have learned some good things, and in it I have seen some things that I thought were not entirely satisfactory, for instance, the " Wheel Chart " in the May number,—at least, as adapted to country roads. A wheel with only $\frac{1}{16}$ inch dish, has certainly no strength beyond simply the tenons of the spokes, and then only ¼ inch dodge.

This Chart will probably do for city use, but not for the country, where the roads are rough and uneven. A wheel with the dish and dodge given in the Chart would be of little use here in the country, with only a ⅞ or ¾ inch spoke; the spokes would soon break at the tenons, as there is nothing to hold them but simply the strength of the wood of the spoke; or if the spoke would not break, the tenons would soon get loose in the hubs.

A hub only 6 inches in length is not a very good thing, as it is much harder on a spindle than a hub of 7 inches. It may do very well on even roads, but not on our country roads. The hub acts on the spindle like a lever, and the shorter the fulcrum the more power you have; and so with short hubs.

My ideas for a light wheel are as follows : Hub, 7 inches long by 3¾ inches in diameter; 1 or 1$\frac{1}{16}$ inch spokes; rims, 1 or 1¼ deep by ⅞ inch tread; ⅜ inch dodge, and ¼ inch dish. I think a wheel of this kind will be much safer than to follow the Chart.

I would like to hear from some more country coach-makers on this subject, and if you think the above worth a place in your journal you are at liberty to use it; if not, you will know how to dispose of it.

Yours Respectfully,   DANIEL RUPP.

The Hub, Aug 1882

Tables and Charts

PROPOSED TABLES FOR GRADING SPOKES.

EDITOR OF THE HUB—DEAR SIR : I notice that you sometimes publish lists of prices for work, repair jobs and materials, all of which are of great interest and value to the trade, and the issues containing these price-lists are highly prized, and are generally preserved for reference.

I have, with considerable care, prepared what I consider a *consistent* series of tables for giving prices of spokes, which I think is the first and only thing of the kind ever offered to the trade. There ought to be established a uniform, systematic and consistent price-list for spokes, as well as for all other staple articles, and such a list published by such an authority as *The Hub* would go a great way towards establishing a standard. I beg to suggest the following :

HICKORY SPOKES, 26 INCHES LONG, 60 IN SET.

SIZE.	¾	⅞	1 in.	1⅛	1¼	1⅜	1½	1⅝	1¾	1⅞	2 in.	2⅛	2¼
XXX ..	12 c.	10½	9	9	9	9½	10	10½	11	11½	12	12½	13
XX....	10 "	8	8	8	8	8½	9	9	9½	9½	10	11	12
X......	8 "	7	7	7	7	7	7	7½	8	8	9	10	11
No. 1..	6 "	5	5	5	5	5	5½	6	6		• 7	8	9
No. 2..	4 "	3	3	3	3	3	3½	4	4		5	6	7

WHITE OAK WAGON SPOKES, LONG AND SHORT, 52 IN SET.

SIZE.	1½	1⅝	1¾	1⅞	2 in.	2⅛	2¼	2⅜	2½	2⅝	2¾	3 in.	3¼
XXX...	9 c.	9½	10	10½	11	11½	12	12½	13	14	15	16	17
XX....	8 "	8½	9	9½	10	10½	11	11½	12	13	14	15	16
X	7 "	7½	8	8½	9	9½	10	10½	11	11½	12	13	14
No. 1..	5 "	5½	6	6½	7	7½	8	8½	9	9½	10	11	12
No. 2..	3 "	3½	4	4½	5	5½	6	6½	7	7½	8	9	10

WHITE-OAK SPOKES, 27 INCHES LONG, 28 IN BUNDLE.

SIZE.	1½	1⅝	1¾	1⅞	2 in.	2⅛	2¼	2⅜	2½	2⅝	2¾	3 in.	3¼	3½
XXX..	9 c.	9½	10	10½	11	12	13	14	15	16	18	20	24	27
XX....	8 "	8½	9	9½	10	11	12	13	14	15	16	18	21	24
X	7 "	7½	8	8½	9	9½	10½	11½	12	13	14	16	18	21
No. 1..	5 "	5½	6	6½	7	7½	8	8½	9	10	11	12	13	14
No. 2..	3 "	3½	4	4½	5	5½	6	6½	7	7½	8	9	10	11

EXPLANATORY NOTES.—In the foregoing lists, the first grade, marked XXX, are spokes absolutely perfect and free from sap, and all white. The second grade, marked XX, is the first-class standard quality, the best, with some sap. The third grade, marked X, includes first-class spokes nearly as good as XX. Fourth grade, marked No. 1, soft timber, but perfect in every other respect. Fifth grade, marked No. 2, imperfect, but every spoke fit to use. Culls are sold by the lot for what they will bring.

If you think the accompanying lists will be of service or interest to your subscribers, you may feel at perfect liberty to publish them.

Respectfully yours, JOHN KLAER.

The Hub, Jan 1883

Draught for Light and Heavy Wheels.

We have from time to time received many inquiries in regard to tire-setting, the best way of bending and heating, amount of draught, and the methods employed in setting them. The cheapest and best way of bending the various sizes of steel or iron is to have a good tire-bender, suitable for the size tires to be bent. If for light work only, one of the lightest sizes is sufficient; but when both light and heavy work are made, it is necessary to have one for heavy tires also.

Welding steel tires is a practice in which perfection can be attained only by experience; but, with a little help from a brother blacksmith who has had some experience, and with the aid of some welding compound, any smith can soon succeed in making a good weld. In heating the tires, it should be done uniformly; and the smith will soon be able to judge the expansion obtained with a certain degree of heat. Soaking the wheel in water after the tire is on, is another bad practice, good workmen letting them cool off gradually, and as a result they are more uniform and regular. An examination of a wheel that has been cooled too suddenly, will show daylight shining through in some places between the tire and rim, the open spaces in some cases being sufficiently large to allow running the blade of a knife through; and other places will be cramped, or, in other words, the rim forced either sidewise or downward.

CARRIAGE WHEELS.

AMOUNT OF DRAUGHT ON PLAIN WHEELS FOR STEEL TIRES.

Sizes of Steel Tires.	Diameter of Front Wheels.	Amount of Draught.	Diameter of Back Wheels.	Amount of Draught.
5/8 x 2/32	3 ft. 6 in.	2/32 in.	4 ft.	2/32 in. scant.
11/16 x 3/32	“ “	3/32 “	“	2/32 “
3/4 x 1/8	“ “	1/8 “	“	3/32 “ scant.
13/16 x 5/32	“ “	5/32 “	“	1/8 “
7/8 x 6/32	“ “	6/32 “	“	1/8 “ full.
15/16 x 7/32	“ “	7/32 “	“	5/32 “
1 x 1/4	“ “	1/4 “ full.	“	5/32 “ full.
1 1/8 x 9/32	“ “	9/32 “	“	6/32 “
1 1/4 x 10/32	“ “	10/32 “	“	7/32 “
1 3/8 x 3/8	“ “	11/32 “	“	1/4 “
1 1/2 x 14/32	“ “	3/8 “	“	9/32 “
1 5/8 x 1/2	“ “	13/32 “	“	10/32 “
1 3/4 x 18/32	“ “	14/32 “	“	11/32 “
1 7/8 x 5/8	“ “	15/32 “	“	3/8 “
2 x 22/32	“ “	1/2 “	“	13/32 “

AMOUNT OF DRAUGHT ON PATENT WHEELS FOR STEEL TIRES.

Sizes of Steel Tires.	Diameter of Front Wheels.	Amount of Draught.	Diameter of Back Wheels.	Amount of Draught.
5/8 x 2/32	3 ft. 6 in.	2/32 in.	4 ft.	2/32 in. scant.
11/16 x 3/32	“ “	2/32 “ full.	“	4/32 “
3/4 x 1/8	“ “	4/32 “	“	3/32 “ scant.
13/16 x 5/32	“ “	1/8 “ scant.	“	3/32 “
7/8 x 6/32	“ “	1/8 “	“	3/32 “ full.
15/16 x 7/32	“ “	5/32 “	“	1/8 “
1 x 1/4	“ “	6/32 “	“	1/8 “ full.
1 1/8 x 9/32	“ “	7/32 “	“	5/32 “
1 1/4 x 10/32	“ “	1/4 “	“	5/32 “ full.
1 3/8 x 3/8	“ “	9/32 “	“	6/32 “
1 1/2 x 14/32	“ “	10/32 “	“	7/32 “
1 5/8 x 1/2	“ “	11/32 “	“	1/4 “
1 3/4 x 18/32	“ “	3/8 “	“	9/32 “
1 7/8 x 5/8	“ “	13/32 “	“	10/32 “
2 x 22/32	“ “	14/32 “	“	11/32 “

continued next page

A very important subject is the draught to be given to both iron and steel tires, plain or patent wheels, of various sizes and diameters. Good judgment on the part of the smith is a necessary requisite in all cases. One wheel will have more dish than another; the wood in one wheel may be softer than the other, and the softer the wood is, the more it will give, and the more draught be required. A wheel having less dish than another, requires more draught. This will show the importance the amount of draught assumes in relation to the various conditions of wheels. This is more difficult to determine in re-tiring old wheels; some have more or less dish, and may also be loose in the hub, while others dish backward.

As an instance, take a plain wheel 4 feet in diameter, $\frac{5}{8}$ inch tread, steel tire, $\frac{5}{8}$ inch wide by $\frac{1}{8}$ inch thick; if the wheel is well made, from good timber, the front spokes driven straight, with no opening in the rim-joints, and the rims well down to the shoulders, we give scant $\frac{1}{16}$ inch draught. If the diameter of front wheels is 3 feet 8 inches, and the other conditions are the same as above, the draught

would be $\frac{1}{8}$ inch; that is, the tire should be made $\frac{1}{16}$ inch smaller than the circumference of the wheel. Suppose we have a heavier wheel, plain hub, tread of rim $1\frac{7}{8}$ inches, tire 2 inches wide by $\frac{11}{32}$. The front wheels, 3 feet 6 inches in diameter, would require $\frac{1}{2}$ inch draught; and the back wheels, 4 feet in diameter, only $\frac{11}{32}$ inch draught. The smaller the wheel is in diameter, the stiffer it is, and the more draught is required. A sulky wheel the same size as published in plate

No. 55 of the fashion-plates of the October number, tire $\frac{5}{8}$ by $\frac{1}{8}$ inch steel, and the diameter 4 feet 6 inches, would require only $\frac{3}{4}$ inch draught.

To make our explanation more clear, we give with this article a mathematical scale for the various amounts of draught on different sizes and diameters of wheels. The first table is for carriage-wheels with steel tires, plain and patent, 3 feet 6 inches diameter for the front and 4 feet for the back wheels. The second table is for wagon-wheels with iron tires, plain and patent, 3 feet 2 inches diameter for the front, and 4 feet 2 inches diameter for the back wheels.

WAGON WHEELS.

AMOUNT OF DRAUGHT ON PLAIN WHEELS FOR IRON TIRES.					AMOUNT OF DRAUGHT FOR PATENT WHEELS FOR IRON TIRES.				
Sizes of Iron Tires.	Diameter of Front Wheels.	Amount of Draught.	Diameter of Back Wheels.	Amount of Draught.	Sizes of Iron Tires.	Diameter of Front Wheels.	Amount of Draught.	Diameter for Back Wheels.	Amount of Draught.
$1\frac{1}{4}$ x $\frac{6}{16}$	3 ft. 2 in.	$\frac{10}{32}$ in.	4 ft. 2 in.	$\frac{1}{4}$ in.	$1\frac{1}{4}$ x $\frac{6}{16}$	3 ft. 2 in.	$\frac{7}{32}$ in.	4 ft. 2 in.	$\frac{4}{32}$ in.
$1\frac{3}{8}$ x $\frac{7}{16}$	" "	$\frac{11}{32}$ "	" "	$\frac{9}{32}$ "	$1\frac{3}{8}$ x $\frac{7}{16}$	" "	$\frac{1}{4}$ "	" "	$\frac{5}{32}$ "
$1\frac{1}{2}$ x $\frac{8}{16}$	" "	$\frac{3}{8}$ "	" "	$\frac{10}{32}$ "	$1\frac{1}{2}$ x $\frac{8}{16}$	" "	$\frac{9}{32}$ "	" "	$\frac{3}{32}$ "
$1\frac{5}{8}$ x $\frac{9}{16}$	" "	$\frac{13}{32}$ "	" "	$\frac{11}{32}$ "	$1\frac{5}{8}$ x $\frac{9}{16}$	" "	$\frac{10}{32}$ "	" "	$\frac{1}{4}$ "
$1\frac{3}{4}$ x $\frac{10}{16}$	" "	$\frac{14}{32}$ "	" "	$\frac{3}{8}$ "	$1\frac{3}{4}$ x $\frac{10}{16}$	" "	$\frac{11}{32}$ "	" "	$\frac{9}{32}$ "
$1\frac{7}{8}$ x $\frac{11}{16}$	" "	$\frac{15}{32}$ "	" "	$\frac{13}{32}$ "	$1\frac{7}{8}$ x $\frac{11}{16}$	" "	$\frac{3}{8}$ "	" "	$\frac{11}{32}$ "
2 x $\frac{12}{16}$	" "	$\frac{1}{2}$ "	" "	$\frac{14}{32}$ "	2 x $\frac{12}{16}$	" "	$\frac{14}{32}$ "	" "	$\frac{3}{8}$ "
$2\frac{1}{8}$ x $\frac{13}{16}$	" "	$\frac{17}{32}$ "	" "	$\frac{15}{32}$ "	$2\frac{1}{8}$ x $\frac{13}{16}$	" "	$\frac{15}{32}$ "	" "	$\frac{13}{32}$ "
$2\frac{1}{4}$ x $\frac{14}{16}$	" "	$\frac{19}{32}$ "	" "	$\frac{1}{2}$ "	$2\frac{1}{4}$ x $\frac{14}{16}$	" "	$\frac{16}{32}$ "	" "	$\frac{14}{32}$ "
$2\frac{3}{8}$ x $\frac{15}{16}$	" "	$\frac{20}{32}$ "	" "	$\frac{17}{32}$ "	$2\frac{3}{8}$ x $\frac{15}{16}$	" "	$\frac{1}{2}$ "	" "	$\frac{15}{32}$ "
$2\frac{1}{2}$ x 1	" "	$\frac{5}{8}$ "	" "	$\frac{18}{32}$ "	$2\frac{1}{2}$ x 1	" "	$\frac{17}{32}$ "	" "	$\frac{16}{32}$ "
$2\frac{5}{8}$ x $1\frac{1}{16}$	" "	$\frac{21}{32}$ "	" "	$\frac{19}{32}$ "	$2\frac{5}{8}$ x $1\frac{1}{16}$	" "	$\frac{18}{32}$ "	" "	$\frac{1}{2}$ "
$2\frac{3}{4}$ x $1\frac{2}{16}$	" "	$\frac{22}{32}$ "	" "	$\frac{5}{8}$ "	$2\frac{3}{4}$ x $1\frac{2}{16}$	" "	$\frac{19}{32}$ "	" "	$\frac{17}{32}$ "
$2\frac{7}{8}$ x $1\frac{3}{16}$	" "	$\frac{23}{32}$ "	" "	$\frac{21}{32}$ "	$2\frac{7}{8}$ x $1\frac{3}{16}$	" "	$\frac{5}{8}$ "	" "	$\frac{18}{32}$ "
3 x $1\frac{4}{16}$	" "	$\frac{3}{4}$ "	" "	$\frac{22}{32}$ "	3 x $1\frac{4}{16}$	" "	$\frac{21}{32}$ "	" "	$\frac{19}{32}$ "

TABLE OF AVERAGE DIMENSIONS OF ENGLISH CARRIAGE WHEELS.

FROM a serial article entitled "Treatise on the Manufacture of Carriage Wheels," by an anonymous writer signing himself "Old Spoke," which has recently appeared in the London *Coach-Builder*, we quote the following interesting table of average dimensions of English carriage wheels :

DRAG OR CLUB COACH.—Weight of body complete, without load, about 21 cwt. Front wheels, 3 ft. 5 in.; hind wheels, 4 ft. 5 in. high ; sawn felloes, $2\frac{1}{2}$ × $2\frac{1}{2}$ in. for 2-in. tires ; spokes, 12 and 14 ; size $2\frac{1}{8}$ × $1\frac{5}{16}$ in.; stocks without hoops, $7\frac{1}{2}$ and 8 × $10\frac{1}{2}$ in.

MEDIUM-SIZE ELLIPTIC-SPRING COACH.—Weight of body, 13 to 14 cwt. Front wheels, 3 ft. 6 in.; hind wheels, 4 ft. 4 in.; felloes, 2 × $1\frac{7}{8}$ in.; spokes, $1\frac{7}{8}$ × $1\frac{1}{4}$ in., 12 and 14 ; stocks, 7 and $7\frac{1}{2}$ × 9 in.

FULL-SIZE ELLIPTIC-SPRING LANDAU.—Weight of body, $11\frac{1}{2}$ cwt. Front wheels, 3 ft. 4 in.; hind wheels, 4 ft. 2 in.; stocks, $6\frac{3}{4}$ and $7\frac{1}{4}$ × $8\frac{1}{2}$ in., spokes, 12 and 14, $1\frac{3}{4}$ × $1\frac{3}{16}$ and $1\frac{1}{2}$ full × $1\frac{1}{8}$, $\frac{3}{4}$ in. dish before tiring felloes, $2\frac{1}{4}$ × $2\frac{1}{2}$ for $1\frac{1}{2}$-in. tire.

SMALL-SIZE ELLIPTIC-SPRING LANDAU.—Weight, about 10 cwt. Front wheels, 3 ft. 1 in.; hind wheels, 3 ft. 8 in.; stocks, 6 and $6\frac{1}{4}$ × 8 in.; spokes, 12 and 14, $1\frac{5}{8}$ × $1\frac{1}{8}$ and $1\frac{1}{2}$ × $1\frac{1}{16}$ in.; felloes, $1\frac{3}{4}$ × $1\frac{5}{8}$ for $1\frac{3}{8}$-in. tire.

LARGE-SIZE ELLIPTIC-SPRING BAROUCHE.—Front wheels, 3 ft. 4 in.; hind wheels, 3 ft. 10 in.; stocks, 6 and $6\frac{1}{4}$ × 8 in.; spokes, 12 and 14, $1\frac{3}{4}$ × $1\frac{3}{16}$ and $1\frac{1}{2}$ × $1\frac{1}{8}$ in. felloes, $1\frac{7}{8}$ × $1\frac{3}{4}$ for $1\frac{1}{2}$-in. tire.

LIGHT ELLIPTIC-SPRING BAROUCHE.—Weight, about 9 cwt. Front wheels, 3 ft. 2 in.; hind wheels, 3 ft. 8 in.; stocks, $5\frac{1}{2}$ and $5\frac{3}{4}$ × $7\frac{1}{2}$ in.; spokes, 12 and 14, $1\frac{1}{2}$ × 1 full and $1\frac{1}{4}$ × 1 bare ; felloes $1\frac{5}{8}$ for $1\frac{1}{4}$-in. tire.

FULL-SIZE ELLIPTIC-SPRING BROUGHAM.—Front wheels, 3 ft. 2 in.; hind wheels, 3 ft. 10 in.; stocks, $5\frac{1}{2}$ and $5\frac{3}{4}$ × $7\frac{1}{2}$ in.; spokes, 12 and 14, $1\frac{5}{8}$ × $1\frac{1}{16}$ and $1\frac{3}{8}$ × 1 in.; felloes, $1\frac{3}{4}$ for $1\frac{1}{2}$-in. tire.

MEDIUM-SIZE ELLIPTIC-SPRING BROUGHAM.—Weight, about $8\frac{1}{2}$ cwt. Front wheels, 3 ft.; hind wheels, 3 ft. 8 in.; stocks, $5\frac{1}{2}$ and $5\frac{3}{4}$ × $7\frac{1}{2}$ in.; spokes, $1\frac{9}{16}$ × $1\frac{1}{16}$ and $1\frac{1}{2}$ × 1 in.; felloes, $1\frac{3}{4}$ for $1\frac{3}{8}$-in. tire.

FULL-SIZE ELLIPTIC-SPRING SINGLE BROUGHAM.—Front wheels, 3 ft. 2 in.; hind wheels, 3 ft. 9 in. Other sizes a shade lighter than for the Medium Brougham, with a $1\frac{1}{4}$-in. tire.

MEDIUM-SIZE ELLIPTIC-SPRING VICTORIA-PHAETON.—Weight, about $6\frac{1}{2}$ cwt. Front wheels, 2 ft. 10 in.; hind wheels, 3 ft. 6 in.; stocks, $5\frac{1}{4}$ and $5\frac{1}{2}$ × $7\frac{1}{2}$ in.; spokes, $1\frac{1}{2}$ full × 1 full, and $1\frac{1}{2}$ bare × 1 in. bare ; felloes, $1\frac{5}{8}$ for full $1\frac{1}{4}$-in. tire.

LARGE-SIZE VICTORIA-PHAETON.—Front wheels, 3 ft. 1 in.; hind wheels, 3 ft. 8 in.; stocks, $5\frac{3}{4}$ and 6 × $7\frac{1}{2}$ in.; spokes, $1\frac{9}{16}$ × $1\frac{1}{16}$ in.; felloes, $1\frac{3}{4}$ light for $1\frac{3}{8}$-in. tire.

MAIL-PHAETON.—Weight, $7\frac{1}{2}$ cwt. Front wheels, 3 ft. 3 in.; hind wheels, 4 ft.; stocks, $5\frac{1}{2}$ and $5\frac{3}{4}$ × $7\frac{1}{2}$ in.; spokes, 12 and 14, $1\frac{9}{16}$ × $1\frac{1}{8}$ full, and $1\frac{1}{2}$ × $1\frac{1}{16}$ in.; felloes, $1\frac{7}{8}$ full × $1\frac{3}{4}$ for $1\frac{3}{8}$-in. tire.

STANHOPE-PHAETON.—Weight, about $5\frac{1}{2}$ or 6 cwt. Front wheels, 3 ft.; hind wheels, 3 ft. 7 in.; stocks, $5\frac{1}{2}$ and $5\frac{3}{4}$ × $7\frac{1}{2}$ in.; spokes, 12 and 14 ; $1\frac{1}{2}$ × $1\frac{1}{16}$ and $1\frac{3}{8}$ × 1 in.; felloes, i$1\frac{3}{4}$ full for $1\frac{3}{8}$-in. tire.

DOG-CART.—Dog-cart wheels are made of various heights and substance, in proportion to the size and requirements of the body, viz.: from 4 ft. 6 in. to 5 ft. high. For instance : Wheels, 4 ft. 6 in.; stocks, $6\frac{1}{2}$ × $8\frac{1}{2}$ in.; spokes, 14, $1\frac{7}{8}$ × $1\frac{5}{16}$ in.; felloes, 2 for $1\frac{1}{2}$-in. tire. Or, wheels 4 ft. 6 in.; spokes, $1\frac{1}{2}$ × $1\frac{1}{4}$ in.; felloes, $1\frac{7}{8}$ × $1\frac{3}{4}$ for $1\frac{3}{8}$-in. tire. Or, wheels, 5 ft. high ; stocks, 7 × 9 in., front hoop, 5 in. diameter ; back hoop, $4\frac{1}{2}$ in.; spokes, 16, 2 × $1\frac{5}{16}$ in.; felloes, $2\frac{1}{8}$ × 2 for $1\frac{5}{8}$-in. tire.

The author of the above table,—which many of our readers will no doubt be interested to compare with similar tables of American wheels, already presented in *The Hub*,—adds the following supplementary remarks. He says :

"The foregoing sizes and proportions are given as an average and not as an arbitrary rule, as many wheelers and carriage-builders have their own carefully-prepared tables of dimensions, compiled after great experience. Carriage-users are at times partial to the heavier description of wheel, on the score of strength and not for appearance ; while others prefer the lighter description, especially if located in a neighborhood where the roads are in good condition. In all cases the heights of wheels are governed by the shape, requirements, and conditions of the vehicle on which they are to be used ; and these, and strength, should be the primary considerations of the wheeler in constructing carriage wheels."

The Hub, June 1887

TWO ENGLISH TABLES OF WHEEL DIMENSIONS.

THE London *Coach-Builder*, in its issue of Jan. 16th, presents an elaborate illustrated review, entitled : "The Manufacture of Wheels by Hand and Machinery," covering seven pages, from which we quote the following instructive tables. They are introduced with these words :

"It is not intended to give a complete description of the course adopted, but the following are examples of the practice. The first is a table of proportions of stocks [hubs] and spokes, and the second a table of proportions of the spokes in use, of which there are three varieties, divided into light, medium and strong, 17 in each class, ranging from the lightest to the heaviest in use. Every member of the wheel is treated in the same manner, and the proportions definitely laid down."

TABLE NO. 1 : PROPORTIONS OF HUBS AND SPOKES.

Spoke.		Diameter of Stock.					
		14 Spokes.		12 Spokes.		10 Spokes.	
Tenon.	Face.	Elm.	Oak.	Elm.	Oak.	Elm.	Oak.
In.	In.	In.	In.	In.	In.	In.	In.
5	9	5.4	5.1	5.1	5	5	5
$5\frac{1}{2}$	10	6.	5.6	5.6	5.5	5.5	5.4
6	11	6.6	6.2	6.2	6	6	5.9
$6\frac{1}{2}$	12	7.2	6.7	6.7	6.5	6.5	6.3
7	13	7.8	7.3	7.3	7	7	6.8
$7\frac{1}{2}$	14	8.4	7.9	7.9	7.5	7.5	7.2
8	15	9	8.4	8.4	8	8	7.7
$8\frac{1}{2}$	16	9.6	9	9	8.5	8.5	8.1
9	17	10.2	9.5	9.5	9	9	8.6
$9\frac{1}{2}$	18	1 .8	10.1	10.1	9.5	9.5	9
10	19	11.4	10.6	10 6	10	10	9.5
$10\frac{1}{2}$	20	12	11.2	11.2	10.5	10.5	9.9
11	21	12.6	11.8	11.8	11	11	10.4
$11\frac{1}{2}$	22	13.2	12.3	12.3	11.5	11.5	10.8
12	23	13.8	12.9	12.9	12	12	11.3
$12\frac{1}{2}$	24	14.4	13.5	13.5	12.5	12.5	11.7
13	25	15	14	14	13	13	12.2

TABLE NO. 2 : PROPORTIONS OF SPOKES.

No.	Light.	Medium.	Strong.
	In.	In.	In.
1	12 x 7.2	12 x 8	12 x 9
2	$13\frac{1}{2}$ x 8.1	$13\frac{1}{2}$ x 9	$13\frac{1}{2}$ x 10.1
3	15 x 9	15 x 10	15 x 11.2
4	16 x 9.9	$16\frac{1}{2}$ x 11	$16\frac{1}{2}$ x 12.4
5	18 x 10.8	18 x 12	18 x 13 5
6	$19\frac{1}{2}$ x 11.7	$19\frac{1}{2}$ x 13	$19\frac{1}{2}$ x 14.6
7	21 x 12.6	21 x 14	21 x 15.7
8	$22\frac{1}{2}$ x 13.5	$22\frac{1}{2}$ x 15	$22\frac{1}{2}$ x 16.9
9	24 x 14.4	24 x 16	24 x 18
10	$25\frac{1}{2}$ x 15.3	$25\frac{1}{2}$ x 17	$25\frac{1}{2}$ x 19.1
11	27 x 16.2	27 x 18	27 x 20.2
12	$28\frac{1}{2}$ x 17.1	$28\frac{1}{2}$ x 19	$28\frac{1}{2}$ x 21.4
13	30 x 18	30 x 20	30 x 22.5
14	$31\frac{1}{2}$ x 18.9	$31\frac{1}{2}$ x 21	$31\frac{1}{2}$ x 23.6
15	33 x 19.8	33 x 22	33 x 24.7
16	$34\frac{1}{2}$ x 20.7	$34\frac{1}{2}$ x 23	$34\frac{1}{2}$ x 25.9
17	36 x 21.6	36 x 24	36 x 27

The Hub, July 1888

The Best Wheels for Light Work.

OMAHA, NEB., Aug. 14th, 1888.

MR. EDITOR:—We lately had a controversy among the wood-workers in our shop on the question of which were the best proportioned wheels for light work. We took a wheel with an inch spoke for guide, but in gathering the opinions pro and con in regard to the shape and sizes of the hubs, shape, size and number of spokes, shape of rims and its sizes, we all differed. Now what we desire is, could you give us a thorough practical explanation as to how a wheel with an inch spoke, should be made? also its sizes and shape of hub, and the reasons why, also the same as regards the spokes and rims? If you will do this we would be greatly benefited thereby, and it would also instruct others at the same time. You will oblige by publishing it in the Apprentice Department of the September number. Yours truly, THREE SUBSCRIBERS.

We infer our correspondents had in discussion wheels used on first-class buggies, requiring the best shape, best make and best known proportions, and we will commence with the hub: its size, proportions, style, and number and sizes of mortises. The question is always as to how light a buggy wheel, for carrying two persons, can be made, and we, of course, will take the lightest size usually made by first-class carriage-builders, 44 by $46\frac{1}{2}$ inches in diameter. The lightest size hub used is $3\frac{1}{4}$ by 6 inches; its diameter, $3\frac{1}{4}$ inches, is measured at the largest part of the hub between the mortises. The reason for making it $3\frac{1}{4}$ inches diameter is to have sufficient wood left for a $\frac{3}{4}$-inch axle, which is generally used for the lightest two-passenger buggies, and the size of the axle is based on these calculations.

Diameter of axle, $\frac{3}{4}$ inch; thickness of box, $\frac{1}{8}$ inch; making the outside diameter of box 1 inch, leaving $1\frac{1}{8}$ inches thickness all around the box. Our rule is to always have the tenon longer than the width of the spokes at the shoulder, it giving the spoke more strength in the hub; as we have an inch spoke, its tenon will be $\frac{1}{8}$ inch longer than its width. But we do not desire the spoke to touch the box, so the space between the box and spoke should be $\frac{1}{16}$ inch scant, so that the tenon, when in the hub on the finished wheel, will be $1\frac{1}{16}$ inches full, which is considered correct. Most builders make the length of the hub for a $\frac{3}{4}$-inch axle 6 inches, but this we think too long, and out of proportion to its diameter; $5\frac{3}{4}$ inches would be right, but this length axle would have to be specially ordered, and the majority of carriage-builders do not like to do this. The diameter of the bands, front and back, should be neither too heavy nor too light, and the best proportionate size is $2\frac{1}{4}$ inches diameter front, and $2\frac{9}{16}$ inches back. These measurements will give the size, and we may add, turn the bands in $\frac{5}{8}$ inch on the hub front and back. This will give the principal proportions, but the appearance of the hub may be spoiled if not turned in good shape or style. Some builders turn them entirely plain, without any beads, which look well, but one very light bead, front and back, relieves the plainness somewhat, and in general it is better liked. The best adapted style for hubs cannot be described except by an illustration.

NUMBER OF MORTISES AND SIZES:—This is an open question with many, fourteen spokes for each wheel being the regular number, but we will give our reasons why twelve spokes are sufficient and better adapted for a well proportioned wheel. Take a hub, $3\frac{1}{4}$ inches in diameter, with fourteen spokes, and you will notice that the walls between the mortises are very weak, even if the size of the mortise is only $\frac{1}{4}$ inch, full but are generally mortised $\frac{5}{16}$ inch scant. Drive the spokes on such a wheel hard, and the last two mortises left for the spokes to be driven are only $\frac{1}{8}$ inch; that is, the other twelve spokes forced the mortises to that size, which will weaken the walls considerably. This is the reason we favor twelve spokes for each wheel for this size. Now the size for the mortises for twelve spokes should be $\frac{5}{16}$ by $1\frac{1}{8}$ inch for 1-inch spokes. The dimensions for the hubs concentrated are as follows: Diameter and length, $3\frac{1}{4}$ by $5\frac{3}{4}$ inches. Bands $2\frac{1}{4}$ by $2\frac{9}{16}$ inches, turned $\frac{5}{8}$ inch for the support of the bands.

SIZE AND PROPORTIONS OF SPOKES:—Width of spokes at shoulder 1 inch; thickness of tenon to suit the mortise of $\frac{5}{16}$ inch thickness. Thickness of body at the heaviest part $\frac{5}{8}$ inch; thickness of shoulder $\frac{1}{2}$ inch full; thickness of throat $\frac{3}{8}$ inch. We prefer 1-inch wide spokes for such a wheel, because it gives more surface for the spokes in the hubs and consequently strengthens it in the hub, and will not become loose so easy; also if a spoke is well throated, as it should be, the wheel is liable to dish too much after tiring, and will have an appearance similar to an umbrella. Throating the spokes properly is of importance; if they are too rigid at the hub it will be injurious, that is the spokes will work themselves loose in the hub. For this reason the spoke should be throated as near the shoulder as possible, although they should not be throated too close, or the shoulder will fly off from the spokes; very good judgment is here required, and it is very seldom we see a well throated spoke on light work. The size of spokes for this wheel, while heavier than usually taken, does not look heavier for a twelve-spoke wheel than lighter ones with fourteen spokes.

The rims for sizes are $1\frac{1}{8}$ by $\frac{7}{8}$ inch full; that is, $1\frac{1}{8}$ inch on the thickest part, $\frac{3}{4}$ inch tread, $\frac{3}{4}$ by $\frac{5}{16}$ inch tire, round edge, and $\frac{7}{8}$ inch, full, deep. We add $\frac{1}{16}$ inch depth to the rim, because there are only twelve spokes for each wheel, and this prevents the rims from sinking in between the spokes; we also give to the spokes $\frac{5}{16}$ inch tenon, because the rim is deeper, and more liable to work on the shoulder of the spokes, as the spokes are heavier than usual and have sufficient shoulder, and it will be better for the shoulder and spoke and rim. Do not round the rims too much, as it weakens them, and will not look much lighter.

These are our opinions about a well proportioned wheel. Now let us hear your objections, and we invite any other wheel-maker to also give us his views on this subject.

The Carriage Monthly, Sep 1888

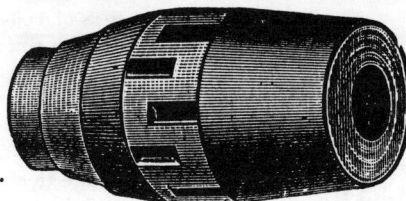

Proportions of Wheels, Tires and Axles.

A well-known carriage builder of western New York asks the following question: "How heavy the wheels should be made that can be driven on a 7/8-inch axle without damage to the axle; that is, without danger of straining the axle. Also, as to the thickest tire that can be put on a wheel with 3/4-inch tread, 1-inch-deep rims. Also, on a 7/8-inch tread, 1 1/16-inch-deep rim, without damage to the wheels."

This is a very important question, and equally instructive to all carriage builders. Will those who have the required knowledge and experience in wheel making, tire setting, and repairing wheels which have been tired with light and heavy tires, that is, with tires out of proportion to the wheels, kindly tell us their experience in the interest of the trade, and we will publish the answers in our July issue. Practical mechanics are aware of the fact that on light and medium carriages proportional sizes in all its parts is a very important matter. If one part is too heavy and the other too light, the light part must give out first before the heavy part will suffer, and this holds good on all its parts on the vehicle, not only on wheels, tires and axles.

The above, from our June issue, has not brought anticipated answers from any of our readers; we therefore make answer.

By proportion of wheels, we understand as follows: shape and size of hubs, shape and size of spokes, and shape and size of rims. This shows that the wheels in the first place must be proportioned regardless of tire and axle, but nevertheless the tires and axles must be in proportion to the wheels.

To give a more complete answer, we reproduce a portion of our Wheel Notes, taken from our Manual of Measurements, which will give our readers a very good idea of the law of proportions of hubs, spokes and rims in connection with tires and axles, as follows:

WHEEL NOTES.

The diameter of the front and back bands in a perfectly proportioned wheel should vary not only with the diameter of the hub, but also with its length, when the diameter is the same; the greater the length, the diameter being the same, the less the diameter of the bands. This is to avoid the cylindrical appearance a hub would assume if the length were increased and bands having the same diameter. The length of a hub of any diameter is too much a matter of local usage and taste to be controlled by the law of proportions.

The thickness of the spoke at the shoulder may be modified and controlled by draftsmen and wheel makers, but must, of course, vary with the number of spokes in the wheel when the hub is the same diameter. The fewer the number the thicker the shoulder. The number of spokes to be used is also a matter of individual or local taste, very difficult to control.

The outside diameter of the axle box at the point where it is under the spoke tenons when the box is set in position, is very important to be considered, for the larger the box, even when the axle is the same, the larger the diameter of the hub must be; and in no case should the length of the spoke tenons in the hub after they are cut off to set the box be shorter than the width of the spoke where it rests on the hub. That is, if in a wheel the width of the spoke at the hub is 1 1/4 inches, the tenons left remaining in the hub after any kind or quality of axle is set, would not be less than 1 5/16 inches, and so in every case the tenon left remaining in the hub should be longer than the width.

The tenon of the spoke at the rim should always be a little less than one-half the width of rim at tread.

The following dimensions are taken from wheels made by the best wheel makers in first-class carriage shops:

WHEELS WITH HUBS THREE INCHES IN DIAMETER.

Exterior diameter of wheels .... 42 by 46 inches.
Exterior diameter of hubs .... 3 inches.
Length of hubs .... 5 1/2 inches.
Mortises of hubs .... 14/16 full by 1/4 inch.
Diameters of bands of hubs .... 1 15/16 by 2 3/8 inches.
Width of spokes at square end .... 13/16 inch.
Thickness of spokes at square end .... 1/2 inch full.
Thickness of throat .... 3/8 inch full.
Thickness of body at center .... 19/32 inch.
Thickness at rim tenon .... 19/32 inch scant.
Number of spokes, front and back .... 12 and 12.
Thickness and depth of rims .... 15/16 by 1 inch.
Width of tread .... 3/4 inch.
Stagger .... 5/8 inch.
Round edge steel tire .... 3/4 by 1/8 inch.
Sizes of axles, 3/4 inch on round, and 13/16 inch on square part.
Width of track .... 4 feet.

WHEELS WITH HUBS THREE AND THREE-QUARTER INCHES IN DIAMETER.

Exterior diameter of wheels .... 42 by 46 inches.
Exterior diameter of hubs .... 3 3/4 inches.
Length of hubs .... 6 1/2 inches.
Mortises of hubs .... 1 inch full by 5/16 inch.
Diameters of bands of hubs .... 2 1/2 by 3 inches.
Width of spokes at square end .... 1 1/16 inches.
Thickness of spokes at square end .... 5/8 inch full.
Thickness of throat .... 9/16 inch.
Thickness of body at center .... 11/16 inch.
Thickness of body at top .... 11/16 inch scant.
Number of spokes, front and back .... 14 and 14.
Thickness and depth of rims .... 1 by 1 1/8 inches.
Width of tread .... 7/8 inch.
Stagger .... 1/2 inch.
Round edge steel tire .... 7/8 by 3/16 inch scant.
Sizes of axles, 7/8 inch on round, and 15/16 inch on square part.
Width of track .... 4 ft. 4 in. to 4 ft. 8 in.

WHEELS WITH HUBS FOUR INCHES IN DIAMETER.

Exterior diameter of wheels .... 42 by 46 inches.
Exterior diameter of hubs .... 4 inches.
Length of hubs .... 6 1/2 inches.
Mortises of hubs .... 1 1/16 by 11/32 inch.
Diameters of bands of hubs .... 2 5/8 by 3 1/8 inches.
Width of spokes at square end .... 1 1/8 inches.
Thickness of spokes at square end .... 11/16 inch.
Thickness of throat .... 1/2 inch.
Thickness of body at center .... 3/4 inch.
Thickness of body at top .... 3/4 inch scant.
Number of spokes, front and back .... 14 and 14.
Thickness and depth of rims .... 1 1/16 by 1 3/16 inches.
Width of tread .... 7/8 inch.
Stagger .... 1/2 inch.
Round edge steel tire .... 7/8 by 3/16 inch full.
Sizes of axles, 1 1/8 inches on round, and 1 3/16 inches on square part.
Width of track .... 4 feet 8 inches.

WHEELS WITH HUBS FOUR AND ONE-QUARTER INCHES IN DIAMETER.

Exterior diameter of wheels .... 42 by 46 inches.
Exterior diameter of hubs .... 4 1/4 inches.
Length of hnbs .... 7 inches.
Mortises of hubs .... 1 3/16 by 3/8 inches.
Diameters of bands of hubs .... 2 3/4 by 3 3/8 inches.
Width of spokes at square end .... 1 1/4 inches.
Thickness of spokes at square end .... 3/4 inch full.
Thickness of throat .... 5/8 inch scant.
Thickness of body at center .... 3/4 inch full.
Thickness on rim tenon .... 3/4 inch.
Number of spokes, front and back .... 14 and 14.
Thickness and depth of rims .... 1 1/8 in. scant by 1 1/4 in.
Width of tread .... 1 inch.
Stagger .... 7/16 inch.
Round edge steel tire .... 1 by 1/4 inch.
Sizes of axles, 1 3/16 inches on round, and 1 1/4 inches on square part.
Width of track .... 4 feet 8 inches.

continued next page

Tables and Charts

Proportions of Wheels, Tires and Axles
(continued from previous page)

The wheel maker has learned from practical experience by making new wheels and repairing them afterwards, and also from repairing the wheels made by other first-class carriage builders, that the best stock and first-class workmanship are essential in the manufacture of wheels, and that their proportions should correspond with the sizes of tires and axles. The tire may be too light or too heavy for the wheels, but it is better to have the tire too light than too heavy. If the tire is too heavy, or non-elastic, it strains not only the rims but also the spokes in the hub, and the wheel is sooner worn out than if the tires had been in propotion to the dimensions of the wheels. If the axles are too heavy for the wheels, the wheels must succumb ; and also, if they are too light when the axle is too heavy, it makes it too rigid, and the wheel must make up the difference of elasticity ; besides, too much wood is taken away from the axle box, and the spokes are weakened. If the axle is too light, they will bend in the center and throw the wheels out of plumb and ruin the wheels.

The Carriage Monthly, Sep 1896

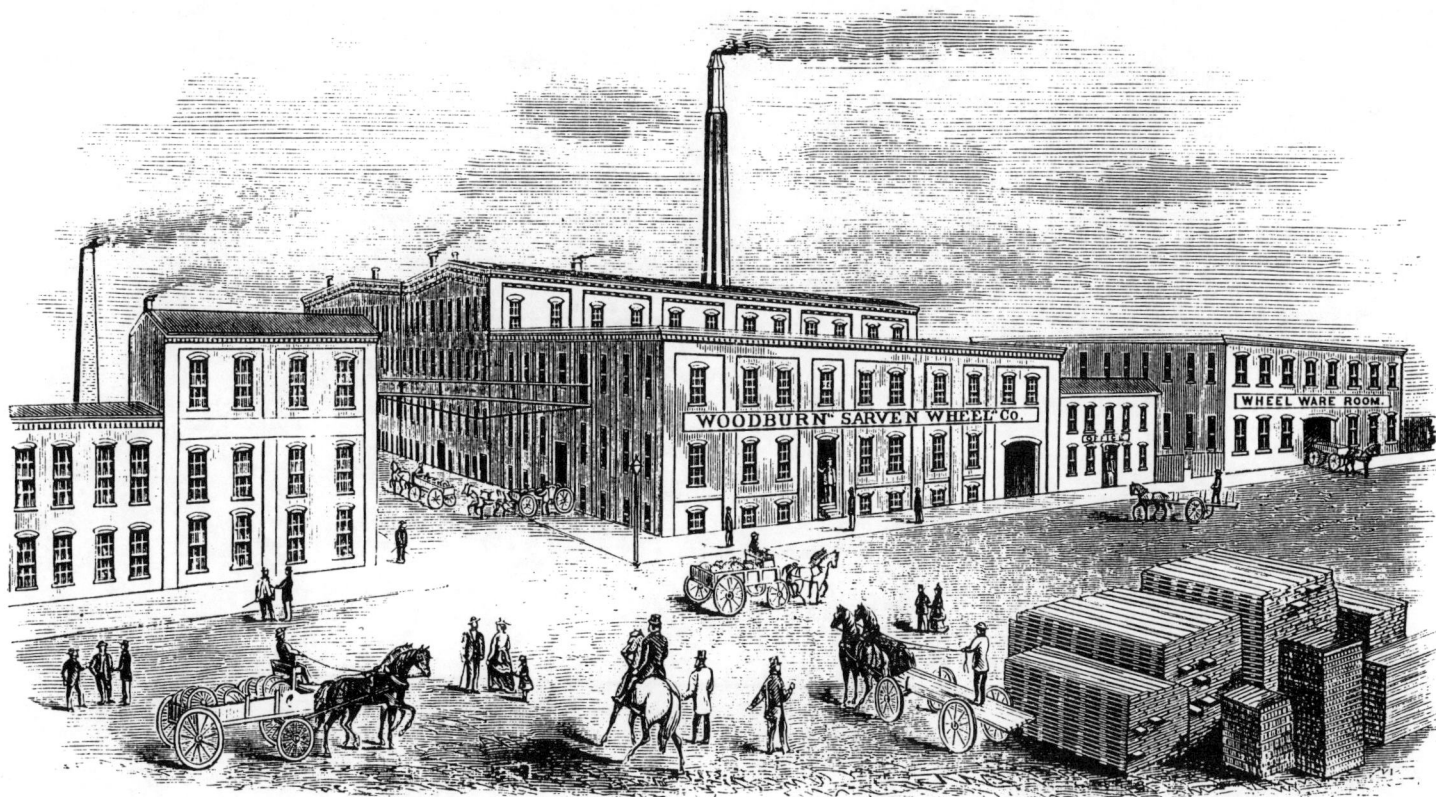